T0142382

Lecture Notes in Networks and Systems   636

The series "Lecture Notes in Networks and Systems" publishes the latest developments in Networks and Systems—quickly, informally and with high quality. Original research reported in proceedings and post-proceedings represents the core of LNNS.

Volumes published in LNNS embrace all aspects and subfields of, as well as new challenges in, Networks and Systems.

The series contains proceedings and edited volumes in systems and networks, spanning the areas of Cyber-Physical Systems, Autonomous Systems, Sensor Networks, Control Systems, Energy Systems, Automotive Systems, Biological Systems, Vehicular Networking and Connected Vehicles, Aerospace Systems, Automation, Manufacturing, Smart Grids, Nonlinear Systems, Power Systems, Robotics, Social Systems, Economic Systems and other. Of particular value to both the contributors and the readership are the short publication timeframe and the world-wide distribution and exposure which enable both a wide and rapid dissemination of research output.

The series covers the theory, applications, and perspectives on the state of the art and future developments relevant to systems and networks, decision making, control, complex processes and related areas, as embedded in the fields of interdisciplinary and applied sciences, engineering, computer science, physics, economics, social, and life sciences, as well as the paradigms and methodologies behind them.

Indexed by SCOPUS, INSPEC, WTI Frankfurt eG, zbMATH, SCImago.

All books published in the series are submitted for consideration in Web of Science.

For proposals from Asia please contact Aninda Bose (aninda.bose@springer.com).

Daria Bylieva · Alfred Nordmann
Editors

Technologies in a Multilingual Environment

XXII Professional Culture of the Specialist of the Future

 Springer

Editors
Daria Bylieva
Department of Social Sciences
Peter the Great St. Petersburg Polytechnic
University
St. Petersburg, Russia

Alfred Nordmann
Institute of Philosophy
Technical University of Darmstadt
Darmstadt, Hessen, Germany

ISSN 2367-3370   ISSN 2367-3389 (electronic)
Lecture Notes in Networks and Systems
ISBN 978-3-031-26782-6   ISBN 978-3-031-26783-3 (eBook)
https://doi.org/10.1007/978-3-031-26783-3

This Springer imprint is published by the registered company Springer Nature Switzerland AG
The registered company address is: Gewerbestrasse 11, 6330 Cham, Switzerland

Contents

Universities as Multilingual Sites for Technological Development

Techniques and Technologies for Multilingual Learning

Introduction: Technologies in a Multilingual Environment

Alfred Nordmann[1(✉)] (iD) and Daria Bylieva[2(✉)] (iD)

[1] Technical University of Darmstadt, Residenzschloss 1, 64283 Darmstadt, Germany
nordmann@phil.tu-darmstadt.de
[2] Peter the Great St. Petersburg Polytechnic University (SPbPU), Polytechnicheskaya, 29, 195251 St. Petersburg, Russia
bylieva_ds@spbstu.ru

Abstract. To introduce the papers from the 22nd conference the *Professional culture of the specialist of the future* they are considered in a broader context of cultural and academic developments. Multilingualism presents a much larger set of questions and challenges than just those of acquiring mastery of many languages. To the extent that the fabric of social life and especially the urban environment is multilingual, everyone inhabits it, regardless of their own knowledge of languages. Discussions within the Philosophy of Technology dramatize this situation if one acknowledges that devices and techniques add further layers of language to an already complex multilingual environment. This dramatic situation is at the same time challenge and opportunity for multilingual learning. This introduction thus provides a conceptual overview of a space that is inhabited by many university students and teachers, asking what they can and must do to gain or regain certain near-extinct species.

Keywords: Multilingualism · Philosophy of technology · Educational practice

By looking at technologies in a mulitingual world, the 22nd conference on the *Professional culture of the specialist of the future* stood at the intersection of various current trends. There is the trend in socio-linguistics towards multilingualism in distinction to mono-, bi- and plurilingualism. There is a trend in the philosophy of technology to consider technology as a kind of language, and language as a kind of technology - which presupposes a broad view of language with an emphasis on its formal, behavioral, performative dimensions. There is a trend in the theory and practice of teaching - not only of language-teaching - that treats the multilingual and multinational classroom as an opportunity for integrated learning. And finally, all this is part of another trend, namely to consider the co-existence of many different languages not primarily as a source of difference and division but as an occasion for peaceful collaboration and cooperation. The interpretation of words and the cultural meaning of things can induce irresolvable conflict, but denizens of a multilingual world are united by the sharing of technologies and thus by their joint participation in linguistic patterns and practices.[1]

---

[1] The journal *Technology and Language* is dedicated to these various aspects. A recent special issue complements this volume specifically [1].

1 Socio-linguistics and Multilingualism

When socio-linguists move towards mulilingualism, they have two sources of inspiration and both of these involve technologies. There is first of all globalization which comes with mobility and migration, tourism and professional assignments. Engineering students from Russia and China might fall in love in Germany. They speak English to each other and their child is a native German plurilingual speaker. There are secondly all the ways in which languages are integrated in our built environment, especially urban infrastructure. Here the focus is not on people who speak different languages in different settings but on everyone who moves among the multilingual signage, who learns to master maps, vending machines, public transportation, or shopping, queuing, dining etiquettes. Accordingly, it is the various natural languages as well as the codes, scripts, or routines of technical artefacts that jointly make up the multilingual condition [2–4, see also Gorohkova et al., this volume]. And the challenge of the multilingual dimension is not how we learn to write, read, and understand the meaning of words and texts, but how we learn to negotiate, orient ourselves and act effectively in a maze of signs and symbols not all of which we fully "understand" (see, for example, the papers by Daineko et al. and by Samorodova et al., this volume). A most obvious example of this would be the many languages of music which coexist in a multicultural society. Quite evidently, people can learn to listen to different musical genres and traditions, following along in song and dance, acquiring a sense of what sounds right and wrong, all of this without explicit knowledge of the principles of composition, without the ability to read and write, produce and properly "understand" music. In our daily lives, this plays out in multilingual settings like modern universities of technology which are the focus of the papers in the third section of this volume.

2 Technology as Language, Language as Technology

Though it is a fairly young field of philosophical questioning, there are many different traditions in the philosophy of technology. In recent years, considerable attention has been paid to human-machine interactions and their transformation through digital technologies. Especially with respect to highly complex simulation models and AI systems, it has become necessary to question human control of technology, asking whether it requires conscious intellectual knowledge of how things work. Often enough, to "know" the working order of things is a matter of tacit or implicit knowledge - it is the ability to anticipate, modulate, and maintain the performance of a black-boxed technical system or device. This is where these philosophical considerations meet up with those of multilingualism. Musical and other works of art share with clockworks and other technical systems that they are composed to form a working order of things. To know a working order is to gain a feeling for the way in which its parts work together to do work or produce an effect, that is, a feeling for the mechanism, organism, or algorithm. This amounts to tacit knowledge of the principles of composition as a language of making and building. Putting together (*compose*) tones according to counter-point, classical harmony, the rules of twelve-tone music, or many non-Western idioms makes for the different languages of music. Similarly, mechanical and electrical engineering, computer programming, architecture and chemistry involve different languages of composition - each a grammar of things - which contribute to the multilingual condition of the modern world [5, 6 and see the contribution by Nesterov in this volume]. Other examples and principled considerations will be found in the first two sections of this volume which concern the languages of technology and how they become salient in the technosphere.

3   Cooperation in the Multilingual Classroom

While socio-linguistics and the philosophy of technology investigate and analyse the multilingual condition, there are also many techniques and technologies that help us successfully navigate a multilingual world. Educational technologies and teaching strategies view the multilingual condition as an opportunity and resource for cooperative and integrative learning. For example, one can learn English as a foreign language not only in classes dedicated to language instruction but also in classes of mechanical engineering - first of all, these classes will often be taught in English to accommodate international groups of students, and secondly, one learns simultaneously the proper ways of putting machine elements together and the proper ways of forming English sentences. This way of learning more than one language at a time promotes students' implicit knowledge of what a language is (see the papers in this volume e.g. by Dubinina et al.). Conceivably, there might be classes where students learn all national languages at once by being shown how to navigate different linguistic cultures with the help of their translation-app. And of course, there are many ways to incorporate project-based learning also within traditional language instruction (see Vorontsova et al., this volume). To the extent that in a multilingual world it is more important to selectively use and deploy language than to become like a native speaker, a totally immersive language experience is not called for anymore - more important than perfect fluency is the ability to skillfully move between languages and to strategically or technologically utilize language. The techniques and technologies for multilingual learning are prominent especially in the fourth section but also throughout this book.

Linguistic difference can bring us together - not just in the classroom but during international sport events or song contests, or international payment systems, shared electric grids, academic peer review and publication practices.[2] A world of mutual technological dependency where we can trustingly share in the same technical routines may well be a peaceful world (compare Zamorev, this volume).

References

1. Bylieva, D., Nordmann, A.: Technologies in a multilingual world. Technol. Lang. 3(3), 1–10 (2022). https://doi.org/10.48417/technolang.2022.03.01

---

[2] From this point of view, English as a shared medium of communication does not signify supremacy of one language over others: German, Russian, Chinese, Indian English are no less different from American English than is British or Australian English. On this account, English is a set of communicative tools that should be judged by the effectiveness of their deployment rather than by fluency in any one local idiom.

2. Aronin, L.: Theoretical underpinnings of the material culture of multilingualism. In: Aronin, L., Hornsby, M., Kiliańska-Przybyło, G. (eds.) The Material Culture of Multilingualism. EL, vol. 36, pp. 21–45. Springer, Cham (2018). https://doi.org/10.1007/978-3-319-91104-5_2

3. Aronin, L.: Multilingualism in the age of technology. Technol. Lang. **1**(1), 6–11 (2020). https://doi.org/10.48417/technolang.2020.01.02

4. Hufeisen, B., Nordmann, A., Liu, W.: Two perspectives on the multilingual condition – linguistics meets philosophy of technology. Technol. Lang. **3**(3), 11–21 (2022). https://doi.org/10.48417/technolang.2022.03.02

5. Nordmann, A.: The grammar of things. Technol. Lang. **1**(1), 85–90 (2020). https://doi.org/10.48417/technolang.2020.01.18

6. Pezzica, L.: Deep technogrammar a wittgensteinian-grammatical approach towards a philosophy of technology. Technol. Lang. **4**(1) (2023). https://doi.org/10.48417/technolang.2023.01.02

The Languages of Technology in Contemporary Culture

The All-Human Space of Transformation: Technology and Culture

Tatiana Bernyukevich$^{(\boxtimes)}$ iD

Moscow State University of Civil Engineering, Moscow, Russian Federation
bernyukevich@inbox.ru

Abstract. The aim of the study is to determine the connection between technology and culture in the context of the transformation of human existence, considered in the philosophical works of Friedrich Dessauer, Oswald Spengler, Pavel Florensky, and the literary works of Alexander Bogdanov and Andrei Platonov. *Methodology and research methods.* The study is based on a comprehensive approach that allows to identify the features of the analysis of the connection between technology and culture in the context of goals set in the works of a number of both philosophers and writers of the first half of the 20th century. *Research results.* An attempt to determine the essence of technology in its relationship with culture is presented in the concepts that underlie the philosophy of technology. Spengler focuses on the metaphysical essence of technology understood as tactics of living. Dessauer reveals the ontological status of technology associated with the formation of God-given forms of Being. According to Florensky, technology as a creative part of Culture is a way of reuniting the human and the divine. *The discussion of the results.* The meanings of life and knowledge can be set by other ways of mastering the world in addition to philosophy, literature is a unique way of familiarizing with the experience of understanding the world. *Conclusions.* Outside the broad ontological and anthropological context of the analysis of technology, it is impossible to set the boundaries of the reorganization of society, determine the current parameters of culture, and predict the future of the man-technology-culture system.

Keywords: Technology · Culture · Creativity · Transformation of the world · Connection between technology and culture · Human and technology

1 Issues of Determining the Role of Technology in Culture and Society

An analysis of works devoted to the role of technics and technology in the development of society and their impact on humans makes it possible for their authors to identify ways to determine the extent of the impact of technics and technology on the human community.

From the perspective of the development of human civilization, the problems of technology are reflected in the works of Oswald Spengler, Lewis Mumford, Karl Theodor Jaspers, Martin Heidegger, Herbert Marcuse, Arnold Joseph Toynbee, Ernest Bloch and

D. Bylieva and A. Nordmann (Eds.): PCSF 2022, LNNS 636, pp. 7–14, 2023.
https://doi.org/10.1007/978-3-031-26783-3_2

other philosophers [1]. Universal understanding of technology in the 20th century is connected with the consideration of global social problems, such as the problems of world wars and local conflicts, the deepening differences in the pace of economic progress of developed and developing countries, the aggravation of environmental problems, etc. These issues are often associated with the crisis of Western culture in general in the philosophers' concepts.

An attempt to determine the essence of technology in its relationship with culture is indicated in the approaches that underlie one of the most popular areas of philosophical knowledge today – the philosophy of technology. This primarily concerns the ideas of Ernest Kapp, Friedrich Dessauer, Peter Engelmeyer, Pavel Florensky, and the concepts of Russian cosmists [2].

Literary works may be equally interesting for the consideration under this question. The anthropological role of technology in the works of the famous writer Andrei Platonov and the artistic embodiment of the ideas of the socio-technical reorganization of society in the utopias of a politician, scientist and publicist, the author of the original theory of tektology Alexander Bogdanov can be discussed.

Therefore, the aim of the study is to determine the connections between technology and culture in the context of the transformation of human existence, considered in the works of Dessauer, Spengler, Florensky, Bogdanov and Platonov.

The study is based on a comprehensive approach that allows to identify the features of the analysis of the connection between technology and culture in the context of goals set in the works of a number of both philosophers and writers of the first half of the 20th century.

2 The System Human—Technology—Culture in the Works of Spengler, Dessauer, Florensky, Platonov and Bogdanov

A large-scale analysis of the phenomenon of global technologization of humanity is presented in the works of Spengler. In "Man and Technics: A Contribution to a Philosophy of Life", he defines the planetary scale of the influence of technology on society and human and emphasizes the fact that at the earliest stages of the development of technology, tendencies were laid to increase its influence on all spheres of public life and on the growing needs of human in technical means. Spengler noted that every invention contains the possibility and necessity of new inventions. The satisfaction of human needs with the help of technology invariably contributes to the generation of other needs. Any victory over nature awakens the desire for other victories. Spengler defined the metaphysical essence of technology as follows: "Technics is the tactics of living; it is the inner form of which the procedure of conflict—the conflict that is identical with Life itself—is the outward expression." [3, p. 457].

The contradictory nature of technics and their place in the life of humanity is due to the fact that, according to Spengler, only human technics are independent of the life of the human species. Within the framework of the existence of life, this is a unique case, since in this case the individual goes beyond the constraints of the species. By technics of his life (life is conscious, intentional, changeable, personal, inventive), a human becomes the only creator of his life among other living being [3, p. 465].

There is an important Spengler's remark regarding the relationship between technics and culture: the inner form of the creative life of human and humanity is culture; therefore, people speak of the creation of culture. Spengler believes that the man's creations are the expression of this being in personal form [3, p. 465]. In this case, the technics of human life is the creation of culture.

Dessauer's research presents a fundamental attempt to determine the ontological status of technology. This status is associated with the formation and implementation of forms of being. Philosopher and theologian Dessauer cinsiders that God endowed Being with "the ability and the law of deployment" in various forms [4]. In fact, human continues to create nature with the help of technology that changes the face of the Earth. The creation of technology, its development and improvement are the implementation of the ability of the world to form, which is given by God and is embodied in the creative and technical activity of man [4].

Culture and technology appear in Dessauer's works in a systemic unity, in particular, he emphasizes that there is no objective cultural property, even of the highest rank, which would not be built by technology, would not be expressed technically, and all the objects of culture are at the same time technical objects [5, p. 174]. In order for the technical to become a way of man's participation in Being, the human himself must be responsible for the technical activity. The philosopher notes that with the help of technology, a person takes fate into his own hands, but if he forgets about it, then accusers of technology are right in calling technology no more than irresponsible functionalism [5, p. 219].

Russian religious philosopher Florensky sees the role of technology in the return of human to Divine existence. A person who has fallen away from God has a way of returning to him through Culture. And it is technology as a creative part of Culture that is the way to reunite the human and the divine. Researchers note the influence of Kapp's concept on Florensky's ideas about the mechanisms for the development of technology and its anthropological role. In 1877, Kapp's "Elements of a Philosophy of Technology: On the Evolutionary History of Culture" that is considered as one of the fundamental works on the philosophy of technology was published. Kapp's idea of organ projection lies in the understanding of technical tools as a projection of the human body, its repetition and embodiment in them. Florensky focuses on the anthropological significance in determining the essence of technology [6]. He believes that the place of human creativity in culture is reflected in the invention of technical means. Florensky, relying on the ideas of E. Kapp, the founder of the philosophy of technology, states that the historical task of technology is to consciously continue its organ projection on the basis of decisions that are given by the "unconscious body-building of the soul." [7, p. 417]. Technology as a conscious "bodybuilding of the soul" in space is the ability to return to the integrity of man, nature, and culture [7, p. 438].

For the famous Russian writer Platonov, technology is associated with many other themes of his work and research such as the transformation of nature and human and the boundaries of these transformations, the essence of creativity and its possibilities, the technical as an expression of the human, etc. [8]. Platonov's interest in technology is connected with his engineering and technical activities; in the 1920s he worked as land reclamation engineer and agricultural electrification specialist. It is also resulted from the undoubted influence of the concepts of Russian cosmism popular at that time, in particular, ideas about transforming the world and overcoming the death of a person as a disadvantage of a natural way of existence [9]. These motifs were vividly reflected in such stories and novels by Platonov of the 1920s and 1930s as "Epifan Gateways", "The Secret Man", "The Ethereal Path", "The Motherland of Electricity", "The Fierce and Beautiful World", "The Foundation Pit", "Chevengur", etc. In the works of Platonov, the problem of the relationship between human and technology found expression in the types of images of the characters. The first type is a genius inventor, an engineer. Such characters include Bertrand Perry ("Epifan Gateways"), Faddey Popov, Mikhail Kirpichnikov and Yegor Kirpichnikov ("The Ethereal Path"), Peter Kreutskopf ("Moon Bomb"). The second type is mechanics, machinists, i.e., "masters". These characters are associated with the image of a "secret person", a kind of naturally gifted master who feels his separation from the world, an "orphan" isolation from it, feels the injustice of being, but is stingy in expressing these feelings and sensations that remain "secret". Mechanic Foma Pukhov, such an intimate person, is the main character of the story "The Secret Man" [10].

The work of a famous Russian revolutionary movement participant, doctor, philosopher, and economist Bogdanov is of intensive research interest today. In the history of Russian and Soviet philosophy, Bogdanov is known as the author of the concept of a special science – tektology. This is the science of the general laws of organization, it is essentially the forerunner of the system approach and cybernetic ideas. In Bogdanov's tektology, defined by him as a universal science of organizational principles, the universe is "an infinitely unfolding fabric of forms of various types and levels of organization, from elements of the ether unknown to us to people's groups and star systems" [11, p. 73].

In the post-revolutionary processes of cultural construction, his ideas of "proletkult" played a certain role too, in addition, he took an active part in organizing a fundamentally new educational institution, the "Proletarian University". In the 1920s he was engaged in research on hematology and gerontology, in 1926 he became director of the world's first Institute for Hematology and Blood Transfusions, which he founded. These studies end tragically for Bogdanov: he died as a result of his own blood transfusion experiment [12–14].

It should be noted that Bogdanov was not only a political figure who developed a project for the improvement of the national economy of the young Soviet republic in 1920, but also a science fiction writer. He is the author of the socio-utopian novels "Red Star" (1908) and "Engineer Manny" (1912).

The novel "Red Star" tells about the journey of the revolutionary Leonid to the planet Mars. During this journey, the character not only gets acquainted with the scientific discoveries and technical projects of the Martians, but, no less important, tries to understand the peculiarities of their social structure and relations with each other [15].

Technics and technology on Mars in the novel "Red Star" are based on the use of "negative type matter". The Martians exploit the "radiating substance" they produce in large quantities. They found an electrochemical way to accelerate the decomposition of its elements hundreds of thousands of times and use it as energy resources. Thus, using this energy, the Martians carry out interplanetary flights by aircrafts "etheronefs" [15, p. 24–25].

The economic structure of the Martian society, based on systematic accounting and redistribution of working time, is same impressive as the technical projects: they use special tables of labor statistics, in which they record the number of working hours at enterprises and in other organizations of various industries [15, p. 58–59].

Particular attention in the novel is given to the education of young Martians. The "House of Children" occupies an entire significant and, moreover, the best part of the city with a population of 15–20 thousand people. There are such institutions in all the big cities of the planet, and in many cases, they may form independent cities. The main principle of education and training in the "House of Children" is diversity and successful socialization. To accomplish this, they deliberately unite, accommodate children of all ages together in the "House of Children", and provide them with teachers of different ages and various practical expertise areas [15, p. 63–64]. In general, the whole organization of life on this fantastic Mars is managed simply and reasonably. This finds expression in the spiritual culture of the Martians [15, p. 71–73].

Despite the fact that the novel "Red Star" does not have high artistic value in terms of grand literature, it is an original reflection of the socio-philosophical concept, which combines political ideas and projects for the technical transformation of the world. For this reason, the content of the novel by Bogdanov is close to the utopian constructions of the bright cosmist philosopher, the founder of Soviet rocketry and astronautics K. Tsiolkovsky.

3   Features of the Consideration of Technology's Transformative Role in Philosophy and Literature

An analysis of the utilitarian and pragmatic tasks of technics and technology is not enough to understand the nature of technology, its place in culture and society [16, 17]. Consideration of technology as a way of transforming the world makes it possible to determine the possibilities and parameters of harmonization of relations between nature and man, to define the ways of development of culture, the limits of change in human qualities. Consideration of the interaction of technology and culture in the context of the space of human world transformation is an appeal to the fundamental problem of the relationship between the human and the technological in culture [18, 19].

There are different ways to determine the features of this interaction. For example, within the framework of the metaphysical understanding of the essence of technology, Spengler defines the essence of technics as a tactic of all life as a whole [20, 21]. F. Dessauer makes a fundamental attempt to determine the ontological status of technology, which he associates with the formation and implementation of forms of being [22]. Florensky seeks to determine the anthropological essence of technology, understand technology as part of culture, he believes that with the help of it, culture creates a chance to return to Divine Being [23].

The problem of determining the features of philosophy and literature, their unity and differences, methods of "philosophical criticism", philosophical understanding of literature or consideration of philosophical texts as literary texts are analyzed in a number of studies and publications. According to V.A. Lektorsky, "…philosophy is the comprehension of the ultimate questions of human existence in the world. But, as is known, the meanings of life and knowledge can be set in other ways…", this is the advantage of literature, a unique way of familiarizing with someone else's experience based on imagination, sensation, and empathy. The worlds created by literature and the meanings embodied in them are easily "interiorized", becoming, as it were, part of personal experience" [24].

This also applies to those literary works in which one of the main themes is the question of the place of technology in culture, the relationship between technology and man, the proportion of technology to the human world. Thus, in Bogdanov's utopian novel "Red Star", an attempt to represent an integral system of anthropological, cultural and technological interaction is expressed in an artistic way. The works of Platonov reflect a wide range of ideas and images associated with deep philosophical ideas, contradictory social processes, grandiose engineering and technical projects and determine the connection between technology and man, the potential of technology in the global reorganization of the world in their context [8, 25–27].

4   On the Future of the System Human—Technology—Culture

It should be noted that attempts to philosophically identify the role of technology in the world, its connection with creativity, the anthropological parameters of technology in the first half of the 20th century determine important milestones not only in the development of the philosophy of technology as a philosophical trend, but also in the development of fundamental problems of philosophical knowledge [28]. The problems are as follows: man's place in the world, the creativity of Being and the role of human in it, the relationship between natural and cultural, profane and divine, the genesis of culture and the ways of its formation, the prospective and boundaries of the transformation of the world. There is a certain probability that outside of this broad ontological and anthropological context it is currently impossible to set the boundaries of the reorganization of society and human through technology and technical creativity, to determine the modern parameters of culture that changes its typological properties with a significant influence of technology, to understand the future of the entire integral system of man-technology-culture [29]. To solve these issues, it is necessary to bring back to consideration and update the ideas of the essence and place of technology, which were expressed by famous philosophers and writers of the beginning of the 20th century, such as Spengler, Dessauer, Florensky, Platonov, Bogdanov.

Acknowledgements. The reported study was funded by RFBR, project number 20-011-00462 A, Friedrich Dessauer's Philosophy of Technology: Epistemology and Anthropology of the Realistic Creativity Theory.

References

1. Papchenko, E.V.: Philosophy of technology: history and development paradigms. Humanit. Soc.-Econ. Sci. **6**(103), 25–27 (2018). https://doi.org/10.18522/1997-2377-2018-103-6-25-27

2. Kapp, E., Noiret, L., Espinas, A.V., Kunov, H.: Philosophy of the Machine. The Role of Tools in Human Development. Priboy, Leningrad (1925)
3. Levit, S.Ya. (ed.): Man and Technics: A Contribution to a Philosophy of Life/Culturology. XX Century: Anthology. YUrist, Moscow (1995)
4. Nesterov, A.Y.: Epistemological and ontological problems of the philosophy of technology: "the fourth kingdom" of F Dessauer. Ontolog. Design 3(21), 377–389 (2016)
5. Dessauer, F.: The Dispute About Technology: Monograph (trans. Nesterov, A.Y.) Publishing House of the Samara Humanitarian Academy, Samara (2017)
6. Polovinkin, S.M.: Christian Personalism of the Priest Pavel Florensky. RGGU, Moscow (2001)
7. Florensky, P.: Works in Four Volumes. Mysl, Moscow (2000)
8. Philosophical anthropology of Andrei Platonov. Institute of Philosophy of the Russian Academy of Sciences, Moscow (2019)
9. Antipov, A.A.: Andrei Platonov's philosophy of technology: ideas and images in the space of mutual interpretation. Russ. Humanit. J. 6(2), 145–153 (2017). https://doi.org/10.15643/libartrus-2017.2.4
10. Bernyukevich, T.V.: The human dimension of technology in philosophy and literature: P. Florensky, F. Dessauer, A. Platonov. Semiot. Stud. 1(4), 22–28 (2011). https://doi.org/10.18287/2782-2966-2021-1-4-22-28
11. Bogdanov, A.A.: Tektologiya: (General Organizational Science), vol. 1. Ekonomika, Moscow (1989)
12. Strezhneva, M.V.: A systematic approach in international studies: on the relevance of the theoretical heritage of Alexander Bogdanov. Polis. Politic. Stud. 4, 112–123 (2020). https://doi.org/10.17976/jpps/2020.04.08
13. Balanovskij, V.V.: Alexander Bogdanov: from criticism of science to the practice of life. Solovyov's Res. 3(35), 157–177 (2012)
14. Sidorin, V.V.: Alexander Bogdanov at the institute of scientific philosophy: from the decline of philosophy to scientific monism. Questions Philos. 10, 91–99 (2021). https://doi.org/10.21146/0042-8744-2021-10-91-99
15. Bogdanov, A.A.: Red Star. Association of Artists. Printing, St. Peterburg (1908)
16. Popkova, N.V.: The social nature of technology. Philos. Sci. Technol. 23(2), 49–60 (2018). https://doi.org/10.21146/2413-9084-2018-23-2-49-60
17. Rozin, V.M.: Technique and Sociality: Philosophical Differences and Concepts. Librocom, Moscow (2012)
18. Jones, W.B.: Book review: Thinking about technology: foundations of the philosophy of technology. Bull. Sci. Technol. Soc. 20(5), 405–408 (2000). https://doi.org/10.1177/2F027046760002000508
19. Peukert, W.: General concepts in nanoparticle technology and their possible implication on cultural science and philosophy. Powder Technol. 158(1–3), 133–140 (2005). https://doi.org/10.1016/j.powtec.2005.04.024
20. Terekhov, O.E.: In the shadow of oneself: Oswald Spengler and humanitarian thought of the 20th century. Bull. Tomsk Stat. Univ. Histor. 57, 137–140 (2019). https://doi.org/10.17223/19988613/57/23
21. Skipin, N.S.: O. Spengler's philosophy of technology in the context of global culture. Notebook. Conservatism 4, 106–112 (2020). https://doi.org/10.24030/24092517-2020-0-4-106-112
22. Nesterov, A.Y.: Cosmos and metacosmos in Dessauer's philosophy of technology: inventing the environment. In: Bylieva, D., Nordmann, A. (eds.) PCSF 2021. LNNS, vol. 345, pp. 22–33. Springer, Cham (2022). https://doi.org/10.1007/978-3-030-89708-6_3

23. Martynova, S.A.: Organo- and technoprojections in the philosophy of P.A. Florensky and cinematography by Dziga Vertov. Questions Philos. **6**, 85–96 (2021). https://doi.org/10.21146/0042-8744-2021-6-85-96
24. Philosophy and literature: problems of mutual relations (materials of the "round table"). Questions Philos. **9**, 56–97 (2009)
25. Tulchinskii, G.L.: Digitalization and the anthropodicy problem in Andrei Platonov's Works. Russ. J. Philos. Sci. **63**(5), 53–66 (2020). https://doi.org/10.30727/0235-1188-2020-63-5-53-66
26. Platonov, A.: Intimate man. In: Selected Works, pp. 668–739. Mysl, Moscow (1983)
27. Nesterov, A.Y., Demina, A.I.: Technology and understanding. Technol. Lang. **2**(4), 1–11 (2021). https://doi.org/10.48417/technolang.2021.04.01
28. Bernyukevich, T.V.: The technical and the religious: concepts and contemporary social practices. In: Bylieva, D., Nordmann, A. (eds.) PCSF 2021. LNNS, vol. 345, pp. 105–112. Springer, Cham (2022). https://doi.org/10.1007/978-3-030-89708-6_10
29. Agazzi, E.: Science, technology and philosophy in the cultural identity of nations. Questions Philos. **8**, 5–15 (2022). https://doi.org/10.21146/0042-8744-2022-8-5-15

Modern Technologies for Teaching Languages and Their Role in the Development of a Global Technogenic Civilization

Sergei Mezentsev[✉] [iD]

National Research Moscow State University of Civil Engineering, Yaroslavl Highway, 26, 129337 Moscow, Russia
perevolochnoe@mail.ru

Abstract. This article discusses modern technologies for teaching languages in the context of multilingualism and their role in the development of a global technogenic civilization. It is emphasized that multilingualism is caused by the processes of globalization, migration flows, "mixing" of representatives of different peoples. Preservation of monolingualism is becoming impossible, and the need for multilingualism is increasing. An important component of multilingualism is the large-scale spread of multilingualism and its varieties (bilingualism and trilingualism). The purpose of this article is to study modern technologies for teaching languages. The methodological and theoretical basis of the article is the publications of recent years devoted to multilingualism, the use of diverse technologies, methods and forms of teaching foreign languages. In the course of the study, methods of analysis and synthesis, systematic and interdisciplinary approaches were used, which made it possible to consider and generalize the results of research in various scientific disciplines: linguistics, psychology, sociology, engineering. Information and computer technologies have significantly changed the educational and communication technologies widely used in teaching foreign languages. A detailed description of the conceptual technological principles of computer learning technologies, the development of mobile technologies, mobile telephone communications and devices, applications (Puzzle English, Learning English, Polyglot, etc.) is given. It is argued that the use of computer and mobile technologies has significantly increased the effectiveness of teaching foreign languages. Further development of technologies for teaching foreign languages will inevitably lead to increased social interactions throughout our globe.

Keywords: Multilingualism · Learning technologies · Technogenic civilization

1 Introduction

The world-historical processes rapidly developing after the collapse of the bipolar world and the accomplishment of the information and computer revolution are fundamentally changing the face of the entire modern technogenic civilization and people's life in all corners of our planet Earth. Accelerating migration flows for political, economic, social,

D. Bylieva and A. Nordmann (Eds.): PCSF 2022, LNNS 636, pp. 15–21, 2023.
https://doi.org/10.1007/978-3-031-26783-3_3

cultural, religious, psychological, ethnic, demographic, environmental and other reasons threaten to develop into a "great migration of peoples." Already now we are witnessing the "mixing" of representatives of different peoples in some countries such as the United States of America, Canada, and countries of the European Union. This circle includes more and more countries, including the Russian Federation. This leads to a deepening of contradictions and problems of a linguistic and communicative nature, an increase in tension and conflicts between individuals, social groups and entire nations.

The desire of certain people to preserve their native language, considered as the main sign of national, cultural and religious identity, on the one hand, and, on the other hand, the desire to know other languages for communication are opposite trends in a number of countries around the world. However, globalization continues despite the difficulties that stand in its way, such as Russia's recent refusal to include English in the system of the Unified State Examination (USE) for graduates of secondary schools.

The need for multilingualism is growing. This means that the knowledge of two or more languages is becoming an inevitable matter of present and future generations. An important component of multilingualism is the large-scale spread of multilingualism and its varieties (bilingualism and trilingualism) on the territory of our planet [1]. At the same time, "multilingualism" ("multilingualism") means the use of several languages within the boundaries of a certain social community, primarily, the state, as well as the use by an individual or a social group one or another language, which is chosen in accordance with a specific communicative situation.

Modern researchers of multilingualism increasingly emphasize that the current human society can no longer be characterized as monolingual, since it is rather multilingual. And this is now considered as the norm, not the exception. The principle of "one language - one people" is irrevocably a thing of the past. It is quite natural that multilingualism is most characteristic to multinational states.

Currently, there are the following forms of social multilingualism:

1) State (two or more state languages).
2) Ethnic (languages of the titular nation and other ethnic groups).
3) Emigrant (learning and using the language of the new Motherland while maintaining or losing one's native language).

It should also be noted that the world-historical process, during which there is population's mass migration, requires an increase in the number of multilingual speakers. In this regard, in the vast majority of countries of the world, the technology of teaching various languages is becoming relevant and of particular importance. The main task of education should be the formation of people's consciousness, in which the language should serve as a "repository" of the nation's cultural codes [2] (Fig. 1).

February 21
INTERNATIONAL
MOTHER TONGUE
DAY

Fig. 1. There are many languages in the world, but only the mother tongue is the "soul of the people"

The purpose of this article is to study modern technologies for teaching languages, to identify their role in the development of a global technogenic civilization.

2   Methods

The methodological and theoretical basis of the article are publications of recent years devoted to multilingualism, its development, distribution and expansion on a global scale, the use of various technologies, methods and forms of teaching foreign languages, mainly English [3–5]. In the course of the study, methods of analysis and synthesis, systematic and interdisciplinary approaches were used, which made it possible to consider and generalize the results of research in various scientific disciplines: linguistics, psychology, sociology, engineering.

3   Results

The development of science, engineering and technology led in the 90s of the last century to the accomplishment of the information and computer revolution, which brought to life many new technologies, technical means and devices. First of all, it is necessary to note information and computer technologies that have changed significantly the communication technologies that are widely used in language teaching. The functional properties of computer and communication technologies have increased the potential of the educational process [6].

This is reflected in:

1)   virtually unlimited possibilities for collecting, storing, transferring, transforming, analyzing and applying knowledge;

2) increasing the availability of education, expanding the forms of education, including distance learning, through which the negative impact of the coronavirus pandemic on the educational process was reduced;

3) ensuring the possibility of obtaining continuous education and advanced training during the entire active period of people's lives;

4) development of personality-oriented, additional and advanced education;

5) significant expansion and improvement of the organizational support of the educational process (virtual schools, laboratories, universities, etc.);

6) creation of unified information and educational learning environment not only for one country, but also for the world community as a whole;

7) ensuring the independence of the educational process from the place and time of training;

8) providing the opportunity to choose an individual learning path;

9) improvement of the methodological and software of the educational process;

10) development of independent students' search activity.

These capabilities of computer technology made it possible firstly to develop new teaching technologies - computer technologies, and then, on their basis, improve the educational process [7]. The conceptual technological principles of computer learning technologies are:

1) the principle of consistency, determining the methodology of computer learning technology, which, on the one hand, is based on didactics, psychology and sociology, and on the other hand, on management theory, computer science, systems engineering, ergonomics, design and other areas of science and technology;

2) the principle of modeling the student's learning activities in the information and computer environment;

3) the principle of mediation of subjects' communication of the educational process through the created computer environment and communication technologies that provide the ability to work in on- and off-line modes;

4) the principle of interactivity, which provides training with the help of special means and operational subjects' feedback of the educational process;

5) the principle of adaptability of algorithms for managing the student's cognitive activity;

6) the principle of distribution not only of the teaching material, but also the subjects of the educational process;

7) the principle of openness of the system to the connection of other systems and modules;

8) the principle of variability in the preparation, updating and design of training materials;

9) the principle of invariance, which provides the possibility of creating computer tools for the preparation of training and monitoring courses;

10) the principle of continuity and integration, based on a reasonable combination in educational process of modern information technologies and traditional approaches to the organization of education;

11) the principle of ergonomics, which makes possible to eliminate the possible negative effects of computer hardware and software on the health of the user [8].

The next important step in the educational process, including language learning, was the rapid development of mobile technologies, mobile telephony and devices, the development and downloading of applications (Puzzle English, Learning English, Polyglot, etc.) [9, 10]. As a result of this, a new learning technology has emerged - mobile learning (m-learning), which is understood as a form of organizing an autonomous and individualized educational process, where the main technology is mobile devices, with the help of which students can form and improve language skills and communication competencies without only during classes in the classroom or student lecture hall, but also at any convenient time and place [11].

Mobile learning methods include:

1) programmed learning (work according to a given program);
2) multimedia learning (simultaneous auditory and visual perception);
3) student-centered learning;
4) involvement in active or interactive activities;
5) gamification - the use of game elements in non-game processes (Fig. 2 and 3).

Fig. 2. Learning English at gateway school Malta

Fig. 3. Intercultural learning at EIL Ireland

Based on these methods, mobile learning applications can be divided into groups according to the priority learning method:

1) Professional multimedia applications (texts, images, audio and video materials).
2) Language social networks (language exchange with native speakers).
3) Interactive application-courses (interaction of the student with the application) for learning vocabulary, grammar and other aspects of the language [12, 13].

4 Discussion

The development of computer and mobile technologies - modern and promising learning technologies - has significantly increased the effectiveness of teaching foreign languages

[14, 15]. But this does not mean that the problems with the study of languages have now disappeared [16]. Obviously, new technologies related to artificial intelligence should come to the rescue. Artificial intelligence is one of the most relevant and discussed topics in science, business and the media today. Modern society is already familiar with "intelligent assistants" such as Siri, Alexa and Google Assistant [17]. There are high hopes for artificial intelligence, as it should help not only in learning foreign languages, but also in translating spoken language. He can become an intermediary translator between subjects speaking different languages. It is possible that artificial intelligence can also become a true "comrade" for generation "Z", which cannot imagine its existence without mobile devices [18]. And in addition to this, music that this generation willingly listens to and which contributes to the study of foreign languages [19].

Further development of technologies for teaching foreign languages will inevitably lead to increased social interactions and multiculturalism throughout our globe [20] (Fig. 4). Especially when the space Internet and mobile communications will form a single whole. Then relationships and communications between people will become networked (Fig. 5). And then it will be possible to say that a global technogenic civilization has been formed.

Fig. 4. Communication in social networks   **Fig. 5.** Global network communication

References

1. Ostapenko, P.S.: Multilingualism: problems of definition and main directions of research in modern linguistics. Russ. Humanit. J. **7**(3), 232–240 (2018). https://doi.org/10.15643/libart rus-2018.3.6
2. Kytina, V.V., Ryzhova, N.V.: The phenomenon of polylingualism in a limited language environment. Bull. People's Friendsh. Univ. Russ. Ser.: Quest. Educ. Lang. Prof. **14**(4), 611–620 (2017). https://doi.org/10.22363/2312-8011-2017-14-4-611-620
3. Oveshkova, A.N.: Working with English corpora as a means of promoting student autonomy. Educ. Sci. **20**(8), 66–87 (2018). https://doi.org/10.17853/1994-5639-2018-8-66-87
4. Baranovskaya, T., Shaforostova, V.: Learner autonomy through role plays in English language teaching. J. Lang. Educ. **4**(4), 8–19 (2018). https://doi.org/10.17323/2411-7390-2018-4-4-8-19
5. Ryabkova, V.V.: Mobile devices and applications for the development of language skills (on the example of the English language). Perspect. Sci. Educ. **2**(38), 320–326 (2019). https://doi.org/10.32744/pse.2019.2.24

6. Mamedova, M.A.: Computer technologies in the field of education and their stages of development. Uchenye zapiski universiteta im P.F. Lesgaft. **3**(193), 264–267 (2021). https://doi.org/10.34835/issn.2308-1961.2021.3.p264-267

7. Zheleznyak, I.L., Kaleeva, Z.N., Kraevaya, N.A.L.: Ways to improve the educational process based on the use of computer technology. Mod. High. Sch. Innov. Aspect **13**(1), 61–68 (2021). https://doi.org/10.7442/2071-9620-2021-13-1-61-68

8. Krasilnikova, V.A.: The Concept of Computer Technology Learning. OGU, Orenburg (2008)

9. Danilina, E.K.: Analysis of the experience of introducing mobile technologies for teaching a foreign language at the university level in Russia and abroad. Soc. Soc. Psychol. Pedag. **4**, 1–6 (2018). https://doi.org/10.24158/spp.2018.4.18

10. Arokiasamy, A.R.: A systematic review approach of mobile technology adoption in higher education. Econ. Manag. Sustain. **2**(2), 48–55 (2017). https://doi.org/10.14254/jems.2017.2.5

11. Troshina, Yu.V., Verbitskaya, N.O.: Mobile learning in a foreign language: concept, functions, models. Mod. Probl. Sci. Educ. **3**, 490 (2015). http://bit.ly/2KIXxdi. Accessed 12 Sept 2022

12. Ilyushkina, M.Yu., Sheinkman, A.M.: E-learning in teaching English: content analysis of mobile applications. Stud. Humanit. **2**, 1 (2019). https://st-hum.ru/content/ilyushkina-myu-sheynkman-am-e-learning-v-prepodavanii-angliyskogo-yazyka-analiz-kontenta. Accessed 12 Sept 2022

13. Koroleva, D.: Always online: mobile technology and social media usage by modern teenagers at home and at school. Educ. Stud. **1**, 205–224 (2016). https://doi.org/10.17323/1814-95-45-2016-1-205-224

14. Samosenkova, T.V., Savochkina, I.V., Goncharova, A.V.: Mobile learning as an effective educational technology in the classroom in Russian as a foreign language. Prospect. Sci. Educ. **2**(38), 307–319 (2019). https://doi.org/10.32744/pse.2019.2.23

15. Altaher, A.W.: Performance expectancy with usability relevance for m-learning technology in education. Bull. NSU Ser.: Inf. Technol. **15**(4), 5–13 (2017). https://doi.org/10.25205/1818-7900-2017-15-4-5-13

16. Titova, S.V.: Didactic problems of integrating mobile applications into the educational process. Bull. Tambov Univ. Ser.: Humanit. **21**(7–8), 7–14 (2016). https://doi.org/10.20310/1810-0201-2016-21-7/8(159-160)-7-14

17. Esionova, E.Yu.: Artificial intelligence as an alternative resource for learning a foreign language. Humanit. Soc. Sci. **3**, 155–156 (2019). https://doi.org/10.23683/2070-1403-2019-74-3-155-166

18. Tatarinov, K.A.: Mobile learning of generation Z. Baltic J. Humanit. **8**(2), 103–105 (2019). https://doi.org/10.26140/bgz3-2019-0802-0024

19. Dorairaju, R., Jambulingam, M.: The role of music and M-learning in English: vocabulary gain among tertiary students. J. Lang. Educ. **3**(2), 39–44 (2017). https://doi.org/10.17323/2411-7390-2017-3-2-39-44

20. Taskaeva, E.B.: Multilingualism in the modern world: cultural traditions and areas of research. Bull. Siber. Stat. Univ. Commun. Humanit. Stud. **2**, 50–57 (2017)

The Four-Dimensionality of Utility as a Philosophic Category

Iskender A. Gaparov(⊠)

Samara National Research University, 34, Moskovskoye Shosse, 443086 Samara, Russia
sarov-1@mail.ru

Abstract. The "utility" category is vital for the philosophy of technology as well as the modern thinking and reflecting on the technological systems and devices. That is why it requires the thorough examination. In order to achieve the goal, the four utility dimensions are to be analyzed. Each dimension is to clarify a certain problematic issue. Ontologically, it may be asked as how does "utility" determine being? Gnoseologically, what is the place occupied by "utility" in the area of cognition? Axiologically, what is the place of "utility" among the value primalities? Praxeologically, how does "utility" affect the act? The methodological ground of the upcoming study is the contrastive-comparative analysis of concepts suggested by domestic and West European thinkers for their compatibility with ontological, gnoseological, axiological and praxeological dimensions of the "utility" category. For this purpose, we have to pay our attention to the ideas by Ch. S. Peirce, W. James, F. Schiller, E. Coreth, P. Engelmeyer, F. Dessauer, T. Kotarbinsky, D.I. Dubrovsky, M.S. Kagan, A.N. Ognev. The result of the present research is presented via the determination methods of the utility notion in accordance with the above-mentioned dimensions. Therefore, ontologically, "utility" acts as an effect affirming the subject in his existence, gnoseologically – as a true idea, axiologically – as a gradation of aspirations, whose sum could be represented within a phenomenon of the will, praxeologically – as a rational balance between the means and the goal, with a number of subjective qualities, which constitute the basis for maintaining their interrelation.

Keywords: Utility · Being · Cognition · Value · Activity

1 Introduction

In the age of technological innovation, the need of comprehension and determination of the category of utility becomes significant, since it, while acting as an attribute inherent in the essence of something objective or subjective, sets the reporting systems for managing them. As the concept of reason (a category or an idea), it acts as the principle of implementation allowing to perform coordinated actions aimed at achieving correctly set purposes. In addition, "utility" acts as an expression of engineering. These are the fundamental issues which were directly or indirectly discussed in the works by E. Kapp [1], A. Espinas [2], F. Bon [3], P.K. Engelmeyer [4], F. Dessauer [5] and which still continue to be conceived in the field of: 1) technology – by V.G. Gorokhov [6], V.M. Rozina [7],

P.D. Tischenko [8], N.A. Yastreb [9], I. A. Gerasimova [10], D. G. Podvoyskiy [11]; 2) semiotics – by A.Yu. Nesterov [12], I.D. Nevvazhay [13]; 3) epistemology – by R. Egré and C. O'Madagain [14], G. Carranante [15]; 4) AI (Artificial Intellect) – by A.V. Rezayev and N.D. Tregubova [16]; 5) CSG (Complex System Governance) – by S.V. Keating and P.F. Katina [17]. A significant deficiency of these studies is the one-sidedness in terms of defining the concept of utility, since it is not something closed only by existential, cognitive, value-based or active areas. "Utility" is the result of intersection between the given dimensions, and thus their categorial expression. Therefore, the fundamental purpose of the forthcoming study shall be defined in clearing these dimensions, which, if not analyzed, make "utility" either an empty abstraction or a mental fiction.

There can be four of these dimensions – ontological, epistemological, axiological and praxeological, in accordance with the studies on the categories of ideal, consciousness, faith in the works by D.I. Dubrovsky. His methodological basis (the analysis of category within four dimensions) is the most preferable since it reveals the essence of activity, which was formerly obscured by ethics or aesthetics and thus served as the expression of "beautiful" or "benign". Also, according to D.I. Dubrovsky, these dimensions are no more and no less due to the arrangement (structure) of consciousness, through the prism of which the most part of phenomena serving as the basis of the category derived from them, passes (by which it is conceived). Finally, it should be noted that the four dimensions represent the content of the category, therefore, on the one hand, they "are not reduced to each other, i.e. each of them may not be logically derived from another one or may not be reduced to it (their fundamental value is manifested in it)" [18, p. 59], while, on the other hand, they "are mutually dependent, meaning that consideration of the problem in one categorial plane assumes (admits, and, in a number of cases, even urgently demands) its consideration in the rest" [18, p. 59].

Hence, ontologically, the question regarding "utility" could be: how does "utility" determine existence? Can existence be useful? The answer should be the clarification of the existential characteristics, in whose context, "utility" may influence a cognizing and acting subject. This question could be solved when appealing to the ideas by A. Schütz, Ch. S. Peirce, P.K. Engelmeyer and E. Coreth. Gnoseologically, the issue regarding "utility" is represented as follows: what is the place occupied by utility in the area of cognition? What makes "utility" cognizable? The answer to these questions should be the clarification of differences and similarities between the components ("utility" and knowledge about it) in the works by W. James, F. Dessauer, as well as the forms, in the context of which "utility" could be captured in the area of cognition. Axiologically, the question related to determination of "utility" may sound like "could "utility" be a value?" Is it given as an element of hierarchical organization of value intentions or is it coordinate with value? The answer to the axiological question should be clarification of the difference between "utility" and "value" and the continuum within which their mutual places are conceived, based on ideas by M.S. Kagan, A.N. Ognev and F. Schiller. And the last one, praxeological question could sound as follows: does "utility" act as an active principle? What is it in the context of activity? A possible solution to this issue could be the clarification of initial basics in the teaching by T. Kotarbinsky, which represent the actions (acts) and the limits within which they are rational.

2 The Ontological Dimension of the Utility Category

To clarify the ontological dimension of the utility category means to define the existence as "useful" and look what would follow from this. Hence, the fundamental question of the ontological dimension could be reduced to how does "utility" define the existence? Since that is existing what does not contradict to logical laws, common sense, inasmuch as "utility" must follow these conditions based on the fact that it is an integral part of the existence. In addition, even the negative sense of the notion of inexistence criticized by Parmenides [19] and redefined by him as a kind of the existence as a virtual beginning, leaves an imprint on categories, including "utility". Problematically, it is similar to "inexistence" since it not subjected to sensing and its holistic perception, as well as reflection on it, are impossible. Such a state of affairs, according to A. Schütz, could evidence of the presence of a natural attitude in which a human spends almost his entire life. In it, "the world is from the outset not the private world of the single individual, but an inter-subjective world, common to all of us, in which we have not a theoretical but an eminently practical interest" [20, p. 534]. This world representing a result of stable combination of experimental data is maintained by the interests of interacting subjects, for whom the being is determined by the goal implementation forms. It can be said "a pragmatic motive governs our natural attitude toward the world of daily life" [20, p. 534]. An exception is hopeless situations that require reflection. As an integral aspect of life, they become a problematic point of an indefinite phenomenon requiring an increased attention to life in all its time intervals. The reflective component turns out to be powerless, since this phenomenon is given in the integrity of the current process of action, and "consideration that the inner experiences of our bodily movements, the essentially actual experiences, and the open anticipations escape the grasping by the reflective attitude shows with sufficient clearness that the past self can never be more than a partial aspect of the total one which realizes itself in the experience of its ongoing working" [20, p. 541].

On the other hand, if "utility" cannot be grasped as a phenomenon, then one can try to turn to the transcendental grounds for the possibility of existence and cognition, which allow talking about the phenomenal world. The utility category may be referred to them since it constitutes the reality in line with "truth", "good", "beauty". These grounds were analyzed by P.K. Engelmeyer in his "Philosophy of Technology", where he defined each of them as an aspiration. The sum of aspirations is equal to the will, which allows determining its implementation in a particular direction. In the spirit close to P.K. Engelmeyer, E. Coreth tried to define the ontological dimension of the utility category. His "utility" acts as one of the kinds of existence-conformable goodness of being, whose integral features are attractiveness and undoubted positivity. They make possible the relation of one being to another, if the first one "corresponds to the nature or peculiarity of an aspiring person, "suits" to him (bonum conveniens), i.e. when it is useful or benefits its existence-conformable self-deployment" [21]. Thus, the "utility" should be understood as anything that represents the possibility to strive for goodness, the value of which lies in achievability, i.e. in limitation and peculiarity, since it will not be possible to achieve something that is essentially infinite.

The ontological dimension of the utility category may be tried to be reconstructed differently based on the ideas by D.I. Dubrovsky. Then the utility category could represent something acting on a subject and causing a positive reaction or some variation in response, "a violation of the natural order of things, including the transformation of the horizons of imagination, which, in their turn, indicate opportunities for new technical objects" [22, p. 5]. So, in learning a foreign language, the work of the brain is activated, whose dynamics positively affects the connections of neurons, resulting in their becoming more plastic and less rigid, which is a beneficial factor for the brain. This example also results in the fact that in terms of the ontological analysis of "utility" the researcher should be interested in the forms in which it can be given, what are its functions, and implementation methods. Since "utility" is rather a subjective than objective reality evaluation criterion, the shape taken by "utility" at the subject's level, will be of importance. In this regard, such shape as idea is the most settled one. It is understood as "a fundamental principle or a deep and original thought (theoretical, artistic etc.), which has a powerful systematizing, heuristic, motivating potential" [23, p. 193]. Conditioned by ideal as noumenal world, the ideas in fact exist objectively. Each of them correlates with the thing as a means. Their objectivity reflects the transpersonal forms. This aspect is obvious in the technology phenomenon, which is the form of objectification of "utility". It represents the connection between an idea and its implementation. The absence of the latter results in an ontologic gap suggesting the emergence of fictitious beings.

To complete the clarification of the ontological dimension of the utility category, the ideas of Ch. S. Peirce should be paid attention to. His "utility" is set as an initial ground of the reality producing the rules possessing practical value. And if a purposeful act is not possible without guiding rules, then in the absence of goals, the subject my not understand what is he doing and why. Instead of understanding that this unclarity is merely subjective, an individual begins to imagine that he contemplates the quality of the object as something self-sufficient. For Ch. S. Peirce, this approach is unacceptable. The action should be supported by prerequisites serving as the basis for the benchmarks. "Utility" may act neither the first one nor the second one. It represents their connection by forming the reality that suits the best to a human.

In principle, any version of reality as an "object of finite general belief" [24, p. 321] possesses a number of principles capable of redirecting the consciousness. Thus, in comparing religion with science it is a fact that their methods of knowing the world may be different. Because the things that are taken by religion as a subject of belief, should represent an experimentally confirmed evidence for science. When it comes to their similarities, the prerequisites of belief and doubt should be accepted. The belief is understood as "a habit or rule of action, and a conscious action of a human being is always aimed at achieving some goal, therefore the true belief may be considered the one which may cause a successful action resulting in an attained goal" [24, p. 304]. Belief is connected with the desire reflecting the need representing the line of thought allowing to satisfy the need. The latter may be definable in an expression of "experience of believing" serving as an indicator of a formed habit defining a number of actions and affecting the human attitude to the world in general.

The other thing is when habits based on the belief, start causing harm or conflict with the time. Then they should be doubted, i.e. hesitancy should be shown regarding the future course of action, while taking the irrationality of an act as the truth. The hesitancy may not be the result of doubt, since it is capable to destroy the value coordinate system. The doubt should be productive. It can be such only if "arise from some indecision, however momentary, in our action" [25, p. 290] regarding a particular state of affairs and things, by forcing to explore it until the dissatisfaction is not eliminated. This is how new belief and habit are shaped. Their advantage is that they guide a person in a non-thetical (non-reflexive) way. This is the basis of the connection between doubt and belief, whose unity can be represented in "utility".

The ontological dimension thus defined is not a project, but a pattern called to help in searching for other dimensions. However, the dimension is unthinkable to be presented completed, if its interpretation ways may be different. Here the certainty of existence is the only criterion. If the utility is existent, its being should consist in the positive effect. Since it refers to the implications, the utility is acceptable to be talked about as a connection between the prerequisite and the implication in the ontological dimension within the pragmaticism of Ch. S. Peirce should be paid attention to. In turn, the substitution of the utility category by the idea turns into a transfer of all happening in reality that is sensually accessible to us into the area of the noumenal world or imaginary realities.

3   The Gnoseological Dimension of the Utility Category

In the gnoseological aspect, "utility" according to D.I. Dubrovsky may represent the act of acceptance of a particular content by a subject. Its content should be given in sense and perception, fixed in the form of thinking. Therefrom comes the question of gnoseological dimension, how does "utility" contribute to the cognition process? The answer should be based on its gnoseological grounds. This category may include such concepts as truth, lies etc. A mistake could be the substitution of "utility" by knowledge since the latter is the result of reflection but not its prerequisite. In this connection it can be said that a subject cognizes when he becomes confident that something affects him. For this reason, it would be legitimate to divide "utility" into "utility as a process" and "utility as a result". Consideration of its empirical and theoretical level could allow understanding the forms of interaction between them and the place taken by "utility" in cognition acts. For this purpose, referring to pragmatism by W. James would be inevitable. The ideas of F. Dessauer are also worth referring to for clarification of the modalities within which "utility" may lie.

Adopting a position of nominalists, W. James also believes that any substance is "a spurious idea due to our inveterate human trick of turning names into things" [26, p. 87]. By replacing names by facts, which they designate, nominalists believe that the properties of things should be contained in names; otherwise, the connection between substance and name will not be available for clarification. Particular attention is paid by James to the definition of "Holy Mysteries" in the communion, the properties of which are unchanged, but at the same time it "has become the very body of Christ" [26, p. 88]. The establishing occurs not by connection or change, but symbolically. The sensual properties of the communion remain the same, only the way of determining of a particular thing's substance changes. Such a line of thought is dictated by the teachings of J. Berkeley [27]. Moreover, it is connected with the perception of an idea, which, as the basis of perception, is not conceivable without it. Another side of this approach is the achievements of Ch. Darwin [28], who discovered "biological accidents". They are associated with adaptations, since, on the one hand, they allow substantiating the mechanism of adaptation, and, on the other hand, to introduce a creator providing for the world order in such a way that each phenomenon is given its peculiar place. But according to W. James, for a human it is important that he/she can form such abstract notions, whose value is defined by the direction of his/her activity area. Thus, "were we lobsters, or bees, it might be that our organization would have led to our using quite different modes from these of apprehending our experiences" [26, p. 171]. Since the reversibility is impossible, the sense producing the concepts and the reason acting as the aspect of their synthesis, are left to be satisfied with. These capabilities are the means of logical processing of facts aimed at ordering of sensory impressions. In this case what is meant by "utility" in the gnoseological terms is not a form of being but a connection law (rule), not a "simple" idea but a true idea.

The true idea is understood by W. James as the capability to learn certain rules, laws and sequences through their confirmation in practice. On the contrary, the thing acting a false idea is not subject to verification, therefore it is knowingly erroneous. The difference between "utility" as an idea from "true idea" consists in its modality. If an idea is in reality, it is rightful to be thought both existing and inexisting at the same time. It is similar to the thinking tried by either Plato and naturalists of early 20th century. According to Plato, a thing represents both matter and an idea, and according to the experiments of physicists, a photon is both a particle and a wave. Reality itself is dependent on the subject. It is actualized, i.e. it is given in sensory experience and is related to an individual. A true idea cannot be "put aside" since it represents the tool (vehicle) of a real action. Similarly, "utility" cannot be defined in terms inaccessible for exploration. Its existence is defined both by the availability (within the consciousness) and the absence (outside the consciousness). If the researcher accepts the first paradigm, then he/she becomes a gnoseologist who tests the idea in the light of truth; if he/she accepts both, then he/she is an ontologist, because of trying to affirm the idea by transferring one world into another.

In general, according to W. James, gnoseological transfer is more productive that the ontological one, since the idea passing the check for imbalance, finely becomes useful, i.e. beneficial. Hence, an undeniable fact is that "true ideas would never have been singled out as such, would never have acquired a class-name, least of all a name suggesting value, unless they had been useful from the outset in this way" [26, p. 204] since something not providing for "utility" as a property is destined for remaining an imaginary phenomenon. Therefore, a connection is formed between "truth" and "utility", which is brought under idea. Therefore, something contemplating the process of check for correspondence with the applicability to the need of human. That what fulfils its purpose in experience, is endowed with utility. The idea in itself is either a fiction or a thing in itself. In order to make the idea reality, it should be defined both by the truth and utility.

The antithesis of the above approach, according to which "utility" is defined as a consequence of our knowledge of the world, is the teaching of F. Dessauer. As opposed to W. James, he takes a realistic point of view, whose basis is Platonism. Its essence consists in the diversity of worlds, each interacting with the others. The interaction is determined by features peculiar to a person, since "all the layers of being are connected in it as an existing unity" [29, p. 48]. Thanks to its creative abilities the transfer of ideas from the noumenal world to the sensual one is legitimized. However, the creative process occurs in "four-dimensional" reality which consists of the dimensions of 1) research; 2) inventions; 3) processing; 4) pure forms. The fact of contact between worlds is determined by the finalist worldview, defined in terms of means, tasks and goals. Through the notions of finalism, "the form is imprinted in the subject by a human "creator" based on his ideas about the goal" [5, p. 91]. A technical device, while representing the idea in the sensual form, fulfills the task. Having surrounded himself with technical devices, a person expands the limits of freedom, while revealing opportunities, which were unattainable. But still, the sensual world limited by the laws of nature, imposes prohibitions represented in the form of rules of use and application, resulting in narrowing of the timeless essence, located in the "fourth kingdom" and possessing a variety of parameters, in the course of its implementation because "the establishing of a technical object is performed in all phases such as the formation, structuring, management, actualization of materials and energies into a really existing unity having existence and power" [5, p. 91]. In this way, utility represents the balance between the tasks and goals or acts as a form of reflection of the "world of ideas" as opposed to the "world of things".

Since the balance represents the interaction between two (and more) elements, thinking "utility" without it is practically impossible. However, it will not be possible to make this connection accessible for contemplation due to the insufficient power of the sense organs. What remains is relying on the reason since it has the ability to bring rational categories under general rules that find a place in nature and are imprinted in the idea. This allows determining "utility" on the practical and theoretical side. On the practical side, it means a particular method of implementing the rules, and on the theoretical, their combination. Instrumentally (practically) there can be one rule for the implementation of a technical device, while theoretically, this rule is one of many allowing to connect tasks with goals (one acceptable solution per function). That is why the question why a technical object cannot be created of improvised things, may be answered simply – because there is connection between an idea and its representation. Pragmatists offer not to pay much attention to the goals, but focus closely on the results. To the contrary, F. Dessauer suggests that such an approach is inacceptable, since obtaining the necessary effect would require many years.

4   The Axiological Dimension of the Utility Category

The axiological dimension of "utility" should clarify its value primalities (limits). Such a paradigm is also supported by A.N. Ognev. His "utility" is a super-task of economic activity, which "is carried out as an intelligible postulate of the removal of the precedent of the establishing grounds, whose model situation is presented in the concept of a preserving cause" [30, p. 17]. Some other approach to the axiological dimension of the utility category belongs to M.S. Kagan, who prefers to use the axiosphere as the justification of the value dimension, which is defined by the aspirations and desires, resulting in "utility" and "value" becoming the ways of expression of the existence. The humanism by F. Schiller could be an antithesis to it, which removes the opposition of the utility and value categories for the benefit of the practical efficiency.

M.S. Kagan believes that "utility" contains a subjective and an objective principle, while the value is aimed at the subject only. One can arbitrarily discuss the object as something that correlates with the subject, while losing its visibility. The value, not accessible for contemplation may seem a fiction. To prove its factuality, "utility" is required acting as a pragmatic relation aimed at the implementation of value as a goal-setting paradigm. M.S. Kagan's saying could be indicative here, where he says that "value is a specifically cultural phenomenon unknown to animal life, while utility characterizes the biological level of existence to the same extent as the socio-cultural one, nutrition is useful both for the animal, and for man, and education and training are useful for human life in the same way as nutrition" [31, p. 77]. The issue here is that a narrow definition of value leads to the blurring of the "bearer of value" and "value" in itself, which was noticed by M. Scheler [32], who clarified the difference between the former and the latter, as between an object and meaning for the subject. Values themselves are objective, while speaking about them is subjective. It is determined in the significance of one for the other, while the ""existence of values" is out of touch with the evaluating subject" [5], and "utility" is one of the reactions of the evaluating subject or a response to contact with value. For "utility", it is not the content that is significant, but the form of manifestation, since it is it that sets the tasks and goals, in conjunction with which "utility" will be determined.

The determination of evaluative judgment is also important for considering the initial boundaries of the axiological dimension. As long as the evaluation is the replacement for a set of terms, therefore as it has multiple meanings determined by their correlation with the object/subject. For example, an examination score reflects gnoseological judgment, since it considers not the examiner's attitude to the spiritual world, but the correspondence of knowledge to reality. However, expert evaluation implies the criterion of technology applicability. As we see the things defined in useful evaluations may be independent from the subject implementing the technology as intended. In this case, the creature acting not by the principle of utility but by the principle of impartiality, represents "a thing in itself". However, such an approach is a dead end both in thinking and in action, if not considering the places occupied by "utility" and "value". The axioshpere may be one of the options of their mutual position, which acts as a continuum setting an evaluative coordinate system. The highest level is the value one, and the lowest is the utility one. Their functionality depends on whether there occur the aspirations "in material and practical and (or) biophysiological processes of human existence or they are born by the level of living activity at which human becomes the subject of activity from the object among objects" [31, p. 82], meaning that the conditioning of "value" by "utility" is necessary since if value presents not the aspiration but reflection on its limits, then "utility" is the aspiration setting the directions to the will capabilities aimed at satisfying the needs. The absence of one of the components prevents from value judgement.

The antithesis to axiosphere by M.S. could be represented by humanism by F. Schiller. According to Protagoras, he reduces theoretical cognition to practical one. This action results in the fact that the difference between "utility" and "value" becomes quantitative out of qualitative. The very opposition of categories represents a contradiction. These issues are raised by him in the dialogue of ""Useless" knowledge. Discussions about pragmatism", where while making Plato and Aristotle as the main heroes, F. Schiller tries to rehabilitate Protagoras following which there is "no opposition between practical and speculative wisdom, because the former grows out of the latter and always remains derivative and secondary, and subordinate, and the last" [33, p. 16]. This position denies the self-sufficiency of the speculative reason, because everything that exists is a reflection of ourselves. However, the latter is characteristic not of an individual subject, but of a group of persons concluding an agreement on practical provisions. Their vision must be shared in order to prevent strife and support the common living with well-being and happiness.

Gradually, the difficulty of generalized practical perception turns into isolated provisions losing their connection with reality. They create special forms of value mindset, which begin to dictate the rules of the field of practical experience. For F. Schiller such a state of affairs and things is considered unacceptable, since it would rather destroy the world than change it for the better. In fact, pure contemplation, on which metaphysicians often rely, is the antithesis of practical applicability. It presents a set of postulates and requirements for experience "since we need them, to make it cosmos convenient for living" [33, p. 18]. They may contain the provisions of natural science, mathematics, etc. To separate them from their own applicability means to create a gap between cognition and action, as could be confirmed by the example of Aristotle about the course of training recommended by the Indian gymnosophist, where there is no connection between the guiding truths and the methods defended.

In order to get rid of the difficulties, it is proposed to determine the actual value of theoretical knowledge by its applicability. Outside of this function, such "truth" is potential. After all, the use can exist where there is a desire. In turn, the desire to apply any truth for its intended purpose depends on the presence of a difference in understanding and misunderstanding. If a person is not interested in any state, then it will be aimless and meaningless. If a person wants to comprehend a certain state of affairs, then this means that he prefers to discover the meaning, i.e. explore the situation for possible consequences. If it corresponds to the initial expectations, then it can be called true, but this does not mean that it is absolutely accurately the same. Rather, its validity is taken on faith due to the fact that in some cases it has passed the check for conformity. Hence value is "a seeming accident and by no means a complete guarantee of truth" [33, p. 20], the necessity is the pertinence, and self-evidence "means only that we have stopped asking questions about it, or that we have not yet started to do so" [33, p. 20].

Important for this part of the study is also the question on clarifying the role of useless knowledge, because it directly relates to the criticism of the approach of W. James regarding the understanding of truth. Since truth is understood as a successful option of behavioral acts or deeds, its connection with the completeness of semantic information about the object of cognition is interrupted. Hence follows the duality of truth. It can be thought as: 1) "a class-name for all sorts of definite working-values in experience" [26, p. 68]; 2) "what would be better for us to believe" [26, p. 77]. But in his article "Ethical Basis of Metaphysics" W. Schiller criticizes James for such separation of notions and for his assuming the prerequisite of "pure reason" as one of the possible ones. On the contrary, he believes that this kind of belief is impossible, since what is practically understood as "pure reason" cannot be clarified. The reason is the means in the battle for survival. Its abstractions should have practical value and real application. Otherwise, the reason is "a monstrosity, a morbid aberration or failure of adaptation, which natural selection must sooner or later wipe away" [34, p. 437].

Thus, if the reason is pragmatic, it is possible to talk about "useless" in cases when people are not capable to use or refuse to use something. Also, that is indirectly useful and that could be connected to unjustified time spending, can also be understood as useless. Therefore, what seems "useless" at first glance does not have sufficient grounds to claim the status of "useful". Thus, it follows that what is called value is a "baseless" utility, i.e. the implementation not supported by the grounds. What is usually understood as "utility" is a set of principles, in relation to which it can be noted that either through them something is realizable and beneficially affects life, or it affects life by the fact of presence, proven by faith.

5   The Praxeological Dimension of the Utility Category

In praxeological dimension, "utility" may be uncovered as an activity and intentionality. If so, "utility" should reflect the will, desire, purposefulness, in the form of a principle, not a concept, i.e. something that acts as a preserving cause, and not as a means of conversion, change, or transformation. "Utility" in the praxeological dimension can only be represented in the analysis of actual actions, but not in imaginary actions. Regarding this clarification of the dimension of "utility", in addition to D.I. Dubrovsky, it would be good to refer to the praxeology founder T. Kotarbinsky. In his work "Treatise on Good Work" he raises the question of structure of management activity and the grounds on which it is based. Moreover, in the same place, through the concept of evaluation and degrees of its gradation, he precedes "the study of the problem of the optimization criterion" [35, p. 14], which is closely related to the definition of the concept of utility, by economic cybernetics.

To ensure the best estimation, it is necessary to analyze the concepts reflecting the essence of this process. At the same time, assessments in the field of praxeology should be relied upon as the practical merits of an action from the point of view of its functionality. In this case, the main issue should cover the ways of determining and grading the functionality on the part of acting subjects. Functionality is clarified on the basis of success, supported by action that leads to a consequence, i.e. anticipated goal. Anything hindering the achievement of the goal is declared unsuccessful. Actions that indirectly contribute to it cannot be declared unsuccessful, because extraneous factors may make approaching the goal possible/easier in the future. However, the event may not also be called exactly successful/unsuccessful, since only the behavior is such. The behavior is evaluated on the basis of the intended goals and measures in relation to them. If a positive relationship is formed between them, then such an action can be considered useful, but only that, "which contributes to possibility or simplification of achieving this goal" [35, p. 106]. On the contrary, harmful in relation to the goal is an action preventing, hindering or precluding from achieving it. In turn, what is understood as utility and expediency of action is declared by T. Kotarbinsky to be interchangeable.

Unlike representatives of utilitarianism, T. Kotarbinsky believes that something defined in terms of "utility" or "expediency" should have a degree of "facilitation" gradation, since actions, incl. Their consequences may be declared unequal in relation to each other. Along with the degree of success (utility), there must be accuracy in achieving the goal, since what is associated with a set of actions may differ from the impact on the material and therefore be subject to gradation. So, on the one hand, a document cannot be half-signed, and the nail, in turn, cannot be partially driven. Due to the fact that the first one is not signed, it may not be accepted, and the picture, which rests on a half-driven nail, may fall while causing damage. On the other hand, the enterprise has the right to create a fabric, or to heat a living space guided by the approach to the needs of people. It turns out that, depending on the measure of the success of an action, this may be the degree of proximity to the goal, which is determined based on whether it is represented by an act or a specific product.

The above analysis of the forms of "utility" makes it clear that its connecting elements are means, goals and actions. However, it must be understood that in addition to the form, there is also the content, which implies everything promoting or hindering communication. This content of the "utility" category consists of different concepts, which need to be discussed in more detail.

One of essential concepts is the "precision" meaning the balance between the product and the reference as to detecting the similarities and differences. Depending on the design of the sample, the product itself sometimes needs to consider small fragments and side objectives that could affect the goal. Another notion of the praxeological dimension of utility is "diligence". It can be seen a set of actions aimed at creating conditions that could contribute to a better achievement of the main goal. In addition to diligence, "expertise" should be considered, which involves accounting for additional considerations when treating an object. Unlike diligence, expertise involves not the creation of conditions, but their development, which is supported by purity. "Purity" refers to the degree to which an object lacks features that might not meet the goals. Such traits are the result of careless exposure and insufficient expertise. In addition to purity in the praxeological dimension, it is necessary to note "effectivity", which is aimed at eliminating means that are unnecessary for the goal. In turn, effectivity is supported by "functionality", which is the sum of the advantages inherent in a set of actions. In order to be functional, the complex should have a sequence, as well as a greater speed of movement, less effort during performing actions, approaching to the planned movement, continuity of movement. Thus, functionality reflects the highest degree of automation of actions, which is supported by less subjective fatigue.

The one who possesses functionality performs actions with a certain reliability. The latter presupposes either an objective reliability that, under given conditions, actions will lead to a given goal, or an awareness of objective reliability. Reliability is its synonym. For the most part, it is relative, therefore, it manifests itself in the form of belief, because it considers a finite number of situations. In the context of "utility", rationality of actions is also necessary, which brings the correctness of the judgment under the events. Due to this connection, "a better adaptation of the plan to the material and to all possible circumstances of the action occurs" [35, p. 120].

The inability to consider these concepts, according to T. Kotarbinsky, leads, at best, to erroreous thinking associated with the adoption of an erroneous judgment; at worst, to the generation of an inappropriate impulse. Therefore, in an act, it is preferable to evaluate some or other action for possible errors, and not be guided by irrationality, spontaneity. In the context of "utility", the connection between means and end can only be limited if individuals consider the internal and external circumstances that influence their actions. The inability to separate the objective and subjective principles in an act leads to the fact that "utility" again begins to be replaced by "imaginary pictures".

6 Conclusion

The performed study represented the analysis of four dimensions of utility as a philosophic category. Each of those measurements had to determine the impact on the part of utility. Thus, ontological measurement was called to answer the question of impact of the utility on the existence. The answer to it was the "positive effect" produced on a cognizing or acting subject, i.e. somehow contacting with the internal and external world and thus affirming its existence. The total of positive effects may be reorganized into a natural attitude representing a form of expression of "utility" in the ontological aspect. To the contrary, gnoseological dimension had to answer the question on the role of utility in cognition. This question was answered by the laws or rules of connection represented by ideas. Their purpose lies in connecting different layers of existence. The form of expression of "utility" in gnoseological dimension is thinking. Axiological dimension is called to answer the question on the place of utility among value primalities. It serves the basis of the value mindset as a primary form of significance, therefore the form of expression is the value as the guide to action for achieving vital objectives regarding satisfaction of needs. And, finally, the praxeological dimension answers the question of the influence of "utility" on human activity as follows: "utility" is a form of expression of the entire activity consisting of a set of acts, among which the goals over them are consistent with the measures of their consistent implementation (positively). Therefore, that what was generally taken as "utility" in the classical form under the paradigm of reasonable balance between the goal and the means, has a completely legitimate right to claim a place among other categories, since each of the concepts reflects its existential, cognitive, axiological or praxeological aspects.

Acknowledgements. The reported study was funded by RFBR, within the scope of the scientific project 20-011-00462 A, Friedrich Dessauer's Philosophy of Technology: Epistemology and Anthropology of the Realistic Creativity Theory.

References

1. Kapp, E.: Grundlinien einer Philosophie der Technik. Druck und Verlag von George Westermann, Braunschweig (1877). (in German)
2. Espinas, A.: Les Origines de la technologie: étude sociologique. Paris (1897). (in French)
3. Bon, F.: Über das Sollen und das Gute. Leipzig (1898). (in German)
4. Jengelmejer, P.K.: Theory of Creativity. Librokom, Moscow (2010). (in Russian)
5. Dessauer, F.: Streit um die Technik. Verlag Herder, Freiburg in Breisgau (1959). (in German)
6. Gorokhov, V.G.: Engineering Sciences: History and Theory (History of Science from the Philosophical Point of View). Logos, Moscow (2012). (in Russian)

7. Rozin, V.M.: Technology as a time challenge: study, concept and types of technology. Philos. Cosmol. **19**, 133–142 (2017). (in Russian)

8. Tishchenko, P.D.: Double helix of life. Epistemol. Philos. Sci. **2**(48), 51–53 (2016). (in Russian)

9. Yastreb, N.A.: Rediscovering instrumentalism: the philosophy of technology in the works of the classics of American pragmatism. Tomsk Stat. Univ. J. Philos. Soc. Polit. Sci. **64**, 241–252 (2021). https://doi.org/10.17223/1998863X/64/23

10. Gerasimova, I.A.: Engineering knowledge in the technogenic civilization. Epistemol. Philos. Sci. **55**(2), 6–17 (2018). https://doi.org/10.5840/eps201855222

11. Podvoyskiy, D.G.: "Dangerous modernity!", or the shadow play of modernity and its characters: instrumental rationality – money – technology. RUDN J. Soc. **22**(1), 40–57 (2022). https://doi.org/10.22363/2313-2272-2022-22-1-40-57

12. Nesterov, AYu.: Semiotic Foundations of Technology and Technical Consciousness. Samara Academy of Humanities, Samara (2017)

13. Nevvazhay, I.D.: Classifications of norms in semiotic conception of norms. Human world: normative dimension – 6. In: Proceedings of the International Academic Conference: Proceedings of the International Scientific Conference, Saratov, 27–29 June 2019, pp. 38–49. Saratov: SEI of HE, Saratov State Law Academy Press (2019). (in Russian)

14. Egré, P., O'Madagain, C.: Concept utility. J. Philos. **116**(10), 525–554 (2019). https://doi.org/10.5840/jphil20191161034

15. Carranante, G.: The boundaries of perception: a conceptual engineering perspective. Philosophy. Université Paris Sciences et Lettres, Paris (2020). https://tel.archives-ouvertes.fr/tel-036 62720

16. Rezaev, A.V., Tregubova, N.D.: "Emotional utilitarianism" and the frontiers of artificial intelligence evolvement. Monit. Pub. Opin. Econ. Soc. Change **2**, 4–23 (2022). https://doi.org/10.14515/monitoring.2022.2.2127

17. Keating, C.B., Katina, P.F.: Complex system governance: concept, utility, and challenges. Syst. Res. Behav. Sci. **36**(5), 687–705 (2019). https://doi.org/10.1002/sres.2621

18. Dubrovsky, D.I.: The Basic Categorical Planes of Consciousness. Questions of Philosophy. Philosophy and Science, Moscow (2008). (in Russian)

19. Plato: Parmenides. Hackett Publishing Company, Indianapolis (1996)

20. Schütz, A.: On multiple realities. Philos. Phenomenol. Res. **5**(4), 533–576 (1945). https://doi.org/10.2307/2102818

21. Coreth, E.: Bases of Metaphysics (Translated from German V. Terletsky). Tandem Publ., Kiev (1998). https://gtmarket.ru/library/basis/5760. Accessed 15 Aug 2022 (in Russian)

22. Nesterov, A.Y.: Clarification of the concept of progress through the semiotics of technology. In: Bylieva, D., Nordmann, A., Shipunova, O., Volkova, V. (eds.) PCSF/CSIS -2020. LNNS, vol. 184, pp. 3–11. Springer, Cham (2021). https://doi.org/10.1007/978-3-030-65857-1_1

23. Dubrovsky, D.I.: The Problem of the Ideal. Subjective Reality. Canon+, Moscow (2002). (in Russian)

24. Melville, Y.K.: Charles Pierce and Pragmatism (At the Origins of American Bourgeois Philosophy of the XX Century). Moscow (1968). (in Russian)

25. Peirce, C.S.: How to make our ideas clear? Popul. Sci. Mon. **12**, 286–302 (1878)

26. James, W.: Pragmatism: A New Name for Some Old Ways of Thinking. Barnes & Noble, New York (2003)

27. Berkeley, J.: A Treatise Concerning the Principles of Human Knowledge. LONDON AGENTS. Kegan Paul, Trench, Trubner & Co., Ltd. (1904)

28. Darwin, C.: The Origin of Species. Signet Classics, New York (2020)

29. Dessauer, F.: Human and cosmos (Trans. German A.Yu. Nesterov). Semiot. Stud. **2**(2), 25–49 (2022). https://doi.org/10.18287/2782-2966-2022-2-2-25-49

30. Ognev, A.N.: Axiology in the System of Philosophical Knowledge. Samara University Publishing House, Samara (2017). (in Russian)
31. Kagan, M.S.: Philosophical Theory of Value. Publishing House "Petropolis", Saint Petersburg (2006)
32. Scheler, M.: Gesämmelte Werke. Bern, Munchen (1957)
33. Schiller, F.C.S.: The ethical basis of metaphysics (Translated from English R.G. Iferov. Vox). Philos. J. **18**, 1–23 (2015). https://doi.org/10.24411/2077-6608-2015-00002
34. Schiller, F.C.S.: The ethical basis of metaphysics. Int. J. Ethics **13**(4), 431–444 (1903). https://doi.org/10.1086/intejethi.13.4.2376273
35. Kotarbinsky, T.: A Treatise on Good Work. Economics, Moscow (1975). (in Russian)

Working Knowledge: What Do I Know About How I Do Something?

Alexander Nesterov$^{(\boxtimes)}$ (iD)

Samara National Research University, 34, Moskovskoye Shosse, 443086 Samara, Russia
aynesterow@yandex.ru, phil@ssau.ru

Abstract. The article discusses the works by Alfred Nordmann, Natalia Yastreb published in 2022 at "Semiotic Studies" and devoted to the problem of "working" or "technical" knowledge. Traditionally determined knowledge (as a justified true belief, as subjectively and objectively sufficient positing that something is true) describes the result of internalizing activity of cognition and is aimed at detection of an actual relation between the state of affairs and their language expressions. "Working knowledge" as a concept is introduced by Nordmann and represents a form of knowledge not on the state of affairs as such but on the method of achieving the target result in case of changing the state of affairs. To describe the method of implementation of working knowledge, a concept of epistemic practice is introduced by Yastreb. The article analyses the working knowledge as a technical knowledge in the context of philosophy of technology by Peter Engelmeyer and Friedrich Dessauer. As a result, it is demonstrated that the question about a "thing" as a reference object of the working knowledge is solved within the "three-act" concept. "Thing" as a reference is either an intellectual idea or a rational formation, or a sensuous artefact, i.e. a result of technical activity achieved by a subject at each of three stages, some sum of stages. A semiotic expression of each stage of the three-act is formed, the thing in each case is demonstrated to be the implementation of a semantic rule. As a general conclusion, the formulated by Nordmann task of generating the theory of working knowledge types is defined more precisely as the task of generating the types of epistemic practices in their general semiotic expression.

Keywords: Knowledge · Working knowledge · Epistemic practices · Perception · Semiosis

1 Introduction

In the first issue of 2022 Semiotic Studies, two articles by professor Alfred Nordmann [1] and professor Natalia Yastreb [2] were published, which raised the issue about the essence, structure and functions of working knowledge. The current research on the concept of knowledge in the context of the cognition theory and in the context of the activity theory are not so much abstract academic studies requiring discussion according to Plato [3], Kant [4], Russel [5, 6], as merely practical search for methodologic, general philosophic tools for generally valid technical expression of system of human subjectivity, creativity, eurilogy. The tradition of epistemological studies sets a question about knowledge as follows: "What exactly do I know when I know something?" or "What does 'to know' mean?", "What does the concept of knowledge mean?". It expresses the traditional receptive paradigm of the representational cognition theory defined by the understanding of the cognition process as appropriation by the subject (interiorization, subjectivization) of the state of affairs being external to the given subject and their explanation based on his experience. Cognition starts with a sensuous perception, it is expressed in rational and linguistic activity of thinking and ends with the formation of a concept in mind (intellect) allowing to correlate sensuous experience with its linguistic designation and thereby meaningfully, that is, with an interpretable volume and content in the formed concept, include this concept in the description and explanation systems. Answer to the question or "What does 'to know' mean?" provides for the construction of ontology (distinguishing between external and internal, subject and object, matter and form, etc.) and the theory of the cognitive process connecting the elements of ontological structuring in a way that a subject has true knowledge about the world allowing her to act, that is, to solve the tasks of satisfying needs and – in the terms of the theory of evolution – survive by truly correlating the pictures of her subjective reality with the actual state of affairs in the objectively self-realizing processes of the universe.

The tradition of praxeological studies is much younger and is presently considerably less expressed as compared to the epistemological one. The activity in itself is an antithesis to the cognition, a project process being reverse to reception, that is the situation of creation or change (exteriorization, objectification) of the state of affairs by a subject in the actual world in connection with the need, will, purpose. The purpose of the activity (action, act, creative act) consists not in the description and explanation, not in achieving the level of concept in the representation act, but in the creation of something new (object, state of affairs, expression, thought, idea), in the change of the existing forms of existence, in the application of the concept in the presentation act. The activity starts with an idea, it continues in a rational and linguistic construct and ends with the creation of an object available for the sensuous perception, which executes the target task of activity on changing of some state of affairs.

Answer to the question "What do I know about how I act?" is more complicated than the answer to the question "What do I know about my own knowledge?" There are some reasons. The first one (ontological) consists in the fact that, figuratively speaking, cognition and activity form a circle or a spiral covering all the human activity: cognition as a reception constitutes a basis for the activity as a projection; the results and the processes of projection, while changing the composition of cognizable, become a new basis for cognition bringing new contents for the concepts and thus creating renewed bases of activity. The second one (epistemic) reveals the essence of the question. Cognition as a reception is simultaneously a projection and an activity: direct cognition (intentio recta) in pure form may only be implemented based on inborn unconditioned reflexes, however, it is presently known that cognition procedures executed by human are set by contingent historical norms, which one way or another are appropriated by an individual from the intersubjective, social space of a biologically non-inherited culture, therefore, are a form of indirect cognition (intentio obliqua), reflection. Cognition as an explanation, as correlation of a sensuous impression and a fragment of some system of rational language in a concept requires application of some system of concepts, that is, it represents the activity aimed at explanation. The same is true for solving the descriptive task. The epistemic question about knowledge thus appears as the first-order reflection: by performing the process of cognition as the interiorization of the existing and the explanation of objects or subjects, I – in order to describe these processes– have to "observe the observation" and express, within the framework of one or another ontology, the processes of activity that lead to knowledge about how I cognize and what. Praxeologically, the given question requires second-order reflection: to describe the activity processes bringing a target result, I have to observe the way I observe the "observation of observation" and express the processes of activity that lead to the knowledge of how I act in this cognition.

These complications historically act in such a way that, up to the radical transformation of philosophical knowledge at the turn of the XIX–XX centuries, questions about the structure, forms, and stages of activity as such were largely considered within the framework of mystical-religious and their equitable handicraft traditions. In the XXI century, it is one of fundamental and already fully explored issues of philosophic knowledge capable of being practically useful in solving particular engineering and technical tasks. In the given article by considering the statements of two authoritative modern philosophers, I would like to define a problem of building the general semiotic definition of working, projective in the broad sense knowledge, as an answer to the question stated in the title of the article.

2   Working Knowledge (A. Nordmann)

In his article "Working knowledge or How to express things in works?" professor Nordmann formulates the concept of working knowledge by using a metaphor of "the grammar of expressing the things in reality": "Similarly to the knowledge of facts and by conditioning it, the objective working knowledge, like the knowledge of the fact and as its prerequisite - is the knowledge about how things interact in the works of activity, what is their contribution to the systemic whole of this interaction" [1, p. 17]. His reasoning is built on the analysis of Wittgenstein's diary entries and theses, where the "fact" (facts, according to the Tractatus Logico-Philosoficus, constitute the world), "work of activity" and "proposal" are significant words, the last two represent "something made by man, something by which things are expressed" [1, p. 18].

According to Nordmann, the propositional knowledge expressed in the receptive activity of cognition as a knowledge is contra-distinguished merely by the working knowledge while putting a question on "How are the things expressed in the creation of activity and what does the working knowledge as opposed to the factual knowledge communicate to the experience about things?" [1, p. 18]. This is the direct statement of the question on the difference between the knowledge, the way it is used for statement of true justified beliefs in description and explanation, on the one hand, and knowledge, the way it is used for creating the things, on the other. To develop this issue, the author suggests an original mental experiment consisting in the building a "Tractatus Technico-Philosoficus", which would perform a function for the technical activity similar to that performed by the more than frequently cited "Tractatus Logico-Philosoficus" for the cognition theory and creative theory of language. "How could look this unwritten "Tractatus Technico-Philosoficus" with its working knowledge in contrast to the well known "Tractatus Logico-Philosoficus" with its factual knowledge?. While according to Wittgenstein, the essence of the sentence is the essence of the description, the essence of the world as a limited whole of the facts, the "Tractatus Technico-Philosoficus" would read in the respective place as follows: the essence of the activity work is the essence of creation, is the essence of the world as an interaction of things" [1, p. 19]. The system of facts creating the sense of a particular fact, is opposed to the system of interaction of things, where the essence of a thing is born by the place of merely the interaction system – this thought is confirmed by the examples of technical artefacts called by the author "the works of activity", in particular by pointing to the meaning of a cogwheel in the clock generated by the clock as a system (or whole) of the interaction of things (parts): "Things acquire "significance" only through their relation to my will (Wittgenstein 1984, 15.10.1916), that is through how they are organized technically, e.g. in the clock mechanism [1, p. 18].

Confirmation of the true of knowledge - the most significant moment in the theory of knowledge - in case of factual knowledge and working knowledge, there are distinguished: "Sentences are the descriptions, we understand their meaning when we know the conditions, at which they are true or false [1, p. 19]. […] "At the time when the factual knowledge is obtained by accepting the sentence as a belief, it objectifies itself in the most literal way by the true sentence and thus in relation to correspondence between the structure of the sentence and the given subject content. Meanwhile the working knowledge is obtained through the intellectual or bodily participation in the activity interaction, it objectivates itself in the work and thus in the successful interaction of things." [1, p. 20]. On the one hand, it is the way in which a classic moment of separation of two types of truth was reproduced already by Hegel [7], on the other hand, here the coherent theory is given a dynamic active character, most clearly formulated in the widely known thesis of Marx ("the criterion of truth is practice" [8]), and the status of a passive-contemplative attitude to the fact is fixed for the correspondent one.

The criterion of the success of the interaction of things, the criterion of the performance of the constructed mechanism, the effectively confirmed ability of the artifact to fulfill its target task, as a rule of determining the truth of working knowledge, allows Professor Nordmann to use the formula of working knowledge applicably to technical sciences, to the philosophy of technology, offering a particular solution to the task of identifying the relationship between working (technical) and factual (scientific) knowledge. On the one hand, the opposition of science and technology, determination of the primacy of natural science or its practical implementation is a quasi-problem, if considered from a historical-epistemic point of view: experimental science arises along with the inductive method of Galileo as a specific form of dialogue between man and nature, while technology - as the fulfillment of desires and the satisfaction of needs by creating new forms in the sensuously perceived world - accompanies man from the moment of his appearance as homo sapiens or even earlier. On the other hand, natural laws ontologically precede every technical object in the order of existence, because they are the necessary condition for the possibility, that is the operability and the effectiveness, of any technical solution [9]. Thus, the ratio between the science and technology is a question about (in)proportion of theoretical (scientific, epistemic, receptive-explanatory and receptive-descriptive) and practical (technical, creative, projective) access of man to the universal laws: on relation of craft skills, its universal expression in an intersubjective language (theoretical), intellectual form, about the condition of the possibility of a craft skill, its generally valid expression in inersubjective language (theoretical) rational form, on the condition of possibility of craft skill (that is, the natural law) and expression of this law in the rational form. Nordmann wrote: "The results of research in technical sciences consist more likely not in the testing of hypotheses, but in the success of the functioning of the activity works, where verification of the hypothesis and the success of functioning of the work (each in its own way) are published, posted, presented and depicted in magazine article" [1, p. 20]. He suggests traditionally humanistic, constructivist-leaning solution to this question, while starting counting not from the cosmic extra-human order consisting of immutable laws, but from human interacting with this order, applying it and recognizing it in this application. In anthropocentric ontological model this solution cannot be different: "The fact that factual knowledge is objectified in articulated sentence, and working knowledge - in interaction of things - is the result of structural similarity, correlating the both forms of knowledge not as the oppositions but by ultimately revealing the factual knowledge as a some play form of the working knowledge" [1, p. 20]. And further: "The activity and knowledge on how worlds can be composed into sentences are primary, the possibility of application of an activity work to depict facts for describing the world is secondary [...] Working knowledge on the successful interaction of things and thus also on the construction of mechanisms and of models, therefore, precedes factual knowledge, that is depiction, reflection and theoretical description of the world. This idea has extensive, heuristically valuable consequences: technology precedes science, science is the application of technology and not only in the sense that the scientific activity requires observation technologies, recording technologies, laboratory research technologies. Firstly, we learn from the activity works created by people, what things can do. And further we notice that things seem to interact in the nature in a way they do in our works. And now we talk about these works that they possess another interesting property - a capability to reflect the interaction of the things in nature, i.e. function as an actual knowledge [1, p. 21].

Professor Nordmann finalizes his article by an explication of the activity approach to understanding as an epistemic procedure. For the receptive tradition, cognition in general form is the transition from sign to its meaning, it is the definition by Augustin that, after a linguistic turn of the early XX century and building a double semantic of sense and value according to Frege, is split into two concepts, and, respectively, transition from sign to meaning as an object or a condition of trueness is described as interpretation, and the transition from sign to its sense, that is the detection of a way through which the sign designates or sets the condition of its trueness, is determined as cognition itself. Obviously, hermeneutic procedures are characterized by the same composition of reception and projection as the procedure of cognition in general: the process of cognition is active, its result is connected with achieving the target task of cognition as activity. It is rather difficult to understand while staying within the framework of historical and philological disciplines regardless that both in XVIII and XIX century practical guides on interpretation are composed as technical instructions for the execution of the mere activity. Nordmann shows that the philosophy of technology manages to find the most obvious formula, probably discovered by M. Mersenn and shared by majority of contemporary specialists in technical sciences: "Actually, we only understand those things which can be effected by us in artificial mechanical work of activity; and the nature not created by man is perceived by us only to the extent to which the dynamics observed in it seems to us to be corresponding to the activity interactions of our works. This thought, based on the 'knowledge is power' concept by Bacon, is developed to the present day in the synthetic biology with its slogan "I don't understand the things which I can't create" [1, p. 22].

Nordmann concludes that working knowledge needs typification, detection and differentiation by kinds. But for getting on with the task, Nordmann's conceptualization - as based on the theses of Wittgenstein - lacks terminological apparatus for expressing actually technical aspects of the theory of creation. The author says that other philosophers could be involved in solving this task, and in the third part of this article I would like to cite two authors, the Russian founder of the philosophy of technology Peter K. Engelmeyer [10, 11] and almost forgotten German classic Friedrich Dessauer [9, 12–14], whose ideas allow to advance in the solving of the problems formulated by Nordmann.

3 Epistemic Practices in Technical Activity (N.A. Yastreb)

In her article "How knowledge Becomes a Technical Object: Epistemic Practices in Information Technology", Nayalya Yastreb formulates the current state of the working knowledge in the production sector of economy of information technologies. In conditions of the information revolution of the last decades, transition from the second-order artificial environment to the third-order environment, the appearance and distribution in society of the technologies percepted as close to strong AI capable of independently executing the operations of understanding of man and the activity as such, there occurs a process repeating the deployment of merely technical engineering knowledge in the XIX century. The difference is that in XXI century, the objects of this knowledge are not the procedures of implementing the idea in a material object, but the forms of knowledge on these procedures.

According to the author, technical knowledge is "such a knowledge, which allows obtaining the planned results based on the natural cause-effect relationship. Forms and content of this knowledge may be different, but its focus on achievement of goals of a man and stipulation by natural reasons make it technical. Provided this understanding, technical knowledge acts as a universal phenomenon going beyond the limits of manufacture and use of machinery and mechanisms, and occurring in every area of cognition and activity." [2, p. 11]. This definition is based on the classic approach to technology, e.g. Engelmeyer describes it as the ability to fulfill desires through the laws of nature, Dessauer emphasizes strict correspondence to the laws of nature, expediency and processed state as the required properties of any technical object. The subject of attention of Yastreb is the historical transformation of this classical object of knowledge occurring with the deployment of the information and cybernetic systems for the information collection, analysis and storage.

We have already emphasized that the working or technical knowledge is formed in the second-order reflection. The ability of expressing within a sentence in an intersubjective language those state of affairs that describe and explain the successful process of objectification of an idea, its transfer from the intramental space to the extramental one, occurs together with the development of the scientific and philosophic language ensuring fixing of such a high-order reflection. The occurrence in the first half of XX century, due to the development of cybernetics and control languages which, while executing the semantic rule perform not a descriptive by performative function, allow implementing and fixing the third-order reflection and higher: along with programming languages and automated control feedback systems, technical objects emerge which act in the area of logical grammatic patterns of reason that materialize the new not within the area of sensuously perceived reality but within the framework of thinking and communication. Strictly speaking, this is the situation of the second artificial nature, where the artifacts that create the area of material culture and define the boundaries of its immaterial aspects as an environment, act with their inherent power not only in the sphere of sensuously accessible reality defined by the laws of natural science, but also (as material technical objects with all the necessary properties) in the sphere of language and thinking defined by the laws of grammar, logic, and the not yet entirely clear laws of neurobiology. Merely the historical transformation of technical knowledge is related to the fact that it itself "is produced as a technical object, operated, used, implemented as a technical object, made the requirements of efficiency, high productivity and safety that are usually made to technical objects. The epistemological specifics of the existence of knowledge as a technical object consists in the fact that the object of cognition is a part of knowledge itself" [2, p. 12].

The technical object, whose object is the technical knowledge, may be called the second-order technical knowledge or the technical knowledge characteristic of the second artificial nature [15]. The forms of work with it differ from the first-order technical knowledge. If the condition of the feasibility of the first one is described in Platonist or constructivist ontological models through the relation of the categories of potential and actual, where the potential cosmos may bear the character of the world of ideas, it is theoretically expressed as the domain of predetermined forms of solutions, or the nature of the biologically and chemically detectable sum of neural connections of an individual or groups individuals; the condition for the feasibility of the second one, at first glance, does not require deep metaphysics: information systems contain an objective set of technical solutions accumulated by mankind and fixed in an intersubjective, interpretable form, the management in constructing such systems consists in the proportionality and speed of selection of knowledge as tools for those tasks available with the system user. However, it is obvious that along with the complication of the technical action itself, its essence remains the same, although knowledge of it is proposed by Yastreb to be functionally constituted as the "epistemic practice", that is, such "methods and techniques of working with technical objects, whose application result is an increase in effectiveness of technical knowledge implemented in these objects. The general objective of any epistemic practices of working with technical knowledge is approaching the truth understood as efficiency" [2, p. 12].

The conclusion from this article is commensurate with the conclusion of the Nord-mann's work and sounds as follows: "Technical knowledge is no longer limited by the sphere of constructing and using machinery and mechanisms and shall be understood in the broad sense as such a knowledge which allows obtaining the planned result based on the natural cause and effect relationship. The cognitive activity may be described as a set of epistemic practices, that is, methods of working with technical objects aimed at increasing the effectiveness of the knowledge on which they are built" [2, p. 14].

4 The Question About the Working (Technical) Knowledge in the Context of Semiotics

The working, projective knowledge differs from receptive knowledge not in the method of construction, but in the subject: it is the sum of beliefs of how to act in relation to things to achieve a goal, as opposed to beliefs of how to describe or explain them. The change in the method of constructing the working knowledge is conditioned by the development of cybernetics and is expressed in the emergence of knowledge about knowledge, requiring new forms, methods of working with it, and new epistemic practices. Technical knowl-edge epistemologically and historically precedes the scientific one, a man first learns to act in relation to things, then discovers that the configurations of things are capable of fulfilling some his tasks (satisfying needs, fulfilling dreams), including the intellectual task of explaining and predicting connections between the observed state of affairs and their grammatical and linguistic expressions.

The essential questions that I still have in connection with the technical knowledge defined in the two articles above are related to the concept of things, on the one hand, and to the necessity to transform Plato's and Kant's receptive definition of knowledge, on the other. According to Nordmann, "logics [of technical - A. N.] work is the law of form of organization of things in it. The words have such an organization in a sentence that they can reflect the subject content, and the gears in the time mechanism - so that the result is the uniform movement of the hour and minute hands" [1, p. 22]. It is obvious that "things" as a word and as a gear of a time mechanism are ontologically different, at least in the sense that the "reflection" in the sentence of the subject content, as a subjective process, occurs only for the interpreting subject, and the uniformity of the movement of the hands, as an objective process, is realized regardless of that. Nevertheless, the law of form is the law of the syntactic organization of signs, and it is objective.

The problem of the "thing" cannot be solved in Wittgenstein's terminology, but it ceases to be a problem when addressing to the classical theory of cognition and its reinterpretation in Peter Engelmeyer's philosophy of technology [11]. What is called here a thing as a meaning or an object denoted by a sign of natural language, is referentially revealed in the form of at least three entities for receptive cognition and the same, but in the reverse order, for projective action. These are the following essences: the object of sensuous perception, the object of the language of reason (Verstand, ratio), the concept of the mind (Vernunft, intellectus). Cognition, in its semiotic expression, grasps the essence firstly an object in sign form, then as a subject of linguistic expression in sign form, and, finally, as a concept in sign form. Activity, being a reversal of cognition, i.e. a change in the sequence of its stages, captures an idea in a concept (a predictable image, a phantasm), expresses it in an subject as a linguistic construct of the reason, and finally fulfills it in an object as a sensuously perceived artifact. Every "thing" that is semantically designated or syntactically expressed by the intersubjective language represents one of those six entities. In cases of indirect cognition, the set of reference objects increases in proportion [16].

The essential complexity in defining the designated "thing" is connected, further, with the fact that each stage of cognition and activity is rule-based, it is revealed through the sum of the rules of the material expression of the sign, pragmatic, syntactic and semantic rules. In other words, every object, subject, and concept in receptive and projective semiosis fall under a semantic, syntactic, pragmatic, and material rule being peculiar for each stage. This peculiarity is revealed for projective activity by the theory of creativity, for example, by Engelmeyer or Dessauer, as a methodology of technical action, and namely it constitutes the basic difficulty in the situations of second-order technical action described by Yastreb, since epistemic practice - as a way of managing technical knowledge - is aimed at regulating particular rules of particular stages of semiosis. Practices aimed in the technical knowledge of the second nature to the management of pragmatics of ideas are essentially different from the practices of managing the syntax of things, etc., despite of the fact that, in the practice of management, they are all denoted by the term "innovation management" or the like. It is this complexity that can and should be overcome through solving the task of identifying types of operational knowledge set by Normann; I would add to this formulation the following: tasks of identifying and describing semiotic types of the second-order working knowledge. This task is partially formulated and solved, but I still don't see any approaches to its complete solution.

The second question is rather philosophical than technical and theoretical. To what extent do the standard definitions of knowledge [17] employed in textbooks and discussed at university seminars for centuries reflect the essence of active, technical knowledge, how much are they commensurate with technical knowledge of the second nature? I can only formulate this question, its discussion is far beyond the possibilities of the article. A true opinion with an explanation, a justified true belief, subjectively and objectively sufficient positing that something is true - these are the definitions of the target results of receptively understood cognition. In technical action, the conviction that "the state of affairs is such-and-such" turns into "belief in the feasibility of a dream, idea, project"; suggestion or opinion expressed in intersubjective language – in a design, drawing or plan; the truth of suggesting – in efficiency and workability, that is, the feasibility of the construction. The truth of judgments as the goal of cognition gives way to the usefulness of action as the goal of activity, respectively, if any receptive knowledge is valuable to the extent of its truth, then any projective knowledge is valuable to the extent of usefulness, efficiency.

The concept of knowledge, determined for receptive cognition, expresses and designates the content of reflection. Its semiotic pattern of description may be taken in a simplified version: sensuous perception is interpreted as the fulfillment of the semiotic rule, sets of sentences of rational intersubjective languages – as the fulfillment of the syntactic rule, mind in the form of a precondition for the possibility of a reference – as the fulfillment of the pragmatic rule. To define knowledge as a justified true suggestion, the syntactic rule regulates the linguistic expression of suggestion, the pragmatic rule regulates its non-linguistic, substantive justification, and the semantic rule defines the truth. Such simplified interpretations are local and inaccurate, but very common and characterize not only knowledge, but also, for example, the ratio of types of the theory of truth. Karen Gloy, for example, expressed the idea that the correspondence theory is described by the semantic dimension of semiosis, the coherent theory – by the syntactic dimension, and the utilitarian and other pragmatic theories – merely by the pragmatic one [18].

I suggest abandoning the simplified interpretation of the dimensions of semiosis and understand sensuous perception, reason and mind as independent stages of cognition, each of which being characterized as semiosis, that is, it having pragmatics, syntactics, semantics, material expression. In this interpretation, knowledge, defined for receptive cognition, is a syntactic expression of a complex intellectual construction allowing the subject to correlate the object of perception and the subject of reason. The semantics of this construction is reflexive, and pragmatics is set by the historical norm of expression, characterizing the contingent, artistic or scientific way of fixing reflection. Working knowledge is a syntactic expression of the complex reflexive interaction of the semiosis of mind, the semiosis of reason and the semiosis of perception, combined in the concept of projective activity designating the successful achievement of a result and has the proportionality of the worldviews of the recipient and sender of this expression as a the condition of its feasibility. Within the framework of the semiotic model, knowledge as a whole is defined through the interaction of heterogeneous rules of different layers of cognition and activity, reflexively connected first in the object of knowledge, and then in the way of its expression.

5   Conclusion

The concept of knowledge is the key concept of both the theory of cognition and the theory of activity. Since it turned out that activity as such includes, among other things, the activity of cognition, so that receptive knowledge is a fragment of working knowledge, and scientific knowledge is a fragment of technical knowledge, so the thesis by A. Nordmann on the task of constructing types of working knowledge and the thesis of N. Yastreb on epistemic practices as a form of implementation of working knowledge should be supported. The general task is to find a system of types of epistemic practices in the form of a semiotic pattern of reflexive self-management of the subject's activity. This is a large-scale but essentially understandable and rather feasible task.

Acknowledgments. The reported study was funded by RFBR, project number 20-011-00462 A, Friedrich Dessauer's Philosophy of Technology: Epistemology and Anthropology of the Realistic Creativity Theory.

References

1. Nordmann, A.: Working knowledge or how to express things in works? Semiot. Stud. **2**(2), 16–22 (2022). https://doi.org/10.18287/2782-2966-2022-2-1-16-22

2. Yastreb, N.A.: How knowledge becomes a technical object: epistemic practices in information technology. Semiot. Stud. **2**(1), 10–15 (2022). https://doi.org/10.18287/2782-2966-2022-2-1-10-15
3. Rowe, C.J. (ed.): Plato: Theaetetus and Sophist. Cambridge University Press, Cambridge (2015)
4. Kant, I.: Kritik der reinen Vernunft. Felix Meiner Verlag, Hamburg (1998). (in German)
5. Russel, B.: The Problems of Philosophy, 2nd edn. Oxford University Press, Oxford (2001)
6. Russel, B.: Human Knowledge: Its Scope and Limits, 1st edn. Routledge, London (2009)
7. Hegel, G.W.F.: Enzyklopädie der Philosophischen Wissenschaften im Grundrisse 1830. Erster Teil. Die Wissenschaft der Logik. In: Werke in 20 Bänden, Band 8. Suhrkamp Verlag, Berlin (1986). (in German)
8. Marx, K.: Thesen über Feuerbach. In: Marx-Engels Werke, Band 3. Dietz Verlag Berlin, Berlin (1969). (in German)
9. Dessauer, F.: Mensch und Kosmos. Verlag Otto Walter AG, Olten (1948). (in German)
10. Engelmeyer, P.K.: Creativity Theory. Book House Librokom, Moscow (2010). (in Russian)
11. Engelmeyer, P.K.: Philosophy of Technology. Lan', St. Petersburg (2013). (in Russian)
12. Dessauer, F.: Philosophie der Technik: das Problem der Realisierung. Verlag von Friedrich Cohen in Bonn, Bonn (1928). (in German)
13. Dessauer, F.: Streit um die Technik. Verlag Josef Knecht, F.a.M (1958). (in German)
14. Dessauer, F.: Streit um die Technik. Verlag Herder, Freiburg im Breisgau (1959). (in German)
15. Nesterov, AYu.: Semiotic Bases of Technology and Technological Consciousness. Samara Humanitarian Academy Publishing House, Samara (2017)
16. Nesterov, A.Yu.: Semiotics as methodology and ontology. Semiot. Stud. **1**(1), 6–13 (2021). https://doi.org/10.18287/2782-2966-2021-1-1-6-13. (in Russian)
17. Lebedev, M.V., Cherniak, A.Z.: Analytical Philosophy. Peoples' Friendship University of Russia, Moscow (2004). (in Russian)
18. Gloy, K.: Wahrheitstheorien. UTB, Stuttgart (2004). (in German)

Social Interaction with Non-anthropomorphic Technologies

Daria Bylieva[1]([✉]) [iD], Alfred Nordmann[2] [iD], Victoria Lobatyuk[1] [iD], and Tatiana Nam[1] [iD]

[1] Peter the Great St. Petersburg Polytechnic University (SPbPU), 195251 Saint-Petersburg, Russia
bylieva_ds@spbstu.ru
[2] Technical University of Darmstadt, Residenzschloss 1, 64283 Darmstadt, Germany

Abstract. The discussion about the special position of robots is often based on their similarity to humans. Our research shows that non-anthropomorphic smart appliances also receive a special regard that is not similar to the attitude to ordinary household appliances. We conducted an ethnographic analysis of 229 cases in which users described online interaction with robot vacuum cleaners, a smart home, a smart refrigerator, and other smart appliances. As a result, the largest number of stories received are devoted to robot vacuum cleaners. These are given names, character is attributed to them and activities described in the context of zoomorphism or anthropomorphism. In our opinion, one of the important reasons for this is the ability of smart technologies to communicate and actively interact. After all, the so-called "desire to establish contact", along with performing quite complex functions, causes people to empathise and respond. We propose that the most correct way to describe this relationship is in terms of a game. People realise that they are acting within the framework of special fictional conditions that do not correspond to reality but get fun from it. People tend to perceive condescendingly even the mistakes of some robots and are ready to help. Real irritation is caused by cases when technologies claim to manage people's lives.

Keywords: Human-machine interaction · Technology · Communication

1 Introduction

Technological means have traditionally been considered as tools, but the modern development of technologies turns out to be a challenge for the instrumental approach. The discussion about the role and status of smart technologies in human society is extremely extensive, but the question of the philosophical foundations of this position is no less interesting. What exactly among other modern technologies makes it necessary to give them a specific position that goes beyond the category of objects and tools [1]? Most often, the discussion concerns the status of a social robot that is intentionally designed to be "socially intelligent in a human like way" such that "interacting with it is like interacting with another person" [2]. Also, the basis for considering the possibility of

D. Bylieva and A. Nordmann (Eds.): PCSF 2022, LNNS 636, pp. 47–58, 2023.
https://doi.org/10.1007/978-3-031-26783-3_6

allocating such robots a special status in comparison to other technologies is that they are "designed to act as our companions" [3]. The ground of people's special attitude is the fact of robots pretending to be and act human, where anthropomorphism therefore plays an important role. This contributes to a type of Human-Machine Interaction [4, 5] which is usually associated with "agentic animism," that is the attribution of psychological, perceptual, name, and artefact properties to a target while not attributing biological properties [6]. We can conclude that attributing moral status to robots and the need for "kind" treatment of them is often based on "apparent similarity to humans," and the assumption that "cruel" treatment of a robot may have implications for their future behavior towards people or animals [7].

The growing significance of technologies in public life [8–10], constant interaction with smart technical means requires their understanding beyond the question of the degree of their similarity to humans and separation on human/unhuman. One of the most well-known theories that integrates technology and humans within one network of interactions is Bruno Latour's actor-network theory. Latour states that "the human, as we now understand, cannot be grasped and saved unless that other part of itself, the share of things, is restored to it. So long as humanism is constructed through contrast with the object […] neither the human nor the nonhuman can be understood" [11]. Today, within the framework of new materialism, some researchers suggest using "flat ontology" that includes technological tools (Jane Bennett), others insist on the principle of the animate/inanimate division (Tim Ingold, Eduardo Kohn). Posthumanism's "flat ontology" offers to completely abandon binaries and dualism: human/nonhuman; nature/culture, subject/object; animate/inanimate; natural/artificial. In particular, it proposes to overcome the human-technology opposition, including technological agents in the common network of equal rights.

Thus, we have two approaches to modern advanced technologies. The first is anthropocentrism, in which the social attitude to such technologies as robots is partly explained by the error of perception of people who take the tool for something more. To some extent, it can be considered as a cultural peculiarity that requires an ethical attitude towards one's own kind. In fact, linking the specific status of smart technologies in human society to their pretence to be human narrows the problem. The second one is posthumanism, that recognizes equality in relations to technologies, which of course may be good as a declaration, but ignores initially unequal positions. Humans, as the creators of technology, have greater opportunities to influence technology, which is ignored in the framework of flat ontology. Some researchers seek to find an intermediate option by combining two positions or finding an average between them. Arianne Françoise Conty claims that "technological tools and commodities cannot be separated from human agency, and thus that we should follow Latour in classifying such tools as techno-human hybrids" [12]. Mark Coeckelbergh suggests continuing to consider robots as instruments; but as instruments-in-relation: they are always connected to humans and the social-cultural fields in which they operate [13].

Thus, on the one hand, technologies are tools created for specific purposes. Owners use them and receive benefits, but also, they can be used by third parties for their own purposes. In any case, technologies have purposes and limitations. On the other hand, in the socio-cultural context, robots act as social beings, to which the appropriate norms of interaction are applied. How do users perceive this situation? Are they really deceived by the appearance and habits of smart systems? Do they act under the influence of norms or grammars that prescribe certain behaviour? Or maybe, in the spirit of anthropomorphism, they tend to see their own kind in technical artefacts? We assume that both are equally unlikely. Adults understand they interact with a highly technical tool made by people, who have their own goals and limitations. However, this does not prevent them from interacting with it in a social way, as with animate beings, so called "agentic animism". An inhumane attitude to alive-looking technology is perceived as morally inappropriate (recall the reaction to the video with blows to the Boston Dynamics robot dog). To be sure, a person who beats and curses any device is unlikely to be perceived as a model to follow. But be that as it may, it would be wrong to say that there are established patterns of interaction with modern technology that require the same relationship as toward a living being. These norms are only being formed now, so people have a great degree of freedom. How does the confidence that one is dealing with a soulless machine combine with quite another way of interacting with it? We believe that interaction with modern technology fits the concept of a game in the understanding of Johan Huizinga, that is, a state of half-faith, which is characteristic of both children's play and many adult activities. Accepting the rules of the game and thus a certain frame of reality, a person acts in accordance with these rules/restrictions, fully realising that the accepted reality exists in the manner of "as if".

2   Social Interaction with Technologies

Despite the spread of social anthropomorphic robots in the fields of medicine, services, trade, advertising, nursing, education, the first machines with which people massively enter into regular interaction are not anthropomorphic. The first robots to inhabit people's homes were robot vacuum cleaners.

There are cases of personalised interaction of people with a variety of non-anthropomorphic technological objects: toys, household appliances, cars, and others. This trend took on a new quality, however, when technical instruments or purposeful devices began to communicate in natural language. At the end of the twentieth century, the topic of interaction with computers as social beings was actively investigated [14–16]. Byron Reeves and Clifford Nass note that "computers, in the way that they communicate, instruct, and take turns interacting, are close enough to human that they encourage social responses (…) any medium that is close enough will get human treatment, even though people know it's foolish and even though they likely will deny it afterwards" [14]. There are also a number of studies on the close personal attitude toward military Packbots which retrieve objects from dangerous territory and thus reduce soldiers' risks – these are given them names, they are awarded battlefield promotions, and their death is mourned [17, 18]. As David J. Gunkel notes, this attitude to technology is the opposite of their traditional instrumental interpretation: "it happens in direct opposition to what otherwise sounds like good common sense: They are just technologies—instruments or tools that feel nothing" [19]. Another early, well-researched example of deep emotional inclusion in a relationship with a technical artefact is the egg toy Tamagotchi which implied or suggested rather than embodied the likeness to an animal. There are cases when businessmen postponed business meetings, people had accidents or refused to fly in an airplane due to the whim of their Tamagotchi, not to mention neglecting work or educational duties. The pet's "death" was accompanied by seeing-off with candles, farewell speeches and gravestones, virtual and real cemeteries were organised [20]. We know that some children are able to get very attached to their toys, imagining them as live companions. However, the described cases spread far beyond the preschool age and demonstrate an unusual mass spread and brightness of emotions. The fundamental difference between Tamagotchi and other toys is its adequate feedback to stimulating engagement and its absence (when inattention leads to the "death" of the pet).

Today, the ability of technology to adequately communicate with owners has reached a new level, primarily due to Artificial Intelligence (AI). Natural language as a complex sign system, which seemed to belong exclusively to humans, is being actively mastered by smart technological devices today. Natural language processing by AI means something more than a transition to a more convenient human-machine interaction. It includes speech recognition (sound waves need to be segmented into words, with checking and error correction), semantic web analysis, decomposition of natural language into semantic parts, speech synthesis, dialogue systems, tonality and emotion analysis, user information analysis. The possibility of meaningful communication implies the transition to a new type of relationship.

Cassirer states that language is not just a tool for representation, but a means of making reality, and the same is true of technology [21]. Coeckelbergh considers that culture becomes dependent on our use of language and technologies. By changing those uses and performances, we can change the larger cultural whole – albeit slowly [22].

Coeckelbergh claims that when a person enters direct communication with technology, it is "fully appearing as a quasi-other, being acknowledged as a quasi-other, and being constructed as a quasi-other". The appeal to a "you" turns a technical object into an artificial second-person, which has a claim on me as a social being [23]. The linguistic environment, which is traditionally the prerogative of a person, contains norms prescribing a certain level of respect for those entering into communication. Receiving messages from technical devices, people on the one hand realise the absence on the other side of a thinking being who would serve as a justification for human communication, and on the other hand they are involved in an ordinary communicative interaction, implying an "adequate response". In addition, home smart appliances actually carry out some purposeful activity. Their communication has a subject that has a basis in the physical world. This compares favorably to a variety of virtual assistants which (unless they are connected to other devices) only appeal to a physical reality in which the subject is not actually able to act. The purpose of the virtual assistants is communication itself.

Thus, we have a communication situation in which there is a certain reason for communication. There is a communicative request in a natural language and a technological object that sends it, which does not pretend to be like a human being. In the following, using the example of smart technical means that people have encountered in everyday life, we research linguistically mediated human interactions with non-anthropomorphic technical objects.

2.1 Analysis of Cases of Communication with Smart Technology

The objective of this research is to study communication of humans with smart technology, its goals, content, and specifics. The main method is netnography which allows for qualitative research based on online communication. Supporting functions were performed by some analytical and quantitative methods (statistical processing, content analysis).

The investigation proceeded in two stages. In March 2022 we studied the resources on the Internet that provide access to the largest number of materials for communicating with smart technology, we later analysed the cases selected on these sites. We examined 229 cases representing stories about communication with robot vacuum cleaners, smart scales, smart refrigerators, various smart home elements, etc. Most of the cases are presented in the form of posts or videos, as well as screenshots of correspondence with smart technology, they are dated 2018–2022. Information was collected on such English and Russian-language websites and social networks as https://www.reddit.com/, https://pikabu.ru/, https://habr.com, https://fishki.net/, https://vk.com/, https://www.you tube.com, https://tiktok.com, https://twitter.com/ https://zen.yandex.ru, on a number of news sites - https://www.vedomosti.ru/, https://regnum.ru/. To search for cases on the listed sites, we used 20 keywords: robot vacuum cleaner, smart home, smart appliances, smart kettle, smart refrigerator, smart bracelet, and others. The robot vacuum cleaner appears in 53.9% of the cases where people talk about their communicative interaction with a smart device. It therefore appears to be the most popular source of communication between a person and a non-anthropomorphized thing. We therefore limit the following discussion to just this smart device – except for its recently acquired smartness, it is a familiar and mundane tool which does not exhibit any human features and is most clearly a non-anthropomorphic technology. There are also many stories, however, related to a smart home (16.6%) and a smart kitchen (17.6%) which is mainly represented by a smart kettle, a smart cooker, and a refrigerator. In addition, there are communication options with a range of other technologies like smart bracelets, smart clothes, smart speakers, smart sockets, smart furniture, etc. (11.9%). Our collection of stories that are told about communicative interaction with all these technologies exhibits the same general pattern which we observed regarding the smart vacuum cleaner.

The authors of the stories highlight foremost the "wrong" operation of technology (24%), non-standard use becomes thematic in 19% of the stories. Then there are reports of devices behaving too "smart" (18%). This option is much more common than "a person is forced to help a robot" (2%) or "communication errors" in 13% of cases. The communication with technology takes several forms. What is written on a display in 11% of the of cases, what is uttered, spoken, or said (5%), oral communication or a back-and-forth between a person and a technology (19%), and further variants of verbal communication (10%) such as a person giving a command, and it is executed wordlessly. Still, in more than half of the cases (55%), communication is non-verbal. Most often, the attitude towards technology is neutral (28%) or as towards a non-sentient animate being (28%), sometimes hostile (13%). Less common is the attitude of being an accomplice (15%). In 16% of the cases, the attitude was similar as to a person (equal), a friend, or an animal and generally had a positive connotation.

2.2 Let's Talk to the Vacuum Cleaner

Vacuum cleaners turned out to be the first publicly available robots that appeared in homes. Numerous studies confirm that people behave towards it in a social way [24, 25], giving them names, genders, and personalities [26]. The modern robot vacuum cleaner exhibits autonomous behaviour, builds a plan of the living area and the trajectory of its movement, and sends information about its condition in natural language orally or via a smartphone: the start and end of their cleaning cycles, communicating success, failure and required repairs. The vacuum cleaner turns out to be not just a routine household appliance but has an emotional and entertainment value when the owner accepts the "game" and interacts with it as with a living being. One of the early studies (2007, when vacuum cleaners were able to communicate only by way of sounds) indicated that people often named vacuum cleaners after their favourite sci-fi or other film characters. Sometimes they changed names, for example, from Robocop to Aarnold (not "Arnold") after the Terminator, because the latter seemed to be a better fit to the personality of their robot. Most owners used the pronoun "he" to describe its masculinity as coming from its shape, colour, and a preconception of male-dominance in the realms of technology and machinery. However, some women-owners referred to Roomba as "he" because they liked the idea of having a man do the cleaning for them [25]. In our study, vacuum cleaners turned out to be both male and female. In most cases, the robots were given ordinary human names, although there were exceptions, for example, Belka (after the Russian astronaut dog), Rosey (after a character from The Jetsons), Simpson Mr. Sparkle, Scruffy, the janitor (Futurama), a couple of robots are called Hansel and Gretel. Sometimes light swear words were used as names. People may call their robots stupid, or conversely, affectionately my Smartie.

Unlike social robots and virtual assistants, the possibilities of interaction with robot vacuum cleaners are quite limited. Nevertheless, they are necessary, at least for monitoring and rescue. For example, it is necessary when a robot reports that it is stuck in a narrow place, cannot overcome an obstacle, or has detected a black colour that is perceived by sensors as a hole and the possibility of falling. Sometimes, even a dark shadow can turn out to be such an insurmountable obstacle and a reason for discussion:

> *My Ozmo thinks a hard shadow from sunlight is a wall but only in one direction so it will enter an area but not leave until the sun moves.*

Messages coming to the smartphone from robots asking for help (like "choked on shoelaces") also cause emotions, since on the one hand the reasons are the most trivial, while on the other hand the robot really is having difficulties, asking for support through messaging just as a person would do. For example, the message from a robot about the problem of overcoming a threshold "Herbert requires your attention. Herbert is stuck near the cliff" is accompanied by ironic comments such as "Herbert looks a little dramatic".

Many people watch the robot being cleaned simply as entertainment. The independence of its work as a fact arouses curiosity and not only among those who have acquired it for the first time:

It is so amazing, sometimes watching him go around all kinds of obstacles on the way to the base. In addition, it also cleans cool! And you can chase the cat in manual mode!

There are many cases when old people express a careful and attentive attitude towards robots. They worry that the vacuum cleaner is tired, needs help or rest:

"Grandma locks the robot vacuum cleaner in every room for at least half an hour so that, I quote: 'so that it works properly, and does not do a bad job, then leave for another room.' Once the robot did actually go somewhere on its own, and she called it literally with the same words as she called upon me when I was a child".

Another interesting case is when the mother of the owner of the smart device turned on the light in every room where the robot worked. In all these cases, the robot is not treated like an autonomously working device (such as a washing machine), but receives a lot of attention:

My mother comes in today and complains: "There's a lot to do in the garden, and I also want to wash the floors". So, I suggest taking my robot vacuum cleaner, start it up and go do business. After 15 minutes, she phones and asks: "How can I turn it off?"—"Why?"—"He's tired!" And I check my mobile app and see that the robot has 87% charge. So, I say: "It is not tired! Let it work!" Then I ask my mom: "Did you manage to do something?"—"No, there was no time, my cat and I sat, watched and helped the robot with advice".

The attitude of younger owners towards the robot is often condescending, it compares to attitudes towards children or pets. When describing the robot, they use zoomorphic descriptions (the "rubs against the legs", like a dog its tail it "waves its brushes"), comparing their robotic vacuum cleaner to a cat: *It constantly hides something and loves the tinsel on the Christmas tree. There is also another problem: we have to push him like a thinking cat who doesn't know whether to go for a walk or not.*

The self-shutdown of the robot when it hits a button, for example, on a door limiter, is described as a "strike". And indeed, for the vacuum cleaner to work correctly, the owners have to work hard themselves. This includes collecting all things from the floor: "*I am forced to pick up all of the random things that accumulate on my floor or risk damaging an expensive "pet" that I've grown quite fond of.*" It also includes more serious activities: Since the brushes of the robot vacuum cleaner suffer from long hair, the owner refuses to let her hair grow long. In another house, where there was a small stair step, the owner made a special ramp and shared a video of how the robot vacuum cleaner masters descents and ascents along it. The need to have a floor free of objects is complicated by the presence of children (and dogs) in the house. However, there are parents who use a vacuum cleaner to teach their children to put things in order:

I have a 4-year-old son. We named our robot Captain Turbot, so I get to tell my son, "Make sure you pick that up, so Captain Turbot doesn't get it". It actually works and he picks most things up.

The construction of a plan of flats by a robot and the drawing of a trajectory also serves as a subject of discussion. In the trajectories, owners tend to see certain "suggestive drawings", such as silhouettes of weapons or indecent gestures. Fixation on the plan of mirrors as an entrance to another room (because of LiDAR sensors), causes mystical comments.

My Roomba doesn't understand tall mirrors, and draws them as actual rooms it can never reach. Once it got confused and thought it was in the mirror universe or maybe mirrors are universes that we can never reach.

In addition to the necessary forms of communication, the owners come up with their own original ways of interacting with their vacuum cleaners. American researchers give an example of a video where three Roomba cleaners with Christmas hats were programmed to sing and dance [25]. The practice of "dressing up" robots or attaching stickers with notes, names, or some objects is widespread. A toy gun was attached to one of the vacuum cleaners, which shoots at everything it crashes into.

Fig. 1. Battle of vacuum cleaner

Attaching objects is fun and can be used to organise games and competitions between robots. Another original use of vacuum cleaners in videos is the battle of robots. The rules and goals of the game may be different: Who will collect more dirt, who will pierce the ball that is tied to the other (Fig. 1), who will push the other into a small room, etc. In Russia, robot vacuum cleaners were initially sold with Chinese or English firmware for translation into Russian. But it is curious that different "creative" variants of Russification have appeared. In some of them, the robot uses the greeting "My Lord" and "Your Majesty", in others it greets the "leather bastards". Also, standard phrases are changed to well-known phrases from movies and cartoon films, for example, alluding to the BB8 droid from Star Wars or Winnie the Pooh. Owners of vacuum-cleaners post videos where one hears complaints and insults in the place of standard phrases. For example, the robot says "fuck off, beast, I'm resting" or "I hate this job, why can't these leather bastards clean up themselves". There are also phrases from a cartoon about a Russian *domovenok* (a household spirit) who got into a city house for the first time: "I will work according to my conscience, do not be afraid for the household. Is there a stable?" Sometimes ready-made voice-overs do not seem interesting enough, and the authors of the videos simulate their own "dialogues" with a "stubborn" robot that utters meaningful phrases, answering the owner's questions or overcoming difficulties. It can be assumed that, having experience with bots and virtual personal assistants, people tend to expect a similar form of communication from a robot vacuum cleaner.

2.3   Related Stories About Related Technologies

Beyond vacuum cleaners, there are today many other smart things in the house, and many houses are themselves becoming smart. The most common actions in a smart home include the automatic switching on and off of lights, automatic notification of intrusion, fire, or leakage, as well as the turning on of Internet-accessible home devices. It seems that communication with a smart home can take many forms, and yet, it should be designed so as to exclude communication errors. In non-standard situations a smart home makes decisions independently which can have various consequences. For example, a smart home often behaves too "vigilantly". So, one Twitter user said in a microblog that he could not enter his home. It turned out that his Batman T-shirt confused an automatic facial recognition system. When the camera captures an unknown person on the porch, it locks the entrance and takes a photo of the guest. The smart lock did not recognize Batman and therefore locked him out. In this case, the system error made the user laugh, and he ended up praising the smart home for its vigilance. But communication with a smart home can obviously have more serious consequences.

Interaction with multiple smart technologies can occasion linguistic confusions:

> "We installed a motion sensor in the toilet, now the light turns on by itself and turns off after 15 seconds if the sensor does not detect any movement. But when the sensor turned off the light, and one of the children was still in the toilet, she began to make loud prolonged sounds. The explanation is simple - my mother-in-law has sound sensors in the house and so the children are used to turning on the light with their voice. I had to explain for several days that we have motion sensors 'and it's enough just to move'".

Irritations cause very human responses also in communication with individual technical devices. A user appeared outraged reporting that "*our fridge just emailed us to say we opened its door too many times in the past month*". In this case, the indignation is not because the technology makes a mistake, but because it puts itself in the role of a judge of the humans, evaluating their actions.

This is the flipside of the expectation that smart technologies will promote better, e.g. environmentally conscious behavior, acting as "a partner or coach in the habituation of virtues" [27]. Smart loudspeakers and electronic assistants can help students prepare for an exam or just in the learning process: "*Once I was preparing for a class at my sister's house, sat in the living room, studied and at some moment said out aloud: 'So, the element nickel was discovered...' and immediately heard a voice '... in 1751'. I continued to ask questions, and the system answered me instantly, it was the best preparation for a class in my life."* But the device can do more than serve as an assistant in studies for school. It can create a New Year's miracle by interacting with a smart socket: *"Once we showed our daughter Sonya a New Year's miracle. We stuck a Christmas tree garland into a smart socket, turned off the power, and asked Sonya what we should say to make the Christmas tree light up. The daughter screamed: One-two-three! Christmas tree, light up! And the lights sparkled. Sonya was happy!"*.

As with smart coffeemakers or stoves, these smart home technologies become incorporated into the social life of a family.

3   Conclusion

Smart appliances gradually fill homes, perform household functions and are capable of communication. Establishing interaction with them is different from interaction with ordinary household appliances. People communicate with technology, laugh at its mistakes, share "its tricks" on the web. The analysis of cases shows that complaining or bragging about the achievements of their technology, people really understand perfectly well that the technology does not have an consciousness the situation and does not need to do so. But this does not prevent them from characterising the actions of smart devices as meaningful and reacting accordingly, which we propose to consider as a type of gaming activity.

Without claiming to be anthropomorphic or zoomorphic, the robot vacuum cleaner became the first robot to move into houses and get a special status. Users give the robots names, attribute character, monitor their movements, draw up a house plan. In addition, even the limited ways of robots to communicate in the form of voice messages to a smartphone deserve attention. In earlier studies, people associated vacuum cleaner robots to a greater extent with characters from science fiction works and gave corresponding names from science fiction books and films [25]. Although this topic is not completely exhausted, and some people continue to see "relatives" of robots from fiction, many users have since switched to human names. Elderly people and children demonstrate the greatest involvement in the work performed by the technology. Young people come up with alternative ways of interacting and playing with robot vacuum cleaners, attributing certain intentions to them. They perceive the communication and some mistakes of robots as funny incidents.

The analysis of cases of interaction with non-anthropomorphic technology shows that a special attitude of a person, which is sometimes similar to the attitude to a living being, is not necessarily related to anthropomorphism, but rather to the specifics of communicative interaction. Smart technical devices are described like pets or even as equal communication partners. In our opinion, one of the important reasons for this is the ability to communicate and actively interact. The "desire to establish contact" itself, along with performing quite complex functions, evokes feelings of sympathy. At the same time, it is interesting that vivid negative emotions are caused not so much by the mistakes of technology (which are often perceived condescendingly), as by its claim to "power" when it makes recommendations or performs actions to regulate people's lives.

What then, does all this signify in respect to android robots and social companions? Do they have a special status that sets them off against other smart technologies, is this special status tied to their thinking and speaking or mostly to their looks? In frustration, we sometimes curse even perfectly old-fashioned and mute technologies, thus engaging with them on a human level, pulling the plug, restarting them with a silent prayer that now they will work again. It appears obvious that the very act of addressing technical devices establishes a special relationship, whether these answer or not, whether they look like humans or not. We have proposed that one opens a language-game by addressing a machine – a game quite literally in that it involves a playful suspension of disbelief and a half-believing as if.

References

1. Ullmann, L.: The quasi-other as a Sobject. Technol. Lang. **3**(1), 76–81 (2022). https://doi.org/10.48417/technolang.2022.01.08
2. Breazeal, C.L.: Designing Sociable Rombots. MIT Press, Cambridge (2002)
3. Darling, K.: Extending legal protection to social robots: the effects of anthropomorphism, empathy, and violent behavior towards robotic objects. In: Calo, R., Froomkin, A.M., Kerr, I. (eds.) Robot Law, pp. 213–232. Edward Elgar Publishing (2016). https://doi.org/10.4337/9781783476732.00017
4. Pezzica, L.: On talkwithability. Communicative affordances and robotic deception. Technol. Lang. **3**, 104–110 (2022). https://doi.org/10.48417/technolang.2021.04.10
5. Sætra, H.S.: Social robot deception and the culture of trust. Paladyn. J. Behav. Robot. **12**, 276–286 (2021). https://doi.org/10.1515/pjbr-2021-0021
6. Okanda, M., Taniguchi, K., Wang, Y., Itakura, S.: Preschoolers' and adults' animism tendencies toward a humanoid robot. Comput. Human Behav. **118**, 106688 (2021). https://doi.org/10.1016/j.chb.2021.106688
7. Sparrow, R.: Virtue and vice in our relationships with robots: is there an asymmetry and how might it be explained? Int. J. Soc. Robot. **13**(1), 23–29 (2020). https://doi.org/10.1007/s12369-020-00631-2
8. Serkova, V.: The Digital Reality: Artistic Choice. IOP Conf. Ser.: Mater. Sci. Eng. **940**, 012154 (2020). https://doi.org/10.1088/1757-899X/940/1/012154
9. Pozdeeva, E., et al.: Assessment of online environment and digital footprint functions in higher education analytics. Educ. Sci. **11**(6), 256 (2021). https://doi.org/10.3390/educsci11060256
10. Bykowa, E., Volkova, Ja., Pirogova, O., Barykin, S.E., Kazaryan, R., Kuhtin, P.: The impact of digitalization on the practice of determining economical cadastral valuation. Front. Energ. Res. **10**, 982976 (2022). https://doi.org/10.3389/fenrg.2022.982976
11. Latour, B.: We Have Never Been Modern. Harvard University Press, Cambridge (1993)
12. Conty, A.F.: The politics of nature: new materialist responses to the anthropocene. Theory Cult. Soc. **35**, 73–96 (2018). https://doi.org/10.1177/0263276418802891
13. Coeckelbergh, M.: Three responses to anthropomorphism in social robotics: towards a critical, relational, and hermeneutic approach. Int. J. Soc. Robot. **14**, 1–13 (2021). https://doi.org/10.1007/s12369-021-00770-0
14. Reeves, B., Nass, C.: The Media Equation: How People Treat Computers, Television, and New Media Like Real People and Places. Cambridge University Press, Cambridge (1996)
15. Turkle, S.: The Second Self: Computers and the Human Spirit. Simon and Schuster, New York (1984)
16. Nass, C., Moon, Y.: Machines and mindlessness: social responses to computers. J. Soc. Issues **56**, 81–103 (2000). https://doi.org/10.1111/0022-4537.00153

17. Singer, P.W.: Wired for War: The Robotics Revolution and Conflict in the Twenty-First Century. Penguin Books, New York (2009)
18. Carpenter, J.: Culture and Human-Robot Interaction in Militarized Spaces: A War Story. Ashgate, New York (2015)
19. Gunkel, D.J.: The other question: can and should robots have rights? Ethics Inf. Technol. **20**, 87–99 (2017). https://doi.org/10.1007/s10676-017-9442-4
20. Bloch, L.-R., Lemish, D.: Disposable love. The rise and fall of a virtual pet. New Media Soc. **1**, 283–303 (1999). https://doi.org/10.1177/14614449922225591
21. Cassirer, E.: Form and technology. In: Hoel, A.S., Folkvord, I. (eds.) Ernst Cassirer on Form and Technology: Contemporary Readings, pp. 15–54. Palgrave Macmillan, New York (2012). https://doi.org/10.1057/9781137007773_2
22. Coeckelbergh, M.: The grammars of AI: towards a structuralist and transcendental hermeneutics of digital technologies. Technol. Lang. **3**, 148–161 (2022). https://doi.org/10.48417/technolang.2022.02.09
23. Coeckelbergh, M.: You, robot: on the linguistic construction of artificial others. Technol. Lang. **3**, 57–75 (2022). https://doi.org/10.48417/technolang.2022.01.07
24. Hendriks, B., Meerbeek, B., Boess, S., Pauws, S., Sonneveld, M.: Robot vacuum cleaner personality and behavior. Int. J. Soc. Robot. **3**, 187–195 (2011). https://doi.org/10.1007/s12369-010-0084-5
25. Sung, J.-Y., Guo, L., Grinter, R.E., Christensen, H.I.: "My Roomba is Rambo": intimate home appliances. In: Krumm, J., Abowd, G.D., Seneviratne, A., Strang, T. (eds.) UbiComp 2007. LNCS, vol. 4717, pp. 145–162. Springer, Heidelberg (2007). https://doi.org/10.1007/978-3-540-74853-3_9
26. Sung, J., Grinter, R.E., Christensen, H.I.: Domestic robot ecology. Int. J. Soc. Robot. **2**, 417–429 (2010). https://doi.org/10.1007/s12369-010-0065-8
27. Fröding, B., Peterson, M.: Friendly AI. Ethics Inf. Technol. **23**(3), 207–214 (2020). https://doi.org/10.1007/s10676-020-09556-w

Multilingual Landscapes in Capital Cities of Northern Caucasus, Russia: Study Based on Panoramic Street View Technology

Larisa Gorokhova[1]([✉]) [ID], Elena Doludenko[2] [ID], and Alexandra Gorokhova[3] [ID]

[1] Pyatigorsk State University, Pr. Kalinina, 9, 357532 Pyatigorsk, Russia
gorohova@pgu.ru
[2] Adyghe State University, ul. Pervomayskaya, 208, 385000 Maykop, Russia
[3] Kuban State University, ul. Stavropolskaya, 149, 350040 Krasnodar, Russia

Abstract. The aim of the study was to gain an insight into the linguistic landscapes of the capital cities of seven North Caucasus republics of Russia and to investigate the main factors influencing them. The study employed the modern Street View technology (Yandex Panoramas Service) to have a better understanding of the multilingual nature of the urban areas in the North Caucasus and to provide an overview of the linguistic situation in the region in its current state. The study was based on the quantitative analysis of the collected data which showed that the linguistic landscapes of the selected cities are dominated by the Russian language, yet experience strong English influence, while the local languages are underrepresented in all investigated cities. The paper discusses possible reasons for the revealed features of the linguistic landscape. Further research may include a qualitative inquiry into the nature of multilingual signs to see how multiple languages are combined in the linguistic landscape of the area.

Keywords: North Caucasus · Linguistic landscape · Street view technology · Multilingualism · Local languages

1 Introduction

In the modern world, both major and minor languages have to face the challenges of dynamic development in all spheres of human life. However, the functional dominance of major languages affects the vitality of the local ones in many ways, shifting their functionality to very limited areas of everyday communication, culture, and, partially, education. Language and culture are the major means of ethnic self-identification, the way to preserve and expand the worldview and mentality of the nation. The vitality of a language is certainly an indicator of its functional capability and potential for its sustainability and further development. The vitality of a language depends on a variety of factors, such as the size of an ethnic group, the number of speakers, both constant and occasional; place of residence; household statutes; age distribution of speakers; ethnic purity of marriages; inclusion in school curricula; national identity and ethnic self-identification; state language policy, etc.

© The Author(s), under exclusive license to Springer Nature Switzerland AG 2023
D. Bylieva and A. Nordmann (Eds.): PCSF 2022, LNNS 636, pp. 59–72, 2023.
https://doi.org/10.1007/978-3-031-26783-3_7

The North Caucasus, being one of the most multiethnic and multilingual territories of Russia, is a place where local languages experience all sorts of challenges, from the dominance of Russian as the language of politics and bureaucracy to the increasing influence of English accompanying the processes of globalization and westernization. Together, these factors form the unique linguistic landscape of the North Caucasus, dynamic and multilingual, especially in the vibrant and diverse capital cities.

Researching linguistic landscapes is a common way to gain insight into how multilingual communities function and what experiences they go through. Technological advancement of the recent years has expanded the horizon of linguistic landscape studies, facilitating the task of gathering data as well as processing them.

The present study employs the modern Street View technology to have a better understanding of the multilingual nature of the urban areas in the North Caucasus and to provide an overview of the linguistic situation in the region in its current state.

1.1 Literature Review

The notion of 'linguistic landscape' (LL) may be interpreted differently by different researchers. At the early stage of LL studies there was a tendency of using it referring to more general features of language use within a certain territory, coming close in meaning to such terms as 'linguistic situation', 'linguistic mosaic' or 'linguistic diversity' [10]. However, with the establishment of LL studies as a distinct field of research with specific practices and procedures, the term has gravitated towards a more narrow meaning, which is reflected in the definition given by R. Landry and R. Y. Bourhis. According to them, LL is "the language of public road signs, advertising billboards, street names, place names, commercial shop signs, and public signs on government buildings combined to form the linguistic landscape of a given territory, region, or urban agglomeration" [21, p. 25]. Apart from its informational aspect, LL can be considered "a symbolic marker communicating the relative power and status of linguistic communities in a given territory" [3, p. 8], which makes LL studies a particularly useful tool to gain insight in matters like language policies, ethnic and local identities construction, and minority language vitality. With "any visible display of written language (a 'sign') as well as people's interactions with these signs" [36, p. 423] as its research object, this field aims at understanding "the motives, uses, ideologies, language varieties and contestations of multiple forms of 'languages' as they are displayed in public spaces" [9, p. 4]. As most (though not all) LL research is conducted in urban areas, it can sometimes be referred to as 'linguistic cityscape'.

Van Mensel et al. [36] ascribed the pioneer role in the field to the study on how Israel language policy changes the face of the city with the official language challenging English in public signage, published in 1977 by Rosenbaum et al. [30] From the very early stage, the concept of LL has been closely connected to that of multilingualism, since it is multilingual cityscapes that most often come into the focus of LL research. Issues discussed in such works include minority languages representation [5, 22], distinctions between official and non-official public signage [1], language policies and their impact on communities [11], etc.

Methods and procedures commonly used in LL studies usually do not demonstrate great variation in terms of collecting and coding data. There are, however, some issues that still demand attention.

First, the **unit of analysis** must be clearly defined, since it is not self-evident what can be considered a linguistic object or sign of public use. Some researchers define linguistic object as "any piece of text within a spatially definable frame" [1, p. 66], others stick to "the larger whole of the establishment" [5, p. 71] as their primary unit of analysis, arguing that several signs may belong to a single venue and should hence be considered as a whole. Whether to take into account the size of a public sign is open to discussion, as well as whether one should register the signs on moving objects or not.

The issue of **sampling data** is no less important. Collecting linguistic data from the whole city would require too much time and effort, therefore a certain research site must be chosen. To gather as much evidence as possible and ensure representativity, the main street or one or more shopping streets is usually selected, though, depending on the research purpose, the sampling area may include a public transport axis, shopping malls, highways, neighborhoods or even villages [9, pp. 6–8].

Coding data may include dividing the signs into categories according to their provenance, topic, function, etc. It is important to distinguish between top-down (imposed by the authorities) and bottom-up (placed by commercial enterprises or private entities) signage [3]. Some researchers would opt for more sophisticated coding schemes, including "elements such as how language appears on the sign, the location on the sign, the size of the font used, the number of languages on the sign, the order of languages on multilingual signs, the relative importance of languages, whether a text has been translated (fully or partially)" [10, p. 3]. In researching multilingual cityscapes, it is essential to determine what languages are displayed, and whether the signs are mono- or multilingual. For multilingual signage, it may be useful to distinguish the type of multilingual information arrangement, which, according to M. Reh, can be duplicating, fragmentary, overlapping, or complementary [29].

Modern technology is a great driver for the development of LL studies. This field has benefited greatly from the introduction of smartphones that have facilitated the data-gathering process and made the citizen science approach possible [27]. Special software was designed for georeferencing objects on the research sites, as well as for processing linguistic data [2]. Services such as Google Street View can also be an asset to the researchers, since they "enable users to scout the LL of distant or less accessible areas viewing panoramic images along routes around the world" [28, p. 398]. We believe that such a technology has great potential and can significantly expand the horizons of LL research.

Recent years have seen a number of studies aimed at researching linguistic landscapes in the post-USSR territories, such as Azerbaijan [33], Moldova [24], Kyrgyzstan [23], Kazakhstan [20], as well as in Russia, focusing mainly on the linguistic cityscapes in multiethnic regions, such as Tatarstan [34], Sakha (Yakutiya) [8], and Bashkortostan [31]. The issue of multilingualism in post-Soviet cities cannot be reduced to local languages challenging the Russian language dominance, but reflects complex processes of constructing local identities against the background of globalization and the influence of

English [26]. However, no comprehensive study of the LL in North Caucasus has been produced yet.

1.2   Language Policy in North Caucasus

All the North Caucasus republics of Russia have legally secured the status of their local languages as official languages of them as subjects of the Russian Federation, along with Russian, the state language of the country. Article 68 of the Constitution of the Russian Federation guarantees this right to be exercised by the republics in the work of local governments, public authorities, and state institutions [19]. Moreover, the Russian Federation guarantees to all its peoples the right to preserve their native languages, and to create conditions for their study and development [Ibid]. At the same time, every North Caucasus republic has a similar article in their own constitution. They are all worded in almost the same way, the difference being the name of the language (or languages) officially equaled to Russian in the area:

- Equal state languages in the Republic of Adygea are Russian and Adyghe. (Article 5 of the Constitution of the Republic of Adygheya) [12]
- State languages of the Kabardino-Balkarian Republic throughout its territory are Kabardian, Balkar and Russian. (Article 76 of the Constitution of the Kabardino-Balkarian Republic) [16]
- State languages in the Karachay-Cherkess Republic are Abaza, Karachai, Nogai, Russian and Circassian. (Article 11 of the Constitution of the Karachay-Cherkess Republic) [17]
- The official languages of the Republic of North Ossetia-Alania are Ossetian and Russian. (Article 15 of the Constitution of the Republic of North Ossetia-Alania) [18]
- The state languages of the Republic of Daghestan are the Russian language and the languages of the peoples of Daghestan. (Article 15 of the Constitution of the Republic of Daghestan) [14]
- The official languages in the Chechen Republic are Chechen and Russian. (Article 10 of the Constitution of the Chechen Republic) [13]
- Ingush and Russian are recognized as official languages in the Republic of Ingushetia. Preservation, protection and development of the Ingush language is the responsibility of the state. (Article 14 of the Constitution of the Republic of Ingushetia) [15]

Most of these republics have also adopted additional regulations on the use of languages by issuing specific documents that outline local language policies. Such documents usually include detailed lists of spheres in which the languages of the title nations are to be used, either on obligatory or voluntary basis. Among other sections specifying the use of Russian and local languages in different spheres, there is one regulating their functioning in the names of geographical objects, inscriptions, road and other visual signs which together constitute the linguistic landscape of the territory, e.g.:

Article 19. The procedure for determining the language of geographical names, topographic designations, inscriptions and road signs.

1) In the Kabardino-Balkarian Republic, geographical names of places are written in the state languages of the Kabardino-Balkarian Republic, and, if necessary, in a foreign language.
2) The Kabardino-Balkarian Republic determines the list of territories and objects where geographical names, topographic designations, inscriptions and road signs must be drawn up in the state languages of the Kabardino-Balkarian Republic.
3) Executive authorities ensure the installation of inscriptions of topographic designations and road signs and are responsible for their appearance and maintenance in due order in accordance with the legislation of the Kabardino-Balkarian Republic and international standards [25].

As is clear from the above example, authorities of the North Caucasus republics officially require to duplicate the names of geographical objects, topographic inscriptions and road signs in all the official languages of the region – this is what we call top-down naming. Indeed, the research has shown that official institutions having their offices in the viewed streets usually feature inscriptions in two or more languages, unlike commercial establishments, which constitute the majority of the investigated city landscapes. Those are often private businesses which bear the names given by their owners in the process of the so-called bottom-up naming. The latter does not require obligatory language dubbing, but it supplies the linguistic landscapes with the variety of onyms in a handful of languages.

2 Research Design

This paper analyses the linguistic landscapes in the capital cities of all seven North Caucasus republics. According to the Great Russian Encyclopedia, the territory of the North Caucasus includes Adygea, Dagestan, the Karachay-Cherkess Republic, the Kabardino-Balkarian Republic, North Ossetia-Alania, the Ingush Republic, the Chechen Republic [32, p. 633]. Apart from those capital cities, Pyatigorsk was also included in the research, being the center of the North Caucasian Federal District since 2010 [4, pp. 86–87]. The capital cities were chosen as the research sites because it is where the most commercial and social activity is concentrated so more bottom-up signage can be found. We intentionally leave out top-down signs within the framework of this study because, as it has been mentioned in 1.2, this kind of signage is strictly regulated by the local legislation and is more or less uniform in all the regions in question. The aim of the study is to gain an insight into the linguistic landscapes of those cities and to investigate the main factors influencing them by researching the representation of local languages in the public signage as well as the share of English signs in it. We also see the mission of the present study in documenting the current linguistic situation in the region before any regulations restricting the use of English in public spaces are adopted, which is quite likely to happen in the near future.

The main research question of the present paper is the following: which are the languages displayed in the linguistic landscape of North Caucasus capital cities, and in what proportion?

Our hypothesis was that the linguistic landscapes of the selected cities are dominated by the Russian Language, yet experience strong English influence, while the local languages are underrepresented.

2.1 Data Collection Procedure

The data collection involved the capital cities of all seven North Caucasus republics:

- Maykop (Adygea);
- Makhachkala (Dagestan);
- Cherkessk (Karachay-Cherkessia);
- Nalchik (Kabardino-Balkaria);
- Vladikavkaz (North Ossetia);
- Magas (Ingushetia);
- Grozny (Chechen Republic);
- and Pyatigorsk as the central city of North Caucasian Federal District.

In each city, two major streets were chosen as the research sites (three streets in Grozny): the main street and the most important pedestrian street/shopping street, based on the information from tourist websites and local citizens surveys (via the internet). In controversial cases, data from "Yandex Maps" [38] were used to determine the suitable streets based on their central location and the number of establishments (shops, restaurants, etc.) on them. Due to the length of some of such streets, only the central and busiest stretches of them were used for collecting data (M.A.Esambayeva Prospect in Grozny - up to the junction with M.G.Gairbekova Street, Magomeda Yaragskogo Street in Makhachkala - up to the junction with Nakhimova Street, Kabardinskaya Street in Nalchik - the pedestrian section).

For gathering data, "Yandex Panoramas" was used, a service by "Yandex Maps", analogous to Google Street View. We opted for the Yandex product because it provides better coverage for the Russian cities and more recent images of them, taken around years 2019–2021.The procedure was the following: the observer was 'moving' first up, then down the street registering all the linguistic objects to their right, except for top-down signage, e.g. belonging to official institutions. This approach allowed for covering both sides of the street.

The total number of linguistic objects in the sampling amounts to 4231, their geographical distribution shown in Table 1. The relatively low number of linguistic objects spotted in Magas is due to the fact that the city itself is much smaller than its counterparts, with only 15279 residents as of January, 1, 2022 [6].

Table 1. Data.

City	Region of Russia	Streets	Number of signs
Cherkessk	Karachay-Cherkess Republic	Lenina (Prospekt) Krasnoarmeyskaya (Ulitza)	614
Grozny	Chechen Republic	V.V. Putina (Prospekt) A.Kh.-A. Kadyrova (Prospekt) M.A.Esambayeva (Prospekt)	528
Magas	Republic of Ingushetia	Idrisa Zyazikova (Prospekt) Zaurbeka Borova (Ulitza)	44
Makhachkala	Republic of Dagestan	Rasula Gamzatova (Prospekt) Magomeda Yaragskogo (Ulitza)	679
Maykop	Republic of Adygea	Proletarskaya (Ulitza) Krasnooktyabrskaya (Ulitza)	818
Nalchik	Kabardino-Balkarian Republic	Lenina (Prospekt) Kabardinskaya (Ulitza)	435
Pyatigorsk	Stavropol Region	Kalinina (Prospekt) Kirova (Prospekt)	695
Vladikavkaz	Republic of North Ossetia–Alania	Mira (Prospekt) Kirova (Ulitza)	418
Total			4231

The advantages of using Street View technology for gathering linguistic data is that it allows to collect a considerable amount of data within a short time period and is more affordable than field research which is especially important for independent studies. However, this approach has a few limitations, as sometimes the text was difficult to discern, especially on smaller signs, or when the street was too wide. Moreover, in the cities where the Yandex Panoramas filming was done during summertime the trees often block the view, which obviously prevents the observer from registering certain signs. In our study, this was the case with Grozny, where we had to add a third street into the selection to ensure representativity, since there was too much greenery in the first two.

2.2 Data Coding Procedure

The data were labelled according to the following categories:

Signs in Russian: The linguistic objects which include only the text in Russian (or proper names of non-Russian origin spelled according to the Russian language rules) in Cyrillic script.

Signs in English: The linguistic objects which include only the text in English (or proper names of non-English origin spelled according to the English language rules) in Latin script.

Mixed Signs (Russian and English): the linguistic objects that include a combination of Russian and English text (or proper names of non-Russian or non-English origin spelled according to the Russian or English language rules respectively) in any proportion.

Signs in English written in Cyrillic letters: The linguistic objects that feature English text in Cyrillic script.

Signs in Russian written in Latin letters: The linguistic objects that feature Russian text in Latin script (or proper names of Russian origin transliterated into English).

Signs in Local Language: The linguistic objects that feature the text in the language of one of the local ethnic groups, other than Russian (or proper names typical of the local culture).

Mixed Signs (Local Languages and Russian or English): The linguistic objects that include a combination of Russian or English and the language of one of the local ethnic groups, other than Russian (or proper names typical of the local culture).

Other: the linguistic objects that include the text in any languages other than Russian, English or any of the languages spoken by the ethnic groups of this region.

There were, however, a few controversial cases where it was difficult to define to which category the sign must be attributed. For example, "Салон красоты АМИНА" ('Beauty salon AMINA') in Vladikavkaz: 'Amina' is a popular name in the North Caucasus but not specific to any of the local cultures. Moreover, it is under the influence of the Russian language that it has received flection -a. In this paper, such cases are regarded as signs in Russian. The fact that all the North Caucasus languages use Cyrillic script does not make the labelling process any easier either, sometimes making it difficult to differentiate between the text in the local language proper and local realia borrowed by the Russian language. For instance, "Кафе Махсыма" (Cafe Makhsyma) in Nalchik: Makhsyma is a traditional Kabardian drink, yet this word is spelled differently in the local language (махъсымэ), therefore we treated this case as a Kabardian realia assimilated by the Russian Language.

3   Results

The results of the study are presented in Table 2.

It is clear that the majority of signs in all cities considered are monolingual, with the average share of multilingual signs in a North Caucasian city linguistic landscape being 15.27% (including all kinds of combinations of Russian, English and local languages). 1.81% of the signs are technically monolingual, but use a different script (Cyrillic or Latin), which can also be considered a manifestation of multilingualism.

It is also easy to notice how small the share of local languages in public signage is, being only 0.45% on average for all cities (0.9% with mixed signs). Even with Pyatigorsk excluded as a predominantly Russian-speaking city where no local language signs can be found, it is barely higher than 1% (1.08%). It is noteworthy, however, that Pyatigorsk is not the only city for which no local language signs were registered, but the same picture can also be seen in Magas and Nalchik (if not taking mixed cases into account). The cases labelled as 'Other' amount to 1.87% of the data. These include signs in Arabic (though mostly in Latin or Cyrillic transcription), Turkish, Italian, and French. Signs in German, Spanish, Uzbek, Corean, Chinese, and Greek were registered but once.

Table 2. Distribution of linguistic signs in the central streets of capital cities according to the language(s) of the sign.

City	Signs in Russian (%)	Signs in English (%)	Mixed signs (Russian and English) (%)	Signs in English written in Cyrillic letters (%)	Signs in Russian written in Latin letters (%)	Signs in local languages (%)	Mixed signs (local languages and Russian) (%)	Other (%)
Cherkessk	80.13	5.54	11.40	0.65	0.81	0.33	0.00	1.14
Grozny	52.65	18.37	21.97	0.57	1.33	0.76	0.57	3.79
Magas	61.36	13.64	22.73	0.00	0.00	0.00	0.00	2.27
Makhachkala	59.65	18.70	15.46	0.44	1.91	0.88	0.00	2.95
Maykop	78.00	6.60	10.02	1.47	1.34	0.49	0.86	1.22
Nalchik	65.75	12.87	17.93	0.23	0.69	0.00	0.69	1.84
Pyatigorsk	79.28	3.45	14.10	1.29	0.72	0.00	0.00	1.15
Vladikavkaz	67.94	11.96	16.27	0.24	0.24	0.72	1.44	1.20
Total	69.98	10.59	14.82	0.78	1.06	0.45	0.45	1.87
Total (Pyatigorsk excluded)	68.16	11.99	14.96	0.68	1.13	0.54	0.54	2.01

The cases labelled as 'Other' amount to 1.87% of the data. These include signs in Arabic (though mostly in Latin or Cyrillic transcription), Turkish, Italian, and French. Signs in German, Spanish, Uzbek, Corean, Chinese, and Greek were registered but once.

With Russian being prevalent in the linguistic landscape of all the cities, its share in the public signage varies considerably - from 80.13% in Cherkessk to 52.65% in Grozny. Bilingual signs combining Russian and English make up the second largest category in all cities, amounting to 14,82% on average. The share of English signs also varies greatly - from a humble 3.45% in Pyatigorsk to a considerable 18.7% in Makhachkala or 18.37% in Grozny.

Apart from the languages displayed, it may be interesting to explore in what proportion different scripts are represented in the linguistic landscapes of the North Caucasus cities. Figure 1 features the distribution of script types used in the region. As expected,

Cyrillic is accounted for almost ¾ registered cases, with 15% of signs combining Cyrillic and Latin and 13% being purely Latin (see Fig. 1).

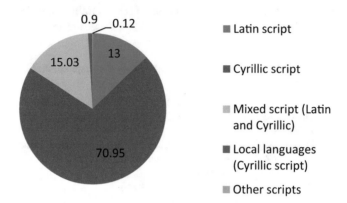

Fig. 1. Diagram showing the distribution of script types used in signs.

Though cyrillic-based, local language scripts form a separate group here because their writing systems are not quite the same as in Russian, with different spelling and additional characters. Other scripts include Arabic, Chinese, Corean, and Greek, though the number of such cases is negligibly small.

4   Discussion

As it was shown in Sect. 3, linguistic landscapes of the North Caucasus capital cities do not contain whatsoever significant manifestations of local language usage. Apparently, this sphere is dominated by the Russian and (very often) English languages. If we leave out top-down naming, the bottom-up commercial names are hardly ever dubbed in a local language. The existing small number of names are based on the local realia (anthroponyms, mythonyms, toponyms, names of local household and cuisine items etc.), with these onyms often originating from the local language but transformed to fit in with the Russian or English orthography.

Why are local languages so poorly represented in the LL of the North Caucasus capital cities? This tendency can be traced back to the times of the Soviet Union when speaking minority languages in cities was strongly discouraged, so they began to be associated with rural life because they were (and still are) mostly used in villages with their traditional lifestyle.

Undoubtedly, a large percentage of the Russian population in all cities (25.01% in Vladikavkaz, 28.77% in Nalchik, 71.29% in Maykop, 54.07% in Cherkessk, and 71.47% in Pyatigorsk) except Grozny (3.3%), Magas (16.23%) and Makhachkala (6.28%) [37], is the reason for the dominance of the Russian language over local languages in many spheres including LL. Despite the official declarations of the equal statuses of Russian and local languages, the sphere of business naming is not strictly regulated allowing for a great deal of freedom in choosing the names and not requiring obligatory translations.

Based on the analyzed data, we may assume that in multilingual communities the local native speakers associate their own language with home and family, while Russian is associated with the officialdom of the state. At the same time, foreign languages (especially English, French and Italian) have the implication for luxury, prosperity, security, and better-off life. As for the use of Arabic in LL, it is largely motivated by Islam being the official religion in most of the North Caucasus republics. Perceived as the language of the Quran prescribing the rules and even small nuances of behavior in every life situation, Arabic is familiar to many local speakers.

Another reason for the underrepresentation of local languages in the LL might be that the business owners do not want to limit the target audience to the speakers of one local tongue when there are several languages spoken in the region, a large percentage of Russians, or an influx of tourists from other regions and abroad. The North Caucasus is the gem of the south of Russia due to the incomparable beauty of its nature, salutary climate, and huge recreational potential. The local governments are now working hard on promoting local tourism, trying to attract domestic and foreign investors and increase the number of visitors, including speakers of other languages.

Finally, the fact that there is no strict division of commercial outlets by languages/nationalities in the investigated capital cities seems to speak in favor of stability in the sphere of interethnic relations in the North Caucasus which is deservedly called a melting pot of nations, cultures, and languages.

Increasing the share of street signs in the local language cannot but become a step towards the promotion of the local languages in the multiethnic areas where Russian is used as lingua franca in everyday communication. It is likely to raise linguistic and cultural awareness of non-title ethnic groups, promote spontaneous language acquisition, and thus contribute to the vitality of the language.

In the recent decades, Russia has seen a massive infusion of foreign vocabulary into all spheres of human activity: politics and economics, information and communication technologies, culture and education. The excessive use of English words has become a distinctive feature of any urban linguistic landscape in Russia, the capital cities of North Caucasus republics being no exception.

One of the reasons is the increase in the number of people who speak and know English in Russia. Bloggers, musicians, DJs, athletes and other media people also contribute to the popularity of English vocabulary.

It should be noted that foreign elements in the public signs are often used merely to catch the attention of potential consumers. Passion for English onyms is a kind of fashion formed in the society under the influence of mass media. The idealized image of Western (primarily American) society with its freedom, technological progress and high standard of living has become a stereotype, especially among younger generations. The names written in Latin script probably seem more stylish and cool (one must admit that fashions and styles mostly come from the West, even if via Moscow); the underlying message of Latin characters in a name might be a hint at the luxury segment meant for those who can afford to travel abroad and buy well-known international brands. The original spelling of international brand names is almost always retained to ensure their recognition by consumers.

Underestimation of local brands, that goes back to the USSR times of total deficit and the cult of imported goods which were always considered to be of better quality, might also be a reason for the overuse of Latin script in LL. Transliteration of this kind is meant to mask the "provinciality" of seemingly uninteresting and bland names of local brands. There is also the legacy of the 1990s when the Soviet economy collapsed leaving the population of Russia with almost no quality goods produced within the country.

The visual difference between Cyrillic and Latin characters involuntarily catches the eye of a recipient: foreignization of a sign attracts the attention, and the resulting defamiliarization creates a mental mark in a reader's memory. Besides, the use of Latin script sometimes helps to unleash the creativity of those who invent LL onyms, as it allows for numerous transcyrillic and multilingual puns, blends, and comic hybrids resulting from mixing two or more languages.

The excessive number of English-based commercial names and signs has been recently heavily criticized on all levels. In March 2022, the State Duma deputy Sultan Khamzayev (by the by a Daghestani himself) addressed the head of the Federal Antimonopoly Service with a letter substantiating his initiative to ban commercial signs in any other languages than those listed as official ones in the area. His reasons were that a sign in a foreign language could provide incomplete or wrong information and mislead the consumers. In fact, the Russian law On Advertising (№38-ФЗ) explicitly bans any information in a foreign language which might be misunderstood and misinterpreted [7]. The document, in its turn, relegates to the regulations of the law On State Language of the Russian Federation [39]. In June 2022, the State Duma outlined a draft law tightening the norms for the use of foreign words and the Latin alphabet in advertising. The new law will require clarifications in the Russian language to any business name containing Latin characters. According to Elena Yampolskaya, the head of the Duma's committee on culture, "in the excessive use of English borrowings there is a plebeian disrespect for one's country and one's native language" [35]. Naturally, the business community of Russia did not welcome the initiative, as the adoption of the law will mean fines and extra expenses on changing the signs or even renaming businesses. However, such initiatives go in line with the current protectionist policies of the Russian government, so the linguistic landscape of local cities is all too likely to face major changes in the next few years.

The present paper highlights the underrepresentation of the local languages in the linguistic landscape of the North Caucasus capitals as well as explores the role of the English language and Latin script in the public signage, suggesting the underlying reasons of the described phenomenon. Yet a quantitative study like this cannot possibly cover all aspects of such a complex and mosaic linguistic landscape. Further research directions may include a qualitative inquiry into the nature of multilingual signs used in the region carefully studying all possible forms of how multiple languages are combined. The research of multilingualism in the North Caucasus can also benefit from contrasting the LL of the capital cities with that of smaller settlements to find out if our findings are something specific only to the urban areas or can be extrapolated to other communities as well.

References

1. Backhaus, P.: Linguistic Landscapes: A Comparative Study of Urban Multilingualism in Tokyo. Multilingual Matters, Clevedon (2007)
2. Barni, M., Bagna, C.: A mapping technique and the linguistic landscape. In: Shohamy, E., Gorter, D. (eds.) Linguistic Landscape: Expanding the scenery, pp. 126–140. Routledge, Abingdon-on-Thames (2009)
3. Ben-Rafael, E., Shohamy, E., Amara, M.H., Trumper-Hecht, N.: Linguistic landscape as symbolic construction of the public space: the case of Israel. In: Gorter, D. (ed.) Linguistic Landscape: A New Approach to Multilingualism, pp. 7–30. Multilingual Matters, Bristol (2006)
4. Bruenok, A.V., Mukhanov, V.M., Pavlinov, P.S.: Pyatigorsk. In: Kravets, S.L. (ed.) Great Russian Encyclopedia, vol. 28. Bolshaya Rossiyskaya Entsiklopediya, Moscow (2015)
5. Cenoz, J., Gorter, D.: Linguistic landscape and minority languages. In: Gorter, D. (ed.) Linguistic Landscape: A New Approach to Multilingualism, pp. 67–80. Multilingual Matters, Clevedon (2006)
6. Federal State Statistics Service. (in Russian). https://rosstat.gov.ru/storage/mediabank/chisl_%D0%9C%D0%9E_Site_01-01-2022.xlsx, Accessed 26 Apr 2022
7. Federal'nyi zakon O reklame (Federal Law on Advertising). https://docs.cntd.ru/document/901971356. Accessed 6 Oct 2022
8. Ferguson, J., Sidorova, L.: What language advertises: ethnographic branding in the linguistic landscape of Yakutsk. Lang Policy **17**, 23–54 (2018)
9. Gorter, D.: Methods and techniques for linguistic landscape research: about definitions, core issues and technological innovation. In: Pütz, M., Mundt, N. (eds.) Expanding the Linguistic Landscape: Multilingualism, Language Policy and the Use of Space as a Semiotic Resource. Multilingual Matters, Bristol (2018)
10. Gorter, D.: Introduction: the study of the linguistic landscape as a new approach to multilingualism. In: Gorter, D. (ed.) Linguistic Landscape: A New Approach to Multilingualism, pp. 1–6. Multilingual Matters, Bristol (2006)
11. Hult, F.M.: Language policy and planning and linguistic landscapes. In: Tollefson, J.W., Pérez-Milans, M. (eds.) Oxford Handbook of Language Policy and Planning, pp. 333–351. Oxford University Press, New York (2018)
12. The Constitution of the Republic of Adygea. (in Russian). https://docs.cntd.ru/document/804932122. Accessed 6 October 2022
13. The Constitution of the Chechen Republic. (in Russian). https://docs.cntd.ru/document/819051373. Accessed 6 October 2022
14. The Constitution of the Daghestan Republic. (in Russian). https://docs.cntd.ru/document/802018919. Accessed 6 October 2022
15. The Constitution of the Republic of Ingushetia. (in Russian). https://docs.cntd.ru/document/720705885. Accessed 6 October 2022
16. The Constitution of the Kabardino-Balkarian Republic. (in Russian). https://docs.cntd.ru/document/720705883. Accessed 6 October 2022
17. The Constitution of the Karachay-Cherkess Republic. (in Russian). https://docs.cntd.ru/document/720705884. Accessed 6 October 2022
18. The Constitution of the Republic of North Ossetia-Alania. (in Russian). https://docs.cntd.ru/document/304200016. Accessed 6 October 2022
19. The Constitution of the Russian Federation. (in Russian). http://www.constitution.ru/. Accessed 6 Oct 2022
20. Kulbayeva, A.: Polycentricity of Linguistic Landscape and Nation-Building in Post-Soviet Kazakhstan. Cent. Asian Aff. **5**, 289–312 (2018)

21. Landry, R., Bourhis, R.Y.: Linguistic landscape and ethnolinguistic vitality. J. Lang. Soc. Psych. **16**(1), 23–49 (1997)
22. Marten, H.F., Van Mensel, L., Gorter, D.: Studying minority languages in the linguistic landscape. In: Gorter, D., Marten, H.F., Van Mensel, L. (eds.) Minority Languages in the Linguistic Landscape, pp. 1–15. Palgrave-MacMillan, Basingstroke (2012)
23. McDermott, A.: The linguistic landscape of post-Soviet Bishkek. Region Regional Stud. Rus. East. Eur. Cent. Asia **8**(2), 227–255 (2019)
24. Muth, S.: Informal signs as expressions of multilingualism in Chisinau: how individuals shape the public space of a post-Soviet capital. Intern. J. Sociol. Lang. **228**, 29–54 (2014)
25. Law about the languages of the peoples of the Kabardino-Balkarian Republic (in Russian). https://docs.cntd.ru/document/802054078. Accessed 6 Oct 2022
26. Pitina, S.: English influence on linguistic landscape of modern Russian cities. Inter. J. Eng. Ling. **10**, 61–68 (2020)
27. Purschke, C.: (T)Apping the linguistic landscape methodological challenges and the scientific potential of a citizen-science approach to the study of social semiotics. Ling. Landscap. **3**(3), 246–266 (2017)
28. Puzey, G.: Linguistic landscapes. In: Hough, C. (ed.) The Oxford Handbook of Names and Naming, pp. 395–411. Oxford University Press, Oxford (2015)
29. Reh, M.: Multilingual writing: a reader-oriented typology with examples from Lira Municipality (Uganda). Intern. J. Sociol. Lang. **170**, 141 (2004)
30. Rosenbaum, Y., Nadel, E., Cooper, R.L., Fishman, J.A.: English on Keren Kayemet street. In: Fishman, J.A., Cooper, R.L., Conrad, A.W. (eds.) The Spread of English, pp. 179–196. Newbury House, Rowley (1977)
31. Sadurov, R.T.: Field research of cultural and linguistic landscape in a multiethnic region. Ecolog. Lang. Communicat. pract. **1**, 23–29 (2020). (in Russian)
32. North Caucasus. In: Kravets, S.L. (ed.) Great Russian Encyclopedia, vol. 29. Bolshaya rossiyskaya entsiklopediya, Moscow (2015). (in Russian)
33. Shibilyev, J.: Linguistic landscape approach to language visibility in post-Soviet Baku. Bilig **71**, 205–232 (2014)
34. Solnyshkina, M.I., Ismagilova, A.R.: Westernization and globalization of the linguistic landscape of Kazan Republic of Tatarstan. XLinguae **8**(2), 36–53 (2015). (in Russian)
35. State Duma clarified the suggestion to ban foreign language signs. (in Russian). https://www.rbc.ru/politics/10/06/2022/62a32d779a7947fa38eecb6a. Accessed 6 Oct 2022
36. Van Mensel, L., Vandenbroucke, M., Balckwood, R.: Linguistic landscape. In: Garcia, O., Flores, N., Spotti, M. (eds.) The Oxford Handbook of Language and Society, pp. 423–449. Oxford University Press, Oxford (2016)
37. All-Russia Population Census, Volume 4. Language competence, citizenship. (in Russian). https://www.gks.ru/free_doc/new_site/perepis2010/croc/vol4pdf-m.html. Accessed 6 Oct 2022
38. Yandex Maps. https://yandex.ru/maps. Accessed 17 June 2022
39. Law on State Language of the Russian Federation. (in Russian) https://normativ.kontur.ru/document?moduleId=1&documentId=12411. Accessed 6 Oct 2022

Conceptions and Cultural Significance of Bilngualism

Mary Danielyan⑩ and Liliia Metelkova(✉) ⑩

Moscow State University of Civil Engineering, Moscow, Russian Federation
{DanielyanMG,MetelkovaLA}@gic.mgsu.ru

Abstract. Arguments in favour of bilingualism in different regions and countries have always been based on political and historical factors. Currently, the advantages of bilingualism are also determined by socio-economic factors. The literature review presented in this article outlines an interdisciplinary approach to bilingualism and summarizes the main findings from more than 50 scientific studies and documents. The studies analyzed in this article cover a wide range of areas of interest in relation to bilingualism, such as psychology, economics, sociology, linguistics, literature, public policy, and political science. Analysis of the many definitions and explanations of bilingualism and biglottism, allows us to say that these terms do not have an unambiguous definition, they can be interpreted both in a broad and in a narrow sense. Many linguists engaged in language contacts believe that bilingualism can be attributed to the phenomenon which demonstrates a change in the language code in a particular speech act, i.e. the ability of an individual to use a non-native language when communicating with the native speakers of this language.

Keywords: Bilingualism · Bilingualism problem · Monolingual

1 Introduction

In modern multicultural society, the phenomena of bilingualism and biglottism are clearly traced. In the era of globalism, the knowledge of two or more languages is a fairly common situation, which partly leads to a mixture of cultures and traditions. However, the question arises, what is considered bilingualism or biglottism?

Biglottism tends to occur in those social societies where there are two or more languages, each with its own social functions. An example of active bilingualism (biglottism) is the linguistic behavior of migrants who use a second language when communicating with the representatives of another language community or at the official level. Children of migrants attending schools with the Russian language used in education process actively use the Russian language at school. However, communication in their families takes place in their native language exclusively. This example demonstrates a partial or fragmentary bilingualism.

For bilinguals who are fluent in two language systems, using two languages alternately is a natural speech behavior. In such a language situation, the question arises - which language is dominant (mother tongue) and which is additional.

D. Bylieva and A. Nordmann (Eds.): PCSF 2022, LNNS 636, pp. 73–79, 2023.
https://doi.org/10.1007/978-3-031-26783-3_8

2 Research Questions

To achieve the goal of our study, it is necessary to answer the following interrelated questions:

Can a person who does not fully speak the language be considered bilingual, or is it a case of biglottism?
Do writers have a linguistic flair in two different language systems?
How often do extralinguistic substitutions occur in bilingual speech?
Can languages learned and used in a human activity process be considered the dominant?

3 Purpose of the Study

Not always the language learned in childhood is dominant in life. The main factor influencing the linguistic behavior of an individual is his cultural socialization. In our study, we propose to determine the degree of influence of the socio-cultural sphere on the use of a particular language, as well as to analyze the phenomena associated with the cultural characteristics of bilingual individuals and their impact on the linguistic personality.

4 Research Methods

The literature review covers more than 50 scientific sources and studies, published journal articles, statistical data, articles in the media and books of Russian researchers.

In addition to the Russian-language review, foreign literature was analyzed, the results of studies for comparative analysis were considered.

The studies analyzed in this article cover a wide range of areas of interest in relation to bilingualism and multilingualism, such as psychology, economics, sociology, linguistics, public policy and political science.

The methods used in the study are diverse, based on statistical and quantitative methods, as well as on a number of qualitative methods, including case studies.

5 Results

5.1 The Influence of Bilingualism on Linguistic Personality

The word "bilingualism" first appears with a certain frequency in the speeches and works of psychologists and teachers. As a rule, they considered bilingualism as a problem that needed to be solved, a disease that needed to be eradicated. Thus, in his book, Isaac Epstein (La pensée et la polyglossia) recognizes polyglossion only in terms of reception. If it is proper for a cultured person to read and understand several languages ("passive polyglossia"), then, according to I. Epstein, in fact, one language is enough to express his thought ("active monoglossy"), which is the main function of language, according to Epstein [1]. Some educators have gone so far as to argue that bilingualism can have a negative impact on a child's intellectual development. This was the thesis presented by two Belgian delegates, Verheyen and Toussaint, at the International Bilingualism Conference held in Luxembourg in 1928 [2]. However, public opinion was not yet ready for this kind of argument.

A typical discrepancy in approaches to the study and use of language can be a comparison of the works of B. Hufeisen - linguist and A. Nordmann – philosopher [3]. The linguist focuses on the use of language and the actual semantic meaning of words. At the same time, the philosopher considers the language and technology of the issue primarily from the point of view of their use, so it is the meaning that becomes the product of use. However, both specialists emphasize the importance of culture and context for the meaning and use of language.

The same dual understanding of bilingualism is sometimes observed in the political sphere. Official bilingualism was often introduced not to promote the development of two languages, but rather "to ensure [citizens'] right to individual monolingualism" [4]. Belgium is a good illustration of such a paradox, conceived by the French-speaking bourgeoisie as a monolingual "trans-ethnic" state and remaining so for almost a century, ignoring the real diversity of languages spoken on its territory [5].

5.2 Bilingualism in Literature

In literature, bilingualism results in a series of monolingual periods: Russian and American in the work of Vladimir Nabokov, Polish and English in the work of Joseph Conrad, Czech and French in the work of Milan Kundera. According to Grutman, after switching from one language to another, authors are often hesitant to return to their former language of writing, as if taboo, feeling a typical sense of guilt, comparing bilingualism with "cohabitation of two languages" [6].

Literally bilingualism means the sequential or simultaneous use of two languages by one author. By the language of writing, we mean the language that a bilingual writer has chosen (or that circumstances have imposed on him) to create his literary works. Thus, the term does not include languages that the author can use in his works for the purposes of plausibility or to convey the local flavor.

It should also be noted that there is a difference between authors who use the two languages of writing, as not all of them are literary bilinguals. The following five categories were proposed by Vallverdú and described by M. Fernández for French-speaking writers:

a) a writer who writes alternately in French and another language;
b) French-speaking writers writing in another language;
c) an allophone writer writing in French;
d) a writer who writes in French but also uses another language for non-literary texts: ideological prose, scientific essays, newspaper articles
e) a writer who writes sequentially in French and in another language (or vice versa), but never simultaneously [7, 8].

5.3 Bilingualism and Biglottism

Translated into foreign languages, "bilingualism" and "biglottism" are absolute synonyms, but in linguistics the terms "bilingualism" and "biglottism" do not have an unambiguous definition, they can be interpreted both in a broad and narrow sense. Many linguists who deal with language contacts believe that biglottism can be attributed to the phenomenon in which there is a change in the language code in a particular speech act. This means that "bilingualism" in the broad sense is the ability of an individual to use a non-native language to one degree or another when communicating with native speakers of this language, and in a narrow sense – fluency in a non-native language in oral and written forms in all spheres of communication.

Thus, V.A. Aurorin, who used a narrow understanding of the term, considered bilingualism "equally fluency in two languages. In other words, biglottism begins when the degree of knowledge of a second language comes close to the degree of knowledge of the first one" [9]. Researchers who explain the phenomenon of bilingualism in a broad sense argue that "the possession of a second language to any extent can be regarded as bilingualism" [10, 11].

U. Weinreich argued that bilingualism or biglottism is precisely "the practice of alternating the use of two languages", and people "who carry out this practice are called bilinguals or the ones with biglottism" [12].

In the era of the USSR, scientists V. V. Vinogradov, Y. D. Desheriev, L. Shcherba, et al., who dealt with the problem of language contacts in a multilingual state and described their interaction, introduced the term (national-Russian) bilingualism, which was explained as any change in the language code.

In modern linguistics, the interpretation of the term "bilingualism/biglottism" sometimes contradicts each other. Thus, according to the theory of G.M. Vishnevskaya, the phenomenon of bilingualism can be divided into two groups: the first is the possession of a non-native (second) language at the level of a native speaker, and the second is the use of a non-native language to varying degrees [13].

Describing language contacts and explaining them through the term "bilingualism", special attention is paid to the quality of knowledge of a non-native language at all levels of the language: phonetic, lexico-grammatical, lexico-semantic and syntactic [14]. Based on this, bilinguals are recognized as those individuals who speak a non-native language at a level that allows it to be used with native speakers of this language. According to the definition of Yu.D. Desheriev, bilingualism is the practice of "alternate active use by an individual of two languages in his speech activity,… And an individual who uses two languages in his speech activity is called bilingual" [15]. Consequently, "bilingualism" in the narrow sense of the term means fluency in two languages: native and additional (non-native).

Analyzing the many definitions and explanations of bilingualism and biglottism, we can say that biglottism is a linguistic behavior that is more necessary for an individual for social interaction, and bilingualism is a natural linguistic behavior under natural language conditions. So, if an individual for interaction in a multilingual society uses a non-native language that he speaks at one level or another, then we can talk about social biglottism. Biglottism is not a natural language behavior, as a second language is mastered by a person in the process of its study and is acquired, studied and used in specific social situations where the use of the language of the majority is required for successful communication. Often such situations are called the phenomenon of partial bilingualism.

Bilingualism is a natural language situation when the code change occurs fluently, the speaker does not experience discomfort and difficulties in communicating, since the degree of proficiency in the native and second language is almost the same.

Often bilinguals themselves find it difficult to answer in a definite way about the place and meaning of the "main" language. As a rule, bilinguals note that the first language is considered to be the language of everyday communication and the "internal" language, although the second language is not characterized by them as a secondary, additional, since both languages were learned in natural linguistic conditions. However, bilingual languages can be contrasted depending on the order of their assimilation and the scope of their use. For the "natural" bilinguals, it is not always easy to determine the dominant and native language: depending on the specific situation, two different languages act as native languages, that is, both their languages are the first ones. With an equal knowledge of two languages, "pure" bilinguals, as a rule, do not switch from one language to another, though such a transition is possible if one of the communicators experiences language discomfort and smoothly switches to the second one.

5.4 Types of Bilingualism

According to the method of mastering and the degree of proficiency in two languages, researchers of language contacts distinguish more than 20 types of bilingualism, which reflect its various manifestations, and also take into account the conditions under which the formation of languages and the spheres of their interaction occur.

One common type of biglottism is "natural" and "artificial". For natural biglottism, a multilingual society is necessary, in which permanent language contacts of speakers of two languages occur, and artificial bilingualism is formed in conditions of monolingualism, with the purposeful formation of skills in using another language system. An example of artificial biglottism is the Russian-national biglottism, formed and functioning on the territory of the former Soviet Union.

L. Shcherba distinguished between a pure type of biglottism and a mixed one. Pure biglottism is characterized by the use of one language in a particular situation, for example, the presence of one language only in the family circle and another – for communication outside the home. A distinctive feature of the mixed type of biglottism is the alternating code change that occurs naturally when communicants do not notice the transition from one language to another.

Sometimes natural bilingualism (biglottism) is equated with everyday bilingualism, meaning that the second language is learned only through the language environment and constant speech practice, and artificial biglottism is called educational, since the second language is learned with the help of educational and methodological rules.

Weinreich, when classifying bilingualism (biglottism), proceeded from the way in which the individual learned a second non-native language. Thus, he singled out the composite bilingualism that takes place in mixed families – the existence of two different autonomous languages for communication; coordinate bilingualism – the use of one particular language in certain language situations (for example, the linguistic behavior of migrants in the host country); subordinational bilingualism, when a second language is learned by explaining grammatical rules in the native language of students [15].

Tréguer-Felten, Truchot, Grenier et Nadeau, speaking of bilingualism, describe the difference between the use of one language as a means of communication in relation to a second language as an element of cultural identity [16–18]. In the first case, the more people use a language, the more useful that language is as a communication tool. Grenier and Nadeau call this principle the "external effects of the language network" [16].

Language, however, conveys not only meaning from a purely terminological point of view, but also moral values and general opinions regarding other social or political phenomena and events as well [19, 20].

The qualitative study of emails between companies with different language profiles by Tréguer-Felten illustrates what can happen when companies of different nationalities use a common language (English) as a simple means of communication and face cultural misunderstandings.

6   Discussion

Summarizing the above, we can conclude that bilingualism is a complex and multifaceted phenomenon that implies the possession of two language systems equally, and the use of two languages is perceived as a natural state. In practical terms these languages have equal functional application. Psycholinguistically, pure bilinguals themselves do not distinguish languages as dominant or complementary. Individuals with biglottism, as a rule, prefer their native language in informal communication. In language practice, biglottism means a dominant language, when switching to a second language, individuals experience some communicative discomfort.

Can bilingualism be considered a natural language situation in the era of globalization? Many countries are considered multilingual in terms of population. Judging by the migration processes taking place everywhere, a large percentage of the world's population is has biglottism or is bilingual, actively using two or more languages in language practice. In a number of European countries, two or more languages (Belgium, Switzerland, Canada) are endowed with the status of the state language. On the territory of the Russian Federation, along with the Russian state language, there are also regional languages that have the status of the state language, for example, Tatar, Bashkir, Crimean Tatar, Yakut, Buryat, Ossetian, Chuvash, etc. Knowledge of several languages makes it possible to educate a tolerant community in which not only different language systems, but also cultural traditions of peoples harmoniously coexist.

1. References

1. Epstein, I.: La pensée et la polyglossie. Payot, Paris (1915). (in Fr.)
2. Chernositova, T.L.: The problem of bilingualism and diglossia in the context of literary and artistic translingualism. Young Sci. **14** (94), 607–609 (2015). https://moluch.ru/archive/94/20937.10.09.2022)
3. Hufeisen, B., Nordmann, A., Liu, A.W.: Two perspectives on the multilingual condition - linguistics meets philosophy of technology. Technol. Lang. **3**(3), 11–21 (2022). https://doi.org/10.48417/technolang.2022.03.02

4. Patrimoine canadien. Avantages économiques du bilinguisme (2016). (in Fr.). https://www. caslt.org/files/learn-languages/pch-bilingualism-lit-reviewfinal-fr.pdf. Accessed 6 October 2022
5. Mackey, W.F.L.: Dialecte et diglossie littéraire. In : Giordan, H.; Ricard, A. (Eds.) Diglossie et Littérature, pp. 19–50. Maison des sciences de l'Homme d'Aquitaine, Bordeaux-Talence (1976). (in Fr.)
6. Grutman, R.: Le bilinguisme littéraire comme relation intersystémique. Can. Rev. Comp. Lit./Rev. Can. Lit. Comparée **17**(3–4), 198–212 (1990)
7. Fernández, M.: Los orígenes del término diglosia: historiade una historia mal contada. Historiographia Linguistica. Historiographia Linguistica **22**(1–2), 163–195 (1995). https://doi.org/10.1075/hl.22.1-2.07fer
8. Vallverdu, F. Sociología y lengua en la literatura catalana, Traduit du catalan par José Fortes Fortes. Cuadernos para el Diálogo, Madrid (1971). (in Fr.)
9. Aurorin, V.A.: Bilingualism and school. In: Problems of Bilingualism and Multilingualism, pp. 49–62. Nauka, Moscow (1972). (in Rus)
10. Filin, F. P.: Modern social development and problems of bilingualism. In: Problems of bilingualism and multilingualism, pp. 13–25. Nauka, Moscow (1972). (in Rus)
11. Desheriev Y.D.: Main aspects of research on bilingualism and multilingualism In: Problems of Bilingualism and Multilingualism, pp. 26–42 Nauka, Moscow (1972). (in Rus)
12. Weinreich, U.: Language contacts: state and problems of research. Kyiv: Vyshcha Shkola, p. 263 (1979)
13. Vishnevskaya, G.M.: Bilingualism and its Aspects. Ivan. State University, Ivanovo (1997)
14. Rosenzweig, W.Y.: Language Contacts: Linguistic Problems. Nauka, Moscow (1972)
15. Weinreich, U.: Monolingualism and multilingualism. Lang. Contacts. New Linguist. **6**, 25–60 (1972)
16. Grenier, G., Nadeau S.: English as the lingua franca and the economic value of other languages: the case of the language of work in the montreal labour market. Université international de Venise, The Economics of Language Policy, pp. 267–312. The MIT Press, Cambridge (2016). https://doi.org/10.7551/mitpress/9780262034708.003.0009
17. Tréguer-Felten, G.: Commun'action ou commun'entente? Un défi linguistique pour les entreprises. Synergies Italie **9**, 47–58 (2013). (in Fr.)
18. Truchot, C.: Internationalisation et choix linguistiques dans les entreprises françaises: entre «tout anglais» et pratiques plurilingues. Synergies Italie, no **9**, 75–90 (2013). (in Fr.)
19. Lepage, J-F., Corbeil, J-P.: L'évolution du bilinguisme français-anglais au Canada de 1961 à 2011. StatistiqueCanada (2013). (in Fr.) https://www150.statcan.gc.ca/n1/pub/75-006-x/201 3001/article/11795-fra.pdf. Accessed 6 October 2022
20. Lévesque, J-M. : Le bilingusime et le revenu du travail. Le revenu et l'emploi en perspective (1989). https://www150.statcan.gc.ca/n1/fr/catalogue/75-001-X19890022277. Accessed 6 October 2022

The Problem of Translating a Literary Text into the Language of the Theater (Based on the Dramaturgy of A. S. Pushkin and N. V. Gogol)

Larisa Tyutelova⬤, Elena Sergeeva⬤, Ksenya Sundukova$^{(\boxtimes)}$ ⬤, and Daria Moroseeva⬤

Samara National Research University, 34, Moskovskoye Shosse, Samara 443086, Russia
{tyutelova.lg,sergeeva.en,sundukova.ka}@ssau.ru

Abstract. The proposed article deals with the problem of translating a literary text into the language of the theater. The study focuses on the dramaturgical material from a turning point in the history of the Russian theater; the plays of two Russian playwrights from the first half of the 19th century, Alexander Pushkin and Nikolai Gogol. This article demonstrates the fundamental changes that took place during this period with the composition of the dramatic text and the organization of the stage space dictated by it. It is shown that Pushkin and Gogol in their dramaturgy used some novel strategies, such as: the free depiction of the hero's movement in space, the change of plans – general and large, and the complex interweaving of verbal and visual information. The literary texts of Pushkin and Gogol require that the main role in the production should no longer be played by the interaction between the actor and the viewer, but by the change of scenes, and the use of onstage and offstage space to convey the subjective view of the hero, who has his own ability to see, perceive, and evaluate. The plays during the first half of the 19th century were only able to overcome these problems by verbal means. That is why the plays of Pushkin and Gogol turned out to be fundamentally untranslatable into the language of their contemporary theater, but it was they who set the trends for the further development of theatrical technologies.

Keywords: Theatrical technologies · The language of the theatre · The language of drama · Theatrical space · Dramatic space · Picture · Stage · Alexander Pushkin's dramaturgy · Nikolai Gogol's dramaturgy

1 Introduction

At one time, Alexander Pushkin called Russian theatre "the courtier" [1, p. 149], since for the first time Russian culture got acquainted with the theater as a European passtime in the 17th century at the court of Alexei Mikhailovich. The Russian theatre (unlike the European one) is not based on folk performances. It, like many things that arose in the field of culture in Russia during the Petrine and post-Petrine era, is the result of an orientation towards the ancient artistic paradigm.

D. Bylieva and A. Nordmann (Eds.): PCSF 2022, LNNS 636, pp. 80–89, 2023.
https://doi.org/10.1007/978-3-031-26783-3_9

Since the time of Aristotle, the two arts, literature and theater, have been seen as inextricably linked. The philosopher, in principle, does not talk about them separately: in his "Poetics" it is a question of the performing arts, for which the text as a system of verbal signs is a subsystem of a holistic theatrical performance along with a system of gestural and musical signs. And for a long time in the development of the European theater, the verbal subsystem did not exist autonomously, nor does it play a leading role.

Until modern times, when the question of authorship arises, European theater had remained an improvisational art. In it, the mask (whose presence on the stage is indicated by the whole system of signs) exists thanks to improvisation and the framework of the role and style tradition that restrains the actor. In the improvisational theater, there is only a script before the start of the performance, and, in the end, the verbal text of the performance is different each time in the process of acting. In improvisational theater there is no canonical literary text, with its subtle play of nuances and intonations.

However, let us repeat, the Russian theater emerged when the European theater ceased to be improvisational. The Russian theater of the 18th century relied on the literary foundation created by the author, who began to realize himself as a personal creator of the reality that appeared in his work. At the same time, the literary text remained the scenario of a theatrical performance, but it was no longer solely the general contours of the story played out on the stage. Thanks to the verbal signs, a visual image of the performance and all its main components, an integral and completed event and its participant, appeared in it. There is a theater in which the playwright began to play the main role. And the theatrical performance becomes the realization of the playwright's intention – a kind of "translation" of a literary text into the language of the performance. U. Eco very precisely characterizes such a translation as an interpretation with a change of matter [2, p. 283]). Let us use this very term "translation" in relation to the situation of transcoding a work from the language of one art into the language of another (moreover, it is quite common in modern science [3, 4]).

2   Formulation of the Problem

The process of reforming Russian drama began during the first half of the 19th century [5, 6]. Russian drama of the 19th century largely used the techniques of the Russian novel, which by that time had already been formed as an original aesthetic phenomenon and even brought up an audience that was bored with watching what the modern theater offered. In this regard, such strategies penetrated into the drama as: a free image of the hero's movement in space, a change of plans – general and a close-up, and a complex interweaving of verbal and visual information. However, all this was fundamentally untranslatable into the language of the theater of that time. Quite often the world that arose in the author's imagination did not coincide with the world that the theater stage created. It seems to us that there were several barriers at once: the vision of what needs to be performed by the actor and their individual capabilities, as well as the lack of theatrical technologies for realizing the playwright's plan.

In fact, the literary texts that appeared in the 19th century required the creation of such technologies which were not yet available for its translation into the language of the theater. Therefore, many of the plays discussed below (as well as, Turgenev's dramas, for example) remained dramas for reading until the end of the 19th century.

The ongoing study is intended to prove firstly that the fundamental untranslatability of the language of the drama from the early 19th century into the language of contemporary theater, and secondly to demonstrate that cases of fundamental "untranslatability" are the very "growth points" due to which new theatrical technologies are born. Our focus is on the composition of a dramatic text and the organization of the stage space dictated by it.

3 Methods

This study uses the analytical methods of non-Aristotelian drama, proposed by the school of historical poetics. The study is carried out within the framework of studying the problems of Russian drama of the 19th century and its theatricality [7–10]. The problems of romanization of drama [11, 12], studies on the organization of text in drama [13], the history of scenography [14–18] are also taken into account.

The texts of two Russian playwrights of the first half of the 19th century, who paid close attention to the features of contemporary theater, were selected for analysis: Alexander Pushkin ("Boris Godunov", "Little Tragedies") and Nikolai Gogol ("The Government Inspector", "Marriage", "Players", "Morning of a Business Man", "Lakeyskaya").

4 Results and Discussion

The texts we are studying are such that they require the rejection of acting theater. The main aesthetic event is the interaction between the viewer and the actor. The actor enters the stage - "appears", and initially the stage is just the space for the event to take place. For its design, nothing is needed: it is generic and therefore it does not have its own content since the viewers' attention is focused on the actor. It is worth emphasizing that in the ancient theater the actor performed on the proscenium – a place in front of the skene, where the actors changed clothes [19]. At the turn of the 19th Century, the skene and proscenium merged creating the stage-box on which the actors performed, creating a single visual image of a person and the space in which they are placed. Thus, the construction of this new stage itself had great potential for the development of theatrical art, but the actor continued to exist on it in accordance with the old canons.

It was the writers, not the theater figures, who were able to see and actualize this potential. The main compositional unit of the text is gradually becoming less of a phenomenon (a change of characters on the stage), but rather a "picture" or a "scene" itself. This is a holistic visual image in which the relationship between space and the person existing in this space has fundamentally changed. It is precisely holistic pictures that should appear before the viewer. And here the laws of painting of the 19th century turn out to be important, in which, as in literature, the principles of interaction between person and the world, of which he is an integral part, are important.

The action moves not due to the appearance of the characters on the stage, but due to the rapid change of scenes that become the compositional units of the work.

So, in Pushkin's tragedy "Boris Godunov" we see a rapid change of scenes in which both the spatial and temporal coordinates change, which is recorded in the remarks:

"Kremlin Chambers (1598, February 20)"; "Night. Cell in the Miracle Monastery (1603)"; "Lithuanian border (1604, October 16)"; "The Plain near Novgorod-Seversky (1604, December 21)". It is worth paying attention to the fact that some scenes are separated by significant temporal and spatial gaps (between the action of the scenes "Kremlin chambers" and "Night. Cell..." four years pass).

In our opinion, the playwright's freedom of dealing with space and time is an attempt to transfer novel strategies to the field of dramaturgy. Therefore, if in some cases it is possible to imagine the theatrical technology of changing Pushkin's scenes, then in most cases this change turns out to be impossible for the theater of the 19th century. Let's illustrate our idea. To convey the change of the first two scenes of the tragedy (from the foreground Vorotynsky and Shuisky ("Kremlin Chambers") to the people in the Red Square in the next scene), in the theater of that time, the actors playing the roles of Vorotynsky and Shuisky could simply go to the side, giving way to crowds. However, the junction between the second and third ("Maiden's Field. Novodevichy Convent") scenes suggests the need to somehow show that the viewer is no longer in the Red Square, but on the Maiden's Field. In order to realize Pushkin's plan, it is not enough to "mark" the place – with the help, for example, of a conditional sign-index, as it was in Shakespeare's theater. The playwright chooses iconic places: These are not just any squares or fields; these are the places from key episodes in Russian history.

Additionally, the style of "Boris Godunov" presupposes a quick change of scenery, which the modern theater of Pushkin cannot do in principle. It is significant that there is not even a set designer in this theater (although there are decorators) and there is no understanding that "Theatrical art is the art of space", as noted by Max Hermann [20, p. 32; 21]).

What possibilities did the theatrical machinery of that time have? Since the Renaissance, telaria were used to change scenery. They allowed only three changes of scenery during the performance. These scenes were extremely generic: a house or a castle, a battlefield, or a dense forest.

Only at the end of the 19th century did scenography respond to the challenges of literature, and then a turning circle appeared. Turns, firstly, set the rhythm of the action, and secondly, the circle can change the theatrical perspective. A revolution in the design and technical equipment of the stage was made possible by electricity: electric mechanisms appeared, as well as new lighting possibilities, film projections, and so on.

Writers who are aware of the impossibility to realize their plan often do not even leave indications that a change of theatrical pictures should take place. So, in Gogol's play "The Morning of a Business Man", official Alexander Ivanovich comes to visit the "businessman" Ivan Petrovich, and after the conversation, he leaves. It would seem that he should disappear from the stage, from the space that is revealed to the viewer and is associated with the central character of the play, Ivan Petrovich. However, Gogol shows the hero "in the lackey", where the viewer seems to have moved along with Alexander Ivanovich. But the playwright does not give special instructions to the theater about this movement; only the reader of the play will know about it:

Александр Иванович. Непременно. (Кланяется.) Катерина Александровна. Прощайте, Александр Иванович! Александр Иванович (в лакейской, накидывая шубу. – emphasis added). Не терплю я людей такого рода.... [22, p. 108]	Alexander Ivanovich. Certainly. (Bowing.) Katerina Alexandrovna. Goodbye, Alexander Ivanovich! Alexander Ivanovich (in the footman's room, throwing on a fur coat. - emphasis added). I can't stand people like that... (*Hereinafter interlinear translation of quotations is made by the authors*):

Another challenge to the theater is to indicate what a specific character in the play sees. For example, the hero of "Boris Godunov" masters, "finishes" the place of the event with the help of individual details that he saw:

...смотри: ограда, кровли, Все ярусы соборной колокольни, Главы церквей и самые кресты Унизаны народом. <...> Послушай! что за шум? Народ завыл, там падают, что волны, За рядом ряд... еще... еще... Ну, брат, Дошло до нас; скорее! на колени! [23, p. 195]	... see: fence, roofs, All tiers of the cathedral bell tower, The heads of churches and even crosses swarmed with the people. <...> Listen! what's that noise? The people howled, there they fall like waves, Behind the next row ... more ... more ... Well, brother, It has come down to us; quicker! Kneel!

In the above example, an image arises not only of the stage, but also of the off-stage space. The theater needs to make sure that this space becomes as alive as possible.

Not only Pushkin, but also Gogol repeatedly take significant events and dialogues out of thse scene. In the episodes of Podkolesin's flight (XXI scene II of the "Marriage" act), Khlestakov's farewell (XVI scene IV of the act of "The Government Inspector"), not a single actor gradually remains on the stage, only the voices of heroes invisible to the viewer are heard, however, the action, saturated with significant details, continues to develop beyond the scene:

Голос городничего. Так по крайней мере чем-нибудь застлать; хотя бы ковриком. Не прикажете ли, я велю подать коврик? Голос Хлестакова. Нет, зачем? это пустое; а впрочем, пожалуй, пусть дают коврик. Голос городничего. Эй, Авдотья! ступай в кладовую: вынь ковер, самый лучший, что по голубому полю, персидской, скорей! <...> Голос Осипа. Вот с этой стороны! сюда! еще! хорошо. Славно будет! (бьет рукою по ковру.) Теперь садитесь, ваше благородие! [24, p. 80]	The mayor's voice. So at least cover it with something; at least with a rug. Do you want me to get a rug? Khlestakov's voice. No, why? There's no need; but, perhaps, let them fetch a rug. The mayor's voice. Hey Avdotya! go to the pantry: get a rug, the best one with the blue field on it, Persian, quickly! <...> Osip's voice. From this side! here! more! Well. (Beats the rug with his hand.) Now sit down, Your Honour!

We also note that in the above example there is a remark that is fundamentally unrealizable, although it is framed as a usual instruction for the actor ("beats the carpet with his hand").

Literature poses new tasks for the theater in the field of mastering the stage space. Due to the monologues of the characters, it is expanded and detailed. Laura in Pushkin's Stone Guest says:

Приди – открой балкон. Как небо тихо;
  Недвижим теплый воздух, ночь лимоном
  И лавром пахнет, яркая луна
  Блестит на синеве густой и темной,
  И сторожа кричат протяжно: "Ясно!.. " [25, p. 327]
	Come, open the balcony doors. How quiet is the sky;
	  Immovable warm air, night
	  smells like lemon and laurel, bright moon
	  Shines on a thick and dark blueness,
	  And the watchmen shout out, drawling: "Clearly! .."

At the same time, only at first glance this remark is "descriptive", it conveys not only the details of the stage space, but also the inner state of the heroine (it is noteworthy that the image of a quiet night is contrasted with the image of Paris, where "the sky is covered with clouds, the cold rain is falling and the wind is blowing", about which Laura says next). And the theater, whether it realizes it or not, is faced with the task of embodying this subjectivity on the stage.

It is important that the development of space occurs from the point of view of each individual hero, who has his own ability to see, perceive, and evaluate. So, the hero of "The Miserly Knight":

(Зажигает свечи и отпирает сундуки один за другим.) Я царствую!.. Какой волшебный блеск! [26, p. 297]	(Lights the candles and unlocks the chests one by one.) I reign!.. What a magical brilliance!

The "magic" shine of the contents of the chests is only for the Baron. The theater should make it clear to the audience that the hero is in the power of his passion and gives magical brilliance to something that objectively does not possess it. It is important that the viewer does not see the contents of the chests. Just as he does not see the contents of Ikharev's box in Gogol's play The Players.

There are cards in this casket, but only one remark informs the reader about this ("unlocks the casket, all filled with decks of cards"). The reader learns about the contents of the box before the hero talks about it. The viewer does not have such an opportunity, but for Gogol it is important that the viewer's vision of the contents of the box arises before Ikharev talks about his perception of the contents.

In order to realize the innovation of the device, it is worth remembering that the contemporary theater of Gogol arises on the basis of a drama in which the problem of subjectivity is simply not posed. The remarks of the heroes of the traditionalist drama speak "about what is", and not about how the hero sees and perceives it. Therefore, there are no theatrical technologies for transferring subjective positions. And Gogol's text already requires them to be designated precisely as subjective.

Some of the details noticed by the characters are not only subjectively evaluated by them but are so small that a "close-up" is required to show them to the viewer. Here is how Sobachkin in Gogol's "Fragment" reads letters addressed to him:

(Вынимает из кармана пучок писем). Ну, хоть бы это, например (читает): "Я очинь слава богу здарова но за немогаю от боле. Али вы душенька совсем позабыли. Иван Данилович видел вас душиньку в тиатере и то пришли бы успокоили веселостями разговора". Чорт возьми! кажется, правописанья нет [27, p. 135]	(Takes a bunch of letters out of his pocket.) Well, at least this, for example (reads): "Thank God I'm very healthy, but I get sick from pain. Or you darling completely forgot. Ivan Danilovich saw you in the theatre darling and you would have come to reassure me with the cheerfulness of the conversation." Damn it! it looks like there is no spelling.

Not all of these errors can be conveyed in oral speech, perhaps the viewer should see these letters without the help of the hero.

Pushkin also refers to the same small details:

Лаура (осматривает тело). Да! жив! гляди, проклятый, Ты прямо в сердце ткнул – небось не мимо, И кровь нейдет из треугольной ранки, А уж не дышит – каково? [25, р. 329]	Laura (examining the body). Yes! alive! look damn, You poked right in the heart - I suppose, probably, not around it, And the blood does not come from the triangular wound, And he doesn't breathe - what is it?

Thus, playwrights, as it were, demand that the theater solve the problem of "focalization", which is more in line with the artistic nature of the novel, rather than drama. Under the pen of Pushkin and Gogol, the very principle of organizing a dramatic text, its tasks and composition are changing.

5 Conclusions

The conducted research shows that the process of novelization of drama, which begins in the 19th century, generates texts that present fundamentally new relations between person and space compared to the established dramatic tradition. Here appears a hero who has the ability to subjectively perceive both the event and the place and time in which this event occurs. New theatrical technologies are required to create the image of the space, time, event, and the hero, which the theater did not yet have owing to the fact that it arose on the basis of mask improvisation.

The text proposed by the theater for translation poses the task of the emergence and development of these technologies. The theater only by the end of the 19th century began to solve the task: new theatrical professions appeared, acting changed. The achievements of modern technology and newly emerging art forms, in particular cinema, come to the service of the theater.

References

1. Pushkin, A.S.: About the folk drama and the drama "Marfa Posadnitsa". In: Complete Works in 10 Volumes, Vol. 7: Criticism and Publicism, pp. 146–152. The Science, Leningrad (1978)
2. Eko, U.: To say almost the same thing. Experiences in translation. Symposium, St. Petersburg (2006)
3. Tsvetkova, M.V.: Is film adaptation also a translation? With reference to two screen versions of Shakespeare's Hamlet. News Ural Feder. Univ. Ser. 2: Humanit. **24**(2), 57–68 (2022). (in Russian) https://doi.org/10.15826/izv2.2022.24.2.024
4. Kirillova, N.B.: Mikhail Bulgakov's macrocosm in the mirror of screen culture. Cult. Art. **4**, 485–497 (2016). https://doi.org/10.7256/2222-1956.2016.4.19550

5. Serman, I.: Pushkin and Griboyedov as the reformers of Russian dramaturgy. Revue Études Slaves **59**(1–2), 213–224 (1987). https://doi.org/10.3406/slave.1987.5624

6. Filonov, E.A.: Nikolai Gogol and his readers in the change of eras. Cult. Text. **2**(41), 53–67 (2020). https://doi.org/10.37386/2305-4077-2020-2-53-67

7. Zakharov, N.V.: The problem of stage performance in the dramaturgy of Shakespeare and Pushkin. Knowl. Underst. Skill. **1**, 359–384 (2015). https://doi.org/10.17805/zpu.2015.1.36

8. Lukov, V.A.: Stage/non-stageness of Pushkin's dramaturgy: expert evaluations. Knowl. Underst. Skill. **3**, 129–151 (2015). https://doi.org/10.17805/zpu.2015.3.11

9. Paderina, E.G.: "Players" of Gogol as artistic critique of the principle of "a well done play" by E. Skriba. In: Gogol's Creativity and European Culture. Fifteenth Gogol Readings, pp. 78–87. Novosibirsk publishing house, Novosibirsk, Moscow (2016)

10. Evdokimov, A.A.: Shakespeare in Gogol's later works. Knowl. Underst. Skill. **1**, 286–296 (2016)

11. Paderina, E.G.: On the correlation of literary and theater strategy in the plots of Gogol's first drama concepts. Syuzhetologiya Syuzhetografiya **2**, 87–95 (2016)

12. Zhuravleva, O.A.: The problem of genre accession of the drama for reading of the XVIII-XIX centuries. Cult. Text. **3**(46), 226–238 (2021). https://doi.org/10.37386/2305-4077c-2021-3-226-238/

13. Golovchiner, V.A., Rusanova, O.N.: Text organisation in drama. Bull. BSU **4**(2), 33–42 (2018)

14. Łarionow, D.: Scenography studies – on the margin of art history and theater studies. Art Inquiry Recherches Sur Les Arts **XVI**, 115–126 (2014)

15. Hann, R.: Theatre, performance and technology: the development and transformation of scenography (2nd ed.). Stud. Theat. Performan. **35**(1), 106–108 (2015). https://doi.org/10.1080/14682761.2014.965595

16. Masters, P.: The history and theory of environmental scenography. Theatre Perform. Design **5**(3–4), 317–319 (2019). https://doi.org/10.1080/23322551.2019.1687910

17. Alisher, A.: Art history dynamics of theater scenography. Natl. Acad. Manager. Staff Cult. Arts Herald. **2**, 170–175 (2021). https://doi.org/10.32461/2226-3209.2.2021.240004

18. von Rosen, A.: Why scenography and art history? Konsthistorisk tidskrift/J. Art Hist. **90**(2), 65–71 (2021). https://doi.org/10.1080/00233609.2021.1923566

19. Polyakov, E.N., Inozemtseva, T.O.: Ancient Greek theater – the history of origin. Bull. Tomsk State Univ. Archit. Civ. Eng. **4**(41), 9–23 (2013)

20. Hermann, M.: Theatrical space-event (1930). In: Fischer-Lichte, E., Chepurov, A.A. (eds.) Theater Studies in Germany: Coordinate System, pp. 31–42. Baltic Seasons, St. Petersburg (2004)

21. Wellington, A.T., Trubochkin, D.V.: Formation of a new theatrical space in the conditions of the temporal development of the theater. Euras. Scien. Associat. **5–7**(63), 535–539 (2020). https://doi.org/10.5281/zenodo.3887159

22. Gogol, N.V.: Morning of a business person. In: Gogol, N.V. (ed.) Complete works: In 14 Volumes, Vol. 5: Marriage; Dramatic Passages and Individual Scenes, pp. 102–108. Publishing House of the Academy of Sciences of the USSR, Moscow (1949)

23. Pushkin, A. S.: Boris Godunov. In: Pushkin, A. S. Complete works in 10 Volumes, Vol. 5: Eugene Onegin. Dramatic Works, pp. 187–285. The Science, Leningrad (1978)

24. Gogol, N.V.: The government inspector. In: Gogol, N.V. (ed.) Complete Works in 14 Volumes, Vol .4: The Government Inspector, pp. 5–95. Publishing House of the Academy of Sciences of the USSR, Moscow (1951)

25. Pushkin, A. S.: The Stone Guest. In: Pushkin, A. S. Complete works in 10 Volumes, Vol. 5. Eugene Onegin. Dramatic Works, pp. 316–350. The Science, Leningrad (1978)

26. Pushkin, A. S.: The Miserly. In: Pushkin, A. S. Complete Works in 10 Volumes, Vol. 5. Eugene Onegin. Dramatic Works, pp. 286–305. The Science, Leningrad (1978)
27. Gogol, N.V.: Fragment. In: Gogol, N.V. (ed.) Complete Works in 14 Volumes, Vol. 5: Marriage; Dramatic Passages and Individual Scenes, pp. 123–136. Publishing House of the Academy of Sciences of the USSR, Moscow (1949)

Dynamic and Systems Features of Fiction Abstracts Discourse from the Functional Linguosynergetics Perspective

Marina Cherkunova[1]([⊠]) [iD], Evgeniya Ponomarenko[2] [iD],
and Antonina Kharkovskaya[1] [iD]

[1] Samara National Research University, 34, Moskovskoye Shosse, Samara 443086, Russia
m.cherkunova@mail.ru
[2] MGIMO University, 76, Prospect Vernadskogo, Moscow 119454, Russia

Abstract. The article is focused on the evolving of the overall meaning of mini-texts viewed as dynamic self-organising systems of senses. The authors consider the semantics of the text as a non-linear functional space created as a result of diverse interactions of all textual elements. Such interactions lead to the emergence of various sense combinations and significant sense accretions on the basis of a limited number of language means. English abstracts to modern works of fiction taken from advertising catalogues of major international publishing houses provide the empirical material for the study. Mechanisms of sense formation within minitexts are analysed in terms of dynamic systems approach with the application of methods of functional linguosynergetics. The authors conclude that the synergistic pragmasemantic effect of the abstracts is underpinned by a number of common principles of their ogranisation, the first one being construction of the texts on the basis of an invariant structural pattern of formally independent blocks. The latter are sequenced into a meaningful structure with the help of para-graphemic means. Semantics of the texts is produced by a limited number of thematically concentrated clusters of words gradually unfolding into a non-linear structure of senses. The text overall pragmatics is actualised through the synergistic interaction of its formal, visual and linguistic elements. Insight into the synergetic processes of sense formation within an advertising abstract can potentially give a clue to identifying universal mechanisms of enhancing the pragmatic effectiveness of such highly demanded textual formats as minitexts.

Keywords: Functional linguosynergetics · Discourse · Self-organising system · Sense space · Small-format text · Minitext · Fiction abstract · Functional attractor · Pragmasemantic relations

1 Introduction

The linguistic definitions of communication are akin to other definitions (e.g. in psychology or computer science) in that this process takes place between two parties as an exchange of information through a signalling system, with the linguistic definition having a rather significant feature – its focus on interaction between humans, while non-linguistic theories of communication refer also to non-humans (animals, components of organisms, plants, software units, etc.) [1–3].

Likewise, the linguistic definitions of discourse as a verbal (either written or spoken) incarnation of communication between humans, have their own peculiarities in comparison with discourse treatment in other sciences. Thus, there are two main variants of discourse description in linguistic theories:

1) "little-D": discourse as "a recorded and transcribed talk or text, where excerpts are used to make scholarly arguments" [4, p. 147], an important premise being that it is enriched with "specific semantic and pragmatic accretions generated in a particular communicative event" [5, p. 5]; in fact it is those sense accretions that turn a text as a finished verbal (written/spoken) construct into full-fledged discourse as an open linguomental system;
2) "big-D": discourse as "complex social practices such as education or business" [4, pp. 147–148], an ever accumulated multitude of thematically correlated texts produced within more or less typical contexts [6], wherein the notion of discourse blends with that of genre.

Hence, discourse analysis is aimed at exteriorizing the linguistic peculiarities of the said communicative social practices, and the mental models incorporated in discourse units [7], which alongside both aforementioned treatments of discourse is relevant for the present study.

On top of that discourse is seen as a phenomenon whose functioning is underpinned by a specific combination of stable systems properties and flexible dynamic ones, which ensues particular research interest in discourse synergistic capacity. To that end methods of functional linguosynergetics appear to make the optimal analytical basis for revealing the discourse features in question.

In view of discourse synergistic properties, another up-to-date linguistic topic considered in this work is the functioning mode of small-format texts (or minitexts), since they embody a most expedient synergistic capacity, i.e. structural and semantic synergism or non-linearity. To disclose this concept in relation to minitexts we are looking into the ability of fiction abstracts to wrap voluminous information of the books they represent in brief wording.

Thus, the authors see the purpose of the present paper in analysing dynamic and systems characteristics of the fiction abstract as a type of discourse actualized through minitexts that vividly illustrate the synergistic capacity of speech as a non-linear phenomenon.

2 The Concise Basics of Functional Linguosynergetics

As is known, the name "synergetics" was introduced in its present meaning by the German scientist Hermann Haken who wanted it to embody the idea of joint coherent operation of open non-equilibrium systems' components, since the Greek συνεργία means cooperative action [8].

The characteristics most commonly attributed to the synergetic vision (including that of speech activity) are holistic (1), nonlinear (2), evolutionary (3) [9–13], which implies:

(1) analysing the object of investigation in its wholeness, i.e. as a system comprising both the constituent elements/parts and various forms of their interaction within the system;
(2) disclosing the object ability to function by the figuratively expressed formula "2 + 2 = 5", i.e. generating the overall effect beyond the pure linear addition of the effects produced by the elements on their own, outside the system;
(3) following the process of system evolution from one stage of its development to another in its striving for the desired attractor, i.e. the purpose of its functioning.

Transferring these basic premises to language and speech study, scholars have elaborated different methods of analysing such issues as self-ogranisation of text "substance" (V. Bazylev, G. Moskalchuk), psycholinguistic properties (I. German, A. Pishchalnikova), hypersense (N. Myshkina, O. Grebenkina), semantic and structural attractors (N. Blaznova, A. Pishchalnikova) and many others.

The works by Evgeniya V. Ponomarenko have laid the groundwork for a specific variant of this paradigm, namely *functional linguosynergetics*, which has been successfully developed in further researches by M. Cherkunova, A. Kharkovskaya, D. Khramchenko, V. Malakhova, A. Radyuk, I. Savina and others [14–16, etc.]. Its distinctive feature lies in the investigation of self-organising processes in the formation of discourse functional (pragmasemantic) plane, the coherent contribution of all language elements to the holistic sense space of any discourse fragment. Within this framework a number of postulates have the leading part:

– discourse is defined as "a self-organising system of senses formed in the text by an aggregate of all (oral and written) verbal means, which synergistically mobilizes their functional potential on the way to the author's communicative purport" [12, p. 359];
– the communicative purport makes the system functional attractor towards which all coherent functioning of discourse components should be directed; otherwise the sense space and the discourse system as a whole will be transformed or ruined;
– the discourse system regulatory mechanism (order parameters) is considered to be a network of functional relations between utterances, with 'functional' meaning 'pragmasemantic';
– the system evolution goes on as a sequence of orderly (when the sense plane is well organized and consistent) and chaotic (when it is imbalanced by incongruent semantic elements) stages;
– the overall sense space is generated as a non-linear synergistic effect, and so on.

Furthermore, we are going to illustrate the applied linguosynergetic approach to the functional peculiarities of small-format texts.

3 Small-Format Text as an Object of Linguistic Research

Small-format texts, or minitexts, containing a limited number of words, up to 600 [17, 18], do not only make a convenient object for a linguistic analysis, but due to their brevity and structural simplicity, serve as the basic structures employed in the formation of larger textual constructs.

According to the classical definition offered by E.S. Kubryakova, a small-format text is the one possessing such significant characteristics as visual brevity, relative completeness and separateness, and characterized by both nominative and pragmacommunicative potential [19]. The majority of the researchers unanimously acknowledge the fact that small-format texts, despite their limited volume, possess all the classical features of textuality such as cohesion, coherence, intentionality, acceptability, informativity, situationality, etc. [20–24]. Thus, the notion of small-format texts can be understood as a taxonomic unit of a supra-genre level, encompassing textual samples of various genres whose main categorical features include brevity of form and the ability to incorporate considerable meaning.

As a special textual type, small-format texts are characterized by a number of systemic features which ensure their pragmatic effectiveness including visual integrity, marked asymmetry of external and internal planes, minimisation of the verbal expression, specific forms of compression, polycode character, etc. The complex pragmatic goal of small texts is achieved through the synergistic interaction of all the textual elements as well as the cognitive plane of the participants of the given communicative situation.

For our purposes we take abstracts to fiction books as a vivid example of functional synergism actualised in texts of limited volume that must nevertheless be rhetorically efficient.

4 Fiction Abstracts Through the Prism of Functional Linguosynergetics

The empirical data of this study includes 1 000 abstracts to fiction published in online advertising catalogues of major international publishing houses such as *Legend Times Group* [25]*; Penguin Group* (including *Ebury, Cornerstone, Penguin Press, Vintage*) [26]*; Princeton University Press* [27]*; Urbane Publications* [28] issued within the period from 2018 to 2022.

The ultimate pragmatic goal of the aforementioned abstracts, in other words – their *functional attractor*, consists in persuading the consumer to purchase the book, hence the dynamics of the discourse functional plane development, the selection, composition, combinatorics and the nature of interaction of the textual elements are determined by the necessity to achieve this goal.

Analysis of the dynamic properties of a discourse system should begin with the description of its equilibrium (i.e., initial, typical) state, since the dynamics of its development are manifested as building up a semantic construct against the background of invariant manifestations of textual organization.

Thus, the invariant model of an advertising book abstract includes the following set of obligatory structural elements:

1) the title, which invariably coincides with the title of the book advertised;
2) the name(s) of the book author(s);
3) the body, averaging 2 or 3 paragraphs, arranged according to the principle of the 'inverted pyramid' with the so-called *lead paragraph* introducing general facts concerning the setting, the characters and the key collision of the story and the subsequent parts providing minor details about the plot;
4) the quotation block, containing one or more emotionally charged positive reviews of the book provided by well-established printed sources, renowned men of letters, etc.;
5) the author's profile (accompanied by a photograph), highlighting some facts related to the author's literary career, his/her professional development, listing the author's literary awards, as well as briefing the readers on the key biographical facts relevant for grasping the message of the novel advertised;
6) the illustration (the photo of the book cover);
7) the technical block containing the information about the publication date, the price, the design, etc.

However, the aforementioned systemic structural elements can be supplemented with some optional textual blocks including those of "Key selling points", "Comparison authors", "Target audience", slogan-like elements, etc. Thus, an average advertising abstract has the following layout:

Fig. 1. An advertising abstract from a publisher's catalogue [25]

The stability of the systemic structural features of the abstracts, their uniformity and repetitive character, establish some kind of an orienting framework for a potential recipient to easily find the information relevant for making their opinion about the book, whereas the variable elements, represented as occasional fluctuations, reflect the current tendencies in the external environment and serve the purpose of adjusting the concrete abstract to the particular conditions of its circulation. What is more, these structural fluctuations serve as the pool of potential trends for further development of the systemic features of this type of text (for more information see Cherkunova, Ponomarenko 2021 [14]).

The characteristic property of the structural elements within the book abstract is their semantic autonomy; the structural blocks are not tied by formal cohesive devices, due to which they can be arranged one after another in an arbitrary fashion, their sequence within the textual structure being open to alteration. At the same time, the final sense construct resulting from the development of the pragmasemantic plane of the text, does not depend on the order in which certain structural and semantic elements are introduced into the system. The overall sense of the discourse system with all the semantic and pragmatic accretions (extensions) arises as a result of complex interaction and mutual influence of all the systemic elements however unimportant they may seem. What is more, the resulting sense construct to a great extent is determined by the specific parameters of the perceiving mind with all its current needs, interests and desires included into the broader context of personal outlook, values, life experiences, etc.

At the same time, each specific abstract contains some kind of a 'navigation route' arranged by the author of the text for potential unfolding of the pragmasemantic textual plane, ultimately oriented towards the successful achievement of the discourse system's functional attractor. To this end, the factor of the visual integrity of the text plays the key role since the trajectory of perception and combination of its structural elements is laid with the account of the visual design of the textual elements.

As can be seen from Fig. 1, the illustration block including the photograph of the book cover makes the largest and brightest element of the textual structure. It creates the visual focus of the text, which is most likely to become a starting point for getting acquainted with the abstract. It is at this stage that the discourse system begins its move towards the functional attractor with the formation of an initial set of the recipient's expectations. Due to the information provided within the illustration block through the interaction of the verbal and visual components, the discourse system is filled with a set of meanings that are minimally necessary to form some primary idea of the book. In the example given the illustration provides the name of the book author (*Jane Isaac*), the title (*Evil Intent*), and also, on the yellow background in the top right corner, both the name of the main character of the book (*Helen Lavery*) and her occupation (*DCI (Detective Chief Inspector)*) are mentioned. Together with the image, featuring a dark forest with a lonely figure moving towards a clearing visible in the fog, the abovementioned information forms a system of meanings including the idea that the advertised work is a detective novel, the main character in which is Police inspector Helen Lavery, who appeared in other works previously published by Jane Isaac. In addition, the individual pragmasemantic accretions may incorporate a positive or negative assessment of the work of the advertised author based on the fact of the reader's acquaintance with her previous novels. At this stage the pragma-semantic plane of the discourse brings about two semantic focus points those of "the author of the book" and "the plot". The load of information perceived by a particular recipient creates the initial cognitive platform for the discourse system development.

Further on, depending on the specific content of the initial pragmatic package created in the mind of the target recipient the unfolding of the functional plane of the discourse system can take several trajectories, firstly, provided that the initially formed set of expectations is not consistent with the interests, needs, requirements of the recipient, the development of the discourse system will stop at this stage since the recipient will not proceed with reading the abstract and will turn to the next one. On the contrary, if the initially formed pragmatic package is consistent with the interests and expectations of the recipient, the following functional element will be built into the discourse system. The potential trajectory of this movement is also directed by the paragraphemic design of the structural blocks. Thus, the second visual focus after the illustration might be provided by the slogan-like element, done in a larger font compared to all other structural blocks.

In all probability, it is this structural element that is likely to continue the formation of the pragmasemantic plane of the discourse system.

The slogan-like phrase, which is formulated like *Thrilling new crime novel from bestselling author*, further develops the two previously introduced semantic foci and verbalizes the genre of the publication (*crime novel*) as well as adds a positive evaluation of the author (*bestselling author*). This information, consistent with part of the expectations previously formed within the illustration block (*A DCI Helen Lavery novel + thrilling new crime novel; Jane Isaac + bestselling author*) expands the discourse system due to the introduction of additional information concentrated around the same focus points.

Further movement of the functional plane of the discourse also involves the expansion of the aforementioned semantic domains while the logic of adding structural blocks is also suggested by the paragraphemic means. One of the possible trajectories of the pragmasemantic plane development involves the transition to the first paragraph of the body block – it is highlighted with the orange colour, which draws attention among other textual parts. In this case, the semantic domain concerning 'the book plot', which previously contained information about the name and profession of the main character (*A DCI Helen Lavery novel + thrilling new crime novel*), is supplemented with the details of the storyline through such verbal operators as *a series of women's bodies, rural Hamptonshire, a pentagram carved on their chests, DCI Helen Lavery, a cat and mouse chase, a murderer*:

When **a series of women's bodies** are discovered in the heart of **rural Hamptonshire** with **a pentagram carved on their chests**, **DCI Helen Lavery** is forced into **a cat and mouse chase** with **a murderer** who ultimately turns the tables and targets her.

At the same time, in the given example the autonomy of the discrete textual parts can be traced within the body block of the abstract as well since its second and third paragraphs are quite independent from the first paragraph both formally and semantically making it possible for the recipient to reduce the body of the text to two, or even to one, first paragraph. The information fundamentally significant for getting a glimpse of the plot is provided within the first paragraph; the subsequent elements only add minor details to the already formed pragmatic package. So, the second paragraph of the body contains verbal operators describing other characters in the book while the third paragraph expands on the key conflict (in the example, these verbal elements are highlighted in bold):

*Meanwhile, she is shocked to discover that **her younger son's new best friend** is the nephew of **organised crime boss Chilli Franks** – the man who **has held a grudge** against Helen's family since her father first put him away in the 1990s.*

*As her **personal and professional lives collide**, Helen finds herself in **mortal danger** as she races **to track down the serial killer** and **restore safety** to the streets of Hampton.*

Thus, the pragmasemantic plane of the discourse system is infused with new meanings which, being combined, form the idea of an unusual and intense plot, non-trivial collision and complex interweaving of human destinies. The semantic system built in this way, being conformed to the expectations and needs of a particular recipient, accelerates its progress towards the functional attractor, that is, to the formation of the desire to purchase the advertised book.

The second option for the development of the discourse after the introduction of the slogan might involve adding the author's profile block, which is physically located in close proximity to the lexical marker that verbalizes this functional segment within the slogan – this line of discourse development is also suggested by the spatial and color design of the structural elements since the author's profile is preceded by the subhead "The Author", which is visually highlighted in coloured font (Fig. 2).

Fig. 2. Spatial correlation of textual blocks within an abstract

From the semantic point of view, the author's profile also exploits the effect of the functional synergy of the discourse and non-linearity of its pragma-semantic space (i.e. when $2 + 2 = 5$) since this structural block contains certain semantic elements the high density of which cumulatively creates an impression of a person who has a lot to share with the reader. In the given example, this impression is formed by the interaction of such semantic elements as:

– the implicit suggestion that the author has the first-hand knowledge of crime and detection sphere, which is achieved through mentioning her husband's profession: *...is married to a serving detective*;
– the mention of the fact that the author lives in the very place where the action of her books takes place: *[they] live in rural Northamptonshire, UK*;
– the reference to a successful previous literary experience: *Evil Intent is Jane's eleventh novel and the fourth in the highly acclaimed DCI Helen Lavery series*;
– the emphasis on her having two pets: *live ... with their dogs, Bollo and Digity*;
– the reference to her accounts in social networks: *Twitter: @JaneIsaacAuthor // IG: @JaneIsaacAuthor.*

The resulting effect of the interaction and mutual reinforcement of the pragmatic potential of the abovementioned elements is that the reader does not merely receive a set of facts concerning the author's biography; instead they get an impression of a flesh-and-blood person who lives an ordinary life next door, adheres to traditional social norms and values while being well-informed on the subject matter and capable of having earned professional recognition; thus, her book promises to be worth buying and reading.

Regardless of the order of the structural and semantic blocks introduction into the discourse system the pragmasemantic plane of the text keeps being reduced to two semantic foci including (1) the author of the book and (2) the plot. The semantic development of the discourse system occurs through gradual expansion of these semantic domains – from the introduction of the most general idea within the illustration and slogan blocks to a relatively detailed image created in the body part of the abstract and the author's profile block. At the same time, the functional attractor can be reached by the discourse system (or, on the contrary, the development of the discourse system can collapse) at any stage of the recipient's acquaintance with the text since the intention to purchase the advertised book can be formed at any of the stages of discourse system development described above, and it depends exclusively on the factor of the addressee, namely, on numerous, for the most part poorly predictable individual parameters in the form of specific requests, desires, expectations, prior experiences, and so on.

For example, if the recipient is a fan of this author and has long been waiting for a new novel to come out, the formation of the discourse system and its reaching the functional attractor can be completed after the recipient has read the title of the abstract and has seen the illustration. Meanwhile for a person who is not interested in the detective genre the formation of the discourse system will also be completed after seeing the title and illustration, albeit without the system reaching the desired attractor since in this case the system of meanings created as the result of the development of the discourse system will be diametrically opposite to the one that has initially been intended, that is it will consist in forming a conviction not to purchase the book.

In case the functional attractor has not been reached by this stage and the discourse system continues its semantic development, its further trajectory might involve addition of the remaining structural components – either the quotation block that occupies the central position on the page and is visually highlighted with the colored font (which is, however, smaller in size than that of the body part) or the block labeled "Key Selling Points".

Within the quotation block, the previously introduced semantic focus points are further developed and expanded; thus, the 'author' domain is supplemented by an expressive and evaluative description of her professional skills: *Isaac is the queen of emotional police procedural, which is located at the top of her game*, while the semantic domain concentrated around 'the plot' is enriched with such densely infused semantic markers as *an array of great characters that really live and breathe, characters who invade your real world, clever plot twists, a compelling, powerful and unforgettable read; the story is an adrenaline rush, fast-paced and original*:

*"Isaac is **the queen of the emotional police procedural**. With **characters who invade your real world**, and **clever plot twists**, Evil Intent is **a compelling, powerful and unforgettable read**." Louise Beech author of I am Dust and How to be Brave.*

*"The story is **a real adrenalin rush, fast-paced and original**, with **an array of great characters that really live and breathe**. Jane really is **at the top of her game!**"* Kate Rhodes author of Hell Bay and Pulpit Rock.

It is worth mentioning that in case the discourse system develops according to the trajectory intended by the author of the abstract, gradual increase in the degree of expressiveness of the information introduced into the discourse system is clearly observed, which is functionally significant. The information is given in the neutral factual manner within the illustration block, then it becomes moderately evaluative in the body part of the abstract, and finally it gets highly evaluative within the quotation block. Thus, gradual mechanical accumulation of information centered around two semantic domains with a constant increase in the degree of expressiveness of the verbal units employed eventually generates a synergetic effect, which consists in a resonant increase in the pragmatic potential of the meaning quanta involved. This, in its turn, contributes to the emergence of pragmasemantic extensions in the form of realization of the book high value and prompts the intention to purchase and read it.

The functional role of the structural element entitled "Key Selling points" also consists in expanding the previously outlined semantic foci, since in fact this block duplicates the information previously introduced into the system:

Key Selling Points:

- **The fourth book** in the **highly acclaimed DCI Helen Lavery series**
- **Commercial crime thriller** by **critically acclaimed crime author** with **an extensive backlist**
- **Amazon #1 bestselling author**

Thus, mechanically an excess in the quantity of information is created, which leads to qualitative changes in the discourse system, namely to the strong implication of the high merits of the book in question. In terms of its linguistic parameters, this structural block is characterized by the highest degree of compression, since the information is introduced in the form of bullet points through non-predicative constructions, which creates quite a concentrated sense space where quanta of meaning come into close interaction boosting each other's pragmatic potential, creating an overall synergistic effect and enhancing the idea of multiple merits of this particular novel.

5   Conclusion

To conclude, the development of the pragma-semantic plane of the discourse system of a small-format text leading to the formation of an integral sense construct and achievement of its functional attractor is ensured by such properties of the text as:

1) its being based on the invariant structural pattern including a strictly fixed number of elements;
2) formal and pragmasemantic independence of the structural elements resulting in multiple trajectories of the possible development of the functional plane of the text;
3) active role of the paragraphemic means in sequencing the developmental logic of the discourse system's functional plane;

4) concentration of the sense structure of the text around a limited number of semantic foci;

5) concentric layered expansion of the semantic foci paralleled by the increase in their emotional and evaluative charge resulting in the creation of a multilayer 'vertical' pragmasemantic structure;

6) overload of the discourse system with semantically homogeneous components leading to the appearance of emergent sense accretions due to the mutual influence of the elements involved;

7) non-linearity of the functional plane of the text in which the overall sense does not equal the sum of the meanings of the elements involved, the asymmetry being achieved due to the emergence of pragmasemantic extensions that arise from the synergistic interaction of all the systemic elements refracted by the consolidating influence of the recipient's mind.

The most essential property of a small-format text consists in the non-linearity of its pragmasemantic plane which, in its final version, cannot be reduced to the mechanical sum of the meanings of separate textual units. The overall meaning of the discourse system goes well beyond that, which is vitally significant for small texts aiming at rendering extensive meaning via a laconic textual form.

References

1. Cobley, P.: Communication: definitions and concepts. In: Donsbach, W. (ed.) The Concise Encyclopedia of Communication. 1st edn., pp. 73–76. Wiley-Blackwell, Hoboken (2015)

2. Crystal, D.: A Dictionary of Linguistics and Phonetics, 6th edn. Blackwell Publishing Ltd., USA (2008)

3. Wickham, H.: Advanced R. R Series, 2nd edn. CRC Press, Taylor & Francis Group, Chapman & Hall's book, Boca Raton, London, New York (2019). https://adv-r.hadley.nz/index.html. Accessed 25 Aug 2022

4. Tracy, K., Craig, R.T.: Studying interaction in order to cultivate communicative practices. Action-implicative discourse analysis. In: Streeck, J. (ed.) New Adventures in Language and Interaction, pp. 145–166. John Benjamins Publishing Company, Amsterdam/Philadelphia (2010)

5. Ponomarenko, E.V., Magirovskaya, O.V., Orlova, S.N.: Introduction: professional discourse in the focus of functional linguistics. In: Malyuga, E.N. (ed.) Functional Approach to Professional Discourse Exploration in Linguistics, pp. 1–20. Springer, Singapore (2020). https://doi.org/10.1007/978-981-32-9103-4_1

6. Baumgarten, S., Gagnon, Ch.: Political discourse analysis from the point of view of translation studies. In: Baumgarten, S., Gagnon, Ch. (eds.) Translating the European House: discourse, ideology and politics – selected papers by Christina Schäffner, pp. 172–207. Cambridge Scholars Publishing, Newcastle (2016). https://doi.org/10.7202/1043961

7. van Dijk, T.A. Introduction: the study of discourse. In: van Dijk, T.A (ed.) Discourse Studies: A Multidisciplinary Introduction, 2nd edn., pp. 1–7. Sage Publications Ltd., Los Angeles, London, New Delhi, Singapore, Washington DC (2011)

8. Haken, H.: Synergetics. Introduction and Advanced Topics. Springer, Berlin, Heidelberg (2004)

9. Khramchenko, D., Radyuk, A.: The synergy of modern business English discourse: holistic approach to teaching unconventional rhetoric. In: INTED 2014 Proceedings 8th International Technology, Education and Development Conference, pp. 6779–6783. IATED, Valencia (2014)
10. Köhler, R.: Synergetic linguistics. In: Köhler, R., Altmann, G. Piotrowski, R.G. (eds.) Quantitative Linguistik. Ein internationales Handbuch. Quantitative Linguistics. An International Handbook, pp. 760–775. Walter de Gruyter, Berlin, New York (2005)
11. Olizko, N.S., Mamonova, N.V., Samkova, M.A.: Semiotic and synergetic methods of text analysis. In: Proceedings of the X International Conference on European Proceedings of Social and Behavioural Sciences EpSBS, vol. 86, pp. 1056–1063. European Publisher, London (2020). https://doi.org/10.15405/epsbs.2020.08.123
12. Ponomarenko, E.V.: Functional properties of English discourse in terms of linguosynergetics. In: 3rd International Multidisciplinary Scientific Conference on Social Sciences and Arts SGEM 2016, SGEM2016 Conference Proceedings, 24–30 August 2016, book 1, vol. 3 Education and Educational Research, pp. 355–362. STEF92 Technology Ltd., Albena, Sofia, Bulgaria (2016). https://doi.org/10.5593/sgemsocial2016/B13/S03.042
13. Wildgen, W.: The relevance of dynamic systems theory for cognitive linguistics. In: Interdisciplinary themes in cognitive language research symposium. University of Helsinki and FiCLA (2005). https://www.researchgate.net/publication/238751644_The_relevance_of_dynamic_s ystems_theory_for_cognitive_linguistics. Accessed 25 Aug 2022
14. Cherkunova, M.V., Ponomarenko, E.V.: Dynamic properties of abstracts to works of fiction in English. Linguist. Polyglot Stud. 7(2), 98–107 (2021). (in Russian). https://doi.org/10.24833/2410-2423-2021-2-26-98-107
15. Kharkovskaya, A.A.: The discourse world of English and Russian bestsellers. Iss. Appl. Linguist. 10, 75–88 (2013). (in Russian)
16. Malakhova, V.L.: Functional properties of possessive nominations and their impact on the pragmatic and semantic systemity of the English discourse. In: 4th International Multidisciplinary Scientific Conference on Social Science and Arts SGEM 2017. Conference Proceedings, book 3. Science and Society, vol. II. Psychology and Psychiatry. Language and Linguistics, pp. 745–752. SGEM, Vena (2017). https://doi.org/10.5593/sgemsocial2017/32/S14.096
17. Korbut, A.Yu.: Text symmetrics as a field of the general text theory [Doctor Disser.]. Altai State University, Barnaul (2005). (in Russian)
18. Kharkovskaya, A.A., Ponomarenko, E.V., Radyuk, A.V.: Minitexts in modern educational discourse. Functions and trends. Train. Lang. Cult. 1(1), 62–76 (2017). https://doi.org/10.29366/2017tlc.1.1.4
19. Kubryakova, E.S. On the text and criteria for its definition. Struct. Semant. 1, 72–81 (2001). (in Russian). http://www.philology.ru/linguistics1/kubryakova-01.htm?ysclid=lb78bmwhq464004991. Accessed 24 Apr 2022
20. Kerans, M.E., Marshall, J., Murray, A., Sabaté, S.: Research article title content and form in high-ranked international clinical medicine journals. Eng. Specif. Purp. 60, 127–139 (2020). https://doi.org/10.1016/j.esp.2020.06.001
21. Malyuga, E.N., McCarthy, M.: Non-minimal response tokens in English and Russian professional discourse: a comparative study. Quest. Linguist. 4, 70–86 (2020). https://doi.org/10.31857/0373-658X.2020.4.70-86
22. Molodychenko, E.N.: Communicative and pragmatic features of 'lifestyle instructions' as an Internet genre in consumer culture. Tomsk Stat. Univ. Bullet. Philog. 57, 79–102 (2019). https://doi.org/10.17223/19986645/57/5. (in Russian)
23. Radyuk, F.V., Ivanova, M.V., Badmatsyrenova, D.A., Makukha, V.S.: Pragma-semantic relations in the structure of a business microblog. Intern. J. Eng. Ling. 9(6), 392–403 (2019). https://doi.org/10.5539/ijel.v9n6p392

24. Vedeneva, Y.V., Kharkovskaya, A.A., Malakhova, V.L.: Minitexts of poetic titles as markers of the English cognitive paradaigm. Train. Lang. Cult. **2**(2), 26–39 (2018). https://doi.org/10.29366/2018tlc.2.2.2

25. Legend Times Group. https://www.legendpress.co.uk/catalogues. Accessed 1 Aug 2022

26. Penguin Group. https://www.penguin.co.uk/company/publishers.html. Accessed 1 Aug 2022

27. Princeton University Press. https://press.princeton.edu/catalogs/literature. Accessed 1 Aug 2022

28. Urbane Publications. https://issuu.com/urbanepublicationslimited/docs/urbane_cat_2020. Accessed 1 Aug 2022

The Influence of Descriptive Language Practices on the Process of External Integration of Corporate Knowledge

Ekaterina Mashina[1,2]([✉]) [iD] and Pavel Balakshin[1] [iD]

[1] ITMO University, Kronverkskiy pr. 49, 197101 Saint-Petersburg, Russian Federation
mashina.katherina@gmail.com
[2] BIOCAD Biopharmaceutical Co, Svyazi str. 34-A, Strelna,
198515 Saint-Petersburg, Russian Federation

Abstract. The article assesses the role of accounting for implicit knowledge in creating corporate knowledge management systems and describes specific approaches to formalizing tacit knowledge based on an educational competence approach using a separate assessment of academic competencies and background knowledge of employees. Particular attention is paid to the issues of the need to ensure multilegality in the description of knowledge elements and their unifying structures.

The paper shows that a significant part of the procedures for automated extraction and formalization of corporate knowledge is based on technologies for working with natural language and assumes that the corporate semantic knowledge structures created at the same time, generated based on the company's production documents and employees' educational programs, are based on the concepts and structures of the language of primary documents. This leads to a complication of the conditions for integrating such a monolingual corporate knowledge management system with semantically structured information about accumulated knowledge objects and the relationships between them, combined using Knowledge Graph technology.

The paper concludes that corporate knowledge, regardless of its origin, can be detected and explicitly described with the help of enriched ontologies, using universal methods of statistical and contextual analysis of arrays of corporate documents, and using multilingual types of representation of basic concepts.

Keywords: Corporate knowledge · Knowledge graph · Ontology · Implicit knowledge · Multilingualism

1 Introduction

The ability to effectively change their production processes and amend the results of their activities in connection with changing trends in the development of technologies and markets is today the most important concrete advantage of a modern enterprise.

D. Bylieva and A. Nordmann (Eds.): PCSF 2022, LNNS 636, pp. 104–119, 2023.
https://doi.org/10.1007/978-3-031-26783-3_11

Therefore, a modern company strives to use as much available information as possible to develop adequate technical and managerial decisions related to making changes to the manufactured product and its manufacturing processes [1].

Corporate knowledge management systems have become a method of solving such problems. These systems assume an integrated approach to the search, systematization, evaluation, and dissemination of information assets of the enterprise.

One of the real breakthroughs in the technology of systematization, reuse, and generation of new knowledge was the creation of the Knowledge Graph technology, which provides users with semantically structured information about accumulated knowledge objects and the relationships between them. Currently, more than 570 million objects and more than 18 billion related facts, as well as semantic relationships between them, are in the public domain.

The gradual expansion by users all over the world of the Knowledge Graph information and semantic database, created as a multilingual environment, gives businesses an excellent mechanism for improving knowledge management in each information environment of the enterprise. However, although the integration processes of knowledge management are increasingly increasing the pace of transformation of management technologies, in the conditions of a particular operating enterprise, such processes may not be as effective for several reasons, the main of which is the significant interdependence of the corporate knowledge management system from the knowledge description language system.

This is because the main prerequisites for the description of heterogeneous knowledge as part of one is the following provisions [2]:

- knowledge is verbal;
- newly revealed knowledge is the development of previously accumulated knowledge, and can be added to previously accumulated knowledge by expanding it;
- specialized elements of corporate knowledge are reflected in the texts of internal corporate documents or arrays of industry and other publicly available data used by employees of the enterprise in their work.

This makes it possible to reduce the actions to create a unified system for describing corporate knowledge to the selection of a set of speech concepts, with the help of which it is possible to describe multidisciplinary arrays of corporate knowledge uniformly and to use methods based on natural language text processing as an analysis tool for their search and systematization.

At the same time, the special effects of the use of artificial intelligence elements related to the semantic processing of natural languages in knowledge management systems are explained not only by the verbal nature of human knowledge but also by the documentation of all business processes of the enterprise.

However, these same circumstances can also harm the integration capabilities of the knowledge management system being created, since it demonstrates the determining influence of language features not only on the semantic structure of knowledge description but also on the entire historically accumulated and structured array of corporate knowledge by previously existing rules.

The creation of an innovative product is a complex multidisciplinary process involving a large number of participants; therefore, the processing of the necessary information arrays for their subsequent use in knowledge management systems is a chain of complex tasks. Therefore, the use of artificial intelligence elements, such as semantic search, natural language processing, and machine learning in knowledge management systems becomes a necessity [3].

In this regard, with an increase in the degree of knowledge integration, the task of studying the features of the linguistic representation of previously accumulated corporate knowledge on the integration capabilities of enterprise information systems, as well as the formation of mechanisms that allow the successful implementation of all the possibilities that multilingualism represents for describing knowledge, is relevant.

The objective of this work is to study the influence of linguistic means of knowledge representation on the overall structure of corporate knowledge from the point of view of enterprise information structures to further improve the efficiency of corporate governance systems by expanding their multilingualism.

2 The Process of Describing Knowledge by Means of Natural Language

From the point of view of computer science, the transition from disparate data to structured industrial knowledge is characterized by a gradual complication of the objects under consideration, and the creation of various relationships between them. At the same time, for a more specific process of formalization of knowledge and skills accumulated by the subjects of production activity (actors) in their education and creative activity at all stages of the production of the final product - the methodological scheme of information and technological redistribution is usually used. As shown in Fig. 1, and describing the scheme of relations between corporate data, information, and knowledge [4] (commonly called the "DIMKC model"), which is a "hierarchy" where each later level adds certain semantic properties to the earlier level.

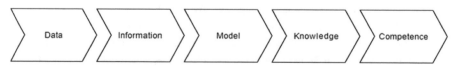

Fig. 1. Methodological scheme of information transformations (information redistribution) arising in the process of knowledge management (DIMKC model)

At the same time with added information technology redistribution, the level of abstraction of the processed material increases, which makes it possible to verbalize the resulting transformation result to an increasing extent. This, in turn, leads to the need to use artificial intelligence methods based on semantic analysis of textual information to analyze the result. Therefore, the creation of specialized methods for the analysis of text documents that generate and fix corporate knowledge is one of the main directions of knowledge management.

The methodological basis for the use of AI elements in the development of knowledge management systems (KM) was laid down by three principal areas of work, which received the conventional names of the American, Japanese, and Scandinavian (European) schools for the study of KM processes [5].

The American KM School of Process Research has focused its attention on the issues of modern enterprise management through the direct influence of knowledge on the business processes of the enterprise [6]. At the same time, the founder of the "American" school of knowledge management, K. Wiig, defined KM processes as an integral part of the process of organizational management of an enterprise. It is based on special methods of describing management information to improve the production and organizational characteristics of the company by ensuring the formal validity of management decisions [7]. At the same time, knowledge management is a complex process that includes the collection, exchange, classification, and formalization of information into knowledge and competencies documented in corporate norms. Later such norms are used for regular training of the employees in all identified production practices, decision support, and planning.

The Japanese KM Process Research School, created by the efforts of Ikujiro Nonaka [8], concentrated its efforts on the research of the processes of knowledge identification and formalization, having developed a spiral model of information transformations SECI, which defines the identification and identification of knowledge as the result of a multi-stage interaction of implicit and explicit knowledge, in which at successive stages: Socialization-Externalization-Combination-Internalization, using there are different types of transformations, the final process of formalization of knowledge takes place. The basics of the methodology developed by the specialists of the "Japanese" school are the procedures for the interaction of the two main concepts of knowledge management theory – explicit knowledge and implicit knowledge, as well as a description of specific methods for identifying implicit knowledge with the subsequent formalization of it into explicit knowledge.

At the same time, a validating assessment of the market for the adequacy of the process of generating corporate knowledge can be conducted, among other things, using the results of the work of specialists from the European School for the Study of Corporate Knowledge Management processes, the founder of which is Karl Erik Sweiby, and which focused its main solutions on measuring the intellectual capital of companies [9]. At the same time, a large amount of work was conducted to determine the value of various intangible assets of the enterprise, such as the value of its brand, the total competence of employees, and reputation. In addition, the Scandinavian school is the author of the description of the general structure and most of the methods used today for assessing corporate intellectual capital and the value of corporate knowledge, as well as the formulation of the idea of the need to change traditional forms of accounting in the direction of their expansion by the possibilities of direct accounting of knowledge (and related indicators) in the final results of the company's activities.

This allows us to present the interaction of all three areas of research on corporate knowledge management processes as a single flowchart of a controlled process with feedback (see Fig. 2), implemented in the form of changes in the valuation of intangible assets [10], arising from the use of newly identified knowledge in the processes of enterprise management, which, is an algorithmic description of a self-learning enterprise.

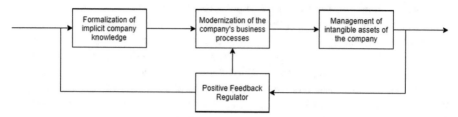

Fig. 2. Presentation of the management of the corporate knowledge management system in the form of a system with positive feedback.

Since any enterprise is interested in the accelerated and justified implementation of newly identified knowledge into the business processes of the enterprise, such feedback will be positive feedback (exacerbating feedback) since the feedback of this type is used in cases where it is necessary to accelerate the reaction to changes in external parameters. However, for this positive quality of positive feedback, any managed system must pay for the possibility of instability in it [11].

The established system of business processes of an enterprise in most cases is characterized by significant inertia [12] and, as a rule, does not involve constant adjustments to compensate for changes in performance evaluation indicators [13]. Therefore, when building this process, special attention should be paid to the creation of the most accurate algorithms for identifying and fixing newly identified and formalized elements of implicit knowledge in the organization's governing documents. At the same time, it is assumed that explicit knowledge can be expressed in a formal form and stored on certain media (in documents, instructions, organized memory sections, etc.) separately from the owner. Whereas implicit knowledge is inherently not formalized and can only exist together with the possessor.

In this regard, it is believed that the mechanisms of identification, fixation, formalization, and subsequent use of implicit knowledge represent a key area of work on the creation of total data management systems based on the principles of artificial intelligence [14, 15], which are also used in the construction of structures of self-learning enterprises.

Since implicit knowledge is quite difficult to transfer from one individual to another, specialized technologies are required for its identification, formalization, replication, and subsequent development. One of the sufficiently productive and valid models of this process is the universal spiral model of SECI knowledge formalization noted above, developed by Ikujiro Nonaka (see Fig. 3).

In the first phase of the process, implicit knowledge is disseminated into the corporate environment of their discussion. In the second phase, there is a process of collective transformation of socialized implicit knowledge spontaneously or consciously organized by a team in the form of a new concept into explicit knowledge. At the next (third) transformation phase, the created concept of new knowledge passes the verification stage, during which the company determines whether the concept created in the previous phase

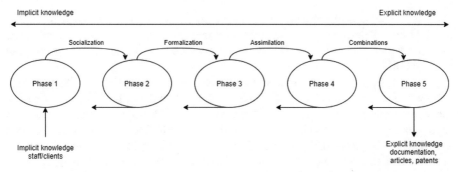

Fig. 3. Universal SECA is a model of the process of interaction of explicit and implicit knowledge, which underlies the creation of corporate knowledge. Phase 1 – dissemination of implicit knowledge; Phase 2 – concept creation; Phase 3 – concept verification; Phase 4 – archetype construction; Phase 5 – knowledge transition to a new level

of the process has the right to exist. In the next phase, the approved concept is transformed into the so-called archetype, which can take either the form of an organizational element (for example, a process or structure), if we are talking about the intangible nature of improvement, or the form of some material prototype if we are talking about the creation of a material object. The last phase of creating explicit knowledge is the distribution of the created prototype across the company's divisions or beyond its organizational boundaries. This procedure conducts the transition of newly identified knowledge to a new level and the exchange of knowledge with the current environment.

Since all stages of the SECI model are characterized by the release of accompanying text documents of strictly defined structures using a single expandable thesaurus, at the end of all stages, a final document describing the newly identified corporate knowledge is created, which from that moment passes into the category of explicit knowledge and is added to the unified structure of corporate knowledge, issued guidelines, production methods, corporate standards, business rules, technologies, drawings, diagrams, information arrays, and other documentary materials, but also the knowledge of the staff, including during training and production activities.

The presented model of the actual transformation of implicit knowledge into explicit, which has proven itself well enough in conducting research, technological developments, and conducting innovative projects, allows for the necessary formalization of a significant part of the non-formalized knowledge that is the subject of research. At the same time, it reveals the presence of significant analogies with the model of launching an innovative object to the market, described in terms of the "Kline-Rosenberg chain model" [16], and with the DIMKC model of information transformations described above, thereby demonstrating the commonality of approaches to the description of information transformation in theoretical computer science, when structuring production processes and when creating corporate knowledge management systems.

At the same time, it should be noted that there is still a wide class of implicit knowledge that requires additional study when creating KM systems, which, with this approach, is quite difficult to detect, but, at the same time, is an essential part of the knowledge used by the company in its activities.

Here we are talking about the so-called personal-behavioral knowledge, which is the result of the influence of many factors that formed the basis of previously accumulated personal experience [17] and the presence of which was first indicated by Michael Polanyi [18].

Thus, it should be borne in mind that the language used to describe knowledge is a system of symbols, and speech (including scientific) is the skill of its use. Moreover, these skills are not always reflected, because a person very rarely notices what he is saying, using constructions in speech that he has inherited in the form of background knowledge. At the same time, the lack of the necessary personal knowledge in production activities can lead to unforeseen situations, since knowledge about production (as well as other) processes cannot be fully formalized in instructions and descriptions, since most of it is stored as "implicit knowledge" in people's memory. Therefore, the process of the formalized description of this implicit knowledge is always the task of the production services of any enterprise.

To avoid a decrease in the universality of descriptions of corporate knowledge associated with the peculiarities of a particular language representation, the process of knowledge formalization already at the stage of creating a corporate knowledge management system should be created based on universal multilingual approaches without focusing on specific languages of representation of knowledge objects.

3   The Information Environment of the Enterprise – as a Repository of Previously Accumulated Formalized Knowledge

From the point of view of knowledge management, any modern enterprise is a complex integrated information Environment (IIE) containing and processing a set of distributed databases describing information about the company's products/services, technical, technological, and organizational environment, processes, and resources of the enterprise necessary for its production and economic activities. The main feature of the IIE of any company is multidisciplinary.

If we take a closer look at the cognitive structure of the IIE parts of any modern company, presented in Table 1, then we can distinguish three main components of the integrated information environment of any business.

The technological part of the IIE, which is the most structured part of IT solutions of any business and has for its purpose data processing aimed at supporting technological production processes of the business; it operates mainly with explicit knowledge presented in analytical models of technological processes, normative technical and technological documentation and the like formalized information in the form of recorded knowledge that allows reliably explaining not only all experience but also predicting future results of technological processes that do not go much beyond the definition of the quantities describing them. This part also includes solutions designed to solve innovative problems that translate implicit knowledge into the field of explicit knowledge. It is obvious that in the "scientific and technological" part of the corporate information environment, the influence of personal-behavioral implicit knowledge is limited only to paradigmatic knowledge [19] (to which the features of the work that characterize a

Table 1. The cognitive structure of parts of the integrated information environment of a modern company

Parameter	Technological part IIE	Economic and organizational part IIE	Part IIE, supporting work with the company's staff/clients
The key meaning-forming objective function	Management of technological processes of product/service creation	Economics and management of the enterprise	Human Resources/Customer Relationship Management
Management object	Information	Company	Person
Integral characteristics of key competencies	Natural sciences and information sciences	Economics and Management	Humanities, pedagogy, psychology, sociology
Centrality of tasks	Information processing	Creating added value	Human interaction
The main type of processed knowledge	Explicit knowledge	Knowledge as intellectual capital	Implicit knowledge

particular scientific community or school may be partially attributed and may not be taken into account in the first approximation).

The economic and organizational part of IIE operates with knowledge in the form of various resources, which include the already mentioned intellectual capital of the company. At the same time, various financial analysis systems and regulatory documents regulating the construction of intra-corporate business processes are used as sources of knowledge. The exceptional importance of the economic and organizational part of the IIE in the corporate knowledge management system is that the knowledge that makes up its solutions is the only feedback for the enterprise that objectively describes the integral results of its activities. At the same time, for the overwhelming number of companies, the result of their activities in the market depends on the behavioral market conditions, in which the determining factor is the behavior of the consumer of goods or services, which must be taken into account, and which is largely determined by personal-behavioral implicit knowledge [20], not verbalizable at this stage.

Part IIE, which supports work with the company's staff/ clients, is responsible for the interaction of individual subjects during the company's functioning and is based on the achievements of pedagogy, sociology, psychology, regulatory information, job distribution, and other documentation regulating HR relations within the company. This part of the IIE operates with a large amount of implicit personal-behavioral knowledge. Therefore, today the sphere of corporate HR resources management and customer relations is (in addition to the economic part) one of the most interesting works on the formalization of implicit personal and behavioral knowledge [20].

At the same time, information sources of corporate knowledge of a particular enterprise, regardless of industrial characteristics, include:

- descriptions of production technologies and business processes,
- methods and corporate standards,
- drawings, technological maps, sketches, diagrams,
- technological and labor documentation,
- regulatory and legal documentation,
- electronic archives and databases of the enterprise,
- information automated systems used in the enterprise,
- organizational structure and job assignments,
- knowledge and skills in the memory of the company's specialists.

Usually, as the most convenient way of structuring elements of corporate knowledge used in various elements of information systems, subject ontologies are used, describing the hierarchy of domain concepts, the connections between them, and the patterns that operate within the framework of the described relationship model [21]. The main advantage of using ontologies, in this case, is the possibility of integrating heterogeneous information based on understandable language tools. At the same time, quick access to its contents, logically justified by some previous agreements, is provided based on a single dictionary of concepts designed to describe documentary content and competence profiles of the company's employees. In addition to this, the applied industrial ontology also acts as a general scheme for integrating disparate corporate data sets.

Since when systematizing corporate knowledge, as a rule, means of semantic analysis of corporate documentary content are used using natural language processing methods, linguistic ontologies are used in industrial information systems. Their distinctive property is the use of natural language words in one resource together with their linguistic properties [22]. At the same time, the main source of concepts of such ontologies is the meanings of linguistic units that should be described by subject branch ontologies.

At the same time, the greatest gain when using such ontologies is achieved by the fact that subject industry ontologies are always built based on a joint understanding of the industry features of key concepts within a certain professional community [23]. In this regard, to increase the possibilities of using ontologies in the communication of professional communities, the formal description of ontology should be expanded by introducing into it the concepts of dictionaries of professional terms describing the features of the field of knowledge under consideration [24].

$$O = <C, R, A, L, F, G, H> \tag{1}$$

where:

$L = L_C \mid L_R$—an ontology dictionary containing a set of lexical units (signs) for concepts L_C and a set of signs for relations L_R;

C—set of ontology concepts;

R—relations between ontology concepts;

A—set of ontology axioms;

F and G are reference functions that link sets of lexical units L_j in L with sets of concepts and relations to which they respectively refer in this ontology. In this case, one lexical unit can refer to several concepts or relations, and one concept or relation can refer to several lexical units;

H - fixes the taxonomic nature of relations (connections), in which the concepts of ontology are connected by non-reflexive, acyclic, transitive relations H from C·C. The expression H (C_1, C_2) means that the concept of C_1 is a sub-concept of C_2.

In cases where a large amount of previously accumulated knowledge is used in the company, excessive specialization of the description may arise, associated with the peculiarities of the procedures for fixing knowledge elements based on the national subject vocabulary. This can significantly complicate further processes of integration of enterprise information systems into comprehensive systems of accumulation and systematization of knowledge.

This process can be observed when conducting work on the integration of elements of information systems created in different countries.

In this regard, the process of creating a uniform approach to the generation of basic ontologies used in corporate information systems based on the principles of multilingualism can be considered one of the ways to improve the efficiency of corporate governance systems.

4   Creating a Unified Approach to the Description of Professional Competencies

From the point of view of the enterprise, the knowledge of the company's employees is not the property of the enterprise, although it is included in the production process. Historically, for a unified description of the knowledge, skills, and competencies of hired employees, the division of employees into professional qualification groups was used, united by fields of activity and considering the level of qualification of such employees. This approach works well for low-skilled workers of mass professions who perform typical actions that do not depend on the specific conditions of their performance.

However, with the general increase in the pace of development and implementation of innovations, such an average approach began to lead to the fact that the costs of retraining an employee in the workplace became unreasonably high and it required the development of methods for quantifying employee knowledge in specific areas of activity. At the same time, from the point of view of the enterprise, the employee's knowledge is formalized.

One of the first successful attempts to determine an employee's knowledge for its subsequent application in corporate knowledge management systems was implemented by Davenport and Prusak [25], who described implicit knowledge as a constantly replenishing combination of personal practical and educational experience, individual values, surrounding contextual information and intuition.

At the same time, if we take into account several works that have significantly concretized the concepts of personal knowledge [26], etc., then the entire volume of implicit (not formalized in company documents) the knowledge that needs to be taken into account when creating a total knowledge management system of the company can be divided into two components:

- educational and competency-based knowledge previously acquired by employees in previous periods through training, the performance of their production duties, or made available to them through communication in the production team [20],

- background knowledge, which is mutually accessible to the speaker and listener knowledge of the subjective features of the real world, which is information that is certainly known to all members of the community that makes up the company's team and is the actual basis of language communication [27].

The next step in the development of an educational-competence approach to the description of corporate knowledge was the proposal to consider the total corporate knowledge, the carriers of which are the company's employees, as a superposition of their competencies in all areas of knowledge. In this regard, the aggregate set of basic competencies for employees of an innovative company involved in the creation of a product can be represented in the form of expression (2):

$$K = \sum_{i=1}^{N} \sum_{j=1}^{M} K_{ij} \tag{2}$$

where N is the number of employees of the company (actors),

M is the total number of competencies accepted for consideration,

K_{ij} is a quantitative description of the j–the competence of the i-th actor.

At the same time, it should be borne in mind that the company's employees not only produce new products/knowledge in the course of work, but also increase their competencies [28], in connection with which the total competence of an employee performing his production duties is proposed as a combination of three components: educational competencies according to certifying documents, competencies obtained during the research, and "social" competencies representing the research competencies of the "co-author environment" of a particular researcher, which can be represented as a sum (3):

$$K = K0 + K1 + K2 \tag{3}$$

where K0 is the educational competence of the actor according to the certifying documents,

K1 – additional competencies of the actor obtained in the course of work,

K2 – "social" competencies, representing the research competencies of his environment.

At the same time, the growth of competencies of both an individual employee and the entire team, which occurred during the execution of the work, can be formally considered by a consistent structural linguistic analysis of the results of his production activities, which is conducted according to a collection of texts authored or co-authored by the actor under study.

At the same time, the task of determining the additional amount of knowledge acquired by a particular employee can be reduced to a consistent modification of the educational ontology of the employee's knowledge by performing the following actions:

- identification of new concepts that an employee uses in "incoming" and "outgoing" production documents,
- expansion of the identified concepts of the initial ontology of specialist knowledge.
- In general, the task of gradual modification of the employee's knowledge ontology can be achieved by sequentially performing the following procedures:

- detection of terms with signs of newly identified concepts for this employee throughout the body of texts under consideration by conducting a frequency analysis of the author's working texts [29],
- finding concepts semantically related to the newly identified term using a comparative analysis of the contextual cosine distance [30],
- building relationships and rules linking newly identified concepts based on semantic relationship extraction methods [31],
- direct expansion of the employee's knowledge ontology [31].

When determining K0, K1, and K2, the difference is only in the types of documents to be analyzed that characterize certain personal competencies. Thus, when determining K0, the corpus of texts describing the educational level of an employee, including the texts of the educational programs he has completed, is subject to analysis. When determining the K1 component, the texts of scientific papers issued by the employee and the texts of his key working documents should be subjected to semantic analysis. When determining the K2 component, the texts of scientific papers written by the co-authors of the employee under study are subject to analysis.

It can be seen from the above that if there are significant differences in the ontological descriptions of the initial training programs (K0), the resulting competencies of employees performing the same work, but using different ontological mappings, explained by the peculiarities of national descriptions of training programs and their composition, can vary greatly. And this can lead to serious "discrepancies" in the description of the employee's current competencies in the future. In this regard, there is also a need for the initial consideration of multilingualism when creating descriptive ontologies of educational programs.

5   Multilingualism in Approaches to Describing the Background Knowledge of the Company's Employees

When constructing methods for identifying implicit background knowledge, a considerable number of researchers propose to build mechanisms for studying and formalizing implicit background knowledge by analogy with studies of background knowledge used in a sufficiently developed linguistic theory of translation [30], which represents the implementation of multilingualism.

Using the methodology of describing background knowledge in the linguistic theory of translation, we will then subdivide implicit background knowledge into four categories:

- personal knowledge related to the naive picture of the world (the so-called everyday knowledge),
- personal cultural knowledge that is closely related to the religious and cultural values surrounding a particular person,
- knowledge of the rules of interpersonal interaction in the situation of performing a particular activity,

- personal verbal competence, expressed in the ability to clearly describe what is happening.

At the same time, another confirmation of the possibility of transferring technologies for studying background knowledge developed in the linguistic theory of translation is also confirmed by the fact that exactly these four categories of implicit personal background knowledge were identified as the main ones considered by buyers when making complex transactions.

Therefore, further, using the standard approaches of the linguistic research of background knowledge in translation, it is possible to determine the most promising areas of identification of the above four categories of implicit background knowledge, among which one can distinguish:

- semantic analysis of the thesaurus of the "everyday oral texts" of the subject, suggesting that the basis of everyday use of language "in live communication" is that language is a person we use intuition, based on previously acquired background knowledge, and not on explicit formalized rules,
- analysis of the contextual features of the argumentation of the subject, presenting arguments, facts, and evidence to explain and strengthen his position or support the other side; since argumentation is always expressed in language in the form of a rather limited set of frames purposefully used for a fairly narrow set of scenarios when evaluating a specially modeled discussion, it becomes possible to fairly objectively evaluate personal background knowledge of interest by several criteria on the ability to rationally weigh arguments, accept them or challenge them,
- analysis of the characteristics of the subject "during examinations", in areas where the rational and formalized knowledge of the subject is not great.

Operational methods used in such studies of implicit background knowledge of employees or candidates for a vacancy can be as diverse types of typical analysis of specialized oral and written storytelling of the subject on a given set of topics, the texts of which can be considered as collections of texts being analyzed. At the same time, as "reference sets" of test requirements for a particular set of background knowledge, text materials collected using the corpus of texts of normalized educational groups, collected with the use of CoP (community of practice) technologies, can be attributed.

6   Conclusion

Since almost any knowledge is verbal in nature, the description of the elements of corporate knowledge always carries the peculiarities of their linguistic representations.

First of all, this is due to the use of linguistic ontologies in industrial information systems for the systematization of the described objects, the distinctive property of which is the use of natural language words together with their linguistic properties in one resource. This, on the one hand, increases the clarity of the created structures for national specialists of a particular industry, on the other hand, significantly complicates the process of integrating international research results into the created corporate knowledge management systems.

In this regard, the transition of corporate information systems from the national to the multilingual nature of the representation of knowledge objects will be able to seriously expand the possibilities of using semantically structured information collected from various sources and technology-oriented Knowledge Graphs in industrial information arrays.

Therefore, the transition to the multilingualism of the representation of knowledge objects can be considered a determining mechanism for the development of corporate knowledge management systems, increasing apart from possibilities of international cooperation but also the competitiveness of individual enterprises.

Of course, the transition to the multilingualism of the representation of corporate knowledge objects will not make a revolutionary breakthrough in the methodologies of formalization of implicit corporate knowledge at a particular enterprise, however, it will allow the "seamless use" of modern world developments in this area.

As a result of the conducted research, the following conclusions can be drawn that are important for the further development of practical methods for assessing and formalizing implicit knowledge in corporate knowledge management systems of companies:

- the lack of considering multilingualism when creating corporate knowledge management systems significantly narrows the possibilities of using modern Knowledge Graph technologies in them, providing access to semantically structured information collected from various sources about accumulated knowledge objects,
- the study of the structural features of the creation of corporate knowledge management systems has shown the key role in this process of elements associated with the processes of detecting implicit corporate knowledge and turning them into explicit knowledge by formalization,
- corporate knowledge, regardless of its origin, can be detected and explicitly described with the help of enriched ontologies, using universal methods of statistical and contextual analysis of arrays of corporate documents, and using multilingual types of representation of basic concepts,
- the most effective method of describing the implicit corporate knowledge of the company's employees as a constantly replenishing combination of personal educational experience, surrounding contextual information, individual values, and intuition is an educational competence approach that divides such knowledge into competencies obtained in the process of professional education, production and co-author experience, as well as Background knowledge, which must be described using structures that allow for multilingualism.

References

1. Shujahat, M., Sousa, M.J., Hussain, S., Nawaz, F., Wang, M., Umer, M.: Translating the impact of knowledge management processes into knowledge-based innovation: the neglected and mediating role of knowledge-worker productivity. J. Busin. Res. **94**, 442–450 (2019). https://doi.org/10.1016/j.jbusres.2017.11.001
2. Kamoun-Chouk, S., Berger, H., Sie, B.H.: Towards integrated model of big data (BD), business intelligence (BI) and knowledge management (KM). In: Uden, Lorna, Lu, Wei, Ting, I-Hsien.

(eds.) KMO 2017. CCIS, vol. 731, pp. 482–493. Springer, Cham (2017). https://doi.org/10. 1007/978-3-319-62698-7_40

3. Buenechea-Elberdin, M., Sáenz, J., Kianto, A.: Exploring the role of human capital, renewal capital and entrepreneurial capital in innovation performance in high-tech and low-tech firms. Knowl. Manag. Res. Pract. 15(3), 369–379 (2017). https://doi.org/10.1057/s41275-017-0069-3

4. Moraga, J., Quezada, L., Palominos, P., Oddershede, A., Silva, H.: A quantitative methodology to enhance a strategy map. Int. J. Prod. Econ. 219, 43–53 (2020). https://doi.org/10.1016/j. ijpe.2019.05.020

5. Pillay, D., Barnard, B.: Entrepreneurship and knowledge management: knowledge requirements, utility, creation, and competency. Expert J. Bus. Manag. 7(1), 44–81 (2019)

6. Wiig, K.: People-Focused Knowledge Management: How Effective Decision-Making Leads to Corporate Success. Elsevier Butterworth Heinemann, Boston (2004)

7. Wiig, K.: Knowledge management: an introduction and perspective. J. Knowl. Manag. 1(1), 6–14 (1997). https://doi.org/10.1108/13673279710800682

8. Nonaka, I., Teece, D.: Managing Industrial Knowledge: Creation Transfer and Utilization. SAGE Publications Ltd, London (2001)

9. Sveiby, K.: The New Organizational Wealth: Managing and Measuring Knowledge-Based Assets. Berrett-Koehler, San Fransisco (1997)

10. Kohlegger, M., Ploder, C.: Data driven knowledge discovery for continuous process improvement. In: North, K., Maier, R., Haas, O. (eds.) Knowledge Management in Digital Change. PI, pp. 65–81. Springer, Cham (2018). https://doi.org/10.1007/978-3-319-73546-7_4

11. Gogan, L., Artene, A., Sarca, I., Draghici, A.: The Impact of Intellectual Capital on Organizational Performance. Procedia Soc. 221, 194–202 (2016). https://doi.org/10.1016/j.sbspro. 2016.05.106

12. Zhanga, Y., Zhanga, M., Luob, N., Wanga, Y., Niu, T.: Understanding the formation mechanism of high-quality knowledge in social question and answer communities: a knowledge co-creation perspective. Int. J. Inf. Manag. 48(10), 72–84 (2019). https://doi.org/10.1016/j.iji nfomgt.2019.01.022

13. Acar, M., Tarim, M., Zaim, H., Zaim, S., Delen, D.: Knowledge management and ERP: complementary or contradictory. Int. J. Inf. Manag. 37(7), 703–712 (2017). https://doi.org/ 10.1016/j.ijinfomgt.2017.05.007

14. Nonaka, I.: Dynamic theory of organizational knowledge creation. Organ. Sci. 5(1), 14–37 (1994)

15. North, K., Kumta, G.: Knowledge Management Value Creation Through Organizational Learning. Springer, Cham (2018)

16. Kline, S., Rosenberg, N.: An overview of innovation. In: Landau, R., Rosenberg, N. (eds.) The Positive Sum Strategy: Harnessing Technology for Economic Growth, pp. 275–307. National Academy of Sciences, Washington (1986)

17. Manville, G., Karakas, F., Polkinghorne, M., Petford, N.: Supporting open innovation with the use of a balanced scorecard approach: a study on deep smarts and effective knowledge transfer to SMEs. Product. Plann. Control 30(10–12), 842–853 (2019). https://doi.org/10. 1080/09537287.2019.1582093

18. Polanyi, M.: Personal Knowledge: Towards a Post-Critical Philosophy. University of Chicago Press, Chicago (1974). https://doi.org/10.2307/2105069

19. Rajapathirana, R., Hui, Y.: Relationship between innovation capability, innovation type, and firm performance. J. Innov. Knowl. 3, 44–55 (2018). https://doi.org/10.1016/j.jik.2017.06.002

20. Silvestre, B., Yîrcă, D.: Innovations for sustainable development: moving towards a sustainable future. J. Clean. Product. 208, 325–332 (2019). https://doi.org/10.1016/j.jclepro.2018. 09.244

21. Kozaki, K., Hayashi, Y., Sasajima, M., Tarumi, S., Mizoguchi, R.: Understanding semantic web applications. In: Domingue, J., Anutariya, C. (eds.) ASWC 2008. LNCS, vol. 5367, pp. 524–539. Springer, Heidelberg (2008). https://doi.org/10.1007/978-3-540-89704-0_36

22. Maksimov, N., Lebedev, A.: Toward a building an ontology of artefact. In: Samsonovich, A.V., Gudwin, R.R., Simões, Ad.S. (eds.) BICA 2020. AISC, vol. 1310, pp. 225–232. Springer, Cham (2021). https://doi.org/10.1007/978-3-030-65596-9_29

23. Casas, M., Perez, M., Rojas, J., Alvarez, J.: Strategic planning model to improve competitiveness for service industry smes using the balanced scorecard. In: Ahram, T., Taiar, R., Colson, S., Choplin, A. (eds.) IHIET 2019. AISC, vol. 1018, pp. 1001–1006. Springer, Cham (2020). https://doi.org/10.1007/978-3-030-25629-6_155

24. Maedche, A., Zacharias, V.: Clustering ontology-based metadata in the semantic web. In: Elomaa, T., Mannila, H., Toivonen, H. (eds.) PKDD 2002. LNCS, vol. 2431, pp. 348–360. Springer, Heidelberg (2002). https://doi.org/10.1007/3-540-45681-3_29

25. Davenport, T., Prusak, L.: Working Knowledge. Harvard Business Review Press, Brighton (1998). https://doi.org/10.1145/348772.348775

26. Lessebr, E., Prusak, L.: Creating Value with Knowledge: Insights from the IBM Institute for Business Value. Oxford University Press, Oxford (2003) https://doi.org/10.1093/019516 5128.001.0001

27. Cheng, Y., Chen, K., Sun, H., Zhang, Y., Tao, F.: Data and knowledge mining with big data towards smart production. J. Ind. Inf. Integr. **9**, 1–13 (2018). https://doi.org/10.1016/j.jii.2017.08.001

28. Li, Y., Song, Y., Wang, J., Li, C.: Intellectual capital, knowledge sharing, and innovation performance: evidence from the Chinese construction industry. Sustainability **11**(9), 2713 (2019). https://doi.org/10.3390/su11092713

29. Dang, C., Le-Hoai, L., Kim, S.: Impact of knowledge enabling factors on organizational effectiveness in construction companies. J. Knowl. Manag. **22**(4), 759–780 (2018). https://doi.org/10.1108/JKM-08-2016-0350

30. Petasis, G., Karkaletsis, V., Paliouras, G., Krithara, A., Zavitsanos, E.: Ontology population and enrichment: state of the art. In: Paliouras, G., Spyropoulos, C.D., Tsatsaronis, G. (eds.) Knowledge-Driven Multimedia Information Extraction and Ontology Evolution. LNCS (LNAI), vol. 6050, pp. 134–166. Springer, Heidelberg (2011). https://doi.org/10.1007/978-3-642-20795-2_6

31. Mashina, E.: Taking into account the specifics of converting highly specialized professional texts to a conceptual series that is understandable to unskilled users when choosing machine translation technologies implemented within a single language. In: Proceedings of the X Congress of Young Scientists, pp. 358–361. ITMO University, St. Petersburg (2021)

**Discerning Languages
in the Technosphere**

Russian Emotional Concepts in the Multilingual Technological Environment

Andrei E. Serikov[(⊠)] ⓘ

Samara National Research University, 34, Moskovskoye Shosse, 443086 Samara, Russia
serikov.ae@ssau.ru

Abstract. What happens to specific linguocultural emotional concepts in the contemporary multilingual technological environment? This issue is discussed on the example of several Russian emotional concepts described by A.Wierzbicka in the framework of the NSM approach. The NSM descriptions are compared to representations of the same concepts in IT applications explicitly related to the theory of basic emotions, and to their representations in RuWordNet. Both NSM descriptions of emotional concepts and their representations in RuWordNet retain their linguocultural specificity, although in different ways. Both of these ways of embedding specific emotional concepts in the multilingual environment are incompatible with the theory of basic emotions, but quite consistent with the Conceptual Act Theory of Emotion, which is more promising from a methodological point of view. If these methods prevail in the future, then the global multilingual environment will become semantically richer, because it will be able to absorb the semantic specifics of individual cultures. On the other hand, the uncritical use in computer applications of universalist ideas associated with the theory of basic emotions creates the illusion that these emotions are indeed universal. That is why, in the multilingual technological environment, the theory of basic emotions can act as a self-fulfilling prophecy. If this scenario dominates in the future, then the global multilingual culture will become semantically more primitive compared to the diversity of cultures that exists today.

Keywords: Linguocultural concepts · Emotions · Semantics · NSM · IT · WordNet

1 Introduction

The famous German linguist Wilhelm von Humboldt suggested that "there resides in every language a characteristic world-view" [1, p. 60]. Therefore, it is possible that peoples who speak different languages feel and think differently. This is due to the fact that "the words of various languages are never true synonyms, even when they designate, on the whole, the same concepts" [1, p. 167]. The Humboldt's idea became the basis for the Sapir-Whorf hypothesis of linguistic relativity. According to Edward Sapir, "the worlds in which different societies live are distinct worlds, not merely the same world with different labels attached" [2, p. 162].

© The Author(s), under exclusive license to Springer Nature Switzerland AG 2023
D. Bylieva and A. Nordmann (Eds.): PCSF 2022, LNNS 636, pp. 123–134, 2023.
https://doi.org/10.1007/978-3-031-26783-3_12

Radically expressed, the hypothesis of relativity would mean that peoples who speak different languages will not be able to understand each other. Sapir's student Benjamin Lee Whorf tried to prove this by contrasting European languages on the one hand, and the languages of American Indians on the other. In particular, he argued that "a Hopi who knows only the Hopi language" does not understand time as it is understood by native English speakers because the Hopi language contains "no words, grammatical forms, constructions or expressions that refer directly to what we call 'time,' or to past, present, or future, or to enduring or lasting, or to motion as kinematic rather than dynamic" [3, p. 57].

But as it turned out later, Whorf did not know the Hopi language well enough to draw such conclusions. In 1983, Ekkehart Malotki published the book "Hopi Time", in which he clearly and in detail demonstrated the presence in the Hopi language of many lexical and grammatical means for expressing time [4]. This failure of Whorf explains why many English-speaking linguists in the late 20th century were skeptical of the hypothesis of linguistic relativity and leaned towards the opposite universalist view, according to which all people think in the same way, regardless of the language they speak.

In the 1950s, Noam Chomsky proposed to understand grammar as a set of computational algorithms [5]. Offering various theoretical models of language, Chomsky always distinguished between deep and surface grammatical structures, internal and external language. According to Robert Berwick and Chomsky, "language evolved as an instrument of internal thought, with externalization a secondary process" [6, p. 74]. From this point of view, mental computations could be understood as internal and therefore independent of the differences of external languages. Based on this idea, the philosopher Jerry Fodor proposed a notion of the language of thought (LOT) that became quite popular among universalists [7]. Steven Pinker used the LOT concept to express the universalist creed in the following words: "People do not think in English or Chinese or Apache; they think in a language of thought" [8, p. 81]. But in practice this means that any idea of English-speaking thinkers is supposed to be easily and without distortion understood by representatives of other cultures. Thus, "since mental life goes on independently of particular languages, concepts of freedom and equality will be thinkable even if they are nameless" [8, p. 82].

However, the fact that mainstream linguists hold the universalist point of view does not prove its validity. Perhaps Stephen K. Levinson provides the most compelling evidence that language can indeed influence perception and thinking. He shows that some languages such as Tzeltal or Guugu Yimithirr force their native speakers to be absolutely oriented in space [9]. According to Anna Wierzbicka, Cliff Goddard and their colleagues, one does not have to choose between radical relativism on the one hand, and universalism on the other. Semantics of different languages and the world-views associated with them do differ, but are not incommensurable. Representatives of different cultures think differently, but the specificity of their thinking can be accurately expressed and conveyed using the Natural Semantic Metalanguage (NSM), developed on the basis of universal semantic primes [10]. Recently, within the framework of the NSM approach, the so-called *minimal languages* have been formed, designed to accurately convey culturally specific meanings in a multilingual environment [11, 12].

Based on the works of such authors as Stevenson and Wierzbicka, it can be argued that the influence of language on perception and thinking, the existence of various specific world-views is a proven fact. But, on the other hand, it is also a fact that today the interaction between representatives of different linguistic cultures is becoming much more intense than in the past. This is largely due to the development and implementation of numerous technologies for automatic natural language processing (NLP), automatic knowledge extraction and database constructing, etc. In this context, the question arises: what happens to specific language concepts and world-views in the multilingual technological environment? Do representatives of different cultures manage to convey to other people the specifics of their perception of the world? Is the multicultural environment that is formed as a result of this richer than individual linguistic world-views? Or vice versa, since most multilingual technologies are developed by people who communicate in English and by default take a universalist point of view, is there some blurring of specific language concepts and the formation of a universal but simplified world-view?

Below, these issues will be considered in relation to Russian emotional concepts. Wierzbicka demonstrates that neither English concepts of supposedly basic emotions such as "anger" or "sadness", nor the concept "emotion" itself are universal. In particular, she suggests semantic descriptions of some specific Russian emotional concepts [13–15]. We will try to find an answer to the question of what happens to these specific concepts in the contemporary technological environment.

In the next section, the current controversy between the theory of basic emotions and the Conceptual Act Theory of Emotion will be discussed. Then, in the third section, Wierzbicka's NSM descriptions of Russian emotional concepts will be given. The fourth section will discuss representation of Russian emotional concepts in IT applications explicitly related to the theory of basic emotions. In the fifth section, the representation of Russian emotional concepts in RuWordNet will be analyzed. In the final section, conclusions will be drawn about the three described options for the embedding of specific linguocultural concepts in the multilingual environment.

2   The Current Controversy on the Theory of Emotions

The psychology of emotions in the late 20th century was dominated by the theory of basic emotions. This theory was based on the ideas of Charles Darwin [16], developed in the middle of the 20th century by Silvan Solomon Tomkins [17] and his students Carroll Ellis Izard [18], Wallace V. Friesen and Paul Ekman [19–23]. The main premises and theses of this theory are as follows: basic emotions are physiological states, innate reactions of the body to certain biologically significant situations; these states may be unconscious or conscious to varying degrees; each basic emotion corresponds to a well-defined facial expression and, possibly, other external signs of behavior; there are dozens of variants of manifestation of the same basic emotion, the so-called *emotion families*, but each of the manifestations contains the mandatory features of this basic emotion; combinations of basic emotions give a range of other, less typical emotions; basic emotions are universal, common to representatives of all human cultures; the outward expression of conscious emotions is regulated by culture, according to the rules of which the expression of some emotions can be emphasized, while the expression of others can be suppressed or masked. There are 6 or 7 basic emotions: *surprise, fear, disgust, anger, happiness, sadness, and contempt.* The success of this theory is associated with the development and application of the Facial Action Coding System (FACS) and the FACS manual created on its basis, first published in 1978 and revised in 2002 [24]. In addition, Ekman and Friesen conducted a number of studies to test the hypothesis of the universality of basic emotions and insisted on its repeated confirmation [25, pp. 23–28].

However, there are approaches that are alternative to the theory of basic emotions. In 1998, James A. Russell and Lisa Feldman Barrett substantiated the hypothesis that affect has two bipolar dimentions: "one of valence (pleasant vs. unpleasant) and one of activation (activation vs. deactivation)" [26, p. 969]. In 1999, they proposed to distinguish between *core affect*, which has such a simple structure, on the one hand, and *prototypical emotional episodes*, which have a more complex structure, on the other. Core affect is only one of the fundamental constituents of emotional episodes, along with other components, such as behavior of certain sort related to a specific object, "attention toward, appraisal of, and attributions to that object; the experience of oneself as having a specific emotion; and, of course, all the neural, chemical, and other bodily events underlying these psychological happenings" [27, p. 806]. The ordinary folk categorization of emotional episodes is based on a natural language words, which is normal, and the basic emotion theorists do the same, which is incorrect because "categories in English are similar – but not identical – to categories found in other languages" [27, p. 806].

In 2009, Russell proposed the notion of psychological construction of emotions based on core affect, contrasting it both with biological (as in the theory of basic emotions) and with social (as totally independent of biological processes) construction. What "gets psychologically constructed are individual token events, which may (or may not) then be classified as emotion, fear, anger and the like by means of a folk concept" [28, p.1267]. Based on these ideas, the Conceptual Act Theory of Emotion were developed, according to which emotional categories do not correlate with individual states of a person, but with integral practical situations, the instant conceptualization of which affects their perception, changes in a person's internal states and corresponding actions. "Over the course of situated activity, numerous modalities and systems in the brain and body respond continually to represent the situation, including exteroceptive perception, interoception, core affect (valuation and salience processes that underlie experiences of pleasure/displeasure and arousal), attention, categorization, executive processing, episodic memory, action, language, reasoning, and so forth" [29, p. 7].

3   The NSM Descriptions of Russian Emotional Concepts

The NSM approach practiced by Wierzbicka contradicts the theory of basic emotions, but is quite consistent with the conceptual act theory of emotion. Wierzbicka convincingly demonstrates that English-language emotional concepts, which are universal in the perspective of the theory of basic emotions, in fact are not. In particular, she analyzes three Russian concepts that are close in meaning to the English concept of *sadness*: *toska, grust'*, and *pečal*.

Here are descriptions of these concepts in NSM:

toska
X thinks something like this:

I want something good to happen
I don't know what
I know: it cannot happen

   because of this, X feels something [15, p. 172].

pečal

(a) X felt something
(b) sometimes a person thinks something like this:
(c) something bad happened
(d) this is bad
(e) if I didn't know that it happened, I would say: "I don't want this"
(f) I don't say this now
(g) because I can't do anything
(h) because of this, this person feels something bad
(i) like people feel when they think something like this
(j) X thought something like this
(k) because of this, X felt something like this
(1) X thought about it for a long time
(m) X felt something bad because of this for a long time.

grust'

(a) X felt something
(b) sometimes a person thinks something like this:
(c) something bad happened now
(d) if I didn't know that it happened, I would say: "I don't want this"
(e) I don't say this now
(f) because I can't do anything
(g) because of this, this person feels something
(h) X felt something like this [13, pp. 14–15].

   Let us compare these descriptions to the description of *sadness*:

sadness

(a) X feels something
(b) sometimes a person thinks something like this:
(c) something bad happened
(d) if I didn't know that it happened I would say: "I don't want this"
(e) I don't say this now
(f) because I can't do anything

(g) because of this, this person feels something bad
(h) X feels something like this [13, p. 9].

Wierzbicka argues that in the meaning of *toska* "elements of something similar to melancholy, something similar to boredom, and something similar to yearning are blended together and are all present at the same time, even though different contexts may highlight different components of this complex but unitary concept" [15, p. 171]. In addition, this Russian word always implies something indefinite, and can have both positive and negative connotations. *Toska* means the yearning for some inexplicable and impossible good, comparing to sadness, which means that something bad happened.

In the case of *grust'* it is important that something bad has happened now which is not necessary in the case of *sadness*. Another difference is that *sadness* implies bad feelings, when *grust'* implies just *some* feelings, not necessarily bad. "Unlike *sadness*, *pečal* has to have a definite cause, it has to imply a negative evaluation of something, as well as a 'bad feeling', and it has to extend in time" [13, pp. 15–16].

In addition, Wierzbicka analyzes two Russian words that are close in meaning to English word *anger*: the noun *gnev* and the verb *serdit'sja*. *Anger* and *gnev* are represented semantically as follows:

anger

(a) X thought something like this about someone:
(b) this person did something bad
(c) I don't want this
(d) I want to do something to this person because of this
(e) because of this, X felt something bad
(f) like people feel when they think something like this.

gnev

(a) X thought something like this about someone:
(b) this person did something bad
(c) I don't want this
(d) if someone does something like this it is bad
(e) I want to do something to this person because of this
(f) because of this, X felt something bad
(g) like people feel when they think something like this [13, p. 20].

The main difference between these concepts is that *gnev* implies ethical evaluation (d), which is not necessary for *anger*. In this respest, *gnev* is closer "to the archaic English word wrath" [13, p. 19], but, unlike it, is more frequent and ordinary in meaning.

Unfortunately, Wierzbicka does not seem to know Russian cognate nouns *zloba* and *zlost'*, which are perhaps very close in meaning to *anger*. She writes that Russian "has no noun corresponding semantically to anger" and therefore she analyzes the verb *serdit'sja* [13, p. 22]. However, the reason for not discussing *zloba* and *zlost'* could be their inner form which has connotations of *bad, evil, malice*. From this point of view, persons who feel *zloba, zlost'* could be negatively evaluated themselves, which in some contexts makes the meaning of these concepts very different from *anger*.

The difference between *serdit'sja* and *anger* is not only that they are different parts of speech. Unlike the meanings of *anger* and *angry*, the meaning of *serdit'sja* implies a more active stance and overt behavioural manifestations of one's feelings:

X serditsja.

(a) X thinks something like this about someone:
(b) this someone did something bad
(c) I don't want this
(d) I want to do something to this person because of this
(e) because of this, X feels something bad
(f) like people do when they feel something like this
(g) X thinks this for some time
(h) because of this, X feels like this for some time
(i) other people can know about this
(j) because X is doing something because of this
(k) like people do when they feel something like this [13, p. 24].

4   IT Applications Related to the Theory of Basic Emotions

The theory of basic emotions is quite popular among information technology (IT) professionals. This is one reason why language-specific emotional concepts analyzed by multilingual computer applications are sometimes interpreted in terms of the putative basic emotions described by Ekman and his colleagues in English. Let us consider two cases where the list of basic emotions is explicitly used in IT applications.

The first case is a project of developing of a multilingual WordNet-Affect. WordNet-Affect is the linguistic resource designed for lexical representation of affective knowledge and created on the basis of Princeton WordNet "through the selection and labeling of the synsets representing affective concepts" [30, p. 1083]. Victoria Bobicev and her colleagues organized a part of WordNet-Affect into six categories, corresponding to the supposed six basic emotions, and translated it into Russian and Romanian. The categories were taken from Ekman's work [31].

As the result, six text files were obtained containing English-language synsets denoting emotions, correlated with Russian and Romanian synsets. There are currently two free downloadable versions available from the Human Language Engineering Laboratory at the Technical University of Moldova. Version 2 consists of 116 synsets for anger, 17 – for disgust, 76 – for fear, 206 – for joy, 96 – for sadness, and 26 – for surprise. Speaking of the Russian concepts discussed above, *toska* and its cognates are included in 4 synsets, *grust'* and its cognates – in 15 synsets, *pečal* and its cognates – in 37 synsets, *gnev* and its cognates – in 17 synsets, *serdit'sja* and its cognates – in 7 synsets. For example, all three concepts *toska, grust',* and *pečal* are elements of a Russian synset corresponding to the English synset < a feeling of thoughtful sadness, melancholy >, both *grust'* and *pečal* are included in a synset corresponding to the synset < emotions experienced when not in a state of well-being, sadness, unhappiness >, the adverbs *gnevno* and *serdito* are parts of a synset corresponding to the synset < with anger, angrily >, etc. [32].

In the multilingual WordNet-Affect, the translation is made in terms of synsets, but synsets are interpreted as special cases of basic emotions, which brings to the semantics something that was not there before, and is not inherent in Princeton WordNet itself. For example, *sadness* is represented in Princeton WordNet as a component of three different synsets:

S: (n) sadness, unhappiness (emotions experienced when not in a state of well-being)
S: (n) sadness, sorrow, sorrowfulness (the state of being sad) "she tired of his perpetual sadness"
S: (n) gloominess, lugubriousness, sadness (the quality of excessive mournfulness and uncheerfulness) [33].

We see that in Princeton WordNet there is no assumption about *sadness* as an umbrella term for the emotion family, but using the multilingual WordNet-Affect for NLP can lead to such an impression. The same can be said for other terms for putative universal basic emotions.

The second case is The Russian Acted Multimodal Affective Set (RAMAS) that was created to use in a Neurodata Lab LLC service for recognizing emotions and heart rate from video and audio. RAMAS consists of multimodal (video, audio, motion and physiology data) recordings, that were made with the help of semi-professional Russian actors. They were given Russian scenarios implying "the presence of one of the six basic emotions (Anger, Sadness, Disgust, Happiness, Fear, and Surprise) in each dialogue or the neutral state" [34, p. 503].

Neither the problem of the possible specifics of Russian emotional situations, nor even the problem of choosing the correct Russian analogues for English emotional categories, was discussed in the article by Perepelkina et al. They don't even say what Russian words they used to translate the names of the English basic emotions. However, this information can be found in other sources. In particular, *sadness* has been translated as *grust'* and *anger* as *zlost'* [35]. As this dataset is used in practical applications for emotion recognition, it reinforces the underlying idea that expression of emotions does not depend on language and culture.

5   Russian Emotional Concepts in RuWordNet

RuWordNet is a part of the Open Multilingual WordNet project, the goal of which is "to link together the existing wordnets created for different languages with an open license" [36, p. 64]. RuWordNet is based on another Russian thesaurus, RuThes, which can be used in most NLP applications, but is not fully compatible with WordNet due to a different structure [37]. RuThes is a bilingual linguistic ontology in which many concepts have names in Russian and English, associated with both Russian and English text entries. Accordingly, one method of linking RuWordNet to WordNet was to semi-automatically match these entries with corresponding WordNet units. Another method was to manually translate those core WordNet concepts that were missing from the English part of RuThes

[36]. As the result, every Russian word that can be found in RuWordNet is represented as a component of synset linked to the corresponding Princeton WordNet English synset or concept.

The question is whether the developers of RuWordNet managed to preserve the semantic specificity of Russian emotional concepts. Let us see what English concepts are associated in RuWordNet with those Russian concepts, the meanings of which were described in Sect. 3.

In RuWordNet, both *pečal* and *grust'* are elements of a Russian synset linked to the English synset <mournfulness, sorrowfulness, ruthfulness> defined as "a state of gloomy sorrow". This Russian synset is designated as a hypernym to another Russian synset that includes *toska* as one of its elements and is related to the English synset <blues, blue_devils, megrims, vapors, vapours> defined as "a state of depression" [38]. However, the Princeton WordNet's hyponyms to <mournfulness, sorrowfulness, ruthfulness> are <woefulness, woe> defined as "intense mournfulness" and <plaintiveness> defined as "expressing sorrowfulness". The Princeton WordNet's hypernym to <blues, blue_devils, megrims, vapors, vapours> is <depression> defined as "a mental state characterized by a pessimistic sense of inadequacy and a despondent lack of activity" [33].

In RuWordNet, there is no hyponymic/hypernymic relationship between *zloba* and *zlost'* on the one hand, and *gnev* on the other. *Gnev* is an element of a Russian synset linked to the English synset <anger, choler, ire> defined as "a strong emotion; a feeling that is oriented toward some real or supposed grievance". *Zloba* and *zlost'* are elements of a synset linked to the synset <malice, maliciousness, spite, spitefulness, venom> defined as "feeling a need to see others suffer". Unfotunatly, *serdit'sja* is an element of a Russian synset which has no link to Princeton WordNet [38].

These examples show that if the translation of emotional synsets is carried out without explicit reference to the supposed basic emotion families, the specificity of Russian concepts is retained. Of course, in this case, it is expressed differently than with the help of NSM tools. The specificity of individual concepts is expressed not through its explicit NSM formulation, but through connotations conveyed by synonyms included in synsets. The synset as a whole has a specific meaning, and this meaning is encoded in a distributed way by including certain concepts in the synset, as well as through the connection between synsets that include the same words and their cognates. This way of expressing the specifics of language categories can be considered as an additional to the NSM. In a sense, it is closer to how the specific meaning of individual words is felt by ordinary native speakers who are not specialists in semantics. Ordinary native speakers do not think about the meaning of words explicitly, intuitively choosing those words that are most appropriate in a given situation. If a person is asked what he meant when he used a certain word, he is very likely to explain it using synonyms of this word that are relevant in this situation. And since this approach is clearly dominant among representatives of the IT community, most of whom are also not specialists in semantics, perhaps the future lies with it.

6 Conclusions

There are at least three options for how language-specific emotional concepts can be embedded in a multilingual environment.

The first way is to ignore the specifics of these concepts. The uncritical use in computer applications of universalist ideas associated with the theory of basic emotions creates the illusion that these emotions are indeed universal. That is why, in the multilingual technological environment, the theory of basic emotions can act as a self-fulfilling prophecy. If this scenario dominates in the future, then the global multilingual culture will become semantically more primitive compared to the diversity of cultures that exists today.

The second way is to express language-specific meanings using NSM explications and minimal languages. This approach seems promising from the point of view of preserving the semantic specificity and richness of individual cultures, but requires special mastery of the appropriate methodology, which is unlikely for those who are not specialists in linguistic semantics.

The third way is to convey specific meanings by translating whole synsets and revealing the relationships between them. This distributed encoding of specific meanings seems more similar to how meanings are expressed by natural language speakers in everyday life, and more natural for specialists in the field of multilingual IT technologies.

The second and third ways can complement each other. In addition, both of them are incompatible with the theory of basic emotions, but quite consistent with the Conceptual Act Theory of Emotion, which is more promising from a methodological point of view. If these methods prevail in the future, then the global multilingual environment will become semantically richer, because it will be able to absorb the semantic specifics of individual cultures.

References

1. Von Humboldt, W.: On Language: On the Diversity of Human Language-Structure and its Influence on the Mental Development of Mankind, trans. Cambridge University Press, Cambridge UK (1988). (Translated by Peter Heath)
2. Sapir, E.: The status of linguistics as a science. In: David Mandelbaum, D. (ed.) Selected Writings of Edward Sapir in Language Culture and Personality, pp. 160–166. University of California Press, Berkeley (1949)
3. Whorf, B.L.: Language, Thought, and Reality. The MIT Press, Cambridge (1956). (Selected Writings of Benjamin Lee Whorf)
4. Malotki, E.: Hopi Time: A Linguistic Analysis of the Temporal Concepts in the Hopi Language. De Gruyter Mouton, Berlin (2011). https://doi.org/10.1515/9783110822816
5. Chomsky, N.: On certain formal properties of grammars. Inf. Contr. 2(2), 137–167 (1959). https://doi.org/10.1016/S0019-9958(59)90362-6
6. Berwick, R.C., Chomsky, N.: Why Only Us: Language and Evolution. The MIT Press, Cambridge (2016). https://doi.org/10.7551/mitpress/10684.001.0001
7. Fodor, J.A.: The Language of Thought. Thomas Y. Crowell Company, New York (1975)
8. Pinker, S.: The Language Instinct: The New Science of Language and Mind. William Morrow, New York (1994)

9. Levinson, S.C.: Space in Language and Cognition: Explorations in Cognitive Diversity. Cambridge University Press, Cambridge (2003). https://doi.org/10.1017/CBO978051161 3609

10. Wierzbicka, A.: "Semantic Primitives", fifty years later. Russ. J. Ling. **25**(2), 317–342 (2021). https://doi.org/10.22363/2687-0088-2021-25-2-317-342

11. Goddard, Cliff (ed.): Minimal English for a Global World. Springer, Cham (2018). https://doi.org/10.1007/978-3-319-62512-6

12. Goddard, C. (ed.): Minimal Languages in Action. Springer, Cham (2021). https://doi.org/10.1007/978-3-030-64077-4

13. Wierzbicka, A.: "Sadness" and "Anger" in Russian: the non-universality of the so called 'basic human emotions'. In: Athanasiadou, A., Tabakowska, E. (eds.) Speaking of Emotions: Conceptualisation and expression, pp. 3–28. Mouton de Gru, Berlin (1998). https://doi.org/10.1515/9783110806007.3

14. Wierzbicka, A.: Emotions Across Languages and Cultures: Diversity and Universals. Cambridge University Press, New-York, Editions de la Maison des Sciences de l'Homme, Paris (1999). https://doi.org/10.1017/CBO9780511521256

15. Wierzbicka, A.: Semantics, Culture, and Cognition: Universal Human Concepts in Culture-Specific Configurations. Oxford University Press, New-York (1992). https://doi.org/10.1017/S0272263100013607

16. Darwin, C.: The Expression of the Emotions in Man and Animals. John Murray, London (1872). https://pure.mpg.de/rest/items/item_2309885_4/component/file_2309884/content. Accessed 15 March 2022

17. Tomkins, S.S.: Affect Imagery Consciousness: The Complete Edition. Springer, New York (2008)

18. Izard, C.I.: Human Emotions. Springer Science + Business Media, New York (1977)

19. Ekman, P., Friesen, W.V.: A new pan-cultural facial expression of emotion. Motiv. Emot. **10**, 159–168 (1986). https://doi.org/10.1007/BF00992253

20. Ekman, P., Friesen, W.V., Tomkins, S.S.: Facial affect scoring technique: a first validity study. Semiotics, **3**, 37–58 (1971). https://1ammce38pkj41n8xkp1iocwe-wpengine.netdna-ssl.com/wp-content/uploads/2013/07/Facial-Affect-Scoring-Technique-A-First-Validity-Study.pdf. Accessed 20 March 2022

21. Ekman, P., Friesen, W.V.: Head and Body Cues in the Judgement of Emotion: A Reformulation. Perc. Mot. Skills, **24**, 711–724 (1967). https://1ammce38pkj41n8xkp1iocwe-wpengine.netdna-ssl.com/wp-content/uploads/2013/07/Head-And-Body-Cues-In-The-Judgement-Of-Emotion-A-Reformulat.pdf. Accessed 20 March 2022

22. Ekman, P.: An argument for basic emotions. Cogn. Emot. **6**(3/4), 169–200 (1992). https://1ammce38pkj41n8xkp1iocwe-wpengine.netdna-ssl.com/wp-content/uploads/2013/07/An-Argument-For-Basic-Emotions.pdf. Accessed 20 March 2022

23. Ekman, P.: Universal facial expressions of emotions. Cal. Ment. H. Res. Dig., **8**(4) 151–158 (1970). https://1ammce38pkj41n8xkp1iocwe-wpengine.netdna-ssl.com/wp-content/uploads/2013/07/Universal-Facial-Expressions-of-Emotions1.pdf

24. Ekman, P., Friesen, W.V., Hager, J.C.: Facial Action Coding System. The Manual on CD ROM. Research Nexus division of Network Information Research Corporation, Salt Lake City UT (2002)

25. Ekman, P., Friesen, W.V.: Unmasking the Face: A Guide to Recognizing Emotions From Facial Expressions. Malor Books, Cambrige (2003)

26. Feldman Barrett, L., Russell, J.A.: Independence and bipolarity in the structure of current affect. J. of Pers. Soc. Psych. **74**(4), 967–984 (1998). https://doi.org/10.1037/0022-3514.74.4.967

27. Russell, J.A., Feldman Barrett, L.: Core affect, prototypical emotional episodes, and other things called *emotion*: Dissecting the elephant. J. Pers. Soc. Psych. **76**, 805–819 (1999). https://doi.org/10.1037/0022-3514.76.5.805

28. Russell, J.A.: Emotion, core affect, and psychological construction. Cogn. Emot. **23**(7), 1259–1283 (2009). https://doi.org/10.1080/02699930902809375

29. Wilson-Mendenhall, C.D., Barrett, L.F., Simmons, W.K., Barsalou, L.W.: Grounding emotion in situated conceptualization. Neuropsychologia **49**(5), 1105–1127 (2011). https://doi.org/10.1016/j.neuropsychologia.2010.12.032

30. Strapparava, C., Valitutti, A.: Wordnet-affect: an affective extension of wordnet. In: 4th International Conference on Language Resources and Evaluation, pp. 1083–1086 (2004) http://www.lrec-conf.org/proceedings/lrec2004/pdf/369.pdf. Accessed 12 April 2022

31. Bobicev, V., Maxim, V., Prodan, T., Burciu, N., Angheluş, V.: Emotions in Words: Developing a Multilingual WordNet-Affect. In: Gelbukh, A. (ed.) CICLing 2010. LNCS, vol. 6008, pp. 375–384. Springer, Heidelberg (2010). https://doi.org/10.1007/978-3-642-12116-6_31

32. Laboratorul de Inginerie a Limbajului Uman. http://lilu.fcim.utm.md/resourcesRoRuWNA.html. Accessed 15 April 2022

33. WordNet: A Lexical Database for English. Princeton University (2010). http://wordnetweb.princeton.edu/perl/webwn?s=&sub=Search+WordNet&o2=&o0=1&o8=1&o1=1&o7=&o5=&o9=&o6=&o3=&o4=&h=000. Accessed 15 April 2022

34. Perepelkina, O., Kazimirova, E., Konstantinova, M.: RAMAS: Russian Multimodal Corpus of Dyadic Interaction for Affective Computing. In: Karpov, A., Jokisch, O., Potapova, R. (eds.) SPECOM 2018. LNCS (LNAI), vol. 11096, pp. 501–510. Springer, Cham (2018). https://doi.org/10.1007/978-3-319-99579-3_52

35. Deineka, D.: It assesses heart rate and breathing from video to understand the client's emotions: the history of the Neurodata Lab's Russian project (2019). https://vc.ru/services/93624-ocenivaet-puls-i-dyhanie-po-video-chtoby-ponyat-emocii-klienta-istoriya-rossiyskoy-razrabotki-neurodata-lab. Accessed 17 April 2022 (in Russian)

36. Loukachevitch N., Gerasimova, A.: Linking Russian Wordnet RuWordNet to WordNet. In: Fellbaum, C., Vossen, P., Rudnicka, E., Maziarz, M., Piasecki, M. (eds.) Proceedings of the 10th Global Wordnet Conference, pp. 64–71. CLARIN-PL digital repository. Global Wordnet Association, Wroclaw (2019). http://hdl.handle.net/11321/718. Accessed 10 April 2022

37. Loukachevitch, N., Dobrov, B.: RuThes linguistic ontology vs. Russian Wordnets. In: Proceedings of Global WordNet Conference GWC-2014, Tartu (2014). https://aclanthology.org/W14-0121.pdf. Accessed 10 April 2022

38. Thesaurus of Russian language RuWordNet. https://ruwordnet.ru/en/search/. Accessed 12 April 2022

Prospects for the Synthesis of the Language of Art, Science and Technology in Russian Science Art

Ivan Aladyshkin$^{(\boxtimes)}$ ⓘ, Natalia Anosova ⓘ, and Olga Noskova ⓘ

Peter the Great St. Petersburg Polytechnic University,
Polytechnicheskaya, 29, St. Petersburg 195251, Russia
aladyshkin_iv@spbstu.ru

Abstract. The paper discusses the problems of integration of science and high-tech art in the interdisciplinary field of science art. It focuses on the changing position of this area of contemporary art in Russia. The authors analyze the innovative pathos of science art and the role of scientific and technical tools used in science art. The idea of science art is based on the development trends in science and art at the turn of the 19th and 20th centuries. Science art is considered to be a consequence rather than a harbinger of the synthesis of the language of art, science and technology, which became apparent in the last century. The growing interest in the phenomenon of science art is mainly due to the specifics of the art market and the demands of the state funds, scientific and educational organizations, as well as the administrative sector that manages them. The crisis of confidence in the new areas of contemporary art, science art especially, activates the search for support in science as one of the most authoritative and influential institutions of modern society. The authors come to the conclusion that science art can be considered to be one of the terms that is used in an effort to combine a number of already existing areas of contemporary art under the same scientific concept.

Keywords: Science art · High-tech art · Transdisciplinarity

1 Introduction

Today, a interdisciplinary area called science art is at the forefront of artistic practices, as it combines the key trends in contemporary art in recent decades. This area is often discussed in the light of overcoming the boundaries between science and art in search for the new forms of artistic expression by means of advanced technologies. The discussions around science art, which started in the international art space back in the 90s of the last century, also came to Russia, which is still lagging behind in the new trajectories of contemporary art, albeit with some delay. This delay was compensated by the intensity of development in the 21st century of science art, the term that was derived from the English language.

In 2012, S.V. Erokhin stated that, despite the intensive development of scientific art, this area remained practically terra incognita for Russian art history and aesthetics,

D. Bylieva and A. Nordmann (Eds.): PCSF 2022, LNNS 636, pp. 135–145, 2023.
https://doi.org/10.1007/978-3-031-26783-3_13

especially for the philosophy of science [1]. Today, the situation has changed essentially. In addition to a number of conferences and collections of materials on the subject, several monographs, more than two dozen scientific papers and dissertations have been published over the past decade. Interest is growing not only in contemporary art centers, but also in scientific and educational institutions, which started educational programs directly or indirectly related to scientific art. The works of S.V. Erokhin, including his monograph "Theory and Practice of Science Art", together with the publications of D. Bulatov and O.E. Levchenko, are in the list of the most important [1–4]. In addition to works on the general issues of science art, there is a growing number of studies dedicated to its internal problems, for example, the analysis of the perception of science art or quality criteria for the works of science art, as well as presentation of experience in preparing scientific and art expositions [5, 6], etc.

When discussing the general problems of the development of high-tech art and new art practices in Russia, it is worth mentioning that, despite the persistence of an ambivalent attitude towards science art, the Russian authors express positive attitude towards new, science based forms of art and dwell on the ways of renovation of scientific knowledge [4, 6, 7]. The main prospect is the possibility of integrating discursive reasoning and intuitive thinking, i.e. two different types of worldview that are still associated with science and art, and their merging implies an appropriate level of technology [8–11].

The purpose of the study is to analyze the prospects for the synthesis of science and art within the framework of science art - one of the main areas in the development of modern artistic practices. The goal determines the tasks of the study:

- to identify the conceptual foundations of science art and its innovative components;
- to analyze the role of scientific and technical materials in science art;
- to study the basic range of value judgments in relation to science art and the tasks declared by its representatives for the renovation of art practices;
- to determine the consistency of the prospects of science art in the light of key development trends, both in the field of contemporary art and scientific knowledge.

2   Words and Things

The growing number of publications devoted to science has not made it quite clear what is referred to as science art. On the contrary, you can get the impression that the vagueness of the phenomenon and the blurring of its boundaries are only intensified. Moreover, the concepts of science art are based on a wide range of similar forms of contemporary art, which makes it reasonable to highlight them in the discussion of art practices [3, 4, 12, 13]. Most often, the definition of the term is replaced by rather abstract maxims about a new interdisciplinary field of synthesis of science and art based on advanced technologies and scientific means [14]. The formulations that reduce the meaning of scientific art to the concept of certain artistic images using modern technologies and scientific developments are no less abstract [15]. However, the term is catching on and the blurred interpretations of the actual relationship between science and art only contribute to its relevance and wider scope of its use.

The pathos of scientific and technical value inherent in scientific art is obvious and discussions about this area of art focus on the development of some synthetic type of thinking of scientists and artists [10]. This requires the description of individual projects, art-objects, programs and exhibition spaces. The formation of an interdisciplinary platform for free combinations and intersections of art and science is illustrated by examples of joint work of scientists and artists in one laboratory. There are numerous examples of such joint work, as there are many opportunities for such cooperation and the forms of their implementation can vary. The Art@Cern project of the European Center for Nuclear Research (CERN) involves a competition of artworks focused on the problems of modern elementary particle physics, the winners of which will receive grants that allow them to work together with the scientists of the center. The experience is indicative of the functioning of the Art Science Labs network, the Symbiotic art and science research laboratory at the School of Human Anatomy and Biology of the University of Western Australia, as well as in other international and national institutions [1, 16]. Description and analysis of specific artistic practices, which are portrayed as harbingers of a synthetic science and art language, often serve as arguments for the prospects and possibilities of its development. With regard to science art, it is easier to define what it implies. Interesting is the fact that in the publications of both international and Russian authors, a projective representation prevails with the analysis of an imaginary phenomenon. The authors, with enthusiasm, point at far-reaching prospects for the renovation of art and science, forms of the joint work of artists and scientists and integration of different types of worldview based on a technological platform adequate to the current realities.

So what is expected from this new union of science and art? If we step back from the predicted birth of a fundamentally new culture and aesthetics, i.e. unfounded forecasts and experiences, which Russian researchers turn to when considering scientific and technical realities, we understand how interconnected the concepts of art and science are. On the one hand, we discuss the adaptation of scientific methods for the implementation of artistic ideas, and on the other hand, we speak about the use of art methods to form new scientific theories. It is assumed that the use of intuitive and aesthetic methods will noticeably strengthen the axiological parameters of modern science and open up new prospects for scientific analysis and comprehension of its results. In the field of art, scientific methods will allow reaching a different level of implementation of complex aesthetic tasks, developing the research potential of artistic practices, while assigning them a scientifically based innovative character. In addition, with the involvement of scientific methods and technologies, comes the transformation of ideas about the works of art, its materiality, its spatial and temporal dimensions, and its involvement in the formation of the viewer's aesthetic experience.

3 Origins

The prospects are clear, but not new, as the term science art and the trends in contemporary art took shape in the last millennium. In fact, all the prospects for the synthesis of science and art mentioned above sound familiar, let alone the associations with science art. It is no wonder that at science art conferences many reports are devoted to the historical examples of the interaction between science and art. Historians of culture and art give a variety of examples, from the specifics of the understanding of art by the ancient Greeks and the features of medieval science to the conclusion that the artists of the Renaissance already associated aesthetics with science [17]. It is hardly worth delving into discussions about scientific art in terms of the syncretism characteristic of antiquity, which unites science, craft and art in the mysterious word "techne". It is well-known that both in the Middle Ages and in modern times, the interpenetration of science and art was sporadic and vague.

Cardinal changes in the ratio of scientific and artistic spaces took place later, at the turn of the 19th–20th centuries, with the appearance of modernism [18]. Then the interconnection of art, science and technology was facilitated by the formation of fundamentally new branches of art - photography and film art. Avant-garde experiments in the first half of the 20th century (Futurism, Dadaism, Bauhaus and Russian constructivism, kineticism) introduced significant scientific and technological elements to all areas of art [19]. In the 1920s, there was also an obvious institutionalization of interdisciplinary areas of science, art and technology. Suffice it to recall the activities of the State Academy of Artistic Sciences or the Bauhaus. The ideas of the synthesis of science and art also received terminological meanings close to science art. For example, in 1939, the composer Edgard Varèse in his lecture spoke of his music as "arsscientia" or "science art" [20]. Then there were the pioneers of cybernetic art, media art and op art, which strengthened the connection between science art and artistic practice. Throughout the last century, the integration of art, science and technology was steadily growing, radically changing the genre structure of the artistic space. By the end of the 20th century, media art, virtual reality, algorithmic painting, robotic sculpture and interactive installations, bio-art and transgenic art became the basic components of contemporary art.

Over the 20th century, the artist acquired the right to use any concepts and means, and any phenomena in his artistic expression. Therefore, while the specifics of the science art of the 20th century was determined by the use of scientific ideas, means and technologies, or physical, chemical, biological, electrical, electromagnetic and any other phenomena to create an artistic image, the lack of original content of art practices was especially obvious. Declarations of overcoming the spatial and temporal restrictions or the involvement of the viewer turning into a co-creator, co-interpreter, co-participant in the creation of art objects, were also represented in various areas of art in the last century. Kinetic art, performance art, happenings, fluxus and video art, conceptual and installation art, interactive cybernetic sculpture and digital art - those areas of art, to a certain extent, stood apart from the spatial and temporal localization of art objects and engaged the viewers in the artistic practices.

Along with the integration of art, science and technology, the change in the basic characteristics of the object and subject of art practices, including the desire to eliminate clear boundaries between them, was one of the most important trends in the cultural dynamics of the entire twentieth century. Largely due to the development of scientific ideas and technologies, the imperative of the procedural nature of art was realized, the measurement of real time became one of the main elements of aesthetic experience, and the viewer turned into an active co-participant of artistic practices. Representatives of science art keep on reproducing modernist aesthetics in their aspirations to review ideas about works of art and the aesthetic experience of the viewer and to expand the spatial and temporal dimensions of art practices. Scientific ideas, concepts and technologies are implemented in art practices, but the concepts of art do not change, which appeals to the principles avant-garde, developed almost a century ago, rather than to postmodern aesthetics. Novelty, scientific character, expansion of the boundaries of artistic expression using the means of non-artistic practices, pathos research, manufacturability, i.e. everything related to science art are the precepts of the avant-garde era. The postulates of the aesthetic platform, which took shape in the first quarter of the last century, are only supplemented by the post-war art practices with the principles of openness, interactivity and dematerialization of art objects. Ultimately, the idea of science art rests on the dilapidated foundation of the avant-garde of a century ago, and there is no conceptual novelty in it.

4 Declarative Synthesis

At the same time, the adherence of the apologists of science art to the principles of the modernist paradigm and the precepts of the avant-garde in any case implies an appeal to novelty, which serves as a justification for the emergence of new areas of art space. But what can science art offer? There is little new in the integration of science and art, and then the point is to reach the integration at a higher level than just the mutual penetration and common use of scientific methods/techniques in art. Science art is proposed as the interdisciplinary area within which the conditions are formed for a fruitful synthesis of discursive thinking and intuitive judgment. This is the space that provides not only the most favorable conditions for the development of the culture of cooperation, but also for the development of a synthetic type of thinking of scientists and artists [9–11]. It all makes sense, even if these prospects are far from being articulated for the first time and are devoid of the charm of novelty. However, the formulations and especially the argumentation of the need for the desired synthesis appeal to the realities of the past rather than the future of science and art.

In most cases, the synthesis of art and science within the framework of science art refers to a rather implicit and an extremely simplified scheme for distinguishing between types of perception with the opposition of its rational and emotional-intuitive invariants. In accordance with the precepts of classical modern European science, a scientist thinks rationally, according to the laws of logic and mathematics and acquired objective knowledge. Yet, an artist, who is guided by intuition and emotional experience, seeks to create a work of art that is far from objectivity, accuracy, consistency, or other criteria of scientific knowledge. This split between art and science is obvious only in relation to the European New Age, when the split was supported by representatives of both camps in every possible way. For a rather short period, the idea of science was really reduced to some kind of extremely rational, pragmatic and practice-oriented activity. Impartial, objective and exact science was opposed to subjective, emotional, far from reality art, which was supposed to be rather irrational and spiritual. The new European utilitarian ideology consolidated the demarcation of science and art, which led to the alienation of artistic activity from socio-economic realities, and scientific activity turned into a servant of progress.

Only in the 21st century, science and art have undergone such transformations that the reference points for the former distinctions (rational/intuitive knowledge, objective/subjective knowledge, etc.) are fairly overdue and do not play a significant role in the demarcation of scientific and artistic practices. To overcome the marked split between science and art today, to focus on the convergence of intuitive and rational perception within one of the leading areas of contemporary art, is not relevant anymore. In the rhetoric of erasing the boundaries of artistic creativity and scientific research, along with the real changes in art practices, there is a play on words and meanings, which in the modern discourse have finally lost their clear outlines. Today it is much easier to postulate a synthesis of science and art than to draw clear lines between them, as well as to answer a simple question - what is the difference between the activities of a modern artist and a scientist. There is an outright demagoguery in the statements about merging of the artist and the scientist, about turning the artist's studio into a laboratory. The statements about the discovery of open spaces for intuitive synthetic judgments in art or the development of a synthetic language of science and art sound especially superficial.

The theme of interdisciplinary expansion of the established traditional boundaries has long been popular both in science and art. The second half of the last century saw an unprecedented expansion of the disciplinary boundaries of professional science and a noticeable lenience of its standards. As a result, science successfully reconciled previously irreconcilable positions and was amazingly perceptive to other forms of cognitive activity traditionally considered non-scientific. Art, in the postmodern paradigm, has erased any restrictions on artistic expression. Thanks to the radical aesthetic innovations of the 20th century, art has been based on conceptual borrowings from the scientific sphere for many decades, likening the game of artistic interpretation to scientific analysis. Indeed, in contemporary art, the constructive type of the creator is dominant, while in science, intuitive insight leading to a specific result is valued.

Modern scientific knowledge is distinguished by the obvious interpenetration and unification into a single whole of the most diverse principles of cognition and representation of the world with the loss of purity of scientific language. Inevitably, the obvious and marked boundaries between the language of science and technology on the one hand, and the language of art on the other, were erased [21–23]. The expansion, weakening, and often disappearance of clear boundaries between science, technology and art only accelerated the information explosion, which deepened the integration processes of the areas under consideration.

Art and science became sources of mutual inspiration long ago. While art systematically expanded its boundaries, turning a scientific experiment into an artistic process, science increasingly longed for the artistic and creative side of the cognitive process. There are many circumstances and grounds for the convergence of scientific and artistic forms of cognition, but in the context of contemporary art, several multidirectional trends stand out: the strengthened role of the creative component in the cognitive act and, at the same time, an increased level of its abstractness and algorithmization. In science and art, productive creativity is increasingly valued, i.e. a creative concept and the ability to implement it without rigorous verification by the classical scientific criteria or standard set of requirements for art objects.

For modern science, which has recognized the hypothetical nature of its results and the unfortunate conventionality of any theoretical constructions being nothing more than exemplary models, the principles of efficiency and functionality have long overshadowed the charm of plausibility and impartiality, truth and objectivity. Art and science have become closer in their relationship with reality, or, more precisely, in their abstraction from reality, operating with symbols and models in its description, dependent on human activity and cognitive activity. Modern science and contemporary art have also converged in unfolding, revealing their models of comprehension/description of reality with the complicity of the observer/viewer, demonstrating the fundamental incompleteness of both scientific knowledge and works of art.

It is difficult to determine when and where exactly scientific discourse became one of the aesthetic platforms, and it is hardly justifiable trying to connect it with any one area of contemporary art [11]. There is no doubt that the growing mutual integration of science and art was taking place long before the articulation of science art. It is undoubtedly happening today to an incomparably larger scale than any particular trend in contemporary art. There are innumerable lines of convergence between science and art in modern realities, and science art only fits into the trends that already exist.

In any case, the design of science art can be described as a consequence of the blurred boundaries between science and art, the inclusion into science of the non-logical, intuitive synthetic components and, accordingly, the development of the means of scientific knowledge in the field of artistic practices. In other words, science art has nothing to overcome, and there is no gap between science and art, much less the opposition of the delimited spheres of intellectual and intuitive representation of the world. Today, it is justifiable to speak only about the residual forms of the once effective demarcation of art and science, the well-known inertia of traditional adherence to long-obsolete distinctions. Science art only captures the final stages of erasing the old boundaries and shifting discussions to the field of artistic practices and the sphere of entertainment. In this sense, the test with the telling title "This is (not) art" [24] is indicative, which was a kind of advertisement organized by the Laboratoria Art & Science Foundation with the support of Kaspersky Lab for the exhibition "Let something else live in me". The exhibition, dedicated to human interaction with non-human agents such as animals, bacteria, plants and technologies, was held in the New Tretyakov Gallery from June 22 to October 10, 2021. In this test, anyone who wanted to try to distinguish museum installations from scientific experiments was invited. The photograph of the described action was attached not to the text of the question, but to the answer. It was done because the description did not give unequivocal grounds to attribute this or that action to an artistic statement or scientific experience, and it becomes much easier to answer correctly using a photograph, since a lot depends on the scene of the action. This again confirms the avant-garde principles rather than a new level of fusion of science and art or the development of a certain scientific and artistic language - everything can be art, any object, process or phenomenon can become an artistic object, if there is a will of the author, or in the context of the exhibition or museum. With science art, the situation is generally the same, and today any scientific ideas and means, any technologies and any experiments carried out with their help, can be represented as contemporary art. As a rule, science art is associated with nature and exact sciences, but there can be situations when Doctor of History writes another serious scientific article in the space of an art gallery, in front of the viewers. The process of writing a scientific article using historical research methods appears as contemporary art, but only within the framework of the exhibition area. If this situation may seem unreal, there is another one, much closer to reality. Let us imagine video recording and broadcasting of laboratory research on screens located in the gallery, using video conferencing technologies, with scientists, artists and viewers, exchanging their opinions. Is it possible to draw conclusions about the synthesis of science and art on this basis? Yes and no.

At present, when the relativity of any demarcations of various areas of human activity is generally recognized, the distinction depends on formal and functional parameters. It is not the means, content or format of the statement that are decisive, but the certified professional training, affiliation with certain institutions, the scope of implementation and application, which, ultimately, determines which area, art or science, this statement should be attributed to. Based on the above examples, everything that happens in an art gallery can justifiably be attributed to art, but work in the laboratory will remain the part of scientific research. This distinction is extremely arbitrary, because it is possible to give examples of cooperation between scientists and artists, the results of which are significant both in scientific and artistic spaces.

5   Art is Art?

Yet, speaking of science art, it must be remembered that initially we are talking about artistic search and experiments, about a new aesthetics, which is closely connected with the philosophy of contemporary art. In the reasoning about an interdisciplinary area of mutual penetration of science and art, there is always some contradiction because it covers not the area of scientific research, but artistic practices with a certain set of tools from modern scientific ideas and technologies, which thematically and conceptually may or may not be related to the field of scientific research. In addition, in most cases, science art remains a rather superficial synthesis of science and art, a kind of play of external forms, a simulation of scientific processes and general scientific imagery. Such contradiction should not be justified by the presentation of art as research, which is increasingly common and opposed by the supporters of the slogan "art is art". In most cases, even if the artistic statement is scientific in its means and partly adjoins scientific knowledge, its cognitive function is secondary and noticeably inferior to the aesthetic and hedonistic goals of artistic expression. Such conclusions come to mind when analyzing almost any modern art practice that involves the use of the latest scientific and technological developments, often in collaboration with researchers.

In the case of science art, no matter how much they talk about a new interdisciplinary field, new artistic practices or new aesthetics, in reality already existing forms and areas of contemporary art are combined. There are many trends and areas in the contemporary art space, and they took shape in the second half of the 20th century. Technological, biological, transgenic, robotic, virtual, information, electronic art and other areas of contemporary art are closely intertwined and it is almost impossible to draw distinction between them. The attempts to combine them were made more than once, and then they received the names of technological art, media art, hybrid art, etc. It is possible that hybrid art, in terms of its content characteristics, guidelines and tasks, is the closest definition of the phenomenon of science art in modern art practices. The representatives of hybrid art operate in the broad areas of art space, trying to overcome the superficial level of using modern technical means or various electronic devices and turning to scientific research.

Scientific imagery can be achieved at all stages of the development of art objects, from research programs, development of research processes and visualization to interpretation of the results. However, it should be noted that representatives of hybrid art, while solving artistic problems, still remain artists operating in the space of contemporary art, only borrowing scientific ideas, solutions and tools.

6 Conclusion

Science art lays claim not so much to the originality of aesthetics as to its relevance. Intellectuality, scientific imagery, innovation and the use of advanced technologies are consistent with the contemporary art, and the interdisciplinary approach is by no means relevant in the scientific field. The relevance of the phenomenon of science art can be explained not by its conceptual foundations that remain extremely vague and devoid of original content, but by the demands of the art market, state funds, scientific and educational organizations, as well as the administrative sector that manages them.

With rare exceptions, the leading areas of contemporary art remain rather localized and limited in their ability to reach a wide audience, remaining largely the field for specialists and researchers. Science art fits well into the aspirations to expand the audience of contemporary art, develop new spaces for solving artistic problems, as well as the search for additional financial and administrative support of contemporary art practices.

The development of new concepts and technologies within the framework of modern aesthetics in the second half of the last century led to growing disillusionment with the possibilities of technical means in the development of artistic practices. Crisis tendencies became obvious at the turn of the century with the growth of public skepticism and criticism of contemporary art, whose representatives, sometimes competing in exposing their own aesthetic attitudes, only contributed to a decrease in the general interest and involvement of the viewers. Moreover, there are still no new constructive and significant concepts that substantiate the forms of modern aesthetics. In fact, the emergence of science art as a new interdisciplinary field was the result of the search not for new expressive means or scientific justification for a whole range of modern art practices, but for their conceptual justification and a kind of scientific alibi.

The weakness of the philosophical and aesthetic justification, the distrust of the viewer activates the search for the support of contemporary art in science as one of the most authoritative institutions of modern society. Although today there is still certain criticism of science, the faith in the omnipotent mind, scientific infallibility, universality and boundlessness of scientific prospects is already in the past. However, the recognition of bias and opportunism in science has not shaken its role and the social trust in it is incomparably higher than in modern art practices. Apparently, there is still a glimmer of hope that scientific research methods and modern technologies and the corresponding categorical apparatus (the language of science) can give credibility to contemporary art and restore not only its relevance, but also the trust of the viewer.

The interest of scientific institutions in the synthesis of art and science, cooperation between artists and scientists has really increased in recent decades, also in Russia. However, science art becomes relevant not only due to the intuitive and aesthetic methods proposed by the scientific community, but in the light of the popularization of science itself. The visualization and artistic interpretation of scientific research are in demand today due to increased public attention to the latest achievements of science. For example, bio-art can be considered in the light of artistic and aesthetic awareness of environmental problems, the use of advanced technologies in art turns out to be the means of their social adaptation, and scientific theories acquire figurative accessibility. Moreover, many areas of modern science are far from visualization, and contemporary art is becoming the means that can bring visual images to extremely abstract theoretical positions. Actually, the main function of the centers of science art that are opening today at leading technical universities and museums is mainly educational. Their main task is to make the "invisible visible" in understanding and mastering rather complex theories and methods of modern science, to draw the attention of young people to scientific problems and certain areas of scientific knowledge.

Ultimately, science art turns out to be just one of the terms used in an effort to single out/combine a number of already existing areas of contemporary art under the denominator of science. The absence of clear boundaries of interpretation and the very definition of the term science art, the variability of interpretations and the diversity of artistic practices that coincide with the idea of science art are not caused by the problems or the ambiguity of theoretical foundations. Science art is devoid of an original conceptual framework or a specific set of art practices, just as the claims declared in this field for a deep synthesis of the language of science, technology and art are devoid of thoroughness. Science art exhibitions prove that some forms of modern integration of science and art are in demand by the wide audience.

Today, it is difficult to say how successful the term science art is in defining the area of mutual penetration of science and art, given the ambiguity and vagueness of its content. It is possible that science art will share the fate of current art, which was understood as ideologically or technically innovative contemporary art with a characteristic mixture of diverse parameters of the aesthetics in the twentieth century. However, current art has remained one of the optional terms in the Russian art space, which is devoid of any specific content.

References

1. Erokhin, S.: Theory and Practice of Scientific Art. Moscow Institute of Energy Security, Moscow (2012)
2. Bulatov, D.: Anthology: evolution of haute couture. Art and science in the era of postbiology, vol. 1–2. Baltic Branch of the Pushkin State Museum of Fine Arts (NCSI), Kaliningrad (2009, 2013)
3. Erokhin, S.: The term "Scientific art" in artistic and scientific discourses. Bull. Cherepovets State Univ. **4–2**, 138–214 (2012)
4. Levchenko, O.: Science-art: problems of terminology. Bull. Russ. State Univ. Philos. Soc. Art Hist. **14**(136), 155–162 (2014)
5. Kazakova, S.: Science art: on the question of the criteria for the quality of a work of modern art. Art Stud. **3–4**, 592–598 (2012)
6. Migunov, A., Erokhin, S., Galkin, D., Gagarin, V.: Scientific art: origins, essence, terminology (based on the materials of the first international scientific and practical conference "Scientific Art", Moscow, Lomonosov Moscow State University, 4–5 April 2012. Bull. Moscow Univ. Ser. 7. Philos. **6**, 96–116 (2012)

7. Dukov, E., Evallier, V.: International scientific conference "Art and Machine civilization." Sci. Telev. **17**(2), 11–32 (2021)

8. Bergantini, L.: Synesthesia in the arts: relations between science, art and technology. In: ARTECH 2019: Proceedings of the 9th International Conference on Digital and Interactive Arts, vol. 17, pp. 225–238. ARS, São Paulo (2019). https://doi.org/10.11606/issn.2178-0447. ars.2019.151267

9. Parusimova, Ya.: On the question of "Scientific art" in modern aesthetic knowledge. Bull. Orenb. State Univ. **11**(186), 244–248 (2015)

10. Shavshina, I.P.: With what artists "eat" science or collaboration as a way to create relevant projects. Creative Modernity **1**(14), 54–58 (2021). https://doi.org/10.37909/978-5-89170-278-3-2021-1009

11. Wilson, S.: Information arts. In: Intersections of Art, Science, and Technology. The MIT Press, Cambridge (2002)

12. Kozhevin, D.: Modern scientific art in Russian artistic practice. Bull. St. Petersburg State Univ. Cult. Art **4–9**, 155–158 (2016)

13. Lorusso, S.: Art and science, art is science. Conserv. Sci. Cult. Herit. **21**, 11–13 (2021). https://doi.org/10.6092/issn.1973-9494/15306

14. Osbourn, A.: Transgressing the disciplines using science as a meeting place: the science, art and writing initiative. In: Stewart, A.J., Mueller, M.P., Tippins, D.J. (eds.) Converting STEM into STEAM Programs. EDSE, vol. 5, pp. 149–167. Springer, Cham (2019). https://doi.org/10.1007/978-3-030-25101-7_11

15. Lvov, A.: Science-art: information technologies in the context of modern aesthetics. Bull. St. Petersburg Univ. Philos. Confl. **31**(3), 59–67 (2015)

16. Gagarin, V., Erokhin, S., Shtepa, V.: International experience of the institutionalization of scientific art. Bull. Tomsk State Univ. **355**, 37–41 (2012)

17. Villaseñor Black, C., Álvarez, M.: Introduction. In: The Future is Now: Reflections on Art, Science, Futurity. Renaissance Futurities, pp. 1–8. University of California Press, Los Angeles (2019). https://doi.org/10.1525/9780520969513-003

18. Popov, D.A.: Formation of science-oriented art: causes and results. Russ. J. Educ. Psychol. **11**, 329–339 (2015)

19. Ershova, N.: Language of art as language of Utopia. Technol. Lang. **1**(1), 28–33 (2020). https://doi.org/10.48417/technolang.2020.01.06

20. The Scientific art of E. Varez (2005). Knowledge-Power **7**(937). https://tech.wikireading.ru/hfTgdbjuUG. Accessed 24 Sept 2022

21. Bylieva, D., Nordmann, A.: Technologies in a multilingual world. Technol. Lang. **3**(1), 1–10 (2022). https://doi.org/10.48417/technolang.2022.03.02

22. Bylieva, D.: Artistic virtual reality. In: Bylieva, D., Nordmann, A. (eds.) PCSF 2021. LNNS, vol. 345, pp. 462–473. Springer, Cham (2022). https://doi.org/10.1007/978-3-030-89708-6_39

23. Bylieva, D., Lobatyuk, V., Ershova, N.: Computer technology in art (Venice Biennale 2019). In: Proceedings of Communicative Strategies of the Information Society (CSIS 2019), p. 18. ACM, Saint-Petersburg (2019). https://doi.org/10.1145/3373722.3373785

24. This is (not) art. https://nplus1.ru/news/2021/06/28/art-or-science. Accessed 24 Sept 2022

American Photography as Technical Art: Ontological, Cultural and Linguistic Aspects

Andrei Vladimirovich Komarov[✉]

Samara National Research University, 34, Moskovskoye Shosse, 443086 Samara, Russia
komarovandrei@yahoo.com

Abstract. The purpose of the present research deals with the interdisciplinary approach implementation regarding such compound phenomenon as photography via clarifying its ontological, cultural and linguistic integral components. The ontological layer presupposes the accomplishment of the existing philosophical concepts' analysis with a focus on visuality, image and the photography's essence, as well as the reasons of the pictorial turn development, which will certainly impact such scientific trends as cognitive linguistics, concept, discourse and anthropocentricity. On the cultural level the author examines the photography, revealing its fine-arts features and points out the art movements of the American photo-art. In the end we examine the American photographic discourse via analyzing the "concept" constructions, actualized within the American photograph's titles and their visual contents. In the present paper the author uses the contrastive-comparative analysis of the philosophical, cultural and linguistic concepts, method of interpretation, methods of discourse and semantic analysis. The author specifies that the ontological, cultural and linguistic peculiarities of the American photography are (being) in the state of interdependence due to their complex multidimensional characteristics: for example, the philosophical "iconic turn" may have influenced the linguistics' cognitive shift, and the lack of knowledge in the sphere of the American culture makes it practically impossible to interpret such photographs as *Migrant Mother, Nipomo, California, 1936* by Dorothea Lange and *The Flatiron, 1902* by Alfred Stieglitz because of their additional metaphoric complexity and the necessity to use specific background knowledge. In the end the author comes to the conclusion, that photographic discourse is a complex communicative phenomenon, a special reflective language, having inherited the Western-European artistic-aesthetic tradition, comprising linguistic and extralinguistic elements, by which American culture is recognized through a technological medium that functions like art – the American photography.

Keywords: American photography · Technical art · Visual arts · Ontology · Culture · The American photographic discourse · Concept · USA

1 Introduction

Photography originated qua technology capturing and preserving the static abstracts of the visual facts via lens, that later gradually transcended its main function, becoming

D. Bylieva and A. Nordmann (Eds.): PCSF 2022, LNNS 636, pp. 146–162, 2023.
https://doi.org/10.1007/978-3-031-26783-3_14

a new type of art. Furthermore, the author claims that photography is to be referred to the semiotic type of arts, because, on the one hand, the photographic sign bears such characteristic as iconicity (visual representation of the reality, visual contents reflecting the interaction of the form and meaning [1, p. 4]), ex altera parte, photography contains the so-called linguistic component, expressed in the form of the title, year and location of the shooting, the author's note and full name, etc. – the expression plane that also needs interpretation and visual-reference.

While analyzing the photography's essence, it is vital to observe the concept of the photographic arts' modelling [2, p. 470]: the visual model proclaims the analogy between camera lens and a human eye as optical devices, whereas the mechanical model postulates the photograph's impossibility of demonstrating a certain scene to a human quite "as we ourselves would have seen it, but it is a reliable index of what was" [3, p. 149]. Ab origine of psychological aesthetics research papers by Gustav Theodor Fechner [4] to contemporary neuroaesthetics, scientists declare that there exists a definite interdependence between human aesthetic experience and the sensation caused by visual stimuli regardless of source, culture, and experience [5, 6], which is supported by activations in specific regions of the visual cortex [7, 8]. As the result, the author defines three integral elements of a photography: the inseparable quality of visuality, its technological nature and objective reality or beingness as the object of representation.

As a consequence of the above-mentioned data, there has been traced the steadily growing interest in the number of works devoted to the philosophy and ontology of photography: [9–12]. We consider such a tendency to be the result of the photograph's qualities influence and its establishment as self-consistent genre of fine arts via creating a new means of relationships with the reality as well as generating its unique language that should have been subjected to conceptualization. Now the photography is able to integrate into the installation in the capacity of a photography sign, the sign referring to the whole complex of meanings connected with the sign. Hereunder, the photography portrays itself, depicting its place within the scope of the cultural space. Moreover, it "plays/performs the genuineness situation", placing reality into a (photo) frame [13, p. 6].

It is worthwhile noting that the root source of the XX century photography explosion is able to be traced even in the first half of the nineteenth century marked by Louis Daguerre and Nicéphore Niépce's invention of a prototype photographic camera [14]. We denote this first stage as a technologic-revolutionary one, because it has presented the medium that has influenced the visuality realm and thereby – the sphere of visual arts. Hence, the technology has stated the implementation process into the art world by the simplification of imagery creation in the second half of the nineteenth century, which resulted in "image revolution" [15]. Yet the aforementioned "image" turn cannot be illustrated only with the expanding significance of the visual paradigm or model of the casual culture. Consequently, this has culminated in new gnoseological cognizance of icons/pictorial indexes within the cultural research domain. Based on the linguo-cognition criticism, "it seeks to promote a visual literacy that has been poorly developed in Western societies since Plato's hostility toward images and logocentrist trends in philosophy" [16, p. 245].

The year of 1992 has been marked by a "pictorial turn", conceptual idea of the American academic, theorist of media, visual art, and literature, William John Thomas Mitchell. His concept manifested the manoeuvre contra the stated hegemony of the verbal language [17]. The complex essence of a shift towards the photographic art has been proved by the "iconic turn" doctrine, that appeared almost simultaneously in the year of 1994. It has been presented by Gottfried Boehm, a German art historian and philosopher, in his essay "Die Wiederkehr der Bilder"/Eng. "The Return of Images" [18]. It is obvious that the purpose of the mentioned research has been the validating of "Allgemeine Bildwissenschaft"/Eng. General Image Science – tactics, counterbalancing general linguistics – in the terms of textual and linguistic means supremacy as part of the linguistic turn, a philosophical phenomenon of the XX century, having used the language as a basis of social practice. This said image knowledge system has been primarily perpetuated in art history, thus it would examine the internal logic of visual representations and obtain coeval analytical admission to visible media, means, practices and values. Nevertheless, the imagery has been perused on a long-term basis concerning their inner significations and connotations. However, the simulacra are not ordinary signs or exempla. They possess the potential of their own that is able to avoid language units' usage. As a result, the iconic turn has preponderated over the pictorial objects study, comprising the integral diapason of visual grasp and cultural spectrum. In general, as it has already been mentioned, the reason of such an interest regarding photographic art may be revealed via particular characteristics of a photography genre that became relevant during the epoch of the image civilization or the civilization of the iconic turn [19].

Most significantly, modern comparative semantics indicates that sign systems and linguae francae exploit the same logical expedients as spoken language, however, the first have plentiful instrumentarium for iconic enhancing. It has been argued that the analysis of iconic symbols in liaison with speech arises eloquent equivalence or symmetry between a spoken utterance and a sign [20]. We suppose, this "image" concept correlates with the author's understanding of the American photographic discourse, interpreted as a complex communicative phenomenon, having inherited the Western-European artistic-aesthetic tradition, comprising linguistic and extralinguistic integral elements, possessing the technological essence and reflecting the infinite being and culture of the United States of America.

2   The Ontology of Visuality and Photography

There is a firmly held conviction that today the modern scientific paradigm is characterized by the increased focus on the visual sources, representations, special methodology used by the representatives of the visual researches. One important reason for this is the growing significance of visuality concerning socio-cultural basics' establishment of the current social experience and social order. The photography, the American photographic art, has stimulated and continues inspiring the changes in the society, increment of knowledge and the evolution of the image interpretation. Whereas the results of the world understanding (world concept) within the new Western-European culture have started relating to the world picture, the portrayal of reality has begun identifying with a picture. The "visual" implicates "depicted", "optical", "estimated by eye", "demonstrative", "image-bearing", "presentational", "viewable", "imaginary" [21, p. 29]. It is becoming more frequent the use of the "turn" metaphor, tracing its origin to Martin Heidegger [22] and the American philosopher Richard Rorty [23], aimed to emphasize the ultra-significancy of results and philosophical status' legitimization of the visual arts researches, as well as to declare the ontological value of the research subject. Notwithstanding the foregoing, there should be the philosophical underpinning of the visual research methodology, as there exists a certain difficulty while interpreting the visual sources and the boarders of visuality as the resource of the socio-economic, cultural, technologically innovative and spiritual knowledge.

In this case we agree with Vyacheslav Vladimirovich Kolodiy, who contends that the significance of the stated above issue is determined by the fact that visualness, represented in the form of photography, is not just a text appendix, not a modern trend, but a primary existence mode of the present-day sociability and culture, basic principle of their configurations' structuring [24, p. 3].

We should concentrate on the fact that special attention has been drawn to the "visual" in the context of the science instrumentarium formulization long before. Hereinafter, we will carry out the analysis of the visual and photography concepts in chronological order.

Aristotle may definitely be called the founder of visualistics. An inclination for sensorial perception he considers as evidence regarding people's inherent pursuance of scholastic attainments. "Anschauungs", regardless of their utility, "are valued for their own sake, and above all – visual impressions, since vision, may be said, is in preference to any other type of perception, not only for the purpose of doing, but also when we are not going to do anything. And the reason for this is that visual sense stimulates our cognition process and reveals a lot of dissimilarities between different things" [25, p. 65].

Having laid the foundation of the western rationalistic philosophy, Aristotle has set out the overall direction for development of the XXI century scientific thought, when it turned out that everything became an image, and the philosophy itself needs theoretical discourse' visualization and even theatricalization for the purpose of meanings' detecting within the exponential growth of events [26, p. 393].

There exist two large dimensions, variously distinguishing the notion of "photography", its essence, qualitative and quantitative characteristics.

The first focus on photography ontology, which may be described as a realistic conceptual idea, identifies the photograph in terms of the exact visual copy of the reality, its objective representation by means of technological supporting aids. In this context the photography is closely connected with memory, and the creator of the picture even though owns subjective view, performs the task of objective capturing.

George Santayana, the famous American philosopher of Spanish origin, was the first scientist who started analyzing the photography phenomenon, being the member of the Harvard photo-club, at the beginning of the twentieth century. The researcher defines photography as a technical means, functioning qua storage of human memory: "The goal of photography – the experience reconstruction, the goal of the art – the experience interpretation. The art is of secondary essence, human life and perception are the primary ones" [27]. In accordance with George Santayana, expressing the positivistic-pragmatic beliefs, the aim of photography is perfecting and broadening the human percipiency, becoming a tool for sense of sight and memory.

Later, in the year of 1913 there has been presented an essay by an American writer Marius De Zayas, proclaiming on the one hand the "primitive" understanding of photography via differentiating this sphere from art: "Art has abandoned its original purpose, the substantiation of religious conception... The Soul of art has disappeared, the body only remaining with us" [28, p. 126]. However, in the end of his work the artist and critic enunciates a doctrine that will have determined the photographic concept development by the present times – acknowledging not only the photography's complexity but also its potential to transform into art.

The second area of focus, developed after having finished photography's promotion as an independent artistic genre, is distinguished by a recognition of an independent reality and poietic/forming power of the photographic images [13, p. 24]. This stage in the evolution of the photographic ontology is associated with the epoch of reflexive modernism.

Wolfgang Kemp argues that since the second decade of the XX century the photography theory has denoted the transition from standardized to categorical comprehending of photography as a medium. The thesis regarding photography being a medium per generis, that is independent of "the arts" interpretations; henceforward, instead of standardized or normative approach we may witness the development of hermeneutic, phenomenological historical and sociological approaches [29].

The XX century is distinguished by the availability of the numerous concepts, identifying the photograph, following no particular pattern. This fact adds additional complexity for creating the typological system of photography's ontological concepts.

However, Siegfried Kracauer, a German writer and cultural critic, recognizing the absolute self-sufficiency of the photo-images universe, states the affinity of the photograph and reality by using the expression "realistic tendency" [30].

The connection between photo-art and reality has been further developed by the American philosopher Stanley Louis Cavell, the representative of post-analytic philosophy. Near the beginning of "The World Viewed", Cavell provides the readers with "a specification of the grammar of the photographic image, specifically in terms of being subject to a range of questions to which there are necessarily answers of a certain form" [31, p. 37]. "You can always ask, pointing to an object in a photograph – a building, say – what lies behind it, totally obscured by it. This only accidentally makes sense when asked of an object in a painting. You can always ask, of an area photographed, what lies adjacent to that area, beyond the frame. This generally makes no sense asked of a painting. You can ask these questions of objects in photographs because they have answers in reality" [32, p. 23–24]. In the second chapter of the "The World Viewed" the Harvard University professor comes to the conclusion that photography is another reality, for the reason that the things are not reflected, but they are presented "themselves". As the result, the route of photography, so as to reach this world and achieve self-hood, in contradiction to the route of modernist painting, may be executed via traversing the "endless presence of subjectivity" [33, p 14].

The comparative analysis of the existing photography definitions, viewing the phenomenon's insight of essence, reveals certain regularities and patterns, enabling us to indicate the below mentioned prospects' system.

The first field of study is called modernist, essentialist or formalist movement. Among the most significant researchers of this scientific area are Walter Benjamin [34], André Bazin [35], Susan Sontag [36], Roland Barthes [10], Hubert Damisch [37], Vilem Flusser [38]. This approach is distinguished by the search for the photography's essence, its identity as medium, its fundamental characteristics and most importantly – its connection with the real world. The representatives of such an approach examine photography essentially, focusing attention on the fact, that during the photographic surveying process the photo-image is created automatically, mechanically and is definitely reproduced or rendered from a technical perspective, that is – not unique.

The second scientific field is known as postmodern or anti-essentialist one. Within the scope of this movement there has been introduced the "representation" notion – that which substitutes the reality, transforming the ontological into symbolic.

Due to the additional complexity of this research area there have pointed out two additional approaches as part of this grand-scale solid scientific field.

Under the frame of *the semiotic approach*, being part of postmodern or anti-essentialist scientific field, philosophers and photo-critics try to answer the question, which way the photography language produces and transfers the meaning, the way photography functions as a means of representation: [39, 40].

The *discourse approach*, as unit of postmodern or anti-essentialist scientific field, specializes in carrying out the analysis not of the certain photography and its functioning principle, but on the survey of a more extensive context, within which it has become possible or actualized. The most distinguished scientists, sharing these views, are Laura Mulvey [41], Rosalind Krauss [42], Roswitha Breckner [43].

The third scientific movement, the post-structuralism, marks the development of the "visual" agenda. At first these social theorists [44–46] have paid attention to the photographic presentation of various historic events and also to the interaction of the photographer and power (authority), to the reflection of different ideologies and control strategies in the sphere of visual representations. Later there has been changed the focus of the researches towards the realm of the critical discourse.

3   American Photography as a Cultural and Arts Medium

"When technology extends one of our senses, a new translation of culture occurs as swiftly as the new technology is interiorized" [47, p. 40]. This thesis, expressed by the Canadian philosopher Herbert Marshall McLuhan, indicates the integral components of the photography – its technological nature, culture reflecting quality and its communicative essence.

Thus, the American photography represents a semiotic type of art, since it contains: name (text embodiment) and visual content requiring interpretation and recording of the objects referring to the USA reality, as well as the so-called value component as far as the names of American photographs, plus their visual embodiments act as an object of research within the second and third part of the given article, and photographs and their names from archives of the largest US museums and art galleries (The Museum of Modern Art in New York, The Danziger Art Gallery, The Library of Congress, The U.S. National Archives) in the amount of 3185 units served as the actual material of research. However, to implement the most precise understanding of photography as a cultural phenomenon, first of all, we need to define such concepts as "cultural symbol", "culture", and "art" as well as determine their degree of correlation between each other.

The cultural symbol is a form of being expressed by pictorial means in art and architecture, intellectual and technological products, everyday items and decorative ornaments, ceremonies, rituals and ordinances, etc.

The symbolism of existence is considerably deeper, diverse and distinct in its semantics as compared to the semantics of sign. While the semantic side of the language sign is its meaning, the images of social being peculiar to the carriers of the respective culture serve as the semantic side of the cultural symbol.

While the language sign serves the communication purposes, the cultural symbol functions as the self-expression of what unites carriers of the same culture, that is methods, contents and sense of their joint being.

The symbols of culture, as opposed to language signs, are communicative signals, in which the "signifier" and "signified" are in the ambiguous mutual reversibility relations. One and the same cultural symbols may be understood and interpreted in different ways. One and the same cultural phenomenon may obtain its multiple symbolic expression. Moreover, cultural symbols can be rightfully attributed to "implicit signals" which can be noticed only by the cultural participant familiar with the cultural symbols. Cultural symbols need not reading but interpreting through "breaking" the cultural code.

The cultural symbols make up a special code that can be called "half-closed" since some part of the culture is slightly open for the direct observation, for example, ethnographic signs – clothes, meals,, music, folk art products, tools, ways of organizing life and activities, family life and even the surrounding nature. Other characteristics remain hidden from the direct observation, and in order to include them into the cultural semiotics, they need to be decoded and interpreted.

In the social and psychological sense, culture represents the collective programming of the individual behavior resulting in setting of rather defined and predictable values, norms, rules, stereotypes and practices in the public and individual consciousness. They unite the society, dictate the ways of reacting to the challenges of the environment, they are passed down from generation to generation; they are formed as a result of social upbringing and allow discriminating between "their own" and "strangers".

Since the culture is formed in the course and as a result of the public being, it can be said that culture and society correlate the same as personality and individual. If the culture is a complex property acquired by the society in the course of social being, the personality is a complex property acquired by an individual in the course of his or her social being. Thus, the social and personal culture are the product of their joint social being [48].

The art is usually considered as a part of culture, while symbolizing the distinctiveness of figurative ways of self-expression of the author and people, the level of achievements of artistic creativity of creators of work of art, life trials, hopes and expectations of people. Art represents an aesthetically sustained reaction of the master to challenges of life and existence. In such a reaction, there is always an attempt to either change a situation or find the way to solve the problem, or distract a person from the experienced difficult feelings. It means that symbols of art transfer, in artistic form, the information on relevant problems experienced by an individual or society as well as on the methods of solving these problems in line with their cultural settings.

Works of art as cultural symbols contain not only the visual or acoustically perceived way of expression, but deep-laid content requiring interpretation. Both visual and acoustic, and meaningful plan of a work of art change depending on the level of social and economic development of the society. Therefore, the plan of expression and the plan of meaning of the symbol of "primitive" or "civilized" art will be noticeably different from each other [49].

Depending on its level, the art may be:

- iconic, i.e. having the figurative form,
- meaningful, i.e. containing artistic ideas,
- semiotic, i.e. suggesting a symbolic code [50].

To sum up the above, the author concludes that the semiotic type of art represents works, which are complex in structure, multipurpose in content and deep in sense, whose cognition requires special expert knowledge in many cases. In other words, the higher the level of artistic form of art, the more diversified interpretation is suggested by them. The American photography represents a semiotic type of art, whose interpretation requires penetrating not only the American cultural environment. To decode the photograph, the depicted historical period (historical aspect), peculiarities of a depicted object, photographer's intentions, his attitude to the created image, creation motive of a considered "frozen fragment of American reality" should be taken into account. The thesis advanced by us allows speaking about the increase of new data and information regarding the research object its complication with the course of time due to the exponential growth of human knowledge, accumulated experience in interpreting the material essences of being, objects of art (cultural products), communicative phenomena of language and extralinguistic content.

Thus, we come to a determination, that American photography is not only the cultural unit, but the object d'art of the visual-semiotic identity, and that is why, the essential properties of the American artistic photography movements are to be denoted.

In the United States of America the XX century has marked the development change not only of the social trends (vectors), but also the paradigms' conflict [51] the presence of which is determined by the so-termed abnormalities. The emerging of these abnormalities is caused by the impossibility of their prevention via using the instrumentarium of the current scientific knowledge. Within the sphere of the photographic discourse the aforementioned thesis is supported by the works of the prominent American photographers referring to the end of the XIX – beginning of the XX century. These photo-artists have influenced the whole national art movement of the following centenary. Therefore, the choice of the depiction traditions of the certain reality objects has been influenced in large measure by the holistic character and semiotic entity of a still picture. Furthermore, it should be pointed out that photographic work traditions have their origins in paintings creation, however, they evolutionize rapidly, searching for new artistic forms of self-actualization.

On the one part, the impact of painting on the photo-art is observed via usage the closed painting composition, characteristic for the definite classic painting genres – landscape, portrait and still life: *The Flatiron, 1902 (Alfred Stieglitz)*; *Still Life, 1916 (Bernard Horne)*; *Miss Koopman, Fashion, 1934 (George Hoyningen-Huene)* [52].

This distinguishing feature is reflected in the American Pictorialism, aesthetic movement, that has been born in the United States of America during the last decades of the XIX century. Pictorialism unites sentimental, pastoral, literary and escapist tendencies.

As the cameras have decreased in size, and photographic reagents have become more compact, in the USA there has been a growing number of people who have not just treated photography as hobby but seriously immersed in this comparably new field of art. Like landscape painters, they have gone to fields, out of town to photograph picturesque forest streams, gardens, houses, bucolic farms: *Landscape, 1890 (Charles S. Bradford)* [51, p. 83]. It should be particularly noted that in the US at the end of the nineteenth century, two partially overlapping directions, which have emerged in the framework of the discourse of Pictorialism photography, have been developing. The first type of the mentioned artistic photographic directions has differed in that picturesque scenes of nature of have been reproduced by amateur photographers. The second, narrower direction has represented the practice of a relatively high level of elitism, a concept of pictorialism, which has been further transformed into the artistic photography having an ideological connection with the aesthetics of European modernism. It has been due to the progressive development of the second branch of pictorialism, its orientation towards the progressive modernism and search for new artistic forms of implementation of the artistic potential of photographers, that American photographers have acquired that peculiar value component and received a pass to artistic museums by the end of the nineteenth century. Establishment of a new type of art has become fixed at the actual level – the objects of new art at the junction of eras have been put to artistic galleries. Furthermore, the second concept of pictorialism, its dualistic nature, initially has set the further dynamics of the American photographic discourse. One of the most important American Pictorialism photographers of the first half of the twentieth century, who has devoted his life to the development of photography, has been Alfred Stieglitz.

For early works by Alfred Stieglitz have been created by him already at the end of nineteenth century upon his returning to the USA from Europe, a distinctive object of the depiction have been the city streets – *Winter–Fifth Avenue, New York, 1893 (Alfred Stieglitz)* [53]. It is worth noting that for two decades of the heyday of pictoralists, whose works have been characterized by the presence of a traditional "aesthetic" object of depiction as well as impressionist softening of the optical pattern, Stieglitz has created photographs, which, admittedly, have been realistic to the fullest extent – without any manipulations with a photographic plate, which have been so widespread among pictorialists. Alfred Stieglitz has produced a significant impact on the development of a new language of modernism through the photo-chanting of machines/technology and the urbanistic environment – the building New York streets, skyscrapers, ferries, airplanes and liners *The Steerage, 1907, Alfred Stieglitz* [54]. This photograph has been influenced by cubism, which has started penetrating into all spheres of art from 1905. Of course, the photographer has been able not create abstract forms as the artist has done, however, he or she has been able to "detect" them by means of "constructing" an object of image in a way emphasizing abstract forms of a particular scene/object of photograph.

The necessity of recognizing photography as a new object of art has anticipated the emergence of the so-called independent group of photographers known as the "The Photo-Secession", which included Alfred Stieglitz as well. The main purpose of the group has been the recognition of photography as an independent type of art, as well as in overcoming its "mechanical" nature. Photographic shoot in this case has been interpreted as a hand-made object. Thus, clear-focus photographs have been ignored. Instead, photographers have experimented with the gum dichromate process, which has softened the photograph surface through some kind of "blurring" as well as by using the golden or silver toning: *Blessed Art Thou amongst Women, 1899 (Gertrude Käsebier)* [55]. The mentioned photograph depicts the surrounding world but the accent is made to rendering the spirit of the depicted scene. Photograph is not a mirror of reality anymore, but it represents a reflection of author's subjectivism serving to the "beautiful".

The new aesthetics of 1910–1920 has taken all the positive features of the "direct" photography (minimum of manipulations with photo shoots): the two-dimensional nature of photography has been proclaimed, which inevitably has resulted in emphasizing of the abstract form: *Vortograph, No. 3, 1916–17 (Alvin Langdon Coburn)* [56]. In taking this photograph, the author has been inspired by the works made in the style of "cubism" as well as by the application of particular technical means. Alvin Coburn has used kaleidoscope attached to the camera lens while creating series of images with a symmetric pattern of light and shadow. Such photographs or "vortographes" are referred to the technological futurism.

4 The American Photographic Discourse and America Concept

The state of the contemporary science is described in terms of the active objects' change, being in the focus of linguistic research, via the turn or shift towards the study of language qua a means of the human cognition of the surrounding world, towards the examination of the language units' functions in the process of communication and in inextricable connection with extralinguistic reality. The linguists' recognition of the dependence of a concrete language performance on the complex of extralinguistic conditions open possibilities for the new description methodology regarding various types of communication.

In connection therewith, the modern linguistics and the cross-disciplines emphasize their attention on such a notion as discourse. Discourse analysis becomes one of the basic description methods of the human life spheres (political discourse, legal discourse, mass media discourse) [57].

While undertaking the further examination we should point out the "discourse" integral qualities. It has become especially important in the light of the modern studies that consider the notions "text" and "discourse" to be the synonyms either inadvertently or deliberately. As a result, such papers describe predominantly the text structure and text qualities. We find such an approach to be inaccurate, because "text" and "discourse" are not synonyms. It is obvious that text and discourse are comparable to each other in terms of size and the number of the presented language units, but actualization and problem solution stated by the author are accomplished on the far more complicated discourse level. While fulfilling the discourse analysis we proceed from the text in terms of the material dimension, but further on we reach the level of the author's solving the stated problem: his or hers implied purpose of the act of speech, "twisting the edges" of the used language means. As a result, we interpret the discourse as a communicative phenomenon, which includes both linguistic and metalanguage (or extralinguistic) characteristics of a photo-art unit.

The author claims that the development of the discourse theory has been the result of the forthcoming of the linguistic anthropocentric paradigm, having returned the "measure of all things" status to a human being. We argue that this trend has been the result of the intense interest in the cognitive research, dealing with the human perception and cognition as the principal qualities of homo sapiens. The establishment of the cognitive approach in the second half of the XX century has proclaimed the development of the multilevel theory of meaning – cognitive semantics, the characteristic feature of which is the transcendency of the linguistic knowledge and the shift towards the metalinguistic sphere as well as the identifying the role of this knowledge in the process of the linguistic meanings' formation and the meaning of the utterance [58, p. 21]. The distinguishing feature of the cognitive semantics is the dynamic approach regarding the meaning, wherein it is recognized less as a structure of hierarchized seme, than as a unity, formed during the cognition process oriented on the surrounding world [59, p. 3]. Therefore, conceptualization is a natural process of the reality cognition, executed by an individual. For this purpose, the representatives of the lingua-cognitive focus area (Kubriakova E.S., Boldyrev N.N., Sternin I.A., Popova Z.D., Babyshkin A.P.) analyze the concept as an increment of the structured knowledge [60, p. 90] with reference to the abstract of the world, mental unit, which purpose is carrying out the intellectual process [61, p. 35].

The researchers, working within the scientific field of linguacultural approach (Wierzbicka A., Stepanov Yu.S., Karasik V.I., Vorkachev S.G., Slyshkin G.G.), perceive the concept as a culturally notable mental entity, existing in the consciousness of the cultural society and actualized inside the boundaries of the language sphere. While lingua-cognitive approach of the concept essence presupposes the progression, beginning from the individual consciousness towards the actuality, the linguacultural approach – from culture in the direction of individual consciousness [62]. That is why, the lingua-culturologists differentiate concepts according to the affiliation with this or that culture, contradistinguishing the concepts – cultural, national and social [63, p. 77].

The author considers it necessary to create a common synthetic definition of the "concept" notion: concept is a semantic entity of a large-scale abstractedness, distinguished by linguacultural specificity, valuable component and possessing the complex structure in the form of the core and the periphery. The author contends, that a concept may be used as a cognitive means so as to examine the linguistic, cultural, and linguo-social peculiarities of a discourse.

The concept structure is represented in the field form, containing the core and the periphery, represented in the form of numerous lexical units, characterized by a different degree of semantic generality. The system structural stability is arranged due to the integral and differential seems or features.

The American photographic discourse due to its essential qualities is understood by the author as a complex communicative phenomenon, reflecting the culture of the United States of America on the linguistic and sign levels (composite visual contents) and comprising the following integral qualities:

1) visuality (direct display of reality);
2) individual-authorship (special techniques and effects within the lensing process);
3) semiotic specificity (photography, a versatile index, that is to be clarified, immerging not only into the linguo-cultural environment, but also obtaining additional knowledge);
4) technical nature (the attempt to avoid subjectiveness due to mechanical recording).
5) cultural loading (the reflection of the US values and culture-bound items).

In the present study the author intends to use the concept system, as a cognitive means, for the purpose of revealing the linguistic and cultural peculiarities of the American photographic discourse space over a period of the XX century. For this purpose there will be introduced the lingua-synergetic approach, the nominative paradigm dynamics of the American photographs' titles, implying long-term system monitoring (one hundred years). In the furtherance of this goal we use the America meta-concept or modus concept, functioning on the principle of a cluster model. The term "cluster model" has been introduced by George Lakoff [64]. This concept is composed of several other concepts. Taken together they are united into one cluster. The present cluster is endowed with integral nature, integrating more or less typical elements. The actualization of a such a compound semantic formation as America meta-concept may be traced via verbalization of its 2 major elements or microconcepts – *Location* and *Human*, that constitute together 58% of the total America concept lexical representatives (excerpts/samples) in the amount of 3185 American photographs' titles and their visual actualizations (Location – 39%, Human – 19%).

The *Location concept* is the most massive in terms of its lexical representatives' volume – 39% of the total amount of excerpts/samples. The discriminant mark of the Location Concept is its frequency interaction with all the possible stated concepts (even those, which we have not studied) with a view to ensuring the intensification of authenticity by means of transferring the "geographical bearings" of the shot, the place of its origin. This tendency may be viewed at the beginning as well as in the second half of the XX century (Fig. 1; Fig. 2).

The *Human concept* takes the second place regarding the number of the lexical representatives – 19% of the total amount of excerpts/samples. The present concept actualization in the second half of the twentieth century differs from its representation in the second half of twentieth century.

Thus, in the first part of the XX century there may the viewed the titles of the photographs, where either a girl or a woman is portrayed. However, these photographs' titles lack a reference to a major image-object. Photographers use impersonal constructions or just avoid mentioning the principal portrayal object – Blueberries and a Peach, Trimming Currency. As the result, there has been observed the belittling of the woman's

Fig. 1. Crew of **New York**. Blue fish fishermen About 1900. Records of the Fish and Wildlife Service. Photographer Unknown

Fig. 2. Ad Reinhardt Hanging Paintings in His Studio, **New York City**, 1966 John Loengard

social significance at the beginning of the of the XX century. Therewith, her role in the frame of the photography's compositional structure may be reduced to the background (field) or the interior feature (Fig. 3; Fig. 4).

Fig. 3. Some blueberries and a peach. 1902. (111-AGD-104) Miles Brothers

Fig. 4. Trimming currency. 1907. Records of the Public Buildings Service (121-BA-361B) Photographer Unknown

It ought to be noted that in the second half of the XX century the situation changes. The human, his/her personality hold a prominent place in the American linguistic world view. Special emphasis is made on the services he/she renders to the society and the state. Therefore, proper nouns begin to be marked with additional significance (John F. Kennedy, Mrs. Gordon Cooper, Cindy Sherman). Moreover, there should be emphasized the lexical marker *Nude*. Its frequency of usage witness the search for the new creative forms and lack of fear, despite facing the modification of the existing original photographic canonicity (Fig. 5; Fig. 6).

Summarizing the above-mentioned data, we come to the conclusion, that the dynamics of the America meta-concept within the boundaries of the XX century photographic discourse is notable for the intensification of authenticity (the high qualitative characteristics of Location concept), anthropocentricity (the influence of Human concept) and the increasing of personification level due to the enhancing interest in real US personalities.

Fig. 5. Wynn Bullock. **Nude** in Forest, 1950   **Fig. 6.** Cornell Capa. **John F. Kennedy**, First Cabinet Meeting, 1961

5   Conclusion

Summarizing everything above-mentioned, it is worth mentioning, that the American photography, being a technological art, because of its attempt to avoid subjectiveness due to mechanical recording, has in possession the valuable component, which relates the photograph to culture and even to fine-arts product. The presence of the valuable component is proved by American culture portrayal as well as by displaying the photographs under analysis in the museums and galleries. Finally, the technological origin of the photo-art is proved via "becoming real using the idea or archetype", epoch-making materialization, recognition and functioning among people [65, p. 397].

Moreover, the American photography is considered to be the semiotic culture, as it contains the special hidden code that should be interpreted using special knowledge. A person, who does not know English language, will hardly be able to explain the photography *Migrant Mother, Nipomo, California, 1936* by Dorothea Lange. Here we deal with the specific background cultural knowledge: the Great Depression as an economic crisis in the United States of America at the beginning of the century, internal migration, thousands of hoboes searching for work, etc. Another photograph *The Flatiron, 1902* by Alfred Stieglitz is marked by a high degree of metaphoricity, as it depicts the first skyscraper, the symbol of New York, having epitomized the Chicago architecture school conception, which really resembles a flatiron device.

The American photographic discourse is interpreted by the author as a complex communicative phenomenon, having inherited the Western-European artistic-aesthetic tradition, comprising linguistic and extralinguistic integral elements, possessing the technological essence and reflecting the infinite being and culture of the USA.

Finally, the present research has shown that the development of the America meta-concept, accepted qua cognitive construction, within the boundaries of the XX century photographic discourse, is distinguished by the intensification of authenticity (the high frequency ratio characteristics of Location concept), anthropocentricity (the influence of Human concept) and the increasing of personification level due to the enhancing interest in real personalities (astronauts, scientists, actors, politicians, painters, musicians, etc.), the citizens of the United Stated of America.

References

1. Giardino, V., Greenberg, G.: Introduction: varieties of iconicity. Rev. Philos. Psychol. **6**(1), 1–25 (2014). https://doi.org/10.1007/s13164-014-0210-7
2. Sonesson, G.: Semiotics of photography: the state of the art. In: Trifonas, P.P. (ed.) International Handbook of Semiotics, pp. 417–483. Springer, Dordrecht (2015). https://doi.org/10.1007/978-94-017-9404-6_19
3. Snyder, J., Neil, W.A.: Photography, vision, and representation. Critic. Inquir. **2**(1), 143–69 (1975). http://www.jstor.org/stable/1342806
4. Fechner, G.T.: Vorschule der Aesthetik. Breitkopf & Härtel, Leipzig (1876). (in German)
5. Deng, Y., Loy, C.C., Tang, X.: Image aesthetic assessment: an experimental survey. IEEE Signal Proces. Magaz. **34**(4), 80–106 (2017). https://doi.org/10.1109/MSP.2017.2696576
6. Zeki, S.: Clive Bell's "Significant Form" and the neurobiology of aesthetics. Front. Human Neurosci. **7**, 730 (2013). https://doi.org/10.3389/fnhum.2013.00730
7. Ishizu, T., Zeki, S.: The Brain's specialized systems for aesthetic and perceptual judgment. Europ. J. Neurosci. **37**(9), 1413–1420 (2013). https://doi.org/10.1111/ejn.12135
8. Brown, S., Gao, X., Tisdelle, L., Eickhoff, S.B., Liotti, M.: Naturalizing aesthetics: brain areas for aesthetic appraisal across sensory modalities. Neuroimage **58**(1), 250–258 (2011). https://doi.org/10.1016/j.neuroimage.2011.06.012
9. Flusser, V.: Für eine Philosophie der Fotografie, European Photography, Göttingen (1983). (in German)
10. Barthes, R.: Camera Lucida. Reflections on Photography. Hill and Wang, New York (1981)
11. Baudrillard, J.: Im Horizont des Objekts: Photographies 1985–1998. Hatje Cantz Verlag, (2000). (in German)
12. Barthes, R.: The photographic message. In: Image-Music-Text translated by Stephen Heath, pp. 15–32. Fontana Press, Fullham (1977)
13. Savchuk, V.V.: Philosophy of Photography. Saint Petersburg University Publishing House, Saint Petersburg (2005). (in Russian)
14. Kluszczynski, R.W.: Visual revolutions: from the electronic to living imagery (2020). https://doi.org/10.26485/AI/2020/22/1
15. Brandt, R.: Bilderfahrungen – von der wahrnehmung zum bild. In: Maar, C., Burda, H. (eds.) Iconic Turn. Die neue Macht der Bilder, hrsg, pp. 44–54. DuMont, Köln (2004). (in German)
16. Bachmann-Medick, D.: Chapter VII: the iconic turn/pictorial turn. Cultural Turns: New Orientations in the Study of Culture, pp. 45–278. De Gruyter, Berlin, Boston (2016). https://doi.org/10.1515/9783110402988-009
17. Mitchell, W.J.T.: The Pictorial Turn. Essays on Verbal and Visual Representation. University of Chicago Press, Chicago, Picture Theory (1994)
18. Boehm, G.: Die wiederkehr der bilder. In: Ders. (ed.) Was ist ein Bild?, pp. 11–38. Wilhelm Fink Verlag, München (1994). (in German)
19. Boehm, G., Mitchell, W.J.T.: Pictorial versus iconic turn: two letters. Cult. Theor. Critiq. **50**(2–3), 103–121 (2009). https://doi.org/10.1080/14735780903240075
20. Schlenker, P.H.: Iconic pragmatics. Natur. Lang. Linguis. Theor. (JSTOR) **36**(3), 877–936 (2018). http://www.jstor.org/stable/45095233
21. Terentieva, I.N.: Visual, too visual… (on the specificity of the iconic turn in the modern mass media and media-researches). Bullet. Nizhny Novgorod Stat. Techn. Univ. Soc. System. Manag. Com. Technol. **2**, 29–35 (2012). (in Russian)
22. Sheehan, T.: The Turn. In: Davis B. (Ed.), Martin Heidegger: Key Concepts, pp. 82–101. Acumen Publishing (2009). https://doi.org/10.1017/UPO9781844654475.007
23. Rorty, R.: The Linguistic Turn: Essays in Philosophical Method. University of Chicago Press, Chicago (1967)

24. Kolodiy, V.V.: Visuality as a phenomenon and its influence on the social cognition and practices. [Ph. D. Dissertation]. National research Tomsk Polytechnic University, Tomsk (2011). (in Russian)
25. Aristotle: Metaphysics. In: Volume 1 under the Editorship of Asmus V.F. Four-Volumes Set, Mysl', Moscow (1975). (in Russian)
26. Nilogov, A.S.: Who Creates the Philosophy in Modern Russia, vol. 1. Pokolenie, Moscow (2007). (in Russian)
27. Santayana, G.: Das fotografische und das geistige bild (1905), theorie der fotografie I. In: Eine Anthologie herausgegeben und Eingeleitet von Wolfgang Kemp, pp. 1839–1912. Schirmer/Mosel, München (1999). (in German)
28. De Zayas, M.: Photography and Photography and Artistic-Photography Camramirez Homepage (1913). http://www.camramirez.com/pdf/DI_Week6_PhotoAndArt.pdf. Accessed 25 Aug 2022
29. Kemp, W.: Theorie der fotografie II, pp. 1912–1945. Munchen (1979) (in German)
30. Kracauer, S.: Das ästhetische grundprinzip der fotografie (1970). In: Kemp W. (Hrsg) Theorie der Fotografie III, Munchen (1983). (in German)
31. Moran, R.: Stanley Cavell on recognition, betrayal, and the photographic field of expression. Harvard Rev. Philos. **23**, 29–40 (2016). https://doi.org/10.5840/harvardreview201610113
32. Cavell, S.: The World Viewed: Reflections on the. Enlarged Edition. Harvard University Press, Ontology of Film (1979)
33. Rothman, W., Keane, M.: Toward a reading of the world viewed. J. Film Video **49**(1/2), 5–16 (1997). http://www.jstor.org/stable/20688129
34. Benjamin, W.: A short history of photography homepage, https://academyphotos.ru/library/benjaminkratrayaistoriya.pdf. Accessed 30 Aug 2022
35. Bazin, A., Gray H.: The ontology of the photographic image. Film Q. **13**(4), 4–9 (1960). www.jstor.org/stable/1210183
36. Sontag, S.: On Photography. Farrar, Straus & Giroux, New York (1977)
37. Damisch, H.: Five notes for a phenomenology of the photographic image. Photog. **5**, 70–72 (1978). https://doi.org/10.2307/778645
38. Flusser, V.: Kommunikologie. Fischer (Tb.), Frankfurt (1998). (in German)
39. Burgin, V.: Thinking Photography. Macmillan, Basingstoke (1982)
40. Alvarado, M., Buscombe, E., Collins, R.: Representation and Photography: A Screen Education Reader. Palgrave, Basingstoke, London (2001)
41. Mulvey, L.: Visual pleasure and narrative cinema. Screen **16**(3), 6–18 (1975). https://doi.org/10.1093/screen/16.3.6
42. Krauss, R.E.: Reinventing the medium. Crit. Inquir. **25**(2), 289–305 (1999). https://doi.org/10.1086/448921
43. Breckner, R.: Pictured bodies: a methodical photo analysis. INTER (Interaction, Interview, Interpretation) Biling. J. Qualitat.-Interpret. Soc. Resear. East. Europ. **4**, 125–141 (2007)
44. Foucault, M., Howard, R.: Ceci n'est pas une pipe. October **1**, 7–21 (1976)https://doi.org/10.2307/778503
45. Bourdieu, P.: Photography: A Middle-Brow ArtTranslated into English by Shaun Whiteside. Polity Press, Great Britain, Cambridge (1990)
46. Van Dijk, T.A.: Discourse and manipulation. Disc. Soc. **17**(3), 359–383 (2006). https://doi.org/10.1177/0957926506060250
47. McLuhan, M.: The Gutenberg Galaxy: The Making of Typographic Man. University of Toronto Press, Canada (1962)
48. Hofstede, G.: Culture's Consequences: Comparing Values, Behaviors, Institutions, and Organizations Across Nations, 2nd edn. Sage Publications, Thousand Oaks, California (2001)
49. Hatcher, E.: Art and Culture. Bergin and Garvey, Westport, Connecticut (1999)

50. Milrud, R.: Symbols of Culture. Palmarium Academic Publishing, Germany, Monograph (2012). (in Russian)
51. Orvell, M.: American Photography. Oxford University Press, Oxford, New York (2003)
52. Danziger, J.: American Photographs 1900/2000. Assouline Publishing, New York (2005)
53. The National Gallery of Art Homepage (2022). https://www.nga.gov/collection/art-object-page.35242.html. Accessed 03 Sep 2022
54. The Metropolitan Museum of Art Homepage (2022). https://www.metmuseum.org/art/collection/search/267836. Accessed 03 Sep 2022
55. The National Gallery of Art Homepage (2022). https://www.nga.gov/collection/art-object-page.150204.html. Accessed 03 Sep 2022
56. The Museum of Modern Art Homepage (2022). https://www.moma.org/collection/works/83725. Accessed 04 Sep 2022
57. Kozhina, M.A.: Language Markers of Polydiscoursity in the Literary Text (As Exemplified by the Crime and Punishment Novel by F.M. Dostoevsky). [Ph. D. Dissertation]. National Research Tomsk State University, Tomsk (2012). (in Russian)
58. Bubyreva, Z.A.: The Nomination Cognitive Basics of the Tactile Perception: as Exemplified by the Russian, French and English Languages. [Ph. D. Dissertation]. Belgorod State National research university, Belgorod (2011). (in Russian)
59. Nikitin, M.V.: The Foundations of the Cognitive Semantics. Herzen State Pedagogical University of Russia Publishing House, Saint Petersburg (2003). (in Russian)
60. Kubriakova, E.S.: Abridged Dictionary of Cognitive Terminology. Philological Faculty of Moscow State University Publishing House, Moscow (1996). (in Russian)
61. Popova, Z.D., Sternin, I.A.: Cognitive Linguistics. "AST: Vostok-Zapad" Publishing House, Moscow (2007). (in Russian)
62. Karasik, V.I.: Linguistic Sphere: Identity, Concepts, Discourse. Peremena, Volgograd (2002). (in Russian)
63. Karasik, V.I.: Lingua-cultural concept as the research unit. In: Sternin, I.A. (ed.) Methodological Problems of Cognitive Science, pp. 75–80. Voronezh State University Publishing House, Voronezh (2001). (in Russian)
64. Lakoff, G.: Women. Fire and Other Dangerous Things. The University of Chicago Press, Chicago (1987)
65. Dessauer, F.: Zur philosophie der technik. was ist technik? – wort und wesen. [Nesterov A.Yu. trans.] In: Ontology of Designing, vol. 2(21), pp. 390–406 (2016). https://www.ontology-of-designing.ru/article/2016_3(21)/12_dessauer.pdf (in Russian)

Socratic Paradoxes: Maieutic Technique of Communication

Anton Zamorev⬚ and Alexander Fedyukovsky$^{(\boxtimes)}$ ⬚

Peter the Great St. Petersburg Polytechnic University (SPbPU), Polytechnicheskaya 29, 195251 St. Petersburg, Russia
fedyukovsky@mail.ru

Abstract. The article is devoted to Socratic maieutics as a communication technique aimed at forming a common language with representatives of any hostile ideology. The relevance of this problem is due to the fact that the conflict level in the contemporary society is threateningly high, and traditional solutions (such as violence, compulsion, and the path of mutual concessions) do not produce a reliable result. The aim of this work is to present Socrates' doctrine as an alternative technique of resolving cross-cultural conflicts, which, on the contrary, requires neither violence, compulsion, nor mutual concessions. Accordingly, we study those principles of Socrates' doctrine which, until now, have been interpreted in philosophy as paradoxes and, therefore, have had no social understanding, not to mention their applying. Initially, two most famous Socrates' formulae are examined: "I only know that I know nothing" and "Virtue is knowledge of virtue". Subsequently, the hidden content of Socrates' doctrine, specified in the synthesis of these two paradoxes, is investigated.

Keywords: Virtue · Justice · Paradox · Golden rule of ethics · Maieutics · Communication · Conflict · Law of contradiction-identity

1 Introduction

The problem of forming a common language between different conflicting cultures has always been relevant. The multilingual and multi-technological world needs a special translation to ensure that everyone is not only able to coordinate with others but to express ourselves or exchange ideas [1]. There are also conflicts in the technological space, for example, between a growing number of vehicles and a limited space, e.g., highways or limited airspace resources [2]. Nevertheless, conflicts between people seem to be the most important.

They have acquired particular significance in our time, when any local conflict on the world stage threatens the outbreak of the nuclear war, and, therefore, the death of civilization. So far, humankind has known only three ways to solve this problem: they are violence, compulsion and compromise, i.e., the path of mutual concessions. Ann Taves writes about the "worlds" created by the people who are subjected to implication adjustments, or overt changes for the resolution of conflicts [3]. But none of these common ways for humankind addresses the problem of internal causes of the conflict. This means that everyone, at best, eliminates, for the time being, its external manifestations, thereby creating a precondition for their further aggravation in the future.

Another way to solve this problem, as we will demonstrate, was opened by Socrates' doctrine. It is aimed at discovering the *imaginary nature* of any conflict. Therefore, at eliminating precisely the internal causes of its occurrence. But the problem is complicated by the fact that Socrates' ideas came to us in scattered quotes. Some of them look so paradoxical that they resemble riddles rather than solutions.

The central problems of this research are two most famous Socrates' paradoxes: the mysterious thesis «I only know that I know nothing» and the equally mysterious doctrine that "virtue is knowledge of virtue". If the former statement of Socrates is suspected to be controversial, the latter one is accused to be tautological. In both cases, the true meaning of these phrases remains unclear for science.

But the most interesting paradox is the third one, which arises when one literally reads the first two statements. Since, if it is true that virtue is knowledge of virtue, and Socrates knows only that he knows nothing, does Socrates himself has virtue? From this point of view, there are two options:

1) the *negative answer* is that Socrates, according to him, knows nothing, which means, in particular, he does not know that there is virtue. Therefore, he has no virtue, by his definition.
2) the *positive answer* is that Socrates has virtue. But then all virtue means the knowledge that Socrates possesses, i.e., the knowledge of ignorance of nothing.

Of these two options, we immediately discard the first one, since it contradicts Socrates' glory as the wisest and most virtuous of the Hellenes. Therefore, the second option remains: to reduce the entire ideal of Socratic virtue exclusively to the realization of his ignorance of nothing. But what can such an ideal provide? And what ethics can be built on its basis? To answer these questions, we will consider each of the paradoxes separately.

2   I Only Know that I Know Nothing

Socrates' phrase "I only know that I know nothing" is widely known, it is quoted constantly and in different contexts [4, 5]. The problem is that Socrates claims to have some knowledge and yet denies that he has it [6]. This paradox is interpreted differently in philosophy. Many, like C. Taylor, see the contradiction in these words and evidence that Socrates ethics reached an impasse [7]. Others believe that there is no contradiction in Socrates' words, since the word "I know" is used here in different meanings.

For example, Gregory Vlastos writes: "When declaring that he knows absolutely nothing he is referring to that very strong sense in which philosophers had used them before and would go on using them long after where one says one knows only where one is claiming certainty. This would leave him free to admit that he does have moral knowledge in a radically weaker sense the one required by his own maverick technique of philosophical inquiry, the elenchus" [8]. Norman Gulley claims, Socrates does not mean what he says: "Socrates' profession of ignorance is merely an expedient to encourage his interlocutors to seek out the truth" [9]. Terry Irwin thinks: "Socrates has renounced knowledge and is content to claim no more than true belief" [10].

In order to discover whose hypothesis is more correct, we will turn to the argument with which Socrates himself tried to explain his statement. As one knows, this argument was a pie chart where all points within the circle denoted truths already known by Socrates. And all points outside the circle, on the contrary, were interpreted as truths, while still unknown to Socrates.

In addition, the points forming the circumference are taken into account: they personify the area of *conscious ignorance*, i.e., the truths unknown to Socrates, but about which he already knows that they are unknown to him. It is assumed that the wider Socrates' undoubted knowledge circle is, the longer its circumference is, i.e., the clearer there is understanding of how much unconscious remains in the world.

The problem is that this gradual growth of the conscious ignorance area does not answer the question at all: how did Socrates move from that to his conclusion: ignorance of *nothing*? At first glance, this conclusion seems to be obtained by mathematical induction, i.e., Socrates simply integrated all the stages of the circumference growth into infinity.

But this is an obvious error: the field of conscious ignorance, according to the condition, is to grow along with the increase of Socratic knowledge circle. And if one expands this range of knowledge to infinity, the area of ignorance will simply disappear from the diagram. In particular, the area of conscious ignorance will disappear, i.e., we will get the circle of knowledge which has an area everywhere and a circumference nowhere.

How, then, did Socrates obtain the opposite conclusion in his reasoning? In order to answer this question, we will refer to the part of the diagram which is discussed the least often. Namely, to the area lying strictly *outside* the circle. Since all truths, at least partially known to Socrates, are to, by definition, either enter the circle of his knowledge, or be located on its circumference, only *completely unknown* truths, about which Socrates knows nothing, are possible *strictly* outside the circle.

Question: Does Socrates admit that some of these truths can refute all his previous knowledge? Consider both answer options:

1) the *negative answer* would mean that Socrates undoubtedly knows something about completely unknown truths. Namely, he undoubtedly knows that none of them can refute his previous knowledge. But this is absurd, because one cannot know anything about completely unknown things, by definition, in particular, about that. Therefore, this option can be excluded.

2) the *positive answer* would mean that Socrates, on the contrary, allows a hypothetical refutation of all his knowledge with unknown truths. But, then, all the previous knowledge of Socrates will *no longer* be undoubted, but only plausible working hypotheses, at best.

3   Who Knows Something Undoubtedly, (S)he Knows Everything

It may be argued that our conclusion is incorrect, since it is permissible to conclude from the definition (in this case, from the definition of "completely unknown truths") only if the objects corresponding to it really *exist*. But, by definition, nothing can be known about *completely unknown* truths. In particular, one cannot know whether they exist or do not. Therefore, the problem of the conclusion correctness is solved in two ways:

1) the *negative solution* would mean that there are *no* completely unknown truths for Socrates. Then, our conclusion from their definition is incorrect, i.e., Socrates can have undoubted knowledge. But, then, the whole field *outside* the knowledge circle will be *empty* in his diagram, for all real truths in it will be included in Socrates' knowledge circle, and he will be omniscient.

2) the *positive solution* would mean that Socrates, on the contrary, allows the presence of completely unknown truths. And, then, our conclusion from their definition is correct. Therefore, in Socrates' diagram the circle of his *undoubted knowledge* will disappear, for they will be only working hypotheses, at best.

Therefore, Socrates' diagram, at least, demonstrates us the following: *we know nothing undoubtedly as long as we do not know everything*. So, in order to undoubtedly know something at least, one need to undoubtedly know *everything*, i.e., to be omniscient and, consequently, to have no doubts about anything.

The similar position, however, was taken by many sophists, e.g., Protagoras of Abdera. According to his doctrine, man is the measure of all things, therefore, for everyone it is true what he undoubtedly believes in at the moment [11]. Obviously, in this definition of truth, every person who undoubtedly *believes* in everything, about which he thinks, is also *omniscient*, i.e., he *undoubtedly knows everything* which is true for him at the moment.

The problem is that Socrates not only did not consider of himself as omniscient, in his discussions he also regularly refuted the imaginary "omniscience" of the Sophists, starting with their leader Protagoras. Therefore, for Socrates personally, only the second option of the two presented was acceptable: to admit that completely unknown truths are and, as a result, may already be a refutation of any knowledge, even if the latter is considered to be proved.

Question: Could Socrates call "knowledge" what, according to his conclusions, is hypothetically refutable? The answer is twofold. On the one hand, in the dialogue "Symposium" Socrates says that even a true opinion *cannot* be called knowledge if its truth is random, i.e., if it is not accompanied by an explanation proving its truth [12]. On the other hand, Socrates admits that evidence does not always remove doubts. Therefore, two kinds of knowledge are allowed:

1) *undoubted knowledge* is a true opinion, which is accompanied by proving explanations and on this basis is recognized as an undeniable truth, i.e., a *dogma*. We will call such knowledge *dogmatic*.

2) *questionable knowledge* is a true opinion, which is also accompanied by proving explanations, but does *not* save from doubts and allows the possibility of revision. It is to this knowledge that Socratic *maieutics* leads us. And that is why we will call his knowledge *maieutic*.

"Maieutics", the art of helping the truth to be born in disputes, was called by Socrates the technique of probing questions, usually asked in order to reveal the ignorance of the interlocutor.

Therefore, the paradox of conscious ignorance can be considered to be solved, since the thesis "I only know that I know nothing" really involves Socrates' using the word "I know" in different meanings. Namely, the words "I know nothing" mean Socrates' lack of undoubted dogmatic knowledge. Therefore, his very *knowledge of this fact* has any longer *not* a dogmatic, but maieutic quality, i.e., a status of the working hypothesis.

Further, it is necessary to determine, how Socrates managed to reduce his entire *ideal of virtue* to this formula? To answer this question, consider the paradox of this ideal.

4 The Golden Rule of Ancient Ethics

The paradox of virtue is associated with the attempts of ancient thinkers to express all moral wisdom in one simple formula so that it is rationally provable. The matter was complicated by the fact that the categories of Good and Evil in different cultures are interpreted differently. Only two of their appointments are common, namely:

1) the *negative purpose* is to guard oneself and other people from acts that directly or indirectly contradict the goals of this culture, and therefore are declared *evil* there and are forbidden.

2) the *positive purpose* is to attract other people to assist in matters that directly or indirectly serve the goals of this culture, and therefore are declared *good* there and are encouraged.

The problem is that none of these purposes is performed if we leave for at least someone the right to *disagree* with our idea of Good and Evil. It is like if we directly permit someone *to rob our most expensive thing*. This is impractical, from the point of view of any possible goal. That is why any morality, however relative in nature, inevitably imposes itself on the world as an *absolute law* claiming *universality*.

For example, the Chinese philosopher Confucius expressed this in the Golden Rule formula "What you do not wish done to yourself, do not do to others". The Jewish prophet Moses expressed the same rule in the commandment "Love your neighbor as yourself". Socrates gave even more precise wording. In the dialogue "Gorgias" [13] in his dispute with Callicles, he argued that the essence of virtue is *justice*, which is the *equality of all before the law*.

It is easy to note that all three formulae express one thing: the *prohibition of double standards in moral matters*, i.e., the requirement to approve one universal standard for oneself and for other people. But is it possible to conclude from here what exactly this standard should be? The answer is ambiguous:

1) the *negative answer* follows from the fact that the Golden Rule allows us to do for others *everything* that we wish for ourselves, i.e., it allows us to decide the fate of other people according to any of our preferences.

2) the *positive answer* also follows, since *none* of us wish other people to decide our fate according to their preferences, not aligning them with ours. Therefore, the Golden Rule, at a minimum, obliges us to *coordinate* our moral ideals with the ideals of our opponents as soon as they come into conflict.

One of the first to recognize this consequence was Confucius, who defined virtuous people as striving to live in harmony with others. Even more precise wording to that was given by I. Fichte, who defined the highest goal of any reasonable society as *complete moral consent among all its possible members* [14]. The problem is that almost *none* of the great moralists of the past considered this ideal to be feasible.

The exception is Socrates only. Although he left neither treatises, nor a clear system of moral commandments, he completely solved the problem of moral consent both theoretically and practically.

5   Solution 1. The Theory of Universal Consent

The theoretical solution to the problem is provided in the dialogue "Symposium", where Socrates confesses his sympathies to Heraclitus' doctrine of the unity of opposites [12]. According to this doctrine, all opposites in nature have initially remained in unity and harmony, like in a lyre or a bow [15]. This means that a conflict between opposing opinions is possible only because of a misunderstanding of their nature.

At the same time, Socrates criticized Heraclitus. But not for the content of his doctrine, which he admired, but for the darkness and inaccuracy in the presentation of thought. Socrates believed that Heraclitus could not understand by the unity of opposites their real coincidence in one sense, but understood by this a certain adjustment [12], i.e., a mutual restriction that allows each of the opposites to be realized within some of their limits of applicability.

Therefore, Socrates was the first to formulate the principle, which was later taken as the basis in the theory of Aristotle's proof. This principle, referred to as the *law of contradiction*, states: "Two conflicting opinions cannot be both true in the same respect, but may be true in different respects". Moreover, any opinion can be called "wrong" only in relation to the system of views where the opposite is true, i.e., the *false system of views is always contradictory*. Therefore, there are only two possible formats for evaluating any conflict:

1) the *negative format* occurs where two subjects apply conflicting opinions in the same respect. In this case, at least one of them is wrong, i.e., it violates the natural limits of the applicability of one's own opinion, which makes the latter contradictory, and therefore impractical.
2) the *positive format* occurs where the two subjects, on the contrary, are equally consistent, i.e., their opinions are not contradictory, and therefore relatively correct. In this case, they are applicable not in one respect, but in different ones, which means that they have no real reason for conflict.

One may ask, how does this second option of agreeing on positions differ from the usual *compromise*? The difference is that *compromise* is the way for both sides of the conflict to derogate from their principles. And this, as a rule, either does *not* suit both sides at all, or arranges them only temporarily, until one of them gains the strength to take revenge. Therefore, this way does not completely remove the essence of the conflict.

On the contrary, according to Socrates, only subjects who are *uncompromising*, i.e., do *not* depart from their principles, are always consistent in their observance; therefore, they do *not* violate the limits of their applicability, that is, they do *not* conflict with each other. And if one violates the limits of the applicability of one's principles, one contradicts oneself and thereby already deviates from one's principles.

This theory may seem like a paradox. But Socrates invariably confirmed it in his disputes with *any* opponents, leading their opinions to external agreement by eliminating internal contradiction.

Hence, it is clear why Socrates, unlike other great moralists, did *not* leave any developed system of moral commandments. The fact is that the principle of *universal moral consent* prohibits the postulation of any moral commandments other than on the basis of the direct or indirect consent of those to whom they are prescribed.

Of course, obtaining the consent of *all* humankind is unthinkable even from the point of view of Socrates' theory. Therefore, it was only about the need to seek agreement with each particular opponent, as a moral conflict was discovered. Such a task for Socrates, on the contrary, was quite solvable through discussion. But for this, Socrates' ethics initially had to acquire the character not so much of a theory as a discussion practice.

6   Solution 2. Truth Birth in the Dispute

Socrates defined the practical solution to the problem as *maieutics*, i.e., the art of helping the truth to be born in disputes through probing questions. According to Socrates, this is the only type of argument suitable for solving the problem of moral *consent*. For this technique assumes that all reasoning is based only on the answers of the opponent, and all conclusions are approved only by the *latter's consent*.

This is with what Socrates' doctrine challenges contemporary culture. In ancient times, all types of argumentation were divided in relation to the opponent into the two: *dialectics* and *rhetorics*. And if Socratic maeutics belongs to the first type, contemporary culture prefers the second one. This can be seen, at least, from the fact that today all types of argumentation are often termed as "rhetorics", i.e., *the art of eloquence*. Although the difference between them is noticeable:

1) rhetorics is a technique of argumentation through the *monologue* in which the host (speaker) oneself expresses all one's arguments without asking the opponent for consent with those. And if the former asks questions, they are only *rhetorical*, i.e., not requiring answers.
2) dialectics is a technique of argumentation through the *dialogue*, in which the host not so much claims as asks and refutes the opponent's opinion only with the latter's own answers, demonstrating their contradiction.

There is no reason to decide which of the two approaches is preferable in general: each of them has its own field of application. For example, *dialectics* is preferable whenever one wishes to make the opponent as conscious as possible. For by asking questions, one does not inspire anything, but simply activates the responder's consciousness. On the contrary, *rhetorics* has an advantage when it is necessary to inspire something to a large crowd or individuals who are in a state of affect. For in these circumstances, the opponents cannot consciously communicate and they perceive the arguments rather emotionally than intellectually.

Therefore, preferring rhetorics to dialectics, contemporary culture aims at introducing people into a state of *rather affect than consciousness*. And in this case, conflict in the society cannot but increase.

Obviously, it cannot be said that dialectics as a technique of probing questions has completely lost its popularity today: it is still actively used in such areas as jurisprudence, pedagogy, psychotherapy, etc. However, within the framework of this technique there are also two opposite directions: *maeutics* and *sophistics*. If Socratic technique belongs to the first type, the contemporary culture prefers the second one. This is evident from the fact that today all types of dialectics are often termed as "sophistics", i.e., the *technique of philosophizing*. Although the difference between them is noticeable:

1) Sophistics is a technique of false philosophizing, for Socrates proved that no one can possess true wisdom. Therefore, "sophists" began to call people who can a) confuse the interlocutor with ambiguous questions; b) catch the former on the inaccurate answers; c) withdraw consciously false conclusions from them and, due to this, they seem to be "wise".
2) Maeutics is a technique of true philosophizing, i.e., the art of helping truth to be born in disputes; at least, that is what Socrates thought. And in order to correspond to this name, the maeutic technique is to differ from sophistics in all the items concerned.

Firstly, maeutics should *not* confuse the responder with ambiguous questions. And since it is impossible to know in advance what exactly the responder will consider "ambiguous", maeutics should provide the responder with the right to counter clarifying questions on his or her part. Maeutics should not limit the responder to answers "yes' or "no", but give the responder the right to clarify in the form of double values of truth. For example, a) both "yes" and "no", but in different respects; b) neither "yes" nor "no" if the subject matter is unrealistic.

Secondly, one cannot catch the responder on the inaccurate answers. This means, it is necessary to give the responder the right to change his or her answers and combine them until (s)he makes a winning combination for oneself. And if the respondent's thesis can be true at least within some limits, then during the conversation, according to Socratic rules, these limits will certainly be discovered.

7 Virtue as Knowledge of Virtue

The unpopularity of maeutics today is due to the fact that the speaker who is arguing, as a rule, aims at imposing a *certain worldview* on the opponent. And if the opponent rejects all the arguments in his or her favor, the leader simply does not know what to do next. But the problem is not the flaws of the technique, but the incorrect target setting of speakers.

For Socrates there was *no* such problem precisely because he did *not* aim at imposing something specific on his opponent. He sought only to help everyone find agreement with both his own views and with others'. It was the inevitability of consent that was the truth born in maieutic disputes. But with such a target setting, Socrates could arrange any outcome of the conversation:

1) the *negative result* occurred in the cases where the opponent's answers were contradictory. This contradiction was tantamount to implicit recognition of its wrongdoing and required correction.
2) the *positive result* occurred in the cases where opponents eliminated all their internal contradictions. But then agreement reigned between them in at least one of the two formats mentioned above: either one directly admits oneself wrong, or both are recognized as right in different respects. In both cases, the reason for moral conflict completely disappears.

The problem is that Socrates did not immediately postulate a definition of virtue which he himself considered sufficient. The technique obliged him to always proceed from the definitions preferred by opponents. It is assumed that Socrates should have received his interpretation from their answers by conclusion. And this meant that Socrates essentially had two definitions of virtue:

1) the *final definition* required to be considered "virtuous" only the lifestyle in which the subject loves justice, strive for agreement with oneself and others, and therefore chooses maieutics as the best way to solve problems. In the dialogue "Gorgias" [13], Socrates receives this definition as a conclusion from the whole discussion.
2) the *original definition* required that only the lifestyle which the opponent oneself *would deliberately prefer as the best for oneself personally* is considered to be "virtuous". In the dialogue "Protagoras" [16] Socrates concludes from this that virtue cannot be preferred only *unconsciously.*

That is how Socrates comes to his paradoxical idea that for perfect virtue there is enough *knowledge of what it is.* For no one, knowing the best way, will consciously choose the worst. Therefore, Socrates considered "the original virtue" to be the desire for this knowledge, i.e., *maieutics* again, and for him "the original sin" was what blocks such a desire in people.

At first glance, there are two such "original sins": either a person is sure that the desire for this knowledge is unnecessary to him or her, or (s)he is sure that this knowledge is impossible, even if necessary. Consider both cases.

8   Sin 1. Dogmatism as a Crime

Dogmatism treats maieutic cognition as *unnecessary.* After all, in order to understand who is right, it is necessary to doubt at least a little about your correctness. Dogmatists, by definition, have no such doubts. Therefore, in any conflict, they prefer to blame the opponent in advance, and rarely need evidence.

At the same time, dogmatics violates the Golden Rule: after all, none of them wants that they themselves were also unfairly accused of wrong. Especially, if this can be followed by some kind of sanctions from the prosecutor.

Most often, dogmatics solves the problem at the expense of the idea of "special moral status", which allegedly allows them personally *not* to observe the Golden Rule and live by double standards. But this idea collapses, *regardless* of what nature they write to their special statuses:

1) the *subjective nature* means that the subject oneself gave a special status by a conscious decision, because (s)he found it *appropriate*, i.e., corresponding to his or her highest goal. And then there is a question: Does (s)he undoubtedly know that his or her true higher goal is just that, and that it corresponds to a special status, and not equal with all people?

2) the *objective nature* means that the subject did *not* give oneself a special status, but received it from certain higher principles which are wise and are not mistaken. And then there is a question: Does the subject undoubtedly know that those principles are wise, and that they did not give the same special status to any of his or her opponents?

Positive answers to all these questions are theoretically possible. But in practice, they are easily crashed by means of the Socratic diagram. Therefore, even a short-lived maieutic conversation is enough for any dogmatic to answer *negatively*, i.e., (s)he would allow *any* possibility in these matters. And then (s)he will have to choose:

1) *either to recognize the special status of all* on the grounds that every reasonable being may have it. But then everyone would presumably have an equal status with all other reasonable beings.

2) *or to deny the special status to all* on the grounds that every reasonable being may *not* have it. But then, especially, everyone's status would be equal with all other reasonable beings.

3) *or to recognize a special status for some* in one's personal preference. But then this status will *no* longer be justified either by the objective wisdom of the higher principles, or by subjective expediency, due to the subject's lack of undoubted knowledge about both.

It is easy to note that in all three cases, the idea of "special moral status" is unfair. Therefore, justice must be interpreted exactly as Socrates interpreted it, i.e., the *equality of all before moral law*. And then dogmatism (as a denial of this equality and as an illusion of "undoubted knowledge" about its infallibility) will be a crime against justice.

9 Sin 2. Skepticism as a Crime

Academic sceptics self-consciously adopted a zetetic form of scepticism directly from the presentation of Socrates in Plato's dialogues [17]. The main source is Plato's dialogue the *Apology*, where there are five Socratic claims that may appears to justify it. Priscilla Sakezles claims that Socrates is skeptical in a certain sense: but he is not dogmatic and self-contradicting skeptical that he is often made out to be [18]. Some scholars have advanced in their own names skeptic interpretations of Socrates [19, 20]. Katja Maria Vogt claims that Socrates advocates the importance of critically examining one's own and others' views on important matters, precisely because one does not know about them [19].

But this Socratic skepticism must be distinguished from *Pyrrhonism*, i.e., from the real skepticism founded by Pyrrho of Elis. The latter, like dogmatism, rejects maieutic practices and related tasks. But *not* so much out of confidence in his rightness, but, on the contrary: out of confidence *in the impossibility of deciding who is right.* After all, if nothing can be known undoubtedly, no solution born in disputes can be considered to be the final truth. Therefore, it is better to supposedly refrain from judgments in any dispute, not to discover anything, but to leave each of the conflict parties in their opinion.

It's a double mistake.

Firstly, Socratic maieutics does not require that decisions be made in the status of "undoubted knowledge" or "final truth". It is enough that the interlocutors answer each other all the questions at least hypothetically. With *any* consistent system of answers, the law of contradiction guarantees the conflict parties firm consent in at least one of the two formats mentioned above.

Secondly, if skeptics object that we also *cannot undoubtedly know* anything about the existence of such guarantees, because we cannot undoubtedly know *anything at all*, then the question will arise: does the skeptic undoubtedly knows *this*? The positive answer to this question is excluded not only by the abbreviated diagram, but also by the condition of the question itself: after all, if you cannot undoubtedly know *anything*, then *this* is also impossible. Therefore, every skeptic must answer negatively, i.e., (s)he must allow both hypotheses in this matter:

1) the firm consent of the conflict parties is impossible.
2) this is quite possible.

But the ideal of justice, according to Socrates, requires only the second hypothesis to be preferred as a "working" one: it is it that ensures moral harmony's coming. Since Pyrrho's skepticism, on the contrary, prefers the first hypothesis, it is also a crime against justice.

10   Original Sin. Knowledge of Good and Evil

Therefore, we found that dogmatism and skepticism are two opposite extremes, leading to a deviation from the original Socratic virtue. But we also discovered that both of them have one common "original sin": the illusion of "undoubted knowledge" of Good and Evil. The difference is that for dogmatists it is the undoubted knowledge of their *own rightness*, whereas for skeptics it is the undoubted knowledge of the *impossibility of discovering who is right*. In both cases, this is a rejection of the desire for moral harmony.

This makes us recall the biblical myth of "original sin", where the progenitors of humankind, Adam and Eve, tasted fruits from the Tree of Knowledge of Good and Evil.

But now we understand that the myth did not mean the true knowledge of Good and Evil, but the illusion of "undoubted knowledge", which leads to one of the extremes in moral conflict and thereby leads us away from virtue.

On the contrary, knowledge of the truth expressed by Socrates "I only know that I know nothing" frees a person from both of these extremes that hinder the search for consent. Therefore, this knowledge is a sufficient condition of Socratic virtue, which was required to be proved.

It is not surprising that representatives of both extremes equally disliked Socrates and his doctrine. Dogmatists accused him of skepticism for doubting everything. And skeptics accused him of dogmatism for not losing hope of discovering who was right. But they especially disliked Socrates for the fact that in practice he always succeeded in this.

11   Results of the Research. Conclusion and Discussion

As a result of the research:

1) It is demonstrated that the imaginary contradiction in Socrates' formula "I only know that I know nothing" is removed by using the word "knowledge" in two different meanings. Namely, the words "I know nothing" mean Socrates' lack of undoubted dogmatic knowledge. But his very knowledge of this fact has a different maieutic quality, i.e., the status of the working hypothesis.
2) It is demonstrated that the imaginary tautology in Socrates' formula "Virtue is knowledge of true virtue" takes place in the original definition of virtue by which Socrates attracted opponents to study. But its result was a meaningful definition of virtue as the desire to live in harmony with oneself and with other people.
3) It is demonstrated that disposal of the "undoubted knowledge" illusion saves a person from both dogmatism and skepticism, i.e., from both extremes that prevent the search for moral consent with oneself or with other people. Therefore, according to Socrates, it is a sufficient condition of original virtue.
4) It is demonstrated that Heraclitus' doctrine of the unity of opposites was transformed by Socrates into a form of doctrine of the original moral harmony of all consistent worldviews. And this means that it is Socratic *maieutics* as a way of mutual deliverance of opponents from internal contradictions that is a true path to perfect virtue.

Being a philosophy in action, this Socratic technique can be applied in various fields to achieve agreement and understanding, becoming a technology for uniting the multilingual and multicultural world. Sławomir Redo claims that it very well lends itself to the implementation of good governance for social justice across the globe [21], the possibilities of its application in educational practice are particularly interesting [22–26].

It remains for us to note that traditional techniques of conflict resolution (such as violence, inspiration and the way of mutual concessions) do not produce a reliable result, as they do not address the internal causes of the problem. Socrates' technology, on the contrary, is aimed at identifying the *imaginary nature* of any conflict, and, therefore, does not require either violence, compulsion, or mutual concessions. And only it can contribute to maintaining sustainable peace and social cohesion in the multilingual world.

References

1. Bylieva, D., Nordmann, A.: Technologies in a multilingual world. Technol. Lang. **3**, 1–10 (2022). https://doi.org/10.48417/technolang.2022.03.01
2. Yang, W., Wen, X., Wu, M., Bi, K., Yue, L.: Three-dimensional conflict resolution strategy based on network cooperative game. Symmetry **14**, 1517 (2022). https://doi.org/10.3390/sym 14081517
3. Taves, A.: Worldview analysis as a tool for conflict resolution. Negot. J. **38**, 363–381 (2022). https://doi.org/10.1111/nejo.12403
4. Hussain, K., Patel, N., Fearfield, L., Roberts, N., Staughton, R.: Mentorship in dermatology: a necessity in difficult times. Clin. Exp. Dermatol. **47**, 622–623 (2022). https://doi.org/10.1111/ced.15071
5. Brull, S.J., Murphy, G.S.: The "True" risk of postoperative pulmonary complications and the Socratic paradox: "I Know that I Know Nothing." Anesthesiology **134**, 828–831 (2021). https://doi.org/10.1097/ALN.0000000000003767
6. Lesher, J.H.: Socrates' disavowal of knowledge. J. Hist. Philos. **25**, 275–288 (1987). https://doi.org/10.1353/hph.1987.0033
7. Taylor, C.: Socrates: A Very Short Introduction. Oxford University Press, Oxford (2001)
8. Vlastos, G.: Socrates' disavowal of knowledge. Philos. Q. **35**, 1–31 (1985)
9. Gulley, N.: The Philosophy of Socrates. St. Martin's Press, Griffin (1968)
10. Irwin, T.: Plato's Moral Theory. Oxford University Press, Oxford (1974)
11. van Ophuijsen, J.M., van Raalte, M., Stork, P.: Protagoras of Abdera: The Man His Measure. Brill, Leiden (2013)
12. Plato: Symposium. Cambridge University Press, Cambridge, UK (2016)
13. Plato: Gorgias. Greenbooks Editore (2020)
14. Fichte, I.: The Vocation of the Scholar. CreateSpace Independent Publishing Platform, California (2017)
15. The Cosmic Fragments by Heraclitus (of Ephesus). Hassell Street Press (2021)
16. Plato: Laches, Protagoras, Meno, Euthydemus. Harvard University Press, Cambridge (1977)
17. Blyth, D.: Plato's Socrates sophistic antithesis and scepticism. Plato J. **19**, 25–42 (2019)
18. Sakezles, P.: What sort of Skeptic is Socrates? Teach. Philos. **31**, 113–118 (2008). https://doi.org/10.5840/teachphil200831214
19. Vogt, K.M.: Belief and Truth: A Skeptic Reading of Plato. Clarendon Press, Oxford (2012)
20. Miller, T.: Socrates warning against misology. Phronesis **60**, 145–179 (2015)
21. Redo, S.: Is Socrates mortal? on the impact of Socratic logic on teaching and learning the united nations crime prevention law. In: Kury, H., Redo, S. (eds.) Crime Prevention and Justice in 2030, pp. 623–635. Springer, Cham (2021). https://doi.org/10.1007/978-3-030-56227-4_30
22. Morabito, M.S., Bennett, R.R.: Socrates in the modern classroom: how are large classes in criminal justice being taught? J. Crim. Justice Educ. **17**, 103–120 (2006). https://doi.org/10.1080/10511250500335726
23. Redo, S., Pływaczewski, E.W., Langowska, A., Alkowski, P.: A Socratic contribution to culture of lawfulness for teaching criminology. Białostockie Stud. Prawnicze **23**, 97–111 (2018). https://doi.org/10.15290/bsp.2018.23.03.08
24. Samorodova, E.A., Belyaeva, I.G., Bylieva, D.S., Nordmann, A.: Is the safety safe: the experience of distance education (or self-isolation). XLinguae **15**, 3–13 (2022). https://doi.org/10.18355/XL.2022.15.01.01

25. Bylieva, D., Hong, J.-C., Lobatyuk, V., Nam, T.: Self-regulation in e-learning environment. Educ. Sci. **11**, 785 (2021). https://doi.org/10.3390/educsci11120785
26. Jianwei, Z.: Different images of knowledge and perspectives of pedagogy in Confucius and Socrates. Complicity Int. J. Complex. Educ. **9**, 1–7 (2012). https://doi.org/10.29173/cmplct 16535

Rendering as Such: Maieutics and Untranslatable Categories

Walker Trimble(⊠) ⓘ

Herzen State Pedagogical University of Russia, St. Petersburg 191186, Russia
parikampi@gmail.com

Abstract. Translation has always been understood as a practice that intends to render compatibility out of incompatibility. Two cultures and modes of thought, two historical ages, are bridged by a conveyance that neither side ever considers to be ideal. The model of translation (source to target) has also been used as a conveyance between categories which were never expected to be juxtaposed. 'Translation' has been used to describe social movements, even changes in the composition of material objects. This essay considers four types of incompatible translation models and argues that they are often used to make things which are not signs appear to behave as though they were. Cognitive science egregiously performs this fallacy by means of a presupposition we call crypto-Cartesianism. The critique of untranslatable categories is applied to computationalist models of brain science. We then show that a different model of compatibility can be helpful to understand the relationship of stimulus and brain activity if we consider the movement of source to target as a maieutic guided by the agency of a conscious subject, a professional (physician or researcher), and a responsive use of technology.

Keywords: Translation theory · Untranslatable categories · Language of Thought · F. Schleiermacher · Awake neurosurgery · Neuroimaging technology

1 Introduction

'*Traduttore, traditore*' is, bar none, the most famous aphorism associated with translation. It sets up the legal conditions of an original sin. The moment the translator conveys their reader from unknowing to knowing they have already betrayed the *original* sense. The fact that the phrase itself remains untranslated means that it is not susceptible to the liar's paradox. We relish its alliteration rolling off the tongue, pulling it out of our cache of useful phrases though the only other words we may know in Italian are '*ciao*' or, perhaps, '*braggadocio*'. The phrase relies on the play of prefixes which serve to superimpose the senses of 'passing over' and 'turning over'. One 'passes over' one sense to another; one 'turns over' one's friend to the enemy. These nice side benefits render the phrase itself untranslatable, even if many other languages do have the same play of prefixes.

The concept of untranslatability often comes to account for the situation of subjective experience. I cannot know if what you call in Russian *toská* is entirely the combination of boredom, distress, and ennui that I do not have a word for in English. Furthermore, I do not know whether your blue is my blue. Qualia – the purported mental correspondents to external things – are always subjective. We transcend their untranslatability by an agreed form of communication based on the presumption that both sets of words are meant to act as elements of communication. I can also come to understand your *toská* by getting to know you. By learning I grow through the praxis of my own subjective lexicon.

One of the earliest treatises on translation comes from Marcus Tullius Cicero's (c. 79–46 BC) *De optime genere oratorum.* Dismissing the value of translation as something only for students (if you were learned you would know the language yourself), he writes that he: "… did not recast debates as a translator, but as an orator, keeping their same meanings but with their forms, their figures, so to speak, in words adapted to our idiom."[1] Of course, he is disingenuous in this. Oratory is never sense over style. His wishes to creatively transform Greek oratory to inform the Latin rhetoric that he has taken into his keeping.

Yet his 'word', or 'form', versus 'sense' has persisted as the most valuable distinction in translation theory. In actuality it is more of a practical distinction. When the translator can (in oratory or poetry) manage a fine rendering of both form and sense, one does not have to be sacrificed for the other. The case is more often that one must compromise between the two. Here better gets the form, there better gets the sense. Thus the work of translation is one where one does one's best to navigate form and sense around the distance between the two languages. Few us have the virtue of Cicero to make a figure that is moulded into a new culture of expression.

A different approach to the problem of form and sense came in the work of the theologian and orientalist Friedrich Schleiermacher (1768–1834). He argued that adaptation to "our idiom" occludes the salutary distance between ways and means of thought. He adopts not so much a textual but what we might now call a cognitive approach. He asks, should could a translator drag his German reader along to make him think like a Roman, or drag a Roman back until he can think like a German [2, p. 49]? He concludes that the first option is best because of the demands put upon the reader. Efforts might be clumsy, infelicitous cribbing of an uncertain level of understanding. But, over time, new foreign ideas and concepts will populate readers' minds. Schleiermacher puts tremendous expectations on translators as forgers of these intentionally rough but coruscating renderings and on the reading public of which he speaks almost like a barbarous nation yearning for change.

The problem with this theory is that there are four possible sources for any element in a translation: the original author, the translator, and the respective languages themselves. If the translator *renders as such* some un- or quasi-translation, the reader can only exclude the target language as the source. A German knows his or her own competencies in German. Not knowing the source language, and not wishing to be burdened by some apparatus appended to the text by the translator, the reader cannot be certain if the

---

[1] Converti enim ex Atticis duorum eloquentissimorum nobilissimas orationes inter seque contrarias, Aeschinis et Demosthenis; nec converti ut interpres, sed ut orator, sententiis isdem et earum formis tamquam figuris, verbis ad nostram consuetudinem aptis. In quibus non verbum pro verbo necesse habui reddere, sed genus omne verborum vimque servavi. Non enim ea me adnumerare lectori putavi oportere, sed tamquam appendere [1, V.14].

'alienating' element came from the author's style or design, were an adaptation by the translator, or were a standard part of the source language itself. In fact, Schleiermacher wants to veer from Cicero's path to pull readers toward an understanding of the alien, untranslatable object which is nowhere near 'our idiom' but the different thought-world of the Other. This is the translation theory of the Romantic age with its fascination with folklore, the sublime and grotesque. By rendering the text as such, the reader is guided along a path of subjective transformation. Any other effort is mere betrayal.

This essay examines forms and senses that go beyond language and yet are set within translation's binary form of [original → target] which we call a 'pairing'. We maintain that outside of semiotics this form is itself a category error and so those who use terms such as 'translation' or 'language' are using the model either as analogy or to gloss over these differences for ideological purposes.

We will begin with classifying four different kinds of untranslatable pairings. We will then go on to show how they apply to the very controversial idea of 'mental representations' and the correspondence between subjective ideas and the things they are supposed to correspond to. This leads to examining the ideological and philosophical crutch which lies at the root of representation theory. Finally, we will see how technology gives us a new philosophical conception of how the biology of mental representation takes place outside of any reference to semiosis but through a maieutics comparable to Schleiermacher's trajectory.

2 Untranslatable Categories

2.1 Bungling Terms

Within the world of discourse, we are told that there are lots of untranslatable things. Russian lawyers often comment that their legal system was based on the German one, yet their language is impoverished of German precision and so is doomed to suffer from another original sin. American philosophers often make the same claim about English and despair for the legacy of the analytic tradition in their poor linguistic soil. It is difficult to imagine generations of Russians and Americans floundering about the world without the proper German words for things. Surely every language has just enough precision for the world it needs. Furthermore, there are things such as the law and philosophy that simply do not exist without the language they are written in. A translation of a law is not even a betrayal, it is a crib, a cheat sheet for non-professionals, or some way for one system to make judgements about another. What is more interesting about these laments is the function they are supposed to perform for Russian jurists and American philosophers. Is it passing the buck? "Adam said, 'It was the woman who gave me to eat." "Eve said, 'It was the snake.'" Perhaps it is the positing of an ideal state itself from which all laws and all philosophy can be imperfect reflections. The bad translation is a foundational myth which allows practitioners some wriggle-room to set up their own meanings. To be a traitor from the onset lets one be the servant of at least two masters, perhaps none.

I have had the tremendous advantage of working on the translations of my own work and finding that the betrayals my translators and I have inflicted against my texts have often led to their improvement. Schleiermacher's Cicero might well wonder at a better turn in a German phrase. Some clarifications required *along the course* of the translation make their way back to a bettered original. From the author's point of view, the impression is that the result is like an arc-shaped slot over a set of gears that allows meanings to slide back and forth and fulfill two functions at once. Nevertheless, the resulting movement of the dials takes place in both languages and is visible as two different sets of meanings, but the inner workings show rounded, versatile boundaries that gracefully oscillate and perform two functions at once. But translation is not primarily an intellectual exercise for bilinguals. What special sense readers with command of only one of the given languages get out of this is hard to reckon with. A text in the only language you can understand is always untranslatable.

Among many accusations of the 'untranslatable' one of the most useful, if neglected, is that of the visionary analyst of conversation Harvey Sacks. He devoted a great deal of his career to determining the categories which allowed a speaker within a conversation to introduce a topic. These turned out to be incredibly varied and complex. Since his materials were often drawn from psychological therapy sessions and suicide hotlines, many of his interlocutors were constantly touching upon the topics of their personal troubles and the things that got them into the positions they were in. Their conversation then entailed techniques for naming who they were and determining the categories in which they fit as patients or members of a group. In this context he noticed the presentation of sets of predicates where the speaker identified themselves with a set of technical terms – what we would now call 'psychobabble'. A teenager in a group therapy session would accuse others of "tearing him down". A patient would say that she did not want a psychiatrist "calling her infantile" [3, p. 200]. Sacks compared these statements to those made by anthropologists, quoting a sentence from the preface of Ronald Berndt's *Excess and Restraint*:

> We were viewed as returning sprits of the dead who had forgotten the tongue of our fathers and wanted to relearn it [in 3, p. 200].

Sacks regards the phrase 'we were viewed' as a framing device used by anthropologists to put a translation of the native subjects' concepts into a predicate. I take him to mean that this is not an actual translation, but the anthropologist's paraphrase of the subjective views of those they study. Sacks considered this predicate to be an asemantic statement. If Berndt stood up at an academic conference in Australia and said "I am the returning spirit of the dead..." he would be taken as mad [3, p. 201]. We might now say that the sentence violates Grice's maxims of communication. Sacks says that this is not a generative statement that one could use to build up other statements. But nor does he say that it is nonsense, which has a philosophical function on its own.

'We were viewed...' serves to set out untranslatable categories of thought. Sacks believed this could be extended to any number of categories. The speaker thus frames out these alien elements as something inserted in speech to perform a function. Especially important was the imposition of incompatible categories over categories that could be understood. Sacks recalls the Freudian programme of forming a science of psychoanalysis. While we may now read Freud's frequent protestations of the 'science' he was creating as quaint, Sacks notes that Freud was challenged with establishing new sets of terminology. One could come up for terms to describe newly discovered phenomena in chemistry or astrophysics and not be concerned that their application would slip into a corrupt lay usage, but the fact that everyone is something of a lay psychologist means a special alienation of language must take place for the precision of since to keep a firm hold on its semantics. Sacks argues:

> The problem is that members [of society in general] take it that such categories – 'manic depressive,' etc. – are additions to a list of categories that exist already, and can be used in just the same fashion that the old ones are usable. They may be better, but they do not otherwise modify the structure of the class of which people come to be seen as members. However, the professional constructing these new categories may take it that one major task he has is somehow to build them so they are unusable in the way that the categories he sees them as replacing were usable. That is, the professionals put it as a programmatic task that they would like to have it that the statement "You're a manic depressive," for example, would be nonsense in ordinary English, i.e., unless said by a therapist [3, p. 202].

He concludes, rightly, that such a project is likely to fail [3 p. 203]. In part this is because these new categories can approximate things which laymen may have noticed in some form but given no name to. Even their lay usage still can acquire its own generative function. I once heard my psychologist friend say "He was being what my husband would call 'hysterical'". If she (the psychologist) had used the term "hysterical" it would have been a diagnosis. Her (sculptor) husband would have meant the term in the lay sense – 'freaking out'. If the lay term had no meaning there would have been no compass within which she would have used the word. So these terms move into a realm of translatability, but through popular betrayal. Through betrayal the aquire another meaning. This is because there is a great deal of power in the ideology of science. Scientific terms give authority to those who use them and may themselves intensify the classes of membership of those to whom they apply. When, 60 years after Sacks, one hears a soccer-mom say "I'm so ADHD!" about her habit of checking the school roster on her smartphone, we can speculate she might be using the untranslatability of a professional diagnosis to accord precision, authority and a certain lack of responsibility for her own behaviour.

2.2 Translatable Matter

Thus there are cases when the untranslatable is born and other cases when untranslatability is thrust upon a set of terms. There may be other cases when translatability would not even be presumed of a certain pairing, for example if the matter has nothing to do with language at all, or even a system of signs. The term 'translational medicine' has an extremely varied and wide set of uses. It roughly refers to the social process of 'translating' scientific discoveries and clinical research into medical practice. This could have been termed 'transferable' medicine, or 'relatable' medicine, or perhaps the best option, 'compatible' medicine. It is difficult to determine what is being translated: knowledge, techniques, information, production methods. The usage seems to have come out of pharmaceutical marketing firms who might have wanted to capture the sense of an operable transfer of something across institutional boundaries that otherwise would have been incomprehensible to each of their respective moieties: researchers, formulation chemists, physicians. In this sense Sacks' 'untranslatable categories' is preserved through its violation: each community enshrines its scientific (and bureaucratic) integrity through incommensurability, yet rather than making them 'compatible', they must be 'translated' so as to preserve that incommensurability. Presumably this in turn preserves the position of the marketing and HR bureaucrats that seem to have come up with the term in the first place. Translation needs translators [4].

At least one can say that these strivings toward social compatibility shift through media that are mostly informational and semiotic. Biology, for its part, often uses 'translation' for things which have no semantic content at all. There translation is often used to name the biochemical processes which account for the formation of complex proteins. That we are not surprised by this shows the extent to which technical terminology has moved out of its preserve into lay usage and back again, each group corrupting the other. Popularization and re-popularization is, obviously, the opposite of the process which troubled Freud and Sacks. Rather than scientific disciplines creating alien terms to mark off new territories of knowledge, biology has popularized terminological shorthand and then adopted it into its own formal language. Of course, physical things cannot be negated or translated like semantic entities. 'Translating' an acid is like 'translating' an orange. It is like demanding that someone provide you with the opposite of a rabbit. The term 'protein synthesis' is perfectly good for biochemistry, 'transfer' is useful. Why genetics needs 'translation' may be a matter of ideology. The four nucleotides that make up the DNA molecule are abbreviated as letters, they bond with each other and are reflected in their opposites in the formation of RNA. In fact the chemical nature of the reactions is no more semiotic than any other chemical process – fire or oxidation. But we have reasons for rendering it as language. The whole processes is termed the genetic *code* as if it were a cypher or system of laws. We have chosen to make this set of chemical interactions into a legal code of life.

Based on what has been reviewed so far, we can read this target category as being one of four things:

1. A semantic system which we have no access to because we don't know the language (Berndt's 'We were viewed as…'),
2. A semantic system which is deliberately concealed (Freud's technical terminology),
3. A non-semantic system which can be converted into one for the purposes of translation ('translational medicine'),
4. A completely non-semantic system for which the term 'translation' is used to give us the impression of some analogical precision and authority ('genetic translation').

2.3 Translatable Thought

Even more elusive and complex is the understanding that the biological processes which make up the brain's activity are also a kind of translation. 'Messages' to and from the brain and the body are 'encoded' in nervous signals. There is also neural 'crosstalk'. Most problematically, the phenomena of the world are translated – encoded – in the brain. This is more than giving the opposite of a rabbit. The rabbit must be isolated from all the stimuli around it (trees, snow) before it is identified, reified into something which could correspond to its neurologically encoded correlate, and then matched with that correlate to inform the organism 'that's a rabbit'. Millenia of epistemological speculation have circled round how this can and cannot take place. When the mind is conceived as a primarily immaterial being, it can take with it some of the immaterial semantic properties of words. Throughout most of human history both things and words were presumed to be immaterial, or that (in Stoicism and Buddhist epistemology) their material nature was subtle and imperceptible.

However contemporary science is built on physicalist presuppositions. In Locke's day, no one could have conceived of the bloody jelly of the brain as being able to contain the self. The discovery of the physiology of nerves by Ramón y Cajal and Golgi around the turn of the 20th century and subsequent discoveries of the statistical ways in which nervous systems work led to an algorithmic explanation of nervous activity. Anatomy and, finally, the neuroimaging techniques that proliferated in the 1990s showed an extraordinarily complex and versatile organ. With some 100 trillion connections it was taken to be 'the most complex place in the universe'.

This has had three deleterious effects on the way we conceive of how thoughts are 'translated' into matter, the last of which we shall take up here. First, the notion that the brain is more complex than the known universe fosters a fallacy of volumes. Since we must conceive of the world, and also a number of things that do not exist in it – the plotline of *War and Peace*, the rules of chess, tomorrow's grocery list – the fact that the brain is more complex comes out quantitatively about right. This is a primitive notion that all things are a matter of containment, like the professor of the Grand Academy of Lagado in *Gulliver's Travels* who declared it was better to carry the referents of the words he used on his back. This presumes that the world is made of a discrete inventory of objects and the brain made up of a discrete catalogue of encodings. Such a view is false for the fact that phenomena are made up of continuities and discontinuities which are objectively hard to define. It is also false because a great deal of the brain is multifunctional and networks are constantly repurposing themselves for different sets of stimuli [5].

Second, this then forms a popular myth (untrue but conceptually productive) that there is a medium of exchange, a set of discrete entities and an encoder which can render one into the other. This makes the untranslatable translatable. Rather than framing, say, the operations of the soul as being beyond any set of processes and uniquely non-generative, it conceives of all phenomena as *a posteriori* exchangeable.

Finally, this pseudo-semiotics is given a pseudo-teleology. Now that the mode of exchange is clear, there needs to be an operator that puts the exchange into effect – a translator. Just as in the other cases, this agent independently operates within the translation machine. This is John Searle sitting in his Chinese Box. In a superb theoretical article, the pioneering developmental psychologist Eleanor Gibson gave this kind of thinking a better characterization:

Dualism beset psychology for hundreds of years until the theory of evolution and a new wave of naturalism brought some healthy changes. Modern psychologists, until recently, have tended to say that the mind-body dichotomy is a philosopher's problem; if we do our job in the laboratory, it will go away so far as we are concerned. But in recent years it has cropped up again and has been the motive for some far from trivial confrontations. The "cognitive revolution" had something to do with the reappearance of the old dualism battle, influenced primarily by nonpsychologists such as Chomsky and Fodor, who have urged a Cartesian philosophy (and consequent psychology) on psychologists working on problems of language, perception, and conceptual thinking. The way the world is perceived must derive from preordained rules and concepts (an analogue of Descartes' innate ideas) that serve as premises for inferences about it (I think that is the substance of what they claim) [6, p. 12].

At once Gibson would seem to be taking up a typical evasive gesture to argue that philosophical problems are not important. Or perhaps, more justly, she wishes to frame them as untranslatable to the realm of experimental psychology. More likely, she means to say that the experimental psychologist frames her enquiry in a fashion so that such questions do not arise. With the sour implication that "non-psychologists" have muddled the field, she continues with an observation of great worth. The "preordained rules and concepts" for Gibson are Chomsky's 'universal grammar' and, respectively, Fodor's 'Language of Thought hypothesis' or 'mentalese'. Implicitly, she sees that the problem of dualism resurfaces because computational rules have been linguistically and logically derived and then set in the place of biological processes. Of course neither Noam Chomsky nor Jerry Fodor would have ever professed themselves to be dualists, they regard all there is about the mind to be in the brain.[2] Yet they arrive at their accounts of language and thought wholly outside any biological accountability. The principles they derive theoretically then serve to operate in the mind like the Cartesian soul. They are immaterial and voluntarist. The ideology around information technology has gone on to serve this notion by creating an artificial parallel with computers. Just as your computer is able to 'act' in a certain way and make calculations on the basis of an internal program, so the brain is *encoded* with certain algorithms that govern the way that it works. Relying on Gibson's critique, I have termed this approach 'Crypto-Cartesianism'. Many of the theoretical difficulties behind such issues as the Hard Problem of consciousness [9] and the nature of discrete inherited behaviours arise from and/or are greatly complicated by crypto-Cartesian presuppositions.

If we view the relationship between the world and what the brain does, or language and what the brain does, as computed representations, we can see why crypto-Cartesianism is very dependent upon an explicit model of translation. There must be a discrete set of stimuli in the world (the source text) and a discrete encoding of that stimuli, qualia, (the target text) and a translator as an algorithmic Cartesian spectator. However, a vast literature of naturalistic epistemology has evolved around the argument that the target is objectively unknowable [8]. We thus are confronted with another untranslatable category. To take the four versions of untranslatable categories, we can reasonably rule out **2**, unless we think of the brain as an encoding machine designed by an ill-spirited god. It is hard also to conceive of **3** as being very likely because the semiotic and biological realms are themselves very different. Thus we are left with **1** and **4**.

---

[2] As a matter of fact, the second edition of Fodor's *Language of Thought* does not ascribe to Cartesianism but sets it alongside his own theory as something of a fellow-traveller [7, pp. 7, 11]. Since Fodor's system regards thinking as only thought and not action, he does not reckon with the possibility that Cartesianism could be supplying thought with *motivation*. Crypto-Cartesianism, *sans* Darwinian instincts, puts both the gas and the ghost in the tank of his thinking machine.

As Gibson implies, crypto-Cartesianism makes presuppositions about how the brain works but does not derive them from biology. Rather, her ecological approach to cognition, developed with her husband J.J. Gibson, used experimental data to arrive at categories and properties of motion through a spatial array and an ethology of objects that are directly visually perceived. The presumption behind these principles, their teleology, as it were, derived from the basic needs of an organism to adapt to and manipulate its environment. Ecological approaches to language cognition much better correspond to brain anatomy than those of Chomsky, but lack the generalizability.

The best representation of **1** is the aforementioned Jerry Fodor's 'Language of Thought Hypothesis'. There he considers that the only way to view thought is as a series of representations. He does not believe that thinking serves action. Thinking is prior to action. Instincts and other thinking that might be outside of ratiocination are irrelevant [7, p. 7]. He does not believe that there could be a real, direct connection between perception and the world. How we get from things to thoughts is not for him a relevant question. His hypothesis does not really engage psychology but the philosophy of representation. The semiotic system is already there because its medium is representation and representation is presumed to be the only valid medium of exchange. Though he would not (in a Berkeleyian fashion) say that there is no means of accessing the world, he simply does not consider that it has anything to do with representation. We may not know the language of mentalese because Fodor is not a psychologist, if he were he would, like Berndt's natives, inform us, but we know that it is a language because it represents and that is what representation is. He does not consider that representation takes other forms – perhaps not the 'Language of Thought' but the 'Dance of Thought', 'Music of thought', 'Sculpture of Thought' or the 'Ritual of Thought' – but it is semiotic and thus is representation. That it *must* be language is a judgement Fodor makes for himself.

This gets at the difference between **1** in Sacks' and Fodor's versions. The statement 'They see us as…' understands two equal cultures which are interlocutors at some level: the anthropologist and his subject, or the doctor and patient, two participants in group therapy. The predicate may be asemantic to one of them, or both, but the nature of the exchange involves communication between two agents. Crypto-Cartesianism requires mediating agents to do the en- and decoding. It needs a translator in contrast to the ecological and holistic theories Fodor dismisses [10]. A clear example of the difference comes from the interesting debate between J.J. Gibson and Ernst Gombrich in 1965. The famous art historian had disputed the ecological psychologist's claim that we may directly perceive distant objects. From his window Gombrich sees things as in a camera, and the projection on his retina is no different. Gibson responded: "I do not accept the eye-camera analogy, since I strongly disbelieve that the retinal image is an image in any proper sense of the term. It cannot be looked at, for there is no seer inside the head to see it" [11].

For crypto-Cartesianism the 'spectator' Gombrich presumes of a mental image, the ghost looking at the view finder, is the immaterial rule standing between stimulus and response. It is not the case that any concept of biological encoding of information is fallacious. Rather it is the manner in which this encoding is 'read' and 'interpreted'. Mentalese defends itself by saying there can be nothing else. But perhaps the logic is wrong, and thus the mechanism is wrong; because the point is not just whether qualia exist, but whether their relations match with the relations they have in the system. That is not a given. If one considers that we must already presume the clear understanding of representation Fodor gives his hypothesis, [7, pp. 20–21] then his claims suddenly become much more modest. The bravado comes from the explicit implication that Fodor discovered the language of thought and can translate it for us. In my opinion, Fodor needed to stick a few more coins into his viewfinder.

It is indisputable that the brain is a biologically determined organ. It is possible, but not necessary that it encodes information in a way that resembles a language or a logic.[3] To employ Canabis' statement "The brain produces thought like the liver produces bile" [12, p. 137], we would say that any encoded information in the brain must be conceived of as having the same level of mediation as an enzyme secreted by the liver. That is to say it mediates to the extent that it serves as the stimulus for another set of behaviours elsewhere in the brain and body, but it does not mediate as a 'message'. When the liver produces bile, or an enzyme, in some sense we can say that it carries information which is transformed or expended through the execution of the emulsification of fat. On the other hand, to speak of translation we need a message which presupposes communication, communication presupposes a sender and receiver and the message which is independent of each.

Thus we can only conclude (and move away from the conundrum of the Hard Problem of Consciousness) by arguing that the 'Language of Thought' is a metaphor and translation can be used analogically as in **4**.

Does this solve the problem? We know it would not satisfy Fodor as he excreted enough bile to emulsify generations of ecological cognitive scientists. Indeed relying on what he lumped together as "pragmatism" and "holism" would not ever get us to logic or language in the forms he and Chomsky [see 13] have demonstrated to be important. Thus we should entertain the possibility that the analogy of translation must work *somewhere* in this biological-semiotic nexus and that it is a more valuable analogy than that between two acids and bases.

Let us consider the possibility that the difference between Alice and Bob, or the source and target in translation theory, relies on some analogy between Cicero's form and sense. It is very likely that the brain represents things at some level and Fodor's version of representation may well be the right one. Perhaps we must then grasp the sense of representation since the form of the biological entity cannot be grasped. What would this sense mean? What does representation mean? We consider representation to mean that which takes the form of something and presents it elsewhere. But that is a form and not a meaning. Then the sense of representation is the *act* of taking one or another form. Representation is its own form of communication, and in this case it may be defined in a rather limited fashion. For example an enzyme can chemically represent the complement of the form of the proteins with which it will interact. When it reacts with the protein, it effects its representation. We can in this sense understand

---

[3] I should state for the record that I do believe there is a relationship between patterns of brain behaviour and language for the simple fact that it would be developmentally and historically strange for them to be utterly unrelated, but there is no evidence to suggest that this relationship is obvious.

representation as the effecting of one form in another place, as the enacting of a form. A token is a representation of a type, and the token can be a good one or a bad one, but if it works at all it invokes the type. This is the case with the sense of a translation. If the aim of translation is form, then the degree of verisimilitude is decisive. But if the aim is the sense, a 'good enough' representation is the one that gets the meaning across. A typical example comes from the law and politics: a country is represented by its president at a state funeral, or it can be represented by the ambassador or the foreign minister. Each of these entities can enter and stand in for the other standing in for the country as an entity. The grain of representation varies: the whole country for the finest grain of resolution, the president, the foreign minister, the ambassador, the flag, seal of state, motorcade. You could have two or three of these latter elements present at one time and representation would still be achieved. Representation may be adequately effected by one of them and it is up to convention to determine what is appropriate. As in chemistry and translation, there is 'good enough' representation in this semiotics: when the reaction takes place and when the meaning gets across.

Thus it may be best to see Fodor's mentalese as a 'good enough' representation of the sense that is a quale or a set of qualia. This soft-pedaling might seem more useful than one might think. First it allows for some of the versatility in the intentionality that Fodor pushes for in the second part of his argument [7, pp. 60–63]. Secondly it allows for the 'processing' of the computational system to work with, refit, and repurpose elements of the representational system, something which suits the multifunctionality of the brain and psycholinguistics [5, pp. 265–272]. Finally, it allows the agency of the organism to participate in its own evaluation of that 'good enough' as part of the process itself and not as a proctor of it. Evaluation takes place not external to the organism but as part of its own activity. It is unlikely that Fodor would have accepted such soft-pedaling as it bears too much of a whiff of pragmatism. The following will show that Fodor needed to hone up his sense of smell.

3   Subjectivity and Alienation: Neural Translation

To see the value of a good enough translation into and out of mentalese, I will take a dramatic leap out of the strictly semiotic world of Fodor and Sacks and into the blood and sinews of the brain and body. This is, after all, in complete conformity with item **4** in our list of untranslatable categories, and we shall pursue it.

First it must be said that a great deal of the contemporary picture we have of the relationship between brain and behaviour comes from imaging technology, especially functional Magnetic Resonance Imaging (fMRI) studies. A particular and popular inter- pretation of these studies is tailor made to arrive at qualia. A set of subjects is monitored while they respond to a stimulus. If enough of them have a similar concentration of activity in a similar area of the brain, that anatomy is said to 'govern' or 'correlate' to that activity. One study entitled "Why do we have a caudate nucleus" argued that activity associated with affection in cats correlates with activity in the caudate nucleus [14]. A similar study argued: "Regions commonly activated by early-stage romantic love (identified in previous studies) and long-term romantic love included the right [ventral tegmental area] and posterior caudate body; bilateral anterior caudate body, mid-insula and posterior hippocampus; and left cerebellum" [15, p. 151]. Love is in the caudate nucleus. It is your caudate nucleus that loves. Romantic love is about everywhere else in the limbic system. Such fuzzy logic smoothed by statistical generality gives crypto-Cartesianism its "little gland", as Descartes famously called his favourite storehouse of the soul, in any part of the brain anatomy where behaviour associated with the object can be associated with its neural correlate. Aside from the triviality of many of these arguments, individual brains have a great deal of variation and so the larger the population of study subjects and the better the statistical value of the results the coarser the grain of individual 'encoding' of a purported 'quale' is likely to be. You might arrive at 'love' but you are not likely to arrive at 'the smell of granny's third dresser drawer, the one with her wedding picture'. Since the essence of qualia are their subjectivity this limits outcomes.

A more important problem is the subjective nature of fMRI readings. Brains are clouds of activity when active and at rest and reading them is more art than science. FMRI-based studies have thus suffered from a crisis of reproducibility. Furthermore, they rely on a signal reading system which renders the vast amounts of atomic measurement data and generalizes upon it. This means that the resolution of the imagery can be over-contoured and so appears finer than it actually is. Thus for brain science and medical practice there is tendency to overstate the claims of such research [16]. Perhaps qualia are too fine to be found if they exist at all.

However there is another possibility. In the early 2000s, the leading neurosurgeon Hughes Duffau developed a method of operating on conscious patients [17, p. 20]. This, in part, had to do with the ramifying growth structure of glial sarcomae. But it also had to do with the inadequacies of neuroimaging and the large individual variation between brains. Now awake surgery is a very common practice. In the 2010s Duffau began to consider awake operations to be a "different philosophy" of neurosurgery [18]. Now "philosophy" in Duffau's rendering of it is for us a bit like Freud's untranslatable categorizations (i.e., 2). He means the term as a general approach to a surgical practice and technique that is something most of us cannot do. I would maintain, however, that the implications of the practice do have philosophical import in the realm of mere speculation.

Let us first briefly describe the procedure. Regular imaging techniques determine the general location of the diseased area which determines the general surgical strategy. A neuropsychologist or speech therapist then determines a set of tests to be administered to the patient during surgery. Since the brain has no pain receptors what is manipulated or removed during surgery does not cause pain. An area around the scalp is given a local anaesthetic to block the pain of the incision. In most cases, during the surgery the patient is fully awake during resection. In some cases the area to be resected is determined on the awake patient and then removed under sedation. As Duffau points out, the principle around determining what and how much brain tissue to remove is utilitarian. One wants to remove as much tissue around the tumour as possible to improve the chances of the cancer not recurring, yet one also wants to do as little damage to the patients' functioning as possible. Once the area around the tumour has been determined, the surgeon begins to apply an electrode around the target area. The mild electrical current 'short circuits' the function of the area to which it is applied and essentially blocks its functioning.

When removed, the functioning returns. This is where the test design is important. The stimulus must match the function of the area touched by the electrode. If at the posing of task A (identifying the meaning of a word, etc.), the area touched by the electrode corresponds to test B, then the identification is not made. There is a risk then that the area test B identifies will be inadvertently removed. Duffau regards (and demonstrates with data from a pool of 700 patients [18, p. 689]) that awake surgical intervention is especially useful in the removal of gliomas. They crawl like spiderwebs up long axional networks and the clear determination of function keeps the surgeon from removing too much tissue along these relatively wide areas.

This is the image in the surgical context that should convey to us something of significance relating to the Hard Problem of consciousness. Doctor and patient are here both consciously walking a path over the patient's brain. The landscape of this path is determined by tests of functionality and we can only regard the features of that landscape to be nothing other than qualia – the subjectively defined neural correspondences to behaviours in the world. But unlike semiotic or syntactic models, or brain imaging techniques, this landscape is mapped by the conscious engagement of patient and doctor together. One may balk at saying that having part of your brain turned off is conscious. But even as Fodor would have to admit, it is even in its negation of activity, intentional. Furthermore, demanding that one has to *think* in the sense that Fodor defines it to engage in mentalese is to impose an *a priori* model of incompatibility (**4**) into this relationship when a set of preexisting pairs is not needed whatsoever. This is not a translation between two pre-existing entities. Determining the nature of the target structure is part of the nature of the procedure.

Awake surgery is not the only type of conscious interaction between brain, patient, technology, and professional. There are few things more inspiring about neuroscience than watching how prosthetic interfaces allow amputees regain the ability to function. If the amputation left some viable nerve tissue projecting from the stump of a limb, a neural interface prosthesis can make use of it. The area enclosed by the interface stimulates and responds to nerve impulses which then have been shown to reenervate part of the tissue of the stump [19]. In other cases the tissue for a different part of the body can be recruited. Nerves once used for the muscles around the shoulder blades can be used for the leg, for instance. Biofeedback methods can be used for those with damage up to the spine, and for some paralytics. This follows from a long-established method where a patient trains themselves to manipulate their brain wave signals to move a cursor on a computer screen. A similar skill has recently been used to control an interface for a patient completely immobilized by ALS through an electrode inserted into his motor cortex [20]. In each of these cases there is a direct, physical relationship between the nervous system (at the level of efferent projections) and a consciously learned system of manipulation. This is not a simple soldering of wires. The prostheses use complex computer interfaces that regularize and soften individual movements, compensating for the lack of fineness that comes out of natural nervous architecture. Even in the most basic forms of connection, when the nerves for a limb are still viable, it takes weeks or months of training to arrive at reasonable functioning.

In a final example, a series of studies were able to construct rough images that purported to be from the dreams of subjects sleeping in an fMRI monitor. This was possible because they had been observed whilst watching particular shapes. A signals analysis was conducted that trained algorithms to recognize the hierarchical firing of visual networks. The subject was then monitored whilst sleeping and the resulting readings run back through the signals data to generate the image. The result was of rather poor quality, but vague, eerie figures emerged. This study was cited to pull back the veil with an objective picture of our subjective lives. Since then a small field has opened up attempting to formalize signals analysis and fMRI to image generation [21, 22].

Each of these examples presents, I would argue, a case for a particular view of the translation of the world, either phenomenal or semiotic, into the subjective activity of the brain. In the first case, a patient aids a surgeon in subjectively identifying the particular brain anatomy that matches the function that patient can perform. The patient orients the doctor over the geography of his or her brain. In the second case both neural interface and patient learn how to work together. While awake brain surgery techniques are used to operate on complex, high-functioning, 'eloquent' parts of the brain, neural interfaces are quite at the other end of the spectrum – highly automatized nervous structures that manage things like the movements of arms and legs. Yet these are not switches between already existing elements. The patient must still learn how to use them, and must still work with the prosthetic interface. The patient and the software learn together to properly determine the responsiveness of the mechanical prosthesis. Thus, in contrast to Fodor's rationalist approach, learning gives us access to the subjective regardless of the level of neuronal sophistication. In the third example, the patient fully trains the software in the functional palette of his brain activity. Training 'translates' digitized activity into visually interpretable signals. It is, in fact, not unreasonable to conceive that such signals could then be 'translated' back into visual images after the training.

We can conclude that these entities as translated in the brain are qualia, if that word is to have any physical referent at all. Yet it is essential to state what these 'translations' are not. They are not the translation of a discrete object in the world into a sign encoded in the brain. A sign (even a token) always aims toward a universal. These qualia are decidedly not universals. They can only be defined for single individuals, other individuals may have different anatomy. They are also only subjectively defined. The physician cannot point to a scan and say where a function is. Were that the case, a particular reading would be enough, and switching on a prosthesis would be like snapping an attachment onto a blender. Contrary to what the ideology of much neuroscience tells us, subjective information can only be precisely defined subjectively.

But this is not a call to neuroscientific skepticism, or neuro-technological Luddism. The above encourages us that the subjective *can* be accessed objectively through a process of *actualization* – learning, doing. While it is wrong to say that qualia do not exist, computationalists are wrong to suggest that they can be accessed by the unidirectional, semiotically determined forms of communication they presume. None of the four models of untranslatability we have considered are adequate because they all posit sources and targets that are complementary units of exchange. Yet exchange is demonstrated to take place and we are at pains to account for at least an analogy to the mechanism by which it does so. With untranslatability evidently violated, the process of the translatable is all that is left.

The problem models of translation face is the presupposition of categories which may not exist. But within Cicero's model is a hidden practical ethos that does not just suggest there is a way to get from source to target, or Alice to Bob, but that provides two different *means* of doing so. He gives preference to sense, but does not reject form. The choice which is best comes in the doing while he never questions that the doing must be done.

I would regard Schleiermacher's sublime contribution to translation theory to be, despite the intervening centuries, the next along this theoretical path. Though he does not say so, translation for Schleiermacher is not a set of codes, it is a maieutic for reader and, in his Romantic aesthetic, for the author as well. This, I would argue, is the best model of translatability we can apply to the 'representation wars' of cognitive science. The subjective can be accessed, but only by walking the path of its determination. This does not mean that we are to create analogies, or propose models that presume categories of conversion. The paradox of modelling complex systems is that the closer the model is to verisimilitude the less valuable it is for explanation. Rather the aim is to set up a system by which the alienation of the reader – the physician, the neural interface, the signals software – is gradually pulled along a path of accommodation and adaptation until the target is reached. It is, of course, a Romantic truism that the learner is transformed by learning and so that object, once learned, is different from what it was at the onset. A brain that has learned to use a prosthetic device is different from one that has not. That is what learning means. But Schleiermacher's model alerts us to the temptation of using analogies of signification to avoid the consequences of what that learning would entail.

Considering the parallels of learning, movement, and translation, one is reminded of the famous story "The Marionette Theatre" by Schleiermacher's near contemporary, Heinrich von Kleist. Romantic alienation abounds in the work of this author, which does not take much effort on the part of a translator to preserve. In the story, one friend confesses to another his love for the marionette theatre. The other, an accomplished dancer, begins a tale meant to support the argument that conscious awareness of what one does is inimical to the gracious performance of the task. He gives several examples, and the narrator chimes in with his own. Marionettes are graceful because they are inanimate and unconscious, a man with a wooden leg can dance better than a ballerina. The story ends with a tale about a bear that can fence better than the most practiced gentleman. Unlike the rest of the examples, mentioned along the course of the conversation, the dancer's last example is strangely detailed: a trip to a distant Russian estate, the jesting rivalry between the young men, and the master who leads the narrator down from the manor house to a bear who parleys his every thrust with its paw. Eerie, exotic, and foreboding, the end has all the elements of one of the *Tales of Hoffman* but with Kleist's characteristic restraint. The reader is gripped by the story for the very fact that the central argument is patently fatuous. Anyone knows that a dancer is more elegant than a marionette, no wooden leg is better than a *pas-de-deux*. Like Baudelaire's famous essay, "On Dolls", the story is about our alienation from our bodies. The bear in the Russian estate is not about what we should be but the shame about what we might be [23].

Perhaps the wonderment at the effectiveness of modern prostheses is the healthier outcome of this Faustian and Cartesian bargain. Rather than seeing that the world of the mind cannot ever connect to the world of the body, we see that their connection can be breached through tools that unite functionality and subjectivity rather than dividing them into unrenderable, or falsely renderable, categories.

4 Conclusion

While the analogies proposed for translation might not ascend to the precision of thought experiments, the increasing tendency to view information in the sense it is conceived in physics sets forth the use of models that derive from linguistic pairings. For example Daniel Dennett long ago noted the subjective element of learning in identifying what he took to be the false properties of qualia: "what counts for an individual as the simple or atomic properties of experienced items is subject to variation with training" [8]. Yet he was unwilling to accept that such a thing as qualia could exist, despite the fact that the foregoing is a fairly good definition of one. Rather than Dennett's own brand of crypto-Cartesianism, one may rather offer as an example the recent work of of Luciano Floridi as a philosophical way forward beyond the confines of language pairings [24]. Floridi concludes that we must see information, and knowledge, not as a lexicon but as a set of affordances between mind and world. This decisively Gibsonian version is closer to the maieutic of Schleiermacher and other Romantic thinkers, including that of Coleridge's theory of perception and the imagination. A maieutic of translation beyond language may serve as a valuable and creative tool for modelling exchanges between semiotic and non-semiotic systems and may aid us in considering communication without presuming some willy-nilly digitization *ab initio*. Digitizing first, we render all things compatible and then compatibility becomes its own philosophical and theoretical *ideé fixe*. A pie in the Matrix does not taste the same. So long as we strive to conceive of heterogeneous things, we are training our senses and our theory forward in ways we fully understand only though the effort. Translation is part of that effort, and the untranslatable as well.

References

1. Cicero. De optime genere oratorum. Documenta Catholica Omnia: Online resource. http://www.documenta-catholica.eu/d_Cicero,%20Marcus%20Tullius%20-%20De%20optimo%20genere%20oratorum%20-%20LT.pdf. Accessed 31 Oct 2022
2. Venuti, L. (ed.): The Translation Studies Reader. Routledge, London (2012)
3. Sacks, H.: Lectures on Conversation, vol. I and II. Wiley-Blackwell, Oxford (1995)
4. History of Translational Medicine. Eupati Open Classroom: Online resource. https://learning.eupati.eu/mod/page/view.php?id=193#:~:text=The%20term%20translational%20medicine%20was,between%20laboratory%20and%20clinical%20research. Accessed 31 Oct 2022
5. Anderson, M.L.: After Phrenology: Neural Reuse and the Interactive Brain. MIT Press, Cambridge (2021)
6. Gibson, E.J.: Introductory essay: what does infant perception tell us about theories of perception? J. Exp. Psychol.: Hum. Percept. Perform. **13**, 151–523 (1987)
7. Fodor, J.A.: LOT 2: The Language of Thought Revisited. OUP, Oxford (2010)
8. Dennett, D.C.: Quining qualia. In: Goldman, A. (ed.) Readings in Philosophy and Cognitive Science. MIT Press (1993)

9. Chalmers, D.J.: Phenomenal concepts and the explanatory gap. In: Chalmers, D.J. (ed.) Phenomenal Knowledge and Phenomenal Concepts: New Essays on Consciousness and Physicalism, pp. 167–194. Oxford University Press, New York (2006)
10. Fodor, J.A., LePore, E.: Holism: A Shopper's Guide. Blackwell, Oxford (1992)
11. Gombrich, E.H., Arnheim, R., Gibson, J.J.: On information available in pictures. Leonardo **4**, 195 (1971). https://doi.org/10.2307/1572214
12. Canabis, P.-J.-G.: Rapports du physique et du moral de l'homme. Gallica, Paris (1802)
13. Chomsky, N.: A review of Skinner's verbal behavior. Language **35**, 26–58 (1959)
14. Villablanca, J.R.: Why do we have a caudate nucleus? Acta Neurobiol. Exp. **70**, 95–105 (2010)
15. Acevedo, B.P., Aron, A., Fisher, H.E., Brown, L.L.: Neural correlates of long-term intense romantic love. Soc. Cogn. Affect. Neurosci. **7**, 145–159 (2012). https://doi.org/10.1093/scan/nsq092
16. Eklund, A., Nichols, T.E., Knutsson, H.: Cluster failure: why fMRI inferences for spatial extent have inflated false-positive rates. Proc. Natl. Acad. Sci. **113**, 7900–7905 (2016). https://doi.org/10.1073/pnas.1602413113
17. Papatzalas, C., Fountas, K., Kapsalaki, E., Papathanasiou, I.: The use of standardized intra-operative language tests in awake craniotomies: a scoping review. Neuropsychol. Rev. **32**(1), 20–50 (2021). https://doi.org/10.1007/s11065-021-09492-6
18. Duffau, H.: Awake mapping is not an additional surgical technique but an alternative philosophy in the management of low-grade glioma patients. Neurosurg. Rev. **41**(2), 689–691 (2017). https://doi.org/10.1007/s10143-017-0937-6
19. Marasco, P.D., et al.: Neurorobotic fusion of prosthetic touch, kinesthesia, and movement in bionic upper limbs promotes intrinsic brain behaviors. Sci. Robot. **6**, eabf3368 (2021). https://doi.org/10.1126/scirobotics.abf3368
20. Chaudhary, U., et al.: Spelling interface using intracortical signals in a completely locked-in patient enabled via auditory neurofeedback training. Nat. Commun. **13**, 1236 (2022). https://doi.org/10.1038/s41467-022-28859-8
21. Ren, Z., et al.: Reconstructing seen image from brain activity by visually-guided cognitive representation and adversarial learning. Neuroimage **228**, 117602 (2021). https://doi.org/10.1016/j.neuroimage.2020.117602
22. Beliy, R., Gaziv, G., Hoogi, A., Strappini, F., Golan, T., Irani, M.: From voxels to pixels and back: self-supervision in natural-image reconstruction from fMRI. In: Advances in Neural Information Processing Systems (2019)
23. von Kleist, H.: On the Marionette Theatre. https://15orient.com/files/kleist-on-the-marionette-theatre.pdf. Accessed 31 Oct 2022
24. Floridi, L.: The Logic of Information: A Theory of Philosophy as Conceptual Design. Oxford University Press, New York (2021)

Visual Culture: A Phenomenological Approach

Vera Serkova[1]([✉]) [ID], Alexandr Ryabov[2] [ID], and Alexander Pylkin[1] [ID]

[1] Peter the Great St. Petersburg Polytechnic University, Polytechnicheskaya 29,
Saint-Petersburg, Russia
henrypooshel@rambler.ru
[2] Saint-Petersburg University of State Fire Service of EMERCOM of Russia,
Moskovskiy prospect, 149, Saint-Petersburg, Russia

Abstract. One of the most important forms of the Multilingual World has always been the language of the visual. The visual world is the basis for the formation of representations of reality and is continuously converted into linguistic structures. The process of such conversion requires special research involving many applied disciplines. Visual studies, whose spontaneous development took place at the end of the last century, became the theoretical basis for combining research efforts in the field of visual studies. The problems of Visual studies are diverse. Contemporary art objects require continuous meaningful correlation, they are open for endless interpretative work. In our opinion, the most important problem of visual studies may be the question formulated in the Kantian spirit how "the visible is possible". The second question can be formulated in a phenomenological way: how the visible is "realized into reality". The multitude of problems contained in this line of research was comprehended in E. Husserl's phenomenology in his theory of spatial syntheses, which Husserl developed in a number of his works. The first part of the article is devoted to the general problems of visual research, and the second part to Husserl's studies that present the problematics of the "visible world" in a new way.

Keywords: Visual turn · Visual studies · Edmund Husserl · Phenomenological approach · Carnal consciousness · Kinesthetic concept

1 Introduction

One important part of Multilingual World is the language of the visual world. In today's culture, the visual image has greatly supplanted the word. Indeed, the "picture" has become an icon of fundamental transformations of modern culture, its turn from the word to the image in the new "post-linguistic reality". As a result of these tectonic shifts visual studies, visual culture as an interdisciplinary, thematically defined field of research that unites many fields of knowledge focused on interpretation of what and how we see, how information embedded in visual images is read and interpreted, how it is transformed during its transmission, how content of the visible is identified and de-identified. In the context of this research topic, the question about the visual is: what does it mean to "see", how the process of seeing is carried out, how can we describe the visible world without this process turning out to be a set of random characteristics. The term fitted well into social theory of culture because, starting from the 20th century, there was not only an active incorporation of art into everyday practice, but also a reverse process of "fabrication" of mass culture works was gaining strength, and the visual environment was saturated with this kind of production. Thus it became clear that contemporary art (in the broad sense of the word) could no longer be seen in terms of its absolute autonomy, as the center of the realm of culture, but as an activity that was always involved in the constitution of everyday life. We continually produce visual images, and continually translate them into linguistic forms. The products of this interaction are like centaurs, with the inseparable duality of picture and verbal content.

The problem field covered by visual culture research is extremely broad, since representatives of a wide variety of disciplinary fields and research schools work together in it. The discreteness of interests and the diversity of analytical programs of visual studies representatives are overcome in some fundamental studies, thanks to the ability to take on the accumulated questions. In one of her comments on the expotential growth of visual studies.

Margaret Dikovitskaya writes: "I envision my task to be to show how visual studies avoids these two ontological perils and negotiates between the Scylla - the lack of a specific object of study - and the Charybdis - the expansion of the field to the point of incoherence. In what follows I offer an overview of this new area of study in order to reconcile its diverse theoretical positions and understand its potential for further research" [1, p. 69]. Margaret Dikovitskaya notes in her last article, devoted to summing up the results of twenty years of visual science development: "Thanks to innovative digital and online technologies, the time-honored practice of 'slow looking' has taken on a new life, visual literacy has enjoyed exponential growth, and the expanded notion of vision has contributed to the development of a multisensory history of art". [2, p. 197]. Her work overcomes the dilemma set by W.J.T. Mitchel, who suggested that every adept of Visual studies should clarify for himself on which pole his interests are concentrated, whether he refers to them as "a theoretical concept or object of research and teaching" [3, p. 165].

Indeed, individual successes in Visual studies are converted into the achievements of individual Visual culture. And the most important thing here is to find a reliable "guide" to this labyrinthine field of contemporary humanism. And perhaps it is Margaret Dikovitskaya, thanks to her deep understanding of the practical nature of visual studies and her efforts to pass on her experience in the practice of brilliantly arranged summer schools on the programs she has developed in the field of visual studies, that this direction is being pursued in her programs.

Academic science, as long as it strives to keep in view all what is commonly called "reality," is forced not only to analyze the ways it coordinates with the latter, but also to integrate mid-level theories interpreting fragments of reality into a coherent system of knowledge.

The efforts of the scientific community to apply a variety of methodologies to research the shifts taking place in culture in the rapidly renewing and visually and informatively saturated space of contemporary culture are understandable. Another impetus for the rapid development of visualism was the expansion of the boundaries and possibilities of anthropologically defined visuality, the transformation of reality into an expanded, hybrid, augmented, virtual one with all the changes in mechanisms of reading, tracing and controlling visual images. The problem field of virtuality unites the research interests of representatives of different disciplines, both natural sciences and humanities. The disciplinary front of visualism is the study of what and how visual perception is represented. The formation of visualism is being carried out by the joint efforts of many researchers, although critical attacks and doubts about the expediency of scientists producing another hybrid of sciences, similar to cultural studies or anthropology, do not cease.

2   The Problematic Field of Visuality

What does Visual studies represent and what are its achievements? The main argument in favor of intensifying efforts in this direction is the fact that visible, perceived reality is becoming increasingly dense, "culturally saturated," informative, multilingual and, consequently, requiring analytical work on processing, assimilation, storage, participation, i.e. special methodological work for understanding the specifics of dealing with visual content. Awareness of the problems and processes with which the modern man deals with every moment determines the problematics of scientific interests in this field. If we talk about specific theoretical issues in this connection, they are diverse, - from research into the ways of encoding and decoding visual information, and the related "culture of vision" - to analysis of perverse forms of information attacks to which bearers of modern culture and recommendations of analysts are exposed [4–10].

The directions of visual culture research can be briefly listed as follows: description of various visual practices (passive and active), analysis of visual images presentation and representation structures in consciousness, use of marginal and universal forms of visual information coding and decoding, development and extinction of high and low genres of visual culture, contextual and universal ways of visual content storage; analysis of visual environment formation sources, problem of relation of art and everyday life; studies of art and everyday life relations.

Essentially, all the studies deal with the enormous role of the visual image, which serves as the basis for what we call "objective reality", the objectivity of which can be questioned by the fact that every person constructs anew and initially unique and individual horizon of visual space by means of his innate ability to realize the world in the forms of its visualization (showing), detection and fixation exclusively in the acts of consciousness, or consciousness. For this reason Plato used the term "vision" as a metaphor for knowledge in general, and the etymological meaning of the word "theory" (θεωρία) as a form of "intelligent vision", knowledge-understanding of things, retains its essential meaning in the basis of visual culture. Visualization of the object world is the basis of the natural human nature, it varies in its contents, depends specifically on what is included in it as its contents, but remains a constant fundamental form of the activity of consciousness. The fundamental areas of visual practice research are the mental twins of visual images, their visualization and the analytics of these processes.

Visual research went in two directions, theoretical and applied. The first direction of research was carried in the analysis of specific visual practices, in descriptive sciences, first of all in art history and art studies, the second out within the framework of philosophical analysis (in phenomenology, analytical philosophy, in cultural studies).

The fundamental point of visual analytics, as James Herbert writes, is the recognition that there should be no preference and privilege to any class of visual images or to any class of people, the bearers of these or those images [11, p. 2]. Keith Moxey, one of the founders and organizers of the field of visual culture, points out that contemporary culture questions the universal meaning of aesthetic value, that is, what since the times of Kant has been regarded as the inalienable meaning of all human activity, as its universal characteristic. The lack of the aesthetic in any object was seen as its defect. Keith Moxey showed in his studies that aesthetic values are not universal and time-dependent, they are transformed according to tastes and values [12, p. 107]. For this reason, the study of the visual sphere is by no means limited to the phenomena of art, in which the aesthetic quality constitutes an essential feature that distinguishes them from the entire sphere of visual objects.

Keith Moxey believes that visual research "opens the door" to the formation of knowledge of such analytical objects, which are saturated with subjectivity as a condition and possibility of their production. This is the essence of the new research strategy - the transfer of knowledge from the position of objectivism to a new methodological basis for describing the manifold structures of subjectivity.

Thus, the problem field of "visual culture" implies the study of visual experience in the broadest sense of the word. Visualistics embraces such abilities that constitute the visual sphere as imagination understood as a way of expanding the field of the directly perceived, identification as the basis for constructing "images of the self", ideation as grasping the essential content of the visible, the work of the unconscious that has no obvious influence on the sphere of consciousness, etc.

So can we answer the provocative question of W.J.T. Mitchell, what is Visual studies [13], this new interdisciplinary hybrid that can link art history with literature, philosophy, studies in film, popular culture studies, sociology and anthropology? He himself admits that visual studies cannot yet be considered a discipline. But what is it, a complex of disciplines, a section of disciplines, or disparate applied research? There are more problems in this field than there are accepted conceptual approaches. The material of visual science, which studies what is supplied by the human eye and what forms the basis for the work of consciousness, is currently emerging spontaneously in interdisciplinary research, and it seems that this state of affairs suits everyone so far. Margaret Dikovitskaya writes: "there is no consensus among its adepts with regard to its scope and objectives, definitions, and methods, there is no consensus among its adepts with regard to its scope and objectives, definitions, and methods". [1, p. 68]. In this connection we would like to point out one more research resource, which would not only supplement the methodological basis of visual research, but, perhaps, put it on the general fundamental research basis.

3 Phenomenological Addition

Phenomenological research has an important, but underestimated role in understanding the mechanisms of visual imagery realization. The visual turn in research strategies, opposed to the linguistic turn, has formed the problem field of visualism. It is common to refer to the studies of M. Merleau-Ponty [14–16]. However, the fundamental foundations of phenomenological analysis of visual mechanisms were laid by the founder of phenomenology Edmund Husserl. An extremely important period in the formation of Husserl's philosophy includes the material published in the 16th volume of Husserliana under the title "Ding und Raum" [17]. At the same time, Husserl develops the theory of reduction in his first book, Ideas Pertaining to a Pure Phenomenology and to a Phenomenological Philosophy [18]. One of the most important themes for Husserl is that of spatial intuition, which he explores from his second book, Ideas 2 [18]. He distinguishes between the space of everyday experience and geometrized (mathematical) experience as perceptions "in space" and perceptions of space itself. However, according to Husserl, neither the visual field nor mathematical constructs directly reveal basic spatial intuitions. What is intuited in spatial experience is not the same as what is described in the mathematical model and what is manifested in the visual picture of perception. Phenomenologically consistent description shows that in these two cases two different meanings of spatiality manifest themselves. According to Husserl, there is a more immediate and more fundamental experience of spatial grasping already incorporated into ordinary visual and everyday perception [19, p. 671]. The most detailed description of Husserl's theory of spatial syntheses of consciousness was given by J. Drummond [20], from modern researchers of this theory we will mention H. Riley, and R. Newell [21, 22].

Husserl bases his analysis of visual structures on the study of elementary quadrilateral eye and head movements (up and down; right and left), which are the basis for the constitution of the visual field. He seeks to understand how the intuition of spatial three-dimensionality is generated, which arises precisely in simple acts of "carnal consciousness". The human eye as a modeling motor structure operates in the mode of planar transformation and limits the possibilities of perception to the limits of the visual field. The world in this reflection is a concentrated, vertically unfolded curvilinear plane at the surfaces of the body, encompassing the body from all sides. Husserl believes that body consciousness models the picture of the world in accordance with the Euclidean model, which admits a certain simplification and essential "incorrectness" in reflecting the picture of the "real" structure of the world of things - visually the world is affirmed as flat, and therefore it is Euclidean geometry that seems "innate," that is, in full accordance with spatial intuitions. Husserl considers it primary in relation to other geometries, because the mathematical intuitions embedded in the axioms of Euclidean geometry initially correspond to the forms of fundamental transformations brought into acts of bodily perception. As already mentioned, Husserl specifically analyzes these elementary and profound forms of intuition, by which he means the immediate contents of the perceived in the syntheses of "carnal consciousness".

The analysis of visual spatial constitution is carried out on the basis of specific acts of consciousness, in which the subjectivities of the visual field, limited by a multiplicity of perspectives related to the point of perception, correlated with the primary impressionistic content of perception, are formed.

It is about reconstructing the consciousness of "available space" in a peculiar (and parallel to temporal) phenomenological series, which, in accordance with the Husserlian tradition, could be defined as the inner consciousness of spatiality.

In the second book, Ideas Pertaining to a Pure Phenomenology and to a Phenomenological Philosophy. Studies in the "Phenomenology of Constitution," the main theme of the study becomes "kinaesthesis," which has the meaning of a spontaneous reaction to a moving and changing source of perception, a bodily correction of the contents of the perceived. The kinaesthesis is a special type of consciousness, which Husserl calls "carnal consciousness", which fixes and directs the processes of bodily correlation. This organistic theory has largely determined the content of French postphenomenological philosophy; it is also reflected in the research of spatial-body problems by such indisputable phenomenologists as M. Merleau-Ponty or Mikel Dufrenne; also the kinesthetic concept has essentially influenced the thinking of those philosophers whose philosophy has an eclectic character, and, in general, their phenomenology is reduced only to the "phenomenological motif". The phenomenological motif of corporeality is particularly evident in the works of postmodern philosophers (Jacques Derrida, Gilles Deleuze, Jean Baudrillard, etc.).

"Carnal consciousness" is not a special kind of consciousness; it represents its basic, fundamental level, underlying all activity. In his second book, Ideas, Husserl describes the experience of spatial constitution, consistently including in the descriptor an ideal situation in which bodily acts are identified at first with the field of "simple" perception, when the perceiver is at absolute rest. Husserl then introduces more and more kinesthetic acts into this simplified scheme - the oculomotor motor coordination of the eye and the cephalomotor kinesthetic coordination of the simplest bodily movements. These types of bodily correction very much change the structure of the original visual field - so much so that the object can appear or disappear depending on the direction of the perceiver's movement. The changes introduced by eye and head movements are the initial forms of bodily interactions on the basis of which perception is constituted, although the very simplicity of the acts analyzed should not give rise to a rebuke of physiological triviality by Husserl's research.

We should not be confused by the fact that three-dimensionality is, as it were, built up or completed by consciousness over the two-dimensional picture of perception, and the act of reflection should not call into question the intuitive nature of spatial constituting as a whole. All spatial intuitions (both two-dimensional and three-dimensional) are primordial and depend only on what field, what horizon, what projection consciousness chooses in its spatial experience. From a phenomenological point of view, painting is as much in line with our natural visual intuition as sculpture is with the two-dimensional picture as natural to our perception as the perceived volume of a figure in three-dimensional projection. Seeing and touching are constitutive forms of perception, full-fledged structures of consciousness, its phases, which phenomenology describes as kinesthetic acts.

Obviously, Husserl's spatial constitution differs from the traditional sensualist theory of sensations and the theory of primary object qualities, because for phenomenological analysis it is the "activity of sensations" whose source lies in the intensional activity of consciousness and not in the material nature of the world that is essential.

The significance of Husserl's kinesthetic theory is that it clarifies the role of corporeality in the overall structure of consciousness, and does not regard perception only as a condition for the activity of consciousness, but as a complete manifestation of it. Husserl's phenomenological description of acts of perception allows him to get rid of the illusion that elementary sensual acts are carried out when consciousness itself is neutralized. For Husserl the sphere of consciousness is total, i.e. there is no single object in which consciousness is neutralized and from which some other "reality" peeks out. Hence Husserl's main objection to the arguments of positivist researchers follows: there are no primary givens (qualities) of the world invested in perception by objective reality, no pure materiality (the hyletic composition of perception). This thought, generally speaking, should be framed tautologically: the world of our knowledge, i.e. the only world that is available to us, is the result of our knowledge.

4 Conclusions

The problematics of the language of visual studies is quite extensive, it is developed on theoretical and applied levels, includes analytics of visual objects and objects functioning, ways of their understanding, translation of meanings, correlation of meaningful content in communication processes. To solve these problems the methods of hermeneutics, structural analysis, logical analysis of relations between objects, meanings and meanings embedded in visual objects are used.

Phenomenological analysis of visual problems is aimed at the research of basic fundamental bases of visual perception, at the forms of spatial intuition, at the description of mechanisms that form and correlate "body consciousness". At the same time Husserl's merit is the development of special terminology, reference concepts that enable him to describe the process of spatial modeling as a fundamental structure of consciousness. Analyzing the "carnal consciousness" that produces visual images, Husserl described them structurally, showing the connection between the subject-content elements of spatial images and the ways they are given (in perception, in fantasy, in representation and in memory). In this sense, his visual theory refers to the analysis of language, which consists of a variety of variant linguistic elements that add up to certain semantic constructions. Husserl was able to translate the statics of the visual picture into the dynamics of a time-varying image. To use an analogy, he not only "multiplied" the stillness of the visual image, but brought its description to the stage of the first experiences in The Lumière brothers. The method of phenomenological descriptor [23] developed by him allows us to show these processes in their formation and to present the processes of visual activity in a completely new perspective. The analytics of the spatial model of perception is indeed Husserl's philosophical discovery.

References

1. Dikovitskaya, M.: Major theoretical frameworks in visual culture. In: Heywood, I. (ed.) The Handbook of Visual Culture, pp. 68–89. Berg, London (2012). https://doi.org/10.5040/978 1474294140.ch-001
2. Dikovitskaya, M.: Visual culture studies: twenty years later. Vis. Stud. **36**(3), 195–197 (2021)
3. Mitchel, W.J.T.: Showing seeing: a critique of visual culture. J. Vis. Cult. **1**(2), 165–181 (2002)
4. Symposium "Visual Studies/Études visuelles: un champ en question." Paris (2011). https://repository.globethics.net/handle/20.500.12424/1583742?show=full https://arthist.net/archive/814/lang=en_US. Accessed 15 Sept 2022
5. Grushka, K., Lawry, M., Chand, A.: Visual borderlands: visuality, performance, fluidity and art-science learning. Educ. Philos. Theor. **54**(4), 404–421 (2022)
6. Elkins, J.: Farewell to Visual Studies (2011). https://arthist.net/archive/814/lang=en_US. Accessed 15 Sept 2022
7. Whitney, D.: A General Theory of Visual Culture. Princeton University Press, Princeton (2011)
8. Bylieva, D.: Artistic virtual reality. In: Bylieva, D., Nordmann, A. (eds.) PCSF 2021. LNNS, vol. 345, pp. 462–473. Springer, Cham (2022). https://doi.org/10.1007/978-3-030-89708-6_39
9. Nordmann, A.: First and last things: the signatures of visualization-artist. Technol. Lang. **2**, 96–105 (2021). https://doi.org/10.48417/technolang.2021.02.10
10. Zhu, Y.: Visualizing the composition: a method for mapping inscription and instruction. Technol. Lang. **3**(2), 127–146 (2022). https://doi.org/10.48417/technolang.2022.02.08
11. Herbert, J.: Visual culture. Visual studies. In: Nelson, R.S., Shiff, R. (eds.) Critical Terms for Art History, 2nd edn., pp. 453–464. University of Chicago Press, Chicago (2003)
12. Moxey, K.: The Practice of Persuasion: Paradox and Power in Art History. Cornell University Press, New York (2001)
13. Mitchell, W.J.T.: Interdisciplinarity and visual culture. In: Art Introduction to Visual Culture. Routledge, London (1999)
14. Reutov, A.S.: Merleau-Ponty's visual phenomenology. Bullet. Perm Univ. Philos. Psychol. Sociol. **4**, 520–527 (2017). (in Russian)
15. Madsen, T.V.: Moving eyes: the aesthetic effect of off-centre pupils in portrait paintings. J. Aesthetics Phenomenol. **6**(1), 59–78 (2019). https://doi.org/10.1080/20539320.2019.158 7966
16. Kang, E., Park, E.J.: Phenomenological transparency through depth of "inside/outside" for a sustainable architectural environment. Sustainability **13**, 16 (2021). https://doi.org/10.3390/su13169046
17. Husserl, E.: Thing and space: Lectures of 1907. Springer, Cham (1997)
18. Husserl, E.: Ideas Pertaining to a Pure Phenomenology and to a Phenomenological Philosophy. First Book: General Introduction to a Pure Phenomenology [Kersten, F., trans]. Nijhoff, Hague (1982 [1913])
19. Husserl, E.: Ideas Pertaining to a Pure Phenomenology and to a Phenomenological Philosophy. Second Book: Studies in the Phenomenology of Constitution. [Rojcewicz, R., Schuwer, A., trans.]. Kluwer, Dordrecht (1989)
20. Drumond, J.: Space. In: Encyclopedia of Phenomenology. Dordrecht, Boston, London (1997)
21. Riley, H., Newell, R.: Egological meets ecological: drawing aspects in perspective(s). Drawing Res. Theory Pract. **6**(2), 307–332 (2021). https://repository.uwtsd.ac.uk/id/eprint/1835
22. Nitsche, M.: Sonic environments as systems of places: a critical reading of Husserl's thing and space. Open Philos. **4**, 136–148 (2021). https://doi.org/10.1515/opphil-2020-0164
23. Serkova, V.A.: Phenomenology of Culture. Polytechnic University Press, St. Petersburg (2010). (in Russian)

Creativity and Life Expectancy in Strategies of Adaptation

Irina Spivak[1,2] (ID), Andrei Zhekalov[1] (ID), Ruslan Glushakov[1] (ID), Vladislav Nyrov[3] (ID), and Dimitri Spivak[4,5(✉)] (ID)

[1] Military Medical Academy, ul. akad. Lebedeva, 6B, St. Petersburg 194044, Russia
[2] St. Petersburg State University, Universitetskaya nab., 7/9, St. Petersburg 199034, Russia
[3] St. Petersburg State Polytechnical University, ul. Polytekhnicheskaya, 29, St. Petersburg 195251, Russia
[4] N.P. Bechtereva Human Brain Institute, Russian Academy of Sciences, ul. akad. Pavlova, 9, St. Petersburg 197376, Russia
`d.spivak@mail.ru`
[5] D.S. Likhachev Russian Institute of Cultural and Natural Heritage, Bersenevskaya nab., 18-20-22, Moscow 119072, Russia

Abstract. The main results of a mass survey of psychological adaptation of 200 normal subjects to unusual or extreme, namely Arctic conditions, are presented along with its molecular biological and genetic correlates. Two main strategies of psychological adaptation to stressful conditions are traced back to the systematic study of polymorphisms of the neurotrophic factor gene (BDNF). The optimal strategy involves the activation of creative performance and tends to occur in subjects with 'stronger' BDNF genotypes. A less effective strategy consists in the activation of psychological defense mechanisms and is proper for subjects with 'weak' genotypes. Since long life expectancy is definitely linked to the 'strong' BDNF genotype, creativity is thus related to longevity in a most plausible way. - By considering the long-term efficacy of different 'dialects' of genetic 'language,' this investigation promotes and constructively pursues the systematic study of inner (genetic) 'multilingualism,' primarily in respect to life expectancy but also involving the functioning of complex psychological capacities such as creativity.

Keywords: Creativity · Longevity · Adaptation · Genetic code

1 Introduction

Longevity studies form an important realm of present-day interdisciplinary scientific research. Based upon family studies, about 25% of longevity may be regarded as owing to hereditary factors; as to the rest, they are currently being related to life style and conscious effort [1]. Systematic study of the genes involved in the regulation of such pivotal organic systems as the serotonin and the dopamine systems, has revealed multiple patterns of their partaking in the prolongation of life expectancy. Since telomere length seems to be indispensable at the present time for ensuring longevity, it is an integral part of this research to trace the association of these gene systems with telomere length. Other genes tend to play a sometimes considerable role in attaining this goal. The neurotrophic factor gene (BDNF) may serve as an instructive example of this. Subjects with its 'weak' (Met/Met) genotype were proven to have higher risk of developing mental disorders, as well as such diseases as diabetes [2, 3]. As to subjects with its 'strong' (Val/Val) genotype, they demonstrated a high probability of enjoying good health [4]. The amount of BDNF in blood serum in humans was demonstrated to be directly linked to telomere length; as a result, subjects with a 'weak' polymorphism were likely to have much lower life expectancy than those with a 'strong' polymorphism [5].

As to psychological factors, in present-day science these are considered a decisive element for attaining a healthy lifestyle and, mutatis mutandis, longevity or 'smart balance' of passive and active adaptive competence and performance – or, speaking in other words, of convergent and divergent thinking [6, 7]. As divergent thinking is directly related to creativity, detecting its input into attaining longevity forms an integral part of longevity studies. Moving in this direction forms the main objective of the present paper.

Reformulating the subject matter of this paper, we feel it to be most constructive referring to the so-called linguistic metaphor, which is quite popular in present-day scientific discourse related to genetic studies [17]. The codes that are applied by different genes in order to store and transmit information and that are related to managing different functions of an organism, may be regarded as different languages. Codes applied by different alleles of the same gene would in this case have to be understood as dialects of the same language. The human genome might be represented along these lines as a realm of inner - or, rather, genetic multilingualism. Thus an important objective of the present paper is the elaboration of methods allowing to assess the long-term efficacy of such dialects, and the level of involvement of higher functions, especially creativity, into their activity.

2 Methods

Two groups of more or less average young male urban professionals, aged 36 ± 11 years, were examined. The main group, comprising 100 subjects, was examined in polar regions, where they had come to live and to conduct working activities, which consisted in operating modern technical gear. The control group comprised 100 subjects as well, all of whom lived and worked in the conditions of much more temperate and mild climate, conducting the same type of professional duties.

Genetic and psychological testing was conducted once for each subject. All of our instructions, as well as the response of subjects, were formulated in Russian, which was the native tongue of all of the members of both our groups. Psychological testing consisted of filling in a form which comprised:

– six different tests, three of which were directed at measuring the level of verbal creativity, while three others were dedicated to figural one. As a result, integral indices of verbal creativity, and of the figural one, were calculated. In testing creativity, we followed a renowned scientific tradition, founded about seventy years ago by Ellis Paul

Torrance and Joy Paul Guilford. A standard Russian version of their tests, elaborated by Elena Tunik, was applied by us in our survey (for general context cf. [8, 9]);
– three questionnaires, directed at measuring the level of psychological tension. The first one of them, widely applied currently in clinical practice, primarily as an express test, was elaborated by Anthony Zigmond and R Philip Snaith, under the title of Hospital Anxiety and Depression Scale [10]. A standard Russian version of the Scale, elaborated by Drobizhev et al., was applied in our research. Not included in this paper are the results of the application of two other inventories, namely, the scale of Psychological Activation, Tension, Comfort, Interest, and Emotional Activation, by Kurgansky and Nemchin; and Scale for Psychological Express Diagnostics of Neuroticization Level, by Wassermann et al.
– a questionnaire elaborated by our research team, directed at the assessment of the level of short-term alterations of consciousness. The questionnaire consisted of 15 items, providing assessment of qualitative alterations of perception patterns, emotional background, cognitive functioning, communicative patterns, and of the contents of dreams. In working upon this part of our agenda, we took into account earlier tests produced by leading research teams in the field [11]. Following this influential trend, we regarded altered states of consciousness as a psychological reserve mechanism, activated sometimes in the service of the ego, primarily in order to cope with stress;
– a set of questionnaires directed at the assessment of the level of activation of psychological defense mechanisms, and of stress coping strategies. These most important psychological resources were briefly reviewed in an earlier paper of ours, dedicated to the same topic of creativity [12]. However, these questionnaires will not be discussed in the text of the present paper.

The biological block of our testing consisted in looking for parallels to psychological functioning primarily at the genetic level. Polymorphic variants of four genes were considered, namely the serotonin-2A receptor gene (5HTR2A), the angiotensin-converting enzyme 1 (ACE1), the α-actinin-3 protein encoding gene (ACTN3), and the neurotrophic factor gene (BDNF). In order to conduct genetic analysis, 5 ml of blood were taken from each subject. 6% EDTA vacutainers were applied produced by Greiner Bio-One company. Total DNA was extracted with the help of a kit designed for the purpose of isolating genomic DNA, produced by Biolabmix company.

Initial results concerning the correlation with creativity of polymorphisms of the four genes were presented in our preliminary report [12]. Based on these results, BDNF was defined as the primary target for further study. As a result, the present paper is focused on the correlation of its polymorphic variants, namely Val/Val, Val/Met, Met/Met, with creativity. Real-time PCR reaction was conducted by us for this purpose, using a DT-Prime (RT-PCR) amplifier, produced by DNA Technology company, and corresponding kits, produced by Sintol company.

Study of telomere length formed another part of the biological block of the present research. Blood samples taken from every subject were applied for this purpose as well. Real-time reverse transcription polymerase chain reaction conducted with the help of the same amplifier was used as the main tool of this study. As a result, relative telomere length (T/S ratio) was detected for each subject, in line with a methodology elaborated previously by N. O'Callaghan, and modified by us especially for the present study [13].

Following this line of analysis, a most accurate assessment of the probability of attaining longevity in each particular case, was acquired [14, 15].

3   Results

Data concerning the distribution of the BDNF genotypes in the European population are available at present, thanks to an extant study undertaken by National Center for Biotechnology Information, USA [16]. Taking into account these data seemed to be constructive, as the size of the groups observed by us, was in fact quite limited (200 subjects all in all). Data for both the main group, observed by us, and the control group, were compared to similar data, characteristic for the European population in general, by means of Pearson's chi-squared test. Results of this testing, presented in Table 1, demonstrate that differences between the main group and the European population in general, were statistically insignificant. The same was true for the difference between our control group, and the European population (the distribution of genotypes in both cases corresponded to the Hardy-Weinberg model, basic for population studies). As a result, we felt authorized to state that both groups observed by us, were representative for the European population in general.

Looking at figures contained in Table 1, especially those concerning the 'strong' (Val/Val) polymorphism, we may nevertheless state that it tended to occur in both our groups much more frequently than in the European population in general (for data, see column 2, Table 1). One might suppose that this tendency owed mostly to the fact that members of both groups had passed rigorous screening, in order to be enlisted. This would explain the prevalence of bearers of the 'strong' polymorphism, which was still not distinguishable in terms of formal statistics, from what was normal for the European population in general.

Table 1. Distribution of BDNF genotype variants by young professionals in normal and stressful conditions, and the European population in general

Group/polymorphism	Val/Val	Val/Met	Met/Met	Calculated chi-squared value
Main group	75,93	18,52	5,56	0,958
Control group	70,10	24,74	5,15	0,986
Average European	64,98	31,26	3,76	

Comment: Val/Val, Val/Met, Met/Met – BDNF genotypes, Average European – average data for population of Europe.

Having tested general psychological state of members of both of our groups, practically all of the indices were part of the respective normal bandwidth. Thus the integral index of neuroticization proved to be normal for practically all members of both groups. One feels authorized to suppose that this tendency also owed to the initial screening, both physiological and psychological, which tended to eliminate weaker candidates. However, some interesting tendencies, taking place in the framework of the norms, may be discerned here. The level of depression may serve as an instructive example (Fig. 1).

Regarding the figure, one has to state that the level of the depression, measured by Zigmond-Snaith scale, which is a standard tool of present-day psychological diagnostics, did not exceed 4 in all cases. Taking into account that the normal level of depression ranges from 0 to 7, one would have to assume that both our groups were quite fine in this respect. However, we see that subjects with the 'weak' BDNF (Met/Met) genotype tended to feel fine in normal conditions (right part, Fig. 1), but became much more depressed when observed under polar conditions (left part, Fig. 1). This difference was statistically significant. On the basis of this result one feels authorized to state that, remaining quite normal, subjects with the 'weak' BDNF genotype tended to react to unusual or extreme conditions much worse than those with the 'strong' (Val/Val) one.

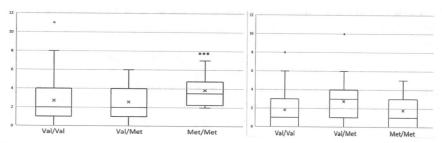

Fig. 1. Depressive attitudes by young professionals in normal and stressful conditions. Comment: left – main group, right – control group; Val/Val, Val/Met, Met/Met (from left to right) – BDNF genotypes, vertical scale – average scores by Hospital Anxiety and Depression Scale; asterisks (***) – statistically significant difference ($p \leq 0.05$).

These regularities are corroborated by the data in Fig. 2 which present the occurrence of short-time qualitative alterations of consciousness. The main trend here consists in sharp activation of this set of psychological processes by subjects with 'weak' BDNF genotype. Comparing the right part of Fig. 2 with its left part, one may state that the level of the corresponding index became practically twice as large under extreme conditions. Thus subjects with 'weak' genotype tended to become not only more depressed

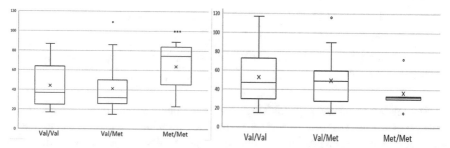

Fig. 2. Occurrence of short-time alterations of consciousness by young professionals in normal and stressful conditions. Comment: left – main group, right – control group; Val/Val, Val/Met, Met/Met (from left to right) – BDNF genotypes, vertical scale – average scores by Altered States of Consciousness Scale; asterisks (***) – statistical significance of difference ($p \leq 0.05$).

when coming to the polar region, but also developed quite pronounced altered states of consciousness. (These features of stress did not, however, exceed the normal bandwidth far enough to prove statistically significant, owing again to their having passed initial rigorous screening.)

Creativity formed another important part of the psychological block of our research. Two regularities were traced back. The first one consisted in the fact that the lowest level of creativity, both verbal or figurative, tended to be linked to the 'weak' (Met/Met) BDNF polymorphism, while its higher levels tended to be related to the 'medium' polymorphism or the 'high' one. In quite a few cases this regularity gained the level of statistical significance. The second regularity consisted in the fact that the level of creativity tended to be higher under polar conditions, especially in subjects with 'medium' or 'high' BDNF polymorphism, most possibly as a result of adaptation to stressful conditions. Figure 3 presents these regularities in an exemplary manner for the case of one of the creativity tests.

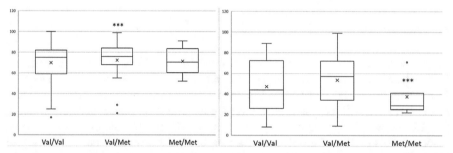

Fig. 3. Level of figurative creativity by young professionals in normal and stressful conditions. Comment: left – main group, right – control group; Val/Val, Val/Met, Met/Met (from left to right) – BDNF genotypes, vertical scale – average scores by Hospital Anxiety and Depression Scale; asterisks (***) – statistically significant difference ($p \leq 0.05$).

In the context of the biological block of our testing measurements of telomere length were obtained for every subject, using in every case sample of blood taken once during the course of psychological testing. Each sample was processed by means of a complex methodology, based upon application of the polymerase chain reaction. As a result, relative telomere length was calculated for every subject observed by us. Following this, mean telomere length was calculated for every polymorphism of every gene involved into our study. New 'telomeric' characteristics of every polymorphism were obtained in this way. Having processed these characteristics, statistically significant correlations were found for only one gene included in our study, namely the neurotrophic factor gene (BDNF). Basic results of this study are presented in Fig. 4. It shows that telomeric length in subjects with 'weak' BDNF polymorphism tends to be much lower (about 1.4 times) than in subjects with either 'strong' genotype or the 'medium' one. This difference is statistically highly significant. Thus the existence of a strong correlation between telomere length and the BDNF genotype ('weak' or 'strong'/'medium') was proven.

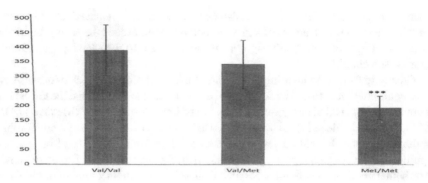

Fig. 4. T/S ratio of telomere length by young professionals in normal and stressful conditions. Comment: Val/Val, Val/Met, Met/Met (from left to right) – BDNF genotypes, vertical scale – average T/S ratio of telomere length; asterisks (***) – statistically significant difference (p $\leq$ 0.05).

4 Discussion

Regarding the results presented above, it would be constructive to divide them into those which correspond to regularities that were already known and those which seem to be quite novel and thus in need of more detailed interpretation. The association of the 'weak' BDNF polymorphism with a poorer state of health, both physical and mental, and with lower life expectancy belongs to the former category, as well as the association of the 'strong' genotype with a more optimal state of body and mind, and with higher life expectancy. The occurrence of states of depression and of episodes of altered consciousness can serve as a plausible example of this kind of results. As demonstrated by Figs. 1 and 2 of the present paper, their possibility is much higher among subjects with 'weaker' BDNF genotype, and much lower for those with its 'stronger' polymorphism. Telomere length may serve as another example. As shown by Fig. 4 of the present paper, it tends to be higher for subjects with 'stronger' genotypes, and much lower among those with 'weak' BDNF genotype; the difference in both cases was calculated significant. As a result, we may state that basic results of this research correspond to those obtained earlier by other research teams quite well.

As to results which may be characterized as novel, this concerns primarily the association of the high level of creative performance with 'stronger' BDNF genotypes (namely, the 'strong' and the 'medium' one), and of the low level of creativity with the 'weak' genotype (for details see Fig. 3). As it was demonstrated in an earlier paper of ours, the activation of psychological reserve mechanisms, especially those providing effective stress coping, tends to be closely linked to the 'weak' BDNF genotype [12]. Taking into account this set of results, it appears that we detected evidence for two quite effective strategies of psychological adaptation to stressful conditions. (1) According to the first of these, people with lower genetic abilities tended to compensate this by activating their psychological defense mechanisms. As to people with higher genetic abilities, they did not need to provide any psychological defense, as their adaptation was already optimal. (2) Thus the second strategy, proper for them, consisted in activation of their creative abilities, if necessary. Reiterating these observations in a short way, evidence for two main strategies were detected by us:

(1) 'stronger' BDNF genotypes (Val/Val, Val/Met) – high level of creativity – longer life expectancy;
(2) 'weak' BDNF genotype (Met/Met) – low level of creativity – shorter life expectancy.

5   Conclusions

As a result of a survey among 200 average people of psychological adaptation to unusual or extreme conditions of labor and life, and of its molecular biological and genetic correlates, the following regularities were detected:

– the neurotrophic factor gene (BDNF) seems to be most appropriate for conducting a detailed study of the interrelation between creativity and longevity;
– subjects with 'stronger' BDNF alleles tend to have higher creative ability, on the one hand, and longer life expectancy, on the other hand. Subjects with 'weaker' genotype tend to be less creative and have a lower life expectancy;
– the existence of two basic strategies of psychological adaptation to stressful conditions seems to be quite plausible. In the framework of the optimal adaptive strategy, subjects with higher genetic abilities cope with stress by means of practicing creativity, both verbal and non-verbal. In the framework of a weaker adaptive strategy, subjects with lower genetic abilities tend to cope with stress by means of activating psychological defense mechanisms;
– by assessing in this manner the long-term efficacy of different 'dialects' of the genetic 'language' (telomere length), and by reconstructing the ways in which these are connected to quite complex psychological phenomena (psychological defenses and creativity) suggests a promising direction for the future systematic study of inner (genetic) multilingualism.

Acknowledgements. This study was sponsored by Russian Foundation for Basic Research, grant 20-013-00121.

The authors are grateful to Ms. A. Trandina, M.D., for having conducted genotyping of all blood samples, and to Mr. P. Shapovalov, for the initial processing of raw data of psychological testing.

References

1. Passarino, G., De Rango, F., Montesanto, A.: Human longevity: genetics or lifestyle? It takes two to tango. Immun Ageing **13**, 12 (2016). https://doi.org/10.1186/s12979-016-0066-z
2. Lau, H., Fitri, A., Ludin, M., Rajab, N., Shahar, S.: Identification of neuroprotective factors associated with successful ageing and risk of cognitive impairment among Malaysia older adults. Curr. Gerontol. Geriatr. Res. **2017**, 4218756 (2017). https://doi.org/10.1155/2017/421 8756

3. Prabu, P., Poongothai, S., Shanthirani, C.S., Anjana, R.M., Mohan, V, Balasubramanyam, M.: Altered circulatory levels of miR-128, BDNF, cortisol and shortened telomeres in patients with type 2 diabetes and depression. Acta Diabetol. **57**(7), 799–807 (2020). https://doi.org/10.1007/s00592-020-01486-9

4. Zhou, J.-X., et al.: Functional Val66Met polymorphism of brain-derived neurotrophic factor in type 2 diabetes with depression in Han Chinese subjects. Behav. Brain Funct. **9**, 34 (2013). https://doi.org/10.1186/1744-9081-9-34

5. Vasconcelos-Moreno, M.P., et al.: Telomere length, oxidative stress, inflammation and BDNF levels in siblings of patients with bipolar disorder: implications for accelerated cellular aging. Int. J. Neuropsychopharmacol. **20**(6), 445–454 (2017). https://doi.org/10.1093/ijnp/pyx001

6. Whitbourne, S.K. (ed.): Special Issue: Successful Ageing. Research in Human Development, vol. 3, no. 2 (2005). https://doi.org/10.4324/9781315799339

7. Walla, C.: A dynamic definition of creativity. Creativ. Res. J. **3**(31), 237–247 (2019). https://doi.org/10.1080/10400419.2019.1641787

8. Kaufman, J., Sternberg, R.: The Cambridge Book of Creativity. Cambridge University Press, Cambridge (2010). https://doi.org/10.1017/CBO9780511763205.029

9. Beghetto, R.A., Corazza, G.E. (eds.): Dynamic Perspectives on Creativity. CTAE, vol. 4. Springer, Cham (2019). https://doi.org/10.1007/978-3-319-99163-4

10. Zigmond, A.S., Snaith, R.P.: The hospital anxiety and depression scale. Acta Psych. Scand. **67**(6), 361–370 (1983). https://doi.org/10.1111/j.1600-0447.1983.tb09716.x

11. Gruzdev, N.V., Spivak, D.L.: An exploratory investigation into the association of neuroticization, cognitive style, and spirituality to reported altered states of consciousness. Int. J. Transpers. Stud. **25**, 56–614 (2006). https://doi.org/10.24972/ijts.2006.25.1.56

12. Spivak, D., Zhekalov, A., Nyrov, V., Shapovalov, P., Spivak, I.: Creativity and its genetic foundations. In: Bylieva, D., Nordmann, A. (eds.) PCSF 2021. LNNS, vol. 345, pp. 72–86. Springer, Cham (2022). https://doi.org/10.1007/978-3-030-89708-6_7

13. Vasilishina, A., Kropotov, A., Spivak, I., Bernadotte, A.: Relative human telomere length quantification by real-time PCR. In: Demaria, M. (ed.) Cellular Senescence. MMB, vol. 1896, pp. 39–44. Springer, New York (2019). https://doi.org/10.1007/978-1-4939-8931-7_5

14. Mayer, S.E., et al.: Cumulative lifetime stress exposure and leukocyte telomere length attrition: the unique role of stressor duration and exposure timing. Psychoneuroendocrin **104**, 210–218 (2019). https://doi.org/10.1016/j.psyneuen.2019.03.002

15. Cabeza de Baca, T., et al.: Chronic psychosocial and financial burden accelerates 5-year telomere shortening: findings from the coronary artery risk development in young adults study. Mol. Psychiatry **25**, 1141–1153 (2019). https://doi.org/10.1038/s41380-019-0482-5

16. RS 6265. In: Submitted SNP(ss) Details: ss182258834 (nih.gov). Accessed 24 Sept 2022

17. Searls, D.: The language of genes. Nature **420**, 211–217 (2002). https://doi.org/10.1038/nature01255

The Language of Smell: Code Failure

Daria Bylieva[(✉)] [ID], Victoria Lobatyuk [ID], Dmitry Kuznetsov [ID],
and Natalia Anosova [ID]

Peter the Great St. Petersburg Polytechnic University (SPbPU), 195251 Saint-Petersburg, Russia
bylieva_ds@spbstu.ru

Abstract. In today's world filled with complex signs and symbols, visual and auditory channels are the most intensive in semiotic terms. The language of smell, associated with the most ancient reactions, is usually considered as secondary and supplementary, and its possibilities for conveying meanings are limited to simple recognition. However, experts have been using the alphabet of smells to convey emotional messages from ancient times to date. The assessment of the role of odors in the modern world became possible due to the Covid-19 pandemic which often involved the loss, change or intensification of the sense of smell. In the course of the study 250 cases were considered, representing the stories associated with the disease and deviations in the perception of odors. The loss of the perception of unpleasant odors makes it impossible to learn about the dangers which cannot be perceived visually like in ancient times (spoiled food, poisoned air, etc.). Phantom interpretation of odors is often unpleasant: people can identify the smells of burning, ammonia, acetone, decomposition, feces, and others, and sometimes the excessiveness of an ordinary smell is unpleasant as well. The change of sign recognition can cause serious consequences for people. Phantom unpleasant odors can result in changes in eating habits and cause problems in communication.

Keywords: Smell · Covid 19 pandemic · Anosmia · Language of smell · Olfactory hallucinations · Phantosmia · Semiotics

1 Introduction

The modern world is a complex semiotic space [1]. People constantly read and decipher the diverse codes that come to them from the outside world. Visual and auditory channels of perception are heavily loaded with symbols. In addition to natural language, people decipher paralingual signs, non-verbal symbols, technical symbols, etc. Unlike vision and hearing, the sense of smell is to a much lesser extent included in the recognition of the sign system of human beings. As a rule, smells are considered to be optional elements, giving an additional dimension to information already perceived through other channels. Recognition of smells involves a fully entwined system including neural operations, corporeal experience, and the cultured environments [2, 3]. Moreover, the language of smell is traditionally regarded as an atavism opposed to the development of human culture. Charles Darwin said that "the sense of smell is of extremely slight service, if any, even to savages, in whom it is generally more highly developed than in the civilized races. Those who believe in the principle of gradual evolution will not admit that this sense in its present state was originally acquired by men. No doubt we inherit the power in an enfeebled and so far rudimentary condition, from some early progenitor, to whom it was highly serviceable and by whom it was continually used" [4]. Sigmund Freud noted: "The diminution of the olfactory stimuli seems itself to be the consequence of man's raising himself from the ground, of his assumption of an upright gait" [5].

Modern technologies are now involved in the process of recognition and generation of odors: Machine olfaction (electronic nose) is used to detect odorant compounds in the air [6, 7]. Moreover, there is a biometric authentication system based on human odor analysis [8]. This apparatus is often referred to as an electronic nose. With the help of technologies capable of the transfer of reality (from cinema to VR), experiments have been carried out to include smells as a special component to enhance the effect of presence and as the object of exhibitions in museums, thus smells penetrate the digital environment [9–11]. Back in the late 1950s, there appeared two competing technologies that allowed sniffing the plot (Smell-O-Vision (Hans Laube) and Aroma Rama (Charles Weiss)). Several devices (iSmell, Scent Dome, Kaori Web, Angewandte Chemie, etc.) were designed to generate a smell from a set of primary recognition codes that could be embedded in web pages, email, television programs, game objects, etc. In 2020, technologies for generating smells for AR or VR reality were developed. For example, in 2022, Swedish scientists from Stockholm and Malmö University included the Nosewise Wine Game in VR, in which users experienced a virtual wine tasting - a specially designed machine that attached to the VR system's controller, and when they lifted the glass, it released a smell of wine [12].

2   The Language of Smell

Smells are natural signs - indices according to Peirce's classification [13]. In ancient times, the understanding of the language of smell was more important for people - the smell could reveal what could not be learned from other sources, visually a lot was hidden, and it was necessary to identify the smell in order to locate food, identify predators and prey, recognize territory, find a receptive mate and identify kin and toxic compounds [14, 15]. Today, unpleasant odors most clearly signal the meaning. As a rule, they act as a warning factor indicating danger or other troubles, such as "food is not fresh", "unhealthy air in the room", "the person is sick or unreliable", etc.

Smells perform the recognition function for the objects, i.e. they can convey a fact of the presence of an object. Therefore, quite often smells are associated with some object (bird cherry, coffee, sea, etc.). In most cases the smell as an identifier coincides with the information obtained visually, and the importance of its perception is not very important for recognition. Among the objects that can be identified by the smell, foods and drinks are of particular importance. There is a connection between touch and taste, delicious smells give a signal about the attractiveness of food and arouse appetite. In this area, a system for verbalization of smells has been developed (for example, there is the Wine Aroma Wheel by Ann C. Noble). In addition to odors denoting a certain category of objects, there are generalizing words that describe a general category of smells (tart, sweet, fresh, etc.) or an emotional impression (soothing, exciting, light, mysterious). However, the language of smell is hard to be translated into verbal terms, and a professional terminology is hardly available. As Morana Alač noted, languages have been marked by their lack of specialized vocabulary to express smells, however, the communicative space of the Internet helps the collaborative verbalization of the language of smell [16].

At the same time, smells can trigger complex psychological mechanisms and cause certain associations and emotions, affecting the psychological or cognitive state of a person, and evoking memories. The olfactory bulb has direct access to the amygdala and hippocampus – the evolutionary "old" parts of the brain, involved in regulating hormones, emotions, and memory, 75% of human emotions are created through smell/aroma [16–18]. Thus, it is implied that smells can give a person an "encrypted message", to which they will respond unconsciously. Unlike the information gathered by other senses, the transmission of olfactory sensory data to the brain is fairly direct [19].

People have been striving to create messages with the language of smell since ancient times. Using the existing alphabet, i.e. well-known odorous substances (musk, herbs, flowers, fruits, spices, resins, ambergris, etc.), the first compositions of odors were created. In the Indus Valley, Ancient Egypt, Mesopotamia, Greece, Rome, Persia, pleasant smells were created for the gods and for the elite. The most noble and rich people could show their status with the help of a smell code, among other things. In Greece and Rome, the smell also served as a sign of prostitutes, marking their body for men [20]. A specific symbolic meaning could be attached to a certain smell: the Nile blue lily is associated with the sign of the pharaoh, roses and violet – with Aphrodite, hyacinth and marjoram – with Hermes. Interaction with the supernatural (various rituals, sacrifices, divination) was accompanied by certain smells, as if opening the entrance to another reality, invisible and non-verbalized. As Suzanne Evans noted, Christianity introduced an epistemological element, attributing to the smell the ability to know the God and understand how the divine shaped the world [21].

Thus, initially fragrances were created both as a message from a person (that is, a person can express himself through a certain aromatic range) and for rites/ambience (to create a certain atmosphere). The same goals for creating messages have remained basic in the language of smell today. Perfumery technologies today make it possible to express oneself with the help of multi-component fragrances. The description of the smell of modern perfumes can be shocking: for example, "Odeur 53" contains, among other elements, the smell of burnt rubber, dust on a hot light bulb, hot metal, a toaster with freshly toasted bread, ink for calligraphy and an electric battery, "Don Xerjoff" contains smells of gunpowder, tobacco, whiskey and burnt sugar, "Muscs Koublaï Khän", in addition to musk, contains notes of civet (a secretion from the glands of animals of the viverrid family) and costus root, which is attributed to the smell of not quite clean hair. Specially created fragrances such as deodorants, air fresheners, food flavorings, etc. should replace or drown out natural smells. Smells can be used to directly affect the psychological or even physiological state of a person. The marketing use of smells is gaining popularity: the smell of space to influence human behavior [22]. In addition, in aroma marketing, the smell of a product can serve as a way to identify a brand [23], the so-called signature scents [24].

For a scent specialist (perfumer, aromatherapist, sensory marketer, etc.) smell is a 'grammar', a system of signs that can be used to create complex harmonies and meanings [25]. For an ordinary person, the recognizable meaning of smells is limited to association with objects, their knowledge of the alphabet of smell is scarce, and there is no possibility of independent generation of a message. At the same time, the influence of those who create the smells of modern civilization is extensive. Rosenbaum wrote that "people in the flavor industry will really be defining what the next generation thinks is strawberry" [26]. Behind the verbally simple attributive message "this is strawberry", there is a complex grammar of smells: "The smell of a strawberry arises from the interaction of at least 350 different chemicals that are present in minute amounts" [27].

3   Impact of Covid19 on Smell Deciphering: Case-Study

Some people lost the sense of smell during/after being afflicted with Covid-19, while others began to perceive smells in different way. Studies show that problems with odor perception were observed in 33 to 98% of COVID-19 patients [28]. These pathologies have received little attention in comparison with other symptoms and consequences of the disease. At the same time, this failure in the information decoding system is a unique case-study that demonstrates the role played by the language of smell in human life.

The purpose of the research was to identify what smell-related changes people most often write about online, sharing their experiences, and what are the changes in their daily life against this background (refusal of food, habitual lifestyle or even depression, difficulties at work and with communication). The main methods in this work include content analysis, statistical data processing and netography.

The study took place in two stages - at the first stage in May 2022, the resources on the Internet were studied, that provided access to the largest number of messages from those who had been ill and lost their sense of smell and at the second stage the cases selected on these sites were analyzed.

During the study, 250 cases were considered, representing stories related to the perception of smells by people who got ill with Covid-19, which was followed by not only the loss of smell, but also the changes in the intensity of smell, the appearance of phantom smells, etc. Most cases are presented by posts dated in 2020–2022. Information was collected on Russian-language sites and social networks such as https://pikabu.ru/, https://fishki.net/, https://vk.com/, https://zen.yandex.ru. To search for cases on the listed sites, the keywords "ansomia", "loss of smell", "phantom smells", "Covid-19 smells disappeared", "change of smells", etc., were used, 20 words in total.

The attitude of the authors of the posts to changes in the sense of smell is negative in most cases (94%) and only in rare cases is it neutral (2%) or positive (4%). Ansomia (40%), when sense of smell is simply lost, and phantom unpleasant odors (42%) are most often shared online, while there are much fewer examples of intensified sense of smell (18%). In 56% of cases, there are related changes in nutrition, in 30% of cases there are stories about cardinal changes in life (including: 8% problems with communication, 4% difficulties at work, 10% depression that required a visit to the doctor, etc.) and 14% of cases there is no change in lifestyle.

3.1   Loss of Smell

Loss of smell is one of the symptoms of Covid-19, which everyone is talking about, saying what it's like to live without being able to use one of the senses: *"The sense of smell disappeared on the fifth day of illness literally overnight. It gave the impression of the lights being turned off suddenly - all the food and the world as a whole suddenly lost a significant part of the existence"*. The loss of understanding of smells is psychologically depressing: *"we started repairing the apartment and fell ill with Covid-19, the smells disappeared, my wife and I don't experience as before the joy of freshly painted walls, because there is no smell of paint, and the world without smells is very sad"*, *"a year or even a little more I lived without any smells, at first there was a feeling of great devastation, for the first two months it seemed as if the world had become less colorful, morally it was difficult because you feel like smelling something as a knee-jerk reaction and try to sniff something. Every time I realized that it was all in vain, but I didn't have depression, there was just some kind of disturbance"*.

The absence of smells as a warning factor can seem dangerous: *"I had Covid-19 two years ago, but the smells never returned, now I try not to tell unfamiliar people that I am devoid of smell, it seems to me that it is easy to use it against me, for example, poison or somehow deceive me"* and can really lead to unpleasant consequences: *"it's hard in everyday life: I was about to give the children rotten soup"*. It turns out that it is the loss of sensitivity to unpleasant odors that is most dangerous: *"All pleasant smells have returned a long time ago. But the disgusting smells such as the one of rotten meat, sour products, and others have not. Before, when I cleaned up the poop after the dog, I felt dizzy, but now I don't feel it. Up to the point that after my dog had been lying next to a dead cat in the forest, I did not feel this stink and even hugged her. Well, I washed her and myself after everyone at home got sick"*.

3.2   Code Failure

Phantom odors, olfactory hallucinations, phantosmia are the names of one condition in which a person feels smells that are not present. Most often this happens to people who have lost their sense of smell during the disease. Researchers confirm olfactory hallucinations in many patients with COVID-19, but there is currently no explanation for their nature [29]. Most often, the phantom odors are unpleasant: people feel smells of burning, ammonia, aciton, decomposition, feces, and so on, although sometimes the obsession of an ordinary smell can also be unpleasant. Olfactory hallucinations are no less intrusive and unpredictable than auditory or visual. They can appear regularly or have a paroxysmal form, lasting from several minutes to several hours and even months. *"The sense of smell is starting to return, but a problem has arisen, everywhere I can smell potatoes baked at the stake. For the first time I felt it on the street, I was surprised who was baking potatoes in the city center, but the smell did not disappear both on the bus and at home"*, *"some kind of sweet chemical smell is haunting, shampoos, soap, linen, food, bus – everything smells like it"*. In some cases, the smell is perceived as a mixture of scents: *"Everything smells about the same. Outside, from the sewer, it smells about the same. A bunch of smells are now mapped into one"*.

A change in the smell of food causes the rejection of a number of products and changing eating habits: *"My taste of cola has changed - a vile, hospital, chemical taste, just disgusting"*. Changes in smells often require changes in diet, a person is forced to abandon foods that give a "dangerous" smell signal: *"All food tasted like the soles of boots, I couldn't smell fragrant sunflower oil and just stopped adding it to food, in general, now I act like that, if something doesn't smell, then I don't eat it"*, *"The worst side effect: three months after the sense of smell returned, suddenly the perception of all protein foods changed. Meat, eggs, milk, and even the human body began to resemble a mixture of expired tomato beans with a large piece of dust. The sensation is not pleasant at all. Fruit tastes like acetone, and ice cream tastes like something sour"*. Quite often, unpleasant odors are caused by protein products, the lack of which can cause serious harm to the body: *"meat tastes like ammonia with rotten meat, minus 5 kg in a month"*. In some cases, some products become attractive against the background of the disgusting smell of ordinary dishes: *"the smell of meat and fish has changed a lot. Fresh vegetables, especially onions, are all disgusting. Because of this, all food seems tasteless. But all sorts of sweets (which I didn't eat before Covid-19) are excellent"*, *"the smells of meat, onions and eggs have changed. From these products I smell a pungent smell of rotten meat, which is generally unbearable. At the beginning, I ate these products through force. Then I realized that it couldn't continue like this and I just had to exclude these foods from the diet, although these foods are my favorite, I just couldn't do without them before, and I used to eat them every day. Now I'll have to change my menu. I tried to 'disguise' the smell with spices, but still I could feel everything. The smell of oatmeal has also changed, but for the better - for some reason it smells sweet, like caramel"*. However, olfactory hallucinations are not necessarily tied to specific products: *"sometimes some products began to smell unexpectedly unpleasant, for example, some vegetable. And it turns out that you simply throw away half of the products, because you can't eat them. All purchases were a kind of lottery"*, *"But right now, as if in a wartime, I'm eating the turnip-tops and thinking about the meaning of life"*.

Women who cook for the whole family had to find solutions not only for themselves. Changes in eating habits also have long-term health or social consequences associated with food sharing rituals. *"An unpleasant, sugary-but-sweet smell began to come from my hair. I dressed it in a bun, but it didn't help. Then I began to feel the same smell from products. As a result, I stopped cooking and eating healthy food. I ordered ready-made food for my family, and I myself just ate oatmeal and honey cake for a month and a half, drank vitamins to keep the nutrient imbalance, but still I got my skin covered with acne"*. *"I used to spend time with my friends in bars and restaurants, but now I feel that all food and alcohol give away the smell of ammonia, I lost almost all my friends within half a year of such a life"*.

Synthesized scents created to convey pleasant odors cease to act: *"perfume smells a mixture of alcohol and a weak fragrance, hand cream smells like some kind of floor cleanser. I normally feel flavors in food, like vanilla, but I don't feel household chemicals, except for air fresheners"*. Moreover, these synthesized scents can often cause a feeling of disgust: *"the smell of rotten fish from shampoos and hair balms"*, *"Creams, gels... The smell seems to be strongly acetic, even causing tears from the eyes, as if corroding the nose and eyes. My husband even freaked out, thinking that I was kidding. First I remove one shower gel, then another, and some creams for/after shaving. I literally could not go into the bathroom. Everywhere, even outside I was haunted by this smell".* The possibility of using scents for self-expression is lost: *"Now I have problems with perfumes. All fresh scents smell like chemistry, a la 'Mister Muscle' or at best, like hairspray".*

A failure in smell recognition creates special problems if a person feels a very unpleasant smell: *"from time to time I feel a putrid smell, the first time I was very scared, I thought someone had been killed and thrown into the elevator shaft, since it was in the elevator when this happened to me for the first time"*, especially if it is people who are perceived as a source of stench: *"a putrid smell emanates from my wife, I endured this for half a year and then was forced to tell it to her, and then I filed for divorce".* Communication problems are one of the most common consequences of wrong recognition of smells. First of all, these are problems with communication at work: *"there was such a strong smell of rotten fish from my girlfriend that I decided to stop communicating, but at work there was a similar incident with a couple of colleagues, and now I am unsociable and prefer to be alone", "I got retired and divorced, unpleasant smells from people brought me to nervous exhaustion, when the whole world seemed to smell of rotten stuff and no one could help you. Then I had to run away from this to an abandoned village and live alone".*

3.3 Recognition of Smell

If it is obvious what a problem the perception of unpleasant smells can be, the intensified smell seems to be perceived as neutral. However, it turns out that this phenomenon is also a consequence of Covid-19.

Though occurring rarely, intensification causes an ambivalent reaction: *"it's like I became a superhero after Covid-19, when my sense of smell returned, I could catch smells at a great distance, now it was not a problem for me to find out what neighbors were having for lunch or what kind of shower gel they used, at first it was funny, but after a month I decided to see a doctor", "today I went out onto the balcony to get some air in self-isolation and felt embarrassed, I could distinguish all the smells from the neighboring house: toilet, tobacco smoke, flowers and much more".*

At the same time, the greatest problems are caused by synthesized odors: *I began to distinguish the smell of perfume at a great distance, if someone nearby smelled heavily, I was getting headaches and nausea, which had never happened before". "I didn't have a loss of smell and taste. On the contrary, it was as if 'amplifiers' were built into them. For example, I found the same sugary-sweet smell in all cheap household cosmetics (shampoos, gels, liquid soap, etc.) that caused slight nausea. I came to the conclusion that this is the smell of the base component of the contents of these bottles and it is the same for different manufacturers".*

A number of professions related to smells are at risk, the problem with the correct decoding of smells can put an end to the work of a perfumer, sommelier, cook, etc. *"I work in a perfume shop, I was very afraid of getting ansomia in a pandemic, this was my phobia but everything turned out differently and my complication is due to the aggravation of the sense of smell, I don't even know what is worse in my work".*

3.4 Technologies of Coping with the Problem

As people share their stories in order to receive social support and assistance, they are also offered advice and guidance on recovery. Advice can range from medical diagnostics and procedures to simple advice. A number of comments are given about the ineffectiveness of going to doctors and taking medications: *"I talked with a lung specialist and an otolaryngologist - they can't help me with anything other than advice! Drink B vitamins and inhale essential oils. (This is normal, there is no cure for this, and I have no complaints to doctors)"*.

The author's explanations of the problem by non-specialists often look like technical metaphors: *"Covid-19 damaged the receptors or nerve cells responsible for smell perception, and the body temporarily repairs the backup line, and then reconnects it back"*. There are also humorous references to the fact that the failure in the perception of smells actually reflects a reality that is hidden from humanity: *"We live in the Matrix. The virus affects the central nervous system and prevents the standard digital signals of the matrix from affecting certain areas of the brain. Because of this, the sense of smell disappears (the first phase of blocking digital signals), and then you begin to smell and feel the taste of the liquid that you are fed through a tube that lies in the capsule. Well, why not transfer really tasty food to batteries?"* Such models correlate with Stanislav Lem's Futurological Congress in the most interesting way. There, virtual reality was chemically modeled, plunging people of the future into a favorable existence against the backdrop of the ongoing apocalypse, while, here, on the contrary, genuine smells supposedly resist digital deception. Despite all the paradoxical nature of the metaphor, it refers to the meaning of the language of smell as more ancient and authentic than the existing visual culture.

The most interesting are the tips for restoring the sense of smell such as the following: it was recommended to inhale substances with a strong smell *"Try to smell pepper paste, the first smell I started to feel after Corona"*, *"My sense of smell restored suddenly when they brought me a whole bag of cilantro and I was cutting it the whole evening. It was just like they turned on the smell"*. In some cases, it is recommended to consciously recall the aroma: *"breathe in bright and familiar aromas, remembering how they smell, what they were. I have Manser red tobacco perfume, which is extremely strong. I brought it to my nose but felt nothing. But I sniffed hard, trying to recall what it was like. After two days, the taste appeared, and then the smell"*. Sometimes it is recommended to use unusual smells to restore the sense of smell (perfume samples, out-of-season fruits, etc.).

4   Conclusion

Smells take a special place in the modern sign system. Despite the development of technological tools for the analysis and creation of smell, the rather modest role and the impossibility of creating messages with the help of smell do not allow us to consider it as a special language. But the loss of smell caused by Covid-19 demonstrated the role which the language of smell plays in the modern world. Though seemingly secondary, the sign system has thus shown its significance even today, in the world where you do not need to smell prey or predators. It turns out that odors in their function of warning about danger dominate over the information obtained in another way. Thus, visual appeal and knowledge of the freshness of the product do not help in overcoming the aversion caused by a phantom bad smell.

Only after losing the sense of smell, we realize the importance of the language of smell. Loss of odor perception affects people more often in a psychological sense, but it can lead to unpleasant and even dangerous consequences when a person cannot recognize the smell warning about danger. The analysis of cases shows that the areas of nutrition and communication turned out to be the most significant in terms of the smell code. Failures in perception of the smells of food and people can lead to the most serious consequences for health, family and social life. It is also interesting that the sharpened ability to identify a variety of odor codes often leads to depression of the owner of such superpower, although it is also possible that in this case people can gradually get used to a space more saturated with various odor codes.

References

1. Nesterov, A.Y., Demina, A.I.: Technology and understanding. Technol. Lang. **2**, 1–11 (2021). https://doi.org/10.48417/technolang.2021.04.01
2. Cerulo, K.A.: Scents and sensibility: olfaction, sense-making, and meaning attribution. Am. Sociol. Rev. **83**, 361–389 (2018). https://doi.org/10.1177/0003122418759679
3. Spackman, J.S., Yanchar, S.C.: Embodied cognition, representationalism, and mechanism: a review and analysis. J. Theory Soc. Behav. **44**, 46–79 (2014). https://doi.org/10.1111/jtsb.12028
4. Darwin, C.: The Descent of Man and Selection in Relation to Sex. Princeton University Press, Princeton (1981)
5. Freud, S.: Civilization and Its Discontents. W. W. Norton & Company, New York (1962)
6. Kochemirovskaya, S.V., Kochemirovsky, V.A.: Laser method of micro-composite materials synthesis for new sensor platforms of an "Electronic Tongue." Technol. Lang. **2**, 16–30 (2021). https://doi.org/10.48417/technolang.2021.02.03
7. Roy, M., Yadav, B.K.: Electronic nose for detection of food adulteration: a review. J. Food Sci. Technol. **59**, 846–858 (2021). https://doi.org/10.1007/s13197-021-05057-w
8. Jirayupat, C., et al.: Breath odor-based individual authentication by an artificial olfactory sensor system and machine learning. Chem. Commun. **58**, 6377–6380 (2022). https://doi.org/10.1039/D1CC06384G
9. Pylkin, A., Serkova, V., Petrov, M., Pylkina, M.: Information hygiene as prevention of destructive impacts of digital environment. In: Bylieva, D., Nordmann, A., Shipunova, O., Volkova, V. (eds.) PCSF/CSIS -2020. LNNS, vol. 184, pp. 30–37. Springer, Cham (2021). https://doi.org/10.1007/978-3-030-65857-1_4
10. Serkova, V.: The digital reality: artistic choice. In: IOP Conference Series: Materials Science and Engineering, vol. 940, p. 012154 (2020). https://doi.org/10.1088/1757-899X/940/1/012154
11. Dozio, N., Maggioni, E., Pittera, D., Gallace, A., Obrist, M.: May I smell your attention: exploration of smell and sound for visuospatial attention in virtual reality. Front. Psychol. **12** (2021). https://doi.org/10.3389/fpsyg.2021.671470
12. Petridou, C.: The sense of smell enters VR gaming world with nosewise odor machine (2022). https://www.designboom.com/technology/sense-smell-vr-gaming-world-nosewise-odor-machine-10-25-2022/. Accessed 1 Sept 2022
13. Peirce, C.S.: Collected Papers, vol. 2. Harvard University Press, Cambridge (1932)

14. Firestein, S.: How the olfactory system makes sense of scents. Nature **413**, 211–218 (2001). https://doi.org/10.1038/35093026

15. Zarzo, M.: The sense of smell: molecular basis of odorant recognition. Biol. Rev. **82**, 455–479 (2007). https://doi.org/10.1111/j.1469-185X.2007.00019.x

16. Alač, M.: We like to talk about smell: a worldly take on language, sensory experience, and the Internet. Semiotica **2017** (2017). https://doi.org/10.1515/sem-2015-0093

17. Paluchová, J., Berčík, J., Horská, E.: The sense of smell. In: Sendra-Nadal, E., Carbonell-Barrachina, Á.A. (eds.) Sensory and Aroma Marketing, p. 33. Wageningen Academic Publishers, Wageningen (2017)

18. Lindstrom, M.: Brand sense: how to build powerful brands through touch, taste, smell, sight and sound. Strateg. Dir. **22**(2) (2006). https://doi.org/10.1108/sd.2006.05622bae.001

19. Lojanica, M.V.: Death smells like strawberries:the olfaction simulacra. Nasle. Kragujev. **43**, 375–390 (2019)

20. Squillace, G.: Perfumes for men, perfumes for women. In: Moore, K.R. (ed.) The Routledge Companion to the Reception of Ancient Greek and Roman Gender and Sexuality. Routledge, London (2022). https://doi.org/10.4324/9781003024378

21. Evans, S.: The scent of a martyr. Numen **49**, 193–211 (2002). https://doi.org/10.1163/156852 702760186772

22. Rimkute, J., Moraes, C., Ferreira, C.: The effects of scent on consumer behaviour. Int. J. Consum. Stud. **40**, 24–34 (2016). https://doi.org/10.1111/ijcs.12206

23. Girona-Ruíz, D., Cano-Lamadrid, M., Carbonell-Barrachina, Á.A., López-Lluch, D., Esther, S.: Aromachology related to foods, scientific lines of evidence: a review. Appl. Sci. **11**, 6095 (2021). https://doi.org/10.3390/app11136095

24. Spence, C.: On the use of ambient odours to influence the multisensory experience of dining. Int. J. Gastron. Food Sci. **27**, 100444 (2022). https://doi.org/10.1016/j.ijgfs.2021.100444

25. van Leeuwen, T.: Introducing Social Semiotics. Routledge, Abingdon (2005)

26. Rosenbaum, R.: Today the strawberry, tomorrow…. In: Klein, N. (ed.) Culture, Curers and Contagion, pp. 80–93. Chandler and Sharp, Novato (1979)

27. Schlosser, E.: Fast Food Nation: The Dark Side of the All-American Meal. Harper Perennial, New York (2005)

28. Kang, Y.J., Cho, J.H., Lee, M.H., Kim, Y.J., Park, C.-S.: The diagnostic value of detecting sudden smell loss among asymptomatic COVID-19 patients in early stage: the possible early sign of COVID-19. Auris Nasus Larynx **47**, 565–573 (2020). https://doi.org/10.1016/j.anl. 2020.05.020

29. Nourchene, K., et al.: Smelling different after COVID-19? Eur. Psychiatry **65**, S514–S514 (2022). https://doi.org/10.1192/j.eurpsy.2022.1309

Term-Metaphors in Construction and Civil Engineering: Based on Metaphorical Nomination of Equipment, Machines and Tools in English and Russian

Elena Morgun$^{(\boxtimes)}$ ⓘ, Daria Burakovaⓘ, Maya Bernavskayaⓘ, and Olga Mikhailovaⓘ

Peter the Great St. Petersburg Polytechnic University, St. Petersburg 195251, Russia
{morgun_ea,burakova_da}@spbstu.ru

Abstract. The research addresses the repertoire of metaphor-induced terms in the specialized domain of engineering. Term-metaphors underlying the language of construction and civil engineering are revealed and scrutinized. We primarily focus on metaphorical nomination of equipment, machines and tools in English and Russian technical terminology. The research is based on Lakoff and Johnson's Conceptual Metaphor Theory. Numerous examples are extracted from specialized technical dictionaries and online glossaries. Metaphorical nomination resides in the nature of every language, it is a lens through which reality is perceived, with metaphors being culturally entrenched patterns reflecting a specific national mentality passed on from generation to generation. Needless to say that technical terms are precise as they designate clearly defined concepts, however, some of them are fundamentally metaphorical in nature. Metaphorical associations in engineering language arise from numerous source domains. The analysis provides insight into the process of meaning encoding/decoding in the sphere of construction and civil engineering, reveals major tendencies (universalities and peculiarities) of metaphorical nominations as well as sheds light into cultural divergence and facilitates understanding of technical vocabularies by foreign languages learners.

Keywords: Term-metaphors · Metaphorical nomination · Civil engineering · Technical terminology · Conceptual metaphor theory

1 Introduction

The language reflects the perception and categorization of the real world around us. In the early 19th century, the Prussian philosopher and philologist Wilhelm von Humboldt (1767–1835) advanced the hypothesis that every language constitutes the outward manifestation of its speakers' (authentic) knowledge, imagination, worldview and culture [1]. These thoughts underwent further development. "Human beings", E. Sapir wrote in 1929, "are very much at the mercy of the particular language which has become the medium of expression for their society" [2, p. 209]. B.L. Whorf, in turn, claimed that:

"the world is presented in a kaleidoscopic flux of impressions which has to be organized by our minds" [3, p. 213]. Thus, language determines thought: processing and understanding reality. People of different cultures perceive and categorize 'reality' differently. Apparently similar concepts could be endowed with different and specific meanings, connotations and images in each language. As A. Wierzbicka puts it: "the meanings of words from different languages don't match […]; they reflect and pass on ways of living and ways of thinking" [4, p. 15]. Naturally, both universal (inter-lingual) and unique (peculiar to a certain language world picture) linguo-cultural components can be found in any language. The imagery entrenched in human minds significantly shapes our perception of reality (both in individual and collective consciousness). The idea is also deeply rooted into the conceptual metaphor theory that postulates that metaphors play a vitally important role in the reflection, perception, understanding of the world [5–7]. They allow to unveil the peculiarities of speech-thinking activity as a stage of cognition and as an interaction and fusion of the individual and collective aspects of knowledge; a conceptual metaphor precedes the linguistic expression. Language is an essential tool for constructing the linguistic world picture, a kind of conceptual layer between the mind (a man's thinking) and the world. Internal semantic structure of a particular language (the language norms, its character and nature) imprints the unique way in which the speakers of a given language view the world. These investigations eloquently show that different languages give different pictures of the world. In this way, language reflects general cognitive processes/cognition [8–10].

Term-metaphors embrace an extremely wide sphere of functioning. "The essence of metaphor is in understanding and expressing one kind of thing in terms of another" [5, p. 5]. Whenever we encounter something new, we subconsciously link it up with something we already know. Thus, each new experience is associated and compared with the most similar thing encountered previously. Term-metaphors used in the engineering context are latent (perceived as linguistically dead), as a rule. In a paradoxical way, they still evoke associations by making salient a certain stereotype in human cognition as reflected in the conceptual and linguistic pictures of the world (language reflects our experience of the world). "We cut nature up, organize it into concepts, and ascribe significance as we do, largely because we are parties to an agreement that holds throughout our speech community and is codified in the patterns of our language" [3, p. 213]. Thus, a variety of different cognitive operations can be employed by people in their effort to make sense of experience.

In recent years, there has been an upsurge in the study of metaphor and metonymy underlying human thought and language in engineering (technical discourse). For extensive collections of papers, see [11–17]. Metaphors, on the one hand, are used for enriching technical and scientific vocabulary; on the other hand, they allow experts in the domain to share and popularize their specialized knowledge [18]. Thus, we make sense of our experience via metaphor and other imaginative structures. The image provides a venue for meaning to occur. The creation of meaning in an image implies that eventually image ceases to be just a copy of an object and begins to serve as an unfolding experience. Images help individuals to make meaning out of their experiences. In this sense, meaning becomes a product of imaginative creativity; even when an experience is described as equally meaningful, identity exists within difference. That is why associations between the individual image and its perception in regard to the nature of and the general qualities ascribed to a particular object (designated by a technical term) may differ in different languages.

Terms are lexical units that designate concepts (pieces of knowledge acting as a kind of aid for carrying out cognitive process) in a specific domain. The cognitive side of a term relates to the underlying conceptualization and modelling the world (a system of thought). Following L.G. Fedyuchenko, the metaphoric term (term-metaphor) is defined "from the standpoint of cognitive terminology as a dynamic cognitive structure, which contains a fragment of the technical picture of the world" [19, p. 77, also 20]. It should also be pointed out that there are a lot of definitions to the language world picture in the present-day linguistics. The current research views this phenomenon as follows: the architecture of the language world picture is naturally encoder-/decoder-specific: reality is conceptualized differently in communities in which a specific language is spoken. However, this process [the process of conceptualization] is of dual character: it discloses both universal (common for different languages) and peculiar (nationally specific) features; native speakers perceive the world through their shared language, its imagery space and inner structure, in particular. It seems to us that the cognitive nature of a technical term-metaphor has so far been insufficiently described in linguistics, therefore, the purpose of our study is to address the repertoire of metaphor-induced terms in the specialized domain of engineering in the light of meaning-making functions: from description to interpretation and imagination. Thus, term-metaphors become the object of special study in terms of their conceptualization, linguistic and cultural equivalency and divergence in English – Russian national world pictures.

2   Corpus and Methodology

A parallel corpus investigation of term-metaphors that are used in the domain of construction and civil engineering and nominate equipment, machines and tools has been performed in the light of cognitive as well as contrastive and cross-cultural approaches. Parallel corpora "give new insights into the languages compared…, can be used for a range of comparative purposes and increase our knowledge of language-specific, typological and cultural differences, as well as of universal features" [21, p. 18, also 16].

The study compares and contrasts term-metaphors in English and their Russian translations from the perspectives of equivalent vs. non-equivalent levels of the units, perceptions and representations of reality. The research features a range of data analysis techniques: continuous sampling method based on statistical foundations, the method of semantic (componential) analysis, description and interpretation of the findings.

The research consisted of four stages. The first stage (I) involved material collection (the random sampling method was used) – we selected English term-metaphors, nominating equipment, machines and tools in construction and civil engineering. These were collected from various specialized technical dictionaries and online glossaries compiled both in the English speaking countries and Russia [22–28]. The main criterion for selecting the units was the presence of a metaphoric constituent element in the term. Then, at the second stage (II), we identified and compiled a systematized corpus of in total about 968 units for the further analysis. At the third stage (III), we examined the equivalents of the term-metaphors in the languages under study (English term-metaphors and their Russian counterparts [27–29]), comparing them to find out their extent of co-incidence. Finally, at the fourth stage (IV), we highlighted the peculiarities recurring in the metaphorical component of the term in the languages under study.

3   Results and Discussion

The analysis of term-metaphors that are used in the domain of construction and civil engineering and nominate equipment, machines and tools shows the overwhelming predominance of HUMAN, NATURE and ARTEFACTS conceptual metaphorical models throughout the technical term corpus; and within each of these metaphorical models distinct semantic paradigms can also be distinguished. For example,

1. HUMAN (523)

    a) face parts: face (82), tongue (25), eye (22), tooth (19), jaw (18), lip (14), mouth (12), hair (11), cheek (8), nose (8), ear (7), eyebrow (2), iris (1).

    E.g.: *adjustable face wrench* – разводной торцевой ключ, *lifting eye* – подъемная проушина, *toothed tool* – многолезвийный/многозубный режущий инструмент, *grab bucket with hinged jaws or teeth* – челюсти/зубья грейферного ковша, *open-jawed spanner* – гаечный ключ с открытым зевом, *bucket lip* – нож ковша (экскаватора), *bent nose pliers* – изогнутые плоскогубцы, *blunt-nose* – тупоносый, *faucet ear* – ушки для крепления трубы к стене, etc.

    Within parts of face: *tooth, jaw, nose, ear* we see the highest level of metaphorical equivalence in the languages analyzed. Figurativeness is built on perceived similarities and associations between the above-mentioned face parts and tools used in construction and civil engineering. This group also reveals unexpected associations and new understanding of term-metaphors in different (English – Russian) languages, for example *eye* is rendered as *проушина*.

    b) parts of human body: head (97), hand (48), arm (25), leg (15), neck (12), body (10), knee (10), foot (8), shoulder (8), finger (5), rib (5), joint (4), toe (3), knuckle (2).

    E.g.: *button-headed/ screw* – винт с круглой/плоско-выпуклой головкой, *hand-operated machine* – ручной станок, *power hand tools* – механизированный инструмент, *gathering arm loader* – погрузчик с загребающими лапами, *arm type* – рычажного типа, *anchor leg* – анкерная лапа, *body* – корпус, каркас, *knee* – коленчатый, *shoulder of lever* – плечо рычага, *finger bit* – зубчатый бур, *joint sealing machine* – машина для герметизации швов, *pile toe* – пята (острие) сваи, etc.

    The large number of "head" metaphors within this group highlights that the head is understood as the top part of the body and, very often, the upper part of a tool. Besides, many mechanisms in construction and civil engineering have the human body model made up of eight major parts: head, body, two arms, two hands and two legs.

c) physical and mental states of a person, activities: dead (8), blind (5), running (3), jumping (2), dancing (1), digestion (1), drunken (1).

E.g.: *dead-smooth file* – *бархатный напильник*, *blind nail* – *гвоздь с уплотненной шляпкой*, *top-running bridge crane* – *опорный мостовой кран*, *jumping formwork* – *подъемно-переставная опалубка*, *dancing* – *дрожание (об индикаторе стрелки прибора)*, *sludge-digestion tank* – *реактор для тепловой обработки осадка сточных вод*, *drunken saw* – *маятниковая (качающаяся) пила*, *etc.*

Motivation of many term-metaphors within this group remains unclear.

d) professions: carpenter (10), conductor (10), banker (1).

E.g.: *carpenter's hammer* – *молоток с гвоздодерными лапками*, *protective conductor* – *заземление*, *banker* – *боек*, *etc.*

Associations within this group, as a rule, are based on a direct link between an instrument and a professional, using this instrument/tool.

2. NATURE (334)

a) names of animals (zoonyms): crane (35), dog (23), ram (15), fish (13), frog (10), horse (8), pig (6), cat (5), caterpillar (5), hog (5), rat (5), duck (3), monkey (3), snake (3), crow (2), donkey (2), sheep (2), squirrel (2), dolphin (1), hawk (1).

E.g.: *crawler crane* – *гусеничный кран*, *dog tongs* – *захватные клещи*, *automatic ram pile-driver* – *строительный механический копер*, *fish tape* – *проволока для протягивания*, *frog* – *крестовина*, *horse scaffold* – *строительные козлы*, *cat ladder* – *стремянка*, *cement hog* – *скребковый цементоперегрузчик*, *rat-file* – *напильник*, *duck's-bill bit* – *ложечный бур*, *monkey wrench* – *разводной ключ*, *crow* – *щипцы*, *etc.*

b) animal body parts: wings (18), tail (16), claw (14), feathers (13), trunk (3), tusk (3), fin (2), legs (2), neck (2), paw (2), tooth/teeth (2), beak (1), fang (1), foot (1), gill (1), horn (1), nose (1).

E.g.: *wing jib crane* – *кран с поворотной стрелой*, *four-wing bit* – *крестообразный бур для вращательного бурения*, *rat-tail* – *тонкий напильник*, *swallow tale* – *в виде ласточкиного хвоста*, *claw hammer* – *молоток с гвоздодерными лапками*, *feather valve* – *лепестковый клапан*, *elephant's trunk* – *гидроэлеватор/хобот для подачи бетонной смеси*, *fin tube* – *труба с ребрами-плавниками*, *dog-leg stair* – *двумаршевая лестница*, *goose/swan neck* – *S-образный*, *cutting paws* – *лапы с режущими кромками*, *dragon's tooth* – *пирсовый гаситель*, *fang* – *раздвоенный конец*, *sheep-foot roller* – *каток с кулачковым валиком*, *bull-nose* – *закругленной формы*, *etc.*

Zoonyms (names of animals) as well as animal body parts in term-metaphors often evoke associations and are rich imagery and connotations. Let us take the example of *crane* – the most productive zoonymic element within this group: *crane* is a tall, long-legged and long-necked bird. In engineering language, a machine used for raising and lowering heavy weights is also a *"crane"*.

c) names of plants (phytonyms): mushroom (4), bush (3), reed (3), rose (3), vine (1),

E.g.: *mushroom construction* – *грибовидная конструкция*, *bush hammer* – *бучарда*, *reed clip* – *проволочный хомут*, *spraying rose* – *разбрызгиватель*, *reed clip* – *проволочный хомут*, *vine-eye screw* – *шуруп с проушиной*, etc.

d) parts of a plant: leaf (6), stem (8), root (2), head (1).

E.g.: *two-leaf grab bucket* – *двухчелюстной грейферный ковш*, *stem bit* – *станок ударно-вращательного бурения*, *root pile* – *буроинъекционная свая*, *mushroom head* – *грибовидная головка*, etc.

Plant names and plant parts rarely evoke associations within the sphere of construction and civil engineering, except for visual resemblance in shape.

e) inanimate nature world: air (15), earth (12), soil (9), stone (8), wind (8), lightning (5), water (5), rain (3), snow (3), star (3).

E.g.: *air-powered cutoff saw* – *пневматическая пила для сухой резки бетона*, *earth mover* – *бульдозер*, *soil-compacting machine* – *машина для уплотнения грунта (грунтоуплотняющая)*, *paving stone laying machine* – *машина для укладки мостовых камней*, *stone drill* – *сверло по камню*, *stone tongs* – *захваты для камней*, *wind-driven generator* – *генератор с ветровым двигателем*, *lightning arrester* – *громоотвод*, *water gauge* – *водяной манометр*, *rain-water leader* – *водосточная труба*, *snow-removal instrument* – *снегоуборочный инвентарь*, *star drill* – *сверло со звездчатым наконечником*, *star-shaped* – *крестовый/звездообразный*.

A very characteristic feature of this group is that inanimate nature world objects as part of term-metaphors tend to mix together ideas (natural phenomena and engineering equipment).

3. ARTEFACTS (111)

a) household items: knife (10), cup (9), ring (8), bucket (7), scissors (6), fork (5), bowl (2), spoon (2).

E.g.: *stopping/putty knife* – *шпатель*, *cup* – *чашевидный/ковшовый*, *ring spanner* – *кольцевой ключ*, *bucket chain excavator on crawlers* – *ковшовый экскаватор на рельсовом ходу*, *paperhanger's scissors* – *обойные ножницы*, *fork wrench* – *вилкообразный/вильчатый гаечный ключ*, *bowl-shaped* – *чашеобразный*, *spoon bit* – *наконечник ложечного бура*, etc.

Household items as parts of term-metaphors: similarity within this group is primarily based on functional resemblance.

b) items of clothing: tie (5), shoe (4), coat (3), ribbon (3), sleeve (3), boot (2), clutch (1).

E.g.: *rod tie* – *стяжной винт*, *pile shoe* – *свайный башмак*, *metal-coated* с *металлическим покрытием*, *ribbon tape* – *рулетка*, *clutch/sleeve* – *муфта*, *boot-shaped* – *в форме сапога*, etc.

c) food: nut (19), sandwich (6), cherry (1), wafer (1).

E.g.: *black nut* – *неточеная гайка*, *ear nut* – *гайка с ушком/барашек*, *sandwich construction* – *многослойная конструкция*, *cherry picker* – *кран мостового типа*, etc.

d) musical instruments: drum (8), accordion (3), piano (2), banjo (1).

E.g.: rotary drum mixer – *мешалка с вращающимся барабаном, accordion type* – *складного типа, piano wire concrete* – *струнобетон, banjo* – *деталь, имеющая форму банджо.*

Items of clothing, musical instruments, gastronomical metaphors are less productive and very often bare visual, perceptual resemblance with a particular object.

Thus, there are universal, present in both languages, as well as unique (language-specific) associations, which are reflected in the metaphorical component of the term.

The analysis of term-metaphors reveals resemblances based on the criterion of shape, position and function as well as contiguous (metonymic) relationships of part/whole, cause/effect, action/result, instrument/action between Source-Target domains. Term-metaphors pervading the civil engineering domain tend to be heavily conceptual in nature and, therefore, rich in imagery and associations as they tend to retain concepts based on a naïve perception of the world. Among the sources of metaphorization in construction and civil engineering terminology, the most productive class is represented by anthroponymically motivated terms. In an overwhelming majority of cases, human-related term-metaphors within the domain under study are based on parts of human body. It seems to us that this can be due to the fact that the body of a human being is initially perceived as a kind of mechanism/machine (mechanized description of the human body) made up of constituent parts: bones, muscles, veins, blood, nerves, etc. and each of these parts performs its own unique function (functional description).

Nature-related term-metaphors are represented by zoonyms, phytonyms and inanimate nature world. Term-metaphors with floral/faunal lexeme in their structure are viewed as essential linguistic elements reflecting the worldview formation through their symbolic meaning. Zoonyms and phytonyms represent the oldest layer of vocabulary in different languages: beliefs in sacred plants and animals, plant and animal manifestations of Gods. Zoomorphic metaphors are the most productive ones in this group. In this way the parallelism between the animal and the human world evokes associations.

Artefact-related term-metaphors include household items, items of clothing, food and musical instruments. The most productive metaphorical model in this domain relates to household items.

Thus, the material analyzed enables us to conclude that term-metaphors constitute an integral part of construction and civil engineering domain. The processes by which technical terms are coined exhibit variations in different languages. While there are undoubtedly universal models of metaphorical nomination and technical terms can be nominated along generally similar lines, cross-linguistic research proves that corresponding (one-to-one) metaphorical equivalents are rather infrequent or only partly identical (e.g. *needle-nose pliers* – плоскогубцы с остроконечными губками, остроносые плоскогубцы, тонконосы, утконосы in Russian).

4 Conclusion

Naturally, the process of nominating the realia of the world around us is forged by the connection between sensory experience (the perception of size, color, form, etc.)

and conceptualization, and metaphor lies at the heart of this process. Thus, people conceptualize their experiences via metaphors. The analysis of engineering language allows us to speak about its metaphorical nature. In particular, the analysis of the terms nominating equipment, machines and tools in the sphere of construction and civil engineering has shown that many of them are regularly subjected to the process of metaphorization. The research revealed that metaphorical associations in engineering language arise from anthropomorphic, zoomorphic, phytomorphic, gastronomical, family and household, etc. source domains that can be arranged into three categories (sources of metaphorization): HUMAN, NATURE and ARTEFACTS.

Term-metaphors nominating equipment, machines and tools in the field of construction and civil engineering do not simply nominate objects; they are often entitled with symbolic meanings that express peculiarities of the people's world outlook since real-world phenomena are firstly perceived and fused by underlying human cognition mechanisms and then expressed in linguistic form. The meanings of many of them are motivated. The identification and comparative analysis of term-metaphors makes it possible through their imagery space to establish intercultural differences within language world pictures. Such phenomena place at our disposal a bewilderingly rich material for cross-cultural and inter-lingual comparison (i.e., linguistic equivalents vs. lack of linguistic equivalents among different languages (native-foreign language dimension) and cultures (native-foreign culture dimension). In this connection further cognitive studies of semantic nature of term-metaphors in order to reveal universal and nationally specific features, concepts in different languages presents a special interest.

References

1. Gumbol'd, V.: Selected Works on Linguistics. Progress, Moscow (1984). (in Russian)
2. Sapir, E.: The status of linguistics as a science. Language 5(4), 207–214 (1929). https://doi.org/10.2307/409588
3. Whorf, B.L.: Language, Thought and Reality: Selected Writings of Benjamin Lee Whorf. Wiley, New York (1956)
4. Wierzbicka, A.: Understanding Cultures Through Their Key Words (English, Russian, Polish, German, and Japanese). Oxford University Press, New York (1997)
5. Lakoff, G., Johnson, M.: Metaphors We Live By. The University of Chicago Press, Chicago (1980)
6. Lakoff, G.: Women, Fire, and Dangerous Things: What Categories Reveal About the Mind. The University of Chicago Press, Chicago (1987). https://doi.org/10.1017/S0272263100008469
7. Kövecses, Z.: Metaphor in Culture: Universality and Variation. Cambridge University Press, Cambridge (2005). https://doi.org/10.1017/CBO9780511614408
8. Littlemore, J.: Applying Cognitive Linguistics to Second Language Learning and Teaching. Palgrave Macmillan, Basingstoke (2009)
9. Littlemore, J.: Metaphoric competence: a language learning strength of students with a holistic cognitive style? TESOL Q. 35, 459–491. https://doi.org/10.2307/3588031
10. Littlemore, J., Krennmayr, T., Turner, J., Turner, S.: An investigation into metaphor use at different levels of second language writing. Appl. Ling. 35(2), 117–144 (2014). https://doi.org/10.1093/applin/amt004

11. Roldán-Riejos, A., Cuadrado, G.: Metaphor and figurative meaning construction in science and technology (English and Spanish). Proc. Soc. Behav. Sci. **212**, 271–277 (2015). https://doi.org/10.1016/j.sbspro.2015.11.348

12. Roldán-Riejos, A., Úbeda-Mansilla, P.: Metaphor use in a specific genre of engineering discourse. Eur. J. Engin. Educ. **31**(5), 531–541 (2006). https://doi.org/10.1080/03043790600797145

13. Roldán-Riejos, A., Moloina, S.: A taste of the technical cuisine: metals and other ingredients. Proc. Soc. Behav. Sci. **178**, 201–206 (2015). https://doi.org/10.1016/j.sbspro.2015.03.181

14. Putri, O.R.U., Zukhrufurrohmah M.Pd.: The ability of the civil engineering students to represent partial derivative symbols as metonymy and metaphor. In: Journal of Physics: Conference Series, vol. 1470, p. 012034 (2020). https://doi.org/10.1088/1742-6596/1470/1/012034

15. Gomes, D., Tzortzopoulos, P.: Metaphors of collaboration in construction. Canad. J. Civil Eng. **47**(2), 118–131 (2020). https://doi.org/10.1139/cjce-2018-0461

16. Carter, E.V., Ionova, V.N.: "Head" metaphors in mechanical engineering (based on the English, German, French and Russian Languages). In: Anikina, Z. (ed.) IEEHGIP 2022. LNNS, vol. 131, pp. 1100–1107. Springer, Cham (2020). https://doi.org/10.1007/978-3-030-47415-7_118

17. Blokh, M., Loseva, O.: Metaphor of negative semantics in scientific and technical text. Bull. Moscow State Region. Univ. Ling. **6**, 6–14 (2017). https://doi.org/10.18384/2310-712X-2017-6-6-14. (in Russian)

18. Raluca, Gh.: Metaphors in technical-scientific texts. In: 1st International Scientific Conference Filko. Conference Proceedings, pp. 147–154. Stip, Macedonia (2016)

19. Fedyuchenko, L.G.: Functions of a metaphor term in a technical text. Bullet. PNIPU. Probl. Ling. Pedag. **3**, 77–86 (2018). https://doi.org/10.15593/2224-9389/2018.3.7. (in Russian)

20. Fedyuchenko, L.: Metaphoric term as translation unit of technical text. In: SHS Web of Conferences, vol. 50, no. 4, p. 01054 (2018). https://doi.org/10.1051/shsconf/20185001054

21. McEnery, A., Xiao, R.: Parallel and comparable corpora: what is happening? In: Rogers, M., Anderman, G. (eds.) Incorporating Corpora: The Linguist and the Translator, pp. 18–31. Multilingual Matters, Clevedon (2007). https://doi.org/10.21832/9781853599873-005

22. Maclean, J.H., Scott, J.S.: The Penguin Dictionary of Building. Penguin, London (1993)

23. Brett, P.: Illustrated Dictionary of Building. Butterworth-Heineman, London (1997)

24. Blockly, D.: The Penguin Dictionary of Civil Engineering. Penguin, London (2005)

25. Gorse, C., Johnston, D., Pritchard, M.: A Dictionary of Construction, Surveying and Civil Engineering. Oxford University Press, Oxford (2012)

26. Barinov, S.M., Borkovskij, A.B., Vladimirov, V.A.: The Comprehensive English-Russian Scientific and Technical Dictionary: in 2 vols, 2nd edn. RUSSO, Moscow (1997). (in Russian)

27. ABBYY Lingvo. http://www.lingvo.ru. Accessed 21 Dec 2021

28. Multitran. https://www.multitran.com. Accessed 21 Dec 2021

29. Rusak, D.A., Ovsinskiy, A.Yu.: The Comprehensive Russian-English Scientific and Technical Dictionary, 4 vols. ETS, Moscow (1996). (in Russian)

Universities as Multilingual Sites for Technological Development

Aspects of Multilingualism in the Transformation of Educational Technologies

Galina A. Dubinina⬡, Larisa P. Konnova⬡, and Irina K. Stepanyan$^{(\boxtimes)}$ ⬡

Financial University under the Government of the Russian Federation, Leningradsky Ave., 49, 125993 Moscow, Russian Federation
ikstepanyan@fa.ru

Abstract. The article gives insight into the way the growing flow of information, its accessibility, and the widespread introduction of digital technologies affect the goals and methods of modern education. The authors indicate that the formation of professional skills, the ability to find and analyze the necessary information, to comprehensively and systematically approach the solution of emerging problems, using modern IT technologies have become the dominant points in today's education. The authors see opportunities for the realization of new goals in the active use of interdisciplinary connections and profile context. The implementation of multilingual technologies in the framework of one discipline is considered. The combination of various disciplines with their own specifics and terminology base creates a special educational environment, which can be considered multilingual. The authors describe their experience of teaching the discipline "Digital Mathematics". Since mathematical language, sublanguage of specialized disciplines, programming languages are used in the learning process, it is proposed to consider such a connection as an interdisciplinary multilingualism. The paper provides a survey of students showing that they consider this approach to teaching useful and effective.

Keywords: Multilingual environment · Interdisciplinary multilingualism · Code switching · Interdisciplinary approach

1 Introduction

The increasing intensity of the flow of information, the rapid development of industry, the digitalization of all spheres of the economy lead to changes in the field of education. First of all, this applies to higher education. University graduates Not only should have a stock of knowledge in the professional field, but also such should possess such skills that will allow them to freely navigate the digital space, track and quickly master developing technologies in order to remain competitive in the labor market. Universities are constantly adapting the educational environment to new realities by introducing new disciplines, updating teaching methods both at lectures and at workshops. Building interdisciplinary connections in the educational process is a current promising trend:

seminars are more and more often conducted by the multidisciplinary teams and, increasingly, by representatives of the real sector of the economy; sessions are more and more often conducted in laboratories and workshops. The use of IT in teaching specialized disciplines and specialized disciplines in a foreign, usually English, language is welcomed. Currently, productive cooperation between teachers of foreign languages and teachers of specialized disciplines has acquired a different dimension. For this purpose, an integrated information environment is widely used [1, 2].

The introduction of interdisciplinary connections, the connection of various disciplines with their own specifics and terminological base, creates a special educational environment that can be considered multilingual.

The generation of modern students is distinguished by mental peculiarity of perceiving information. They are characterized by a fairly high speed of perceiving information, the ability to do several things at once, but at the same time, they reveal short-term memory, inability to retain a large amount of information and difficulty in perceiving large texts [3, 4]. Researchers note the inclination of modern students to more concise, structured materials, preferably presented in the form of various tables and diagrams [5]. Students consider that the main goal of education is the immediate practical application of the obtained knowledge, not realizing that professional knowledge has a cumulative character and, as a result, a delayed effect. Therefore, it is important for teachers to maintain motivation, interest in learning, forming professional skills from the first sessions at the university, which automatically leads to the connection of various fields of knowledge, which means that students are immersed in an interdisciplinary multicultural environment.

Thus, university teachers should dynamically adapt their methods to the educational process. We are no longer talking about the static traditional lectures and seminars of universities of the last century, when the basis of the process lay in the transfer of information from the teacher to students. Even modern approaches to learning, close to interactive communication and cooperation, require correction in order to meet the needs of modern society. Today, a teacher of any profile discipline needs digital communication skills, IT basics, knowledge of English terminology in the professional field, i.e. have the skills to work in a multilingual environment.

The purpose of this study is to summarize the experience of the authors in the use of multilingualism in the educational process in the digital environment, to identify the positive and negative aspects of such pedagogical methods and to identify prospects in their development.

It is worth noting that multilingualism is a term traditionally used by philologists, when several languages are used in the process of some activity.

For instance, philologists most frequently consider bilingualism in the form of dual language instruction in the course of which students master academic disciplines, whose content is taught and assessed in two languages. The lecturer or instructor in a special academic discipline teaches students in two languages, usually English as the language of international communication and the native language. In some cases, teaching international students in English is combined with using the language of the host country in the teaching process, and, if possible, with the native language of a student. The efficiency of such a technology has been proved by vast educational experience and the fact that the drive to higher profilization enhanced the multidisciplinary approach in academic education, especially, when it concerns multilingual professionally oriented training in special disciplines [6].

However, the application of multilingualism is not limited by linguistic issues. Even within the framework of one discipline, it is possible to distinguish such language units as professional language, the language of mathematical formulas, the academic style of information presentation, professional slang, etc. [7], which, in our opinion, are the components of much wider multilingual environment. The article describes the authors' experience in the formation of digital skills among first-year undergraduate students of the Financial University under the Government of the Russian Federation. Interdisciplinary code switching technologies were actively used in the educational process, which contributed to the creation of a multilingual educational environment.

2 Multilingualism in Digital Mathematics

It is common knowledge that multilingualism or polylingualism, most frequently takes place either in the natural or in the educational environment. Many academic disciplines involve dual-language instruction. First and foremost it refers to teaching international students, which is exercised in English as lingua franca and the language of the host country. With the advent of the enhanced academic mobility teaching academic disciplines in English became increasingly popular. This phenomenon is known as English Medium Instruction, which has gained wide use currently. But the situation is not as simple as that, some disciplines, mostly of mathematical cycle, IT or Economics, engage sub-languages of their specialty in combination with international terms in English, Latin, Greek and several other languages. It occurs not only due to natural bilingualism or multilingualism but also due to insufficient language expertise. Nowadays educators often engage in triglossia/trilingualism, consisting of the native language, the first (usually English) and the second foreign languages [8].

The subject of multilingualism has a certain coverage among the pedagogical community, with the conclusion that there is a shortage of complex system-oriented research on how to apply the basic principles of multilingualism in content and language integrated learning to methodological, psychological, pedagogical and methodical concepts.

Surveys presented by [9] attributed multilingualism, to the ability of a person to use several languages; to the alternative use of several languages; to the coexistence of several languages within a social group as well as to the coexistence of several communities in a given geographical area. The above conclusions are in full concord with the authors' perception. The intention to apply multilingualism in the didactic and extracurricular activities coincides with the results of our investigation.

It is well known that mathematics is not as language dependent as other disciplines. The authors admit that students engaged in interdisciplinary training shoe excellent results in, for instance, mathematics because, on the one hand, it is a "universal language", but on the other hand, it abounds in international mathematical terms in English, consists mostly of signs and symbols and is carefully and purposefully designed. The language of mathematics and mathematical symbols is one and the same all over the world, though the people of different countries convert it into their particular spoken language. Mathematics supplies a language for the treatment of the qualitative problems of various sciences. Linguists state that its characteristic features are a limited number of terms and the abundance of examples. Lingvo-didacts call the language of instruction an educational and scientific substyle [10].

Switching between languages within the same discipline used to be criticized among professionals. Especially acute the negative attitude to using the native language would appear in teaching foreign languages (bilingualism and multilingualism were considered to be unprofessional). Today, the use of several languages in presenting the learning material shows good results and contributes to the formation of an inter-disciplinary approach among students. Due to the fact that content and language-integrated learning gained popularity, we extensively use English as a foreign language in our educational activity and base training for mathematics on the professional context, as part of the study of mathematics. We find it necessary to discuss how to make calculations with the help of various programs (Excel, R), while switching to the IT language. We developed and used a glossary of economics terms that combines an economics definition, an international (English) name, a symbolic description, a mathematical formula and calculation codes in R [11].

Our research showed that multilingual approach to teaching academic disciplines, especially in the field of economics, digital mathematics, mathematical modelling of economic processes, and the like, enables a learner to absorb correctly rules and certain other characteristics of language used by surrounding speakers. Some scholars imply that individuals with high language entropy are easier integrated and tend to use multiple languages when a context it demands. [12] We support the above point of view and came to the conclusion that multilinguals immerse themselves in the contexts of a special discipline where several languages are used much better than those who study in the environment where only one language is used.

The present-day enhanced involvement of sub-languages of various disciplines in the educational process is attributable to the tasks facing universities today. Thus, in teaching basic disciplines, it is impossible to do without turning to professional orientation and the use of digital technologies. This fully applies to Mathematics as an academic discipline. With the introduction of powerful digital tools in the financial and economic sphere, the requirements for mathematical training of graduates have increased dramatically. In order to strengthen the mathematical component in the preparation of bachelors of the Financial University, the discipline "Digital Mathematics" was introduced into the programs of all directions. The main objectives of the new discipline are as follows; the formation of students' knowledge about computational methods for implementing mathematical models; the formation of practical skills in the application of computer technologies used in economics and finance.

In parallel with the course of "Digital Mathematics", students study the course of mathematics, which opens up wide opportunities for a deeper understanding of mathematical laws and the practical use of mathematical methods in future professional activities. At the seminars, attention is paid to the formation of professional digital skills, financial computing, as well as the basics of data analysis. The toolkit is the R programming language and the Microsoft Excel spreadsheet editor.

The computer workshop "Digital Mathematics" can surely be called a unique, inter-disciplinary discipline, where multilingualism in teaching is very clearly manifested, since it is impossible to do without referring to the languages of mathematics, eco-nomics, programming and English. In this case, it is difficult to distinguish the main language of the discipline. The program contains both strictly mathematical topics: calculating limits, differentiating functions, working with matrices, etc., and financial calculations on deposits and loans, the study of the production function and the util-ity function. In addition, students get acquainted with the techniques of data analysis: filtering, consolidation, databases. Priority in tasks is given to practice-oriented cases.

An example of such a case is presented below. It demonstrates the idea of using multilingualism in teaching. The case is solved at a workshop on Mathematics by the graphic method and at a computer workshop in EXCEL and R.

CASE:
Three types of raw materials are used for the manufacture of products A and B. The production of one product A requires: raw materials of the first type - 1 kg, the second - 2 kg and the third - 4 kg. The production of one product B requires: raw materials of the first type - 7 kg, the second - 1 kg and the third - 1 kg. Production is provided with raw materials of the first type in the amount of 490 kg, the second type - 200 kg, the third type - 360 kg. The cost of one product A is 90 rubles, product B is 120 rubles. Make an optimal product release plan that ensures maximum profit.

1. **Graphic method.**
   To begin with, the formulation of the problem in the language of unspecified products and raw materials opens up the possibility for the creativity of the teacher as well as students. It is quickly decided what and from what we will produce. At the same time, it is possible to discuss the demand for this product and the availability of raw materials (*sublanguage of professional activity*).
   Next, we proceed to the compilation of a mathematical model by introducing vari-ables and translating the condition of the problem into the language of mathematical symbols.

$$\begin{cases} x + 7y \leq 490, \\ 2x + y \leq 200, \\ 4x + y \leq 360; \\ x \geq 0, y \geq 0; \\ f(x, y) = 90x + 120y \rightarrow \max. \end{cases}$$

   When solving this problem of linear programming, a graphical method is used (Fig. 1), which translates mathematical symbols into the language of *graphic visual-ization*, which, in turn, is actively used in economic analysis. The teacher, explaining the decision, constantly refers to the *sublanguage of professional activity*, because we are talking about important concepts of microeconomics: level curve and objective function's direction of growth.
   Next, the results of the mathematical model

$$\left.\begin{array}{l} x_{max} = 70 \\ y_{max} = 60 \end{array}\right) \Rightarrow f_{max} = f(70;60) = 90 \cdot 70 + 120 \cdot 60 = 13500(\textit{руб.})$$

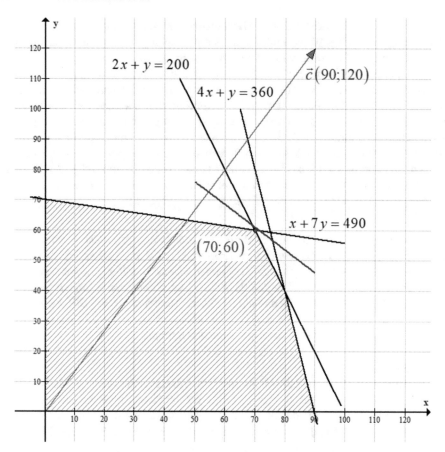

Fig. 1. Solving the case by the graphic method.

must be interpreted based on their task conditions.

2. **Solution in EXCEL.**

According to the terms of the task, students must compile and fill out a screen form using special rules and symbols of the EXCEL software (Fig. 2).

The built-in functions form a separate *language of the EXCEL editor* with syntax and a voluminous array of words formed from phrases either in Russian or in English. The interface, the names of functions and tools will be incomprehensible to the user if he or she does not speak this language. For example, the function "Sum of pairwise products" used to solve this case in the Russian interface is written СУММПРОИЗВ, and in the English interface SUMPRODUCT.

When explaining the solution of the case in EXCEL, the teacher also refers to the mathematical model (mathematical language) and the economic interpretation of the solution (sublanguage of economics).

3. **Solution in R.**

In parallel with solving the case in EXCEL, students are offered a solution in R. The RStudio resource is a wrapper for the *R programming language* and significantly

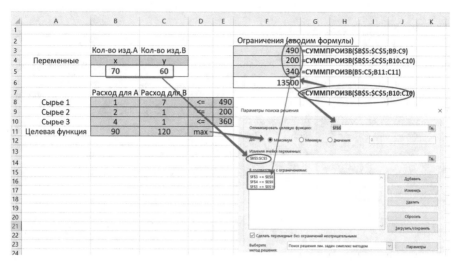

Fig. 2. Solution of the case in EXCEL.

increases the speed of calculations, which allows it to be used for processing large data arrays. Like any programming language, R is a system of data types, operators and codes linked together by a special syntax. R is an international project that uses English for teams.

To solve the case under consideration, students are shown a program consisting of a series of sequential codes in *the English language* (Fig. 3). When explaining the meaning of the commands, the teacher refers both to the mathematical model of the case and to the economic content of the condition, relying on English terms. The solution of the case contains many words known to students, such as make, name, solve, type, etc., as well as new terms and their abbreviations. In this multi-lingual environment, the teacher switches the students' attention from one language to another. Indications of the English-speaking origin of the commands contribute to understanding the work of the program.

The reviewed case study demonstrates the active use of mathematical and economic terminology in both Russian and English in the IT shell. This approach to learning is the concept of the discipline Digital Mathematics. Active switching of students between disciplines, building mutual connections enhances the motivation of students, making the process of obtaining professional knowledge conscious.

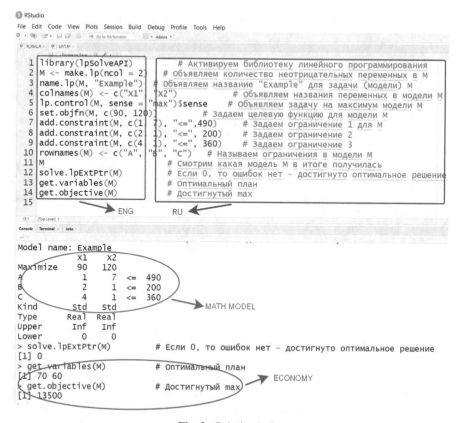

Fig. 3. Solution in R.

3 Pedagogical Technologies in a Multilingual Environment

The multilingual educational environment requires the introduction of special learning technologies into the pedagogical process. One of such technologies is code switching technology. Code switching allows the teacher to more accurately define didactic units and make explicit the connection between the disciplines studied. Using a different code-term allows you to make the teacher's explanations more effective.

According to research [13] the phenomenon of code switching has not only linguistic aspects, but broad interdisciplinary integration possibilities, demonstrating how models of one scientific field are applied to models from other fields. This approach to code switching allows students to more comprehensively and extensively consider professional cases, find new ways to solve them [11].

In their pedagogical practice of teaching students mathematics and digital skills, the authors use the technology of interdisciplinary code switching, using the sublanguages of specialized disciplines, the languages of mathematics and IT, both in Russian and in English. Thanks to this approach, an educational environment is created that can be called multilingual (Fig. 4).

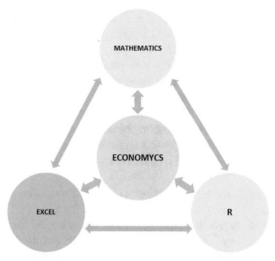

Fig. 4. Multilingual educational environment of digital mathematics as an academic discipline.

At the same time, the first steps in computer practice for students are quite complicated, requiring careful preparation of the teacher for the seminar. But step-by-step students begin to get used to this style of presentation of educational information, and by the end of the course they practically do not notice the moment of switching codes. Strong students successfully solve cases in various resources, independently attracting digital skills and mathematical tools when participating in practice-oriented projects. To consolidate interdisciplinary connections and help students at the first stage of studying the discipline digital mathematics, it is useful to use such a *pedagogical resource* as the *Glossary*. A convenient platform for creating an electronic glossary is LMS Moodle. The glossary in this system can be created by the teacher, and replenished with students.

Together with the students, we developed a glossary of R programming language codes with explanations and examples. As well as a glossary of economic terms that combines terms in Russian language with international names, mathematical formulas and graphical examples in an e-learning course at LMS Moodle. Students' mastery of the Glossary had a positive impact on their academic results when solving professionally oriented cases [11].

Among the pedagogical technologies that contribute to the formation of an interdisciplinary multilingual environment, we can safely name the contextual approach widely used in higher education, proposed by A.A. Verbitsky [14] the idea of which is to consider all the disciplines studied under the prism of forming the necessary professional skills while using the context of the future profession when teaching. This approach contributes to the implementation of early profiling of university education. This is one of the strategic directions in the educational activity of the Financial University. In their practical activity the authors decided to combine the introduction of the contextual approach with the parallel use of modern technology for learning foreign languages, namely,

CLIL (Content and Language Integrated Learning) [15], the so-called content and language integrated training. It allows instructors to strengthen the relationship between the language teaching and the subject content of professional disciplines.

The aforementioned method, CLIL, is of particular interest for this study, since it creates the best opportunities for the formation of a personality capable of independently setting and solving professional tasks, and carrying out effective communication in the professional sphere in the native and foreign languages. The discipline of Digital Mathematics allowed educators to combine these two technologies, using mathematics as a basis – content, digital technologies as a means of implementation and professionally oriented economic context [16]. Such an association has been displaying good results for several years.

One of the confirmations of the mentioned above is the results of a survey of undergraduate students of the second and third courses of the Financial University, where the authors conducted classes in Digital Mathematics during the first year according to the described methodology. The survey was conducted anonymously and was optional, with 101 participants taking part. Students were asked to rate in points from 1 to 5 the usefulness of studying the discipline of Digital Mathematics for

- studying mathematics;
- studying data analysis;
- enhancing their professional skills;
- enhancing their digital skills.

The results of the survey are presented in Table 1.

Table 1. Survey results on the usefulness of digital mathematics.

	0 points	1 point	2 points	3 points	4 points	5 points
Mathematics in the 1st year	1	1	2	6	11	80
Data analysis in the 2nd year	5	2	4	9	27	54
Digital professional skills	0	3	3	17	23	55
Personal digital skills	0	1	1	6	29	64

In our opinion, students highly appreciated both the innovative character of the discipline and the suggested approach to learning. The diagram (Fig. 5) shows the number of students, expressed as a percentage, who rated significance at 4 and 5 points, i.e. as "good" and "excellent".

It should be noted that the respondents were students who had already been able to assess the acquired experience within the framework of specialized disciplines, further study of mathematics and initial work in the future specialty in the internship mode.

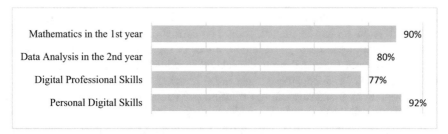

Fig. 5. 4 and 5 points on the usefulness of Digital Mathematics.

4   Conclusions, Challenges and Prospects

The practical experience of teaching the digital mathematics, which combines several areas of knowledge, allows us to draw the following conclusions:

1. Multilingual interdisciplinary environment allows instructors to realize the practical orientation of training, students gain skills in solving practice-oriented cases.
2. Multilingualism stimulates the understanding and conscious inclusion of links between disciplines, both specialized and general scientific orientation, which contributes to an integrated approach to any problems.
3. The creation of a multilingual educational environment is achieved through the combined use of various pedagogical technologies such as contextual approach, contextual learning, code switching, introduction of interactive language resources (glossary, thesaurus).
4. Training in a multilingual environment is characterized by the dynamism and informative contents of workshops, which is in line with the peculiarities of modern students' perception, which positively affects their motivation to achieve academic and professional success.

At the same time, the following problems associated with teaching in an interdisciplinary multilingual environment can be identified. First of all, these are increasing demands on the professional knowledge and skills of teachers. In addition to their disciplines, they must understand the basics of the core disciplines of their university, be proficient in IT, possess a terminological base of related subjects both in their native and English languages. In the research [17, 18] dedicated to modern classifications of academic workers, the need for continuous professional development, inclusion in the world scientific agenda and possession of language skills, as well as communication skills, is noted.

Another urgent problem of modern vocational education is the lack of interdisciplinary educational literature. The need for special textbooks is noted by both students and university professors. First of all, interactive electronic teaching aids, which should be created by a team of teachers from several disciplines in cooperation with IT specialists and psychologists. [19]. Some experience in creating interdisciplinary textbooks has been accumulated today among teachers of foreign languages using CLIL [1, 2].

The creation of an interdisciplinary multilingual environment at universities in cooperation with lecturers and instructors should be carried out by the management of the educational institution. There are a large number of organizational problems, such as changes in curricula, the distribution of teachers' workload, the formation of a timetable for classes, and so on. Today, these issues are still unresolved.

The creation of an interdisciplinary multilingual educational environment and work in it positively affects the academic achievements of students, the formation of their sustainable professional skills. The prospects for such an approach are obvious and are due to the solution of the problems described above. A new aspect of modern education is becoming more and more popular at universities, lecturers and instructors understand the need to apply interdisciplinary multilingualism for the sake of the effectiveness and relevance of teaching sessions. Of particular importance here is the participation of students in various interdisciplinary projects, laboratories and workshops.

References

1. Rzheutskaya, S.Y., Kharina, M.V.: Interdisciplinary cooperation in the integrated information learning environment of technical university. Open Educ. **2**, 21–28 (2017). https://doi.org/10.21686/1818-4243-2017-2-21-28
2. Kobeleva, E.P.: Teaching foreign languages at universities based on cross-disciplinary integration. Int. J. Humanit. Nat. Sci. **11–2**(38) (2019). https://doi.org/10.24411/2500-1000-2019-11756
3. Kraynov, A.: Clip thinking in the context of educational practices: a socio-philosophical analysis. Izvestiya of Saratov Uni. Philos. Psychol. Pedagogy **19**(3), 262–266 (2019). https://doi.org/10.18500/1819-7671-2019-19-3-262-266
4. ZHuravlev, A.L., Nestik, T.A.: Socio-psychological consequences of the introduction of new technologies: Promising Areas Res. Psychol. J. **40**, 35–47 (2019). https://doi.org/10.31857/S020595920006074-7
5. Tenkhunen, P.Y., Eliseeva, Y.A.: Features of perception of educational information by modern students: potential of visual conceptualization. Integr. Educ. **19**, 28–34 (2015). https://doi.org/10.15507/1991-9468.081.019.201504.028
6. Dubinina, G.A., Stepanyan, I.K., Ganina, E.V.: Specificity of dual language workshop in mathematics for foreign entrant students. Espacios. **39** (38), 8 (2018). http://www.revistaespacios.com/a18v39n38/18393808.html Accessed 28 Sep 2022
7. Dubinina, G., Konnova, L., Stepanyan, I.: Technologies for teaching mathematics in a multilingual digital environment. Educ. Sci. **12**(9), 590 (2022). https://doi.org/10.3390/educsci12090590
8. Golovanova, I.I., Lopareva, T.A.: Social conditionality of multilinguism education in educational establishments of the country in the modern period. Int. J. Environ. Sci. Educ. **11**, 1963–1973 (2016). https://doi.org/10.12973/ijese.2016.570a
9. Domilescu, G., Lungoc, C.: Strengthening European identity by promoting multilingualism in education. J. Educ. Sci. **2**, 57–65 (2020). https://doi.org/10.35923/JES.2019.2.05
10. Stycheva, O.A.: Scientific style of speech: issues of didactic support. World Sci. Cult. Educ. **3**, 150–152 (2012). https://cyberleninka.ru/article/n/nauchnyy-stil-rechi-voprosy-didakticheskogo-obespecheniya Accessed 28 Sep 2022
11. Dubinina, G.A., Konnova, L.P., Stepanyan, I.K.: Cross-disciplinary code switching as means of encouraging creativity. In: Bylieva, D., Nordmann, A. (eds.) PCSF 2021. LNNS, vol. 345, pp. 683–704. Springer, Cham (2022). https://doi.org/10.1007/978-3-030-89708-6_56

12. Gullifer, J.W., Titone, D.: Engaging proactive control: Influences of diverse language experiences using insights from machine learning (PDF). J. Exp. Psychol. Gen. **150**(3), 414–430 (2020). https://doi.org/10.1037/xge0000933

13. Isurin, L., Winford, D., Bot, K.: Multidisciplinary Approaches to Code Switching. John Benjamins Publishing Company, Philadelphia (2009). https://doi.org/10.1075/sibil.41

14. Verbitskiy, A.A.: Theory and Technologies of Context Education. MPGU Publ, Moscow (2017)

15. Ting, Y.L.T.: CLIL … not only not immersion but also more than the sum of its parts. Eng. Lang. Teach. J. **65**(3), 314–317 (2011). https://doi.org/10.1093/elt/ccr026

16. Konnova, L.P., Rylov, A.A., Stepanyan, I.K.: Integrative teaching of mathematics as a means of a forming modern economist. Amazonia Investiga **9**(26), 486–497 (2020). https://doi.org/10.34069/AI/2020.26.02.56

17. Efimova, G.Z., Sorokin, A.N., Gribovskiy, M.V.: Ideal teacher of higher school: Personal qualities and socio-professional competencies. Edu. Sci. J. **23**(1), 202–230 (2021). (In Russ.). https://doi.org/10.17853/1994-5639-2021-1-202-230

18. Valisova, A., Subrt, J.: The social competence – significant part of teacher's professional authority. In; Proceedings of the International Conference on Interactive Collaborative Learning (ICL 2015), 20–24 September 2015, Florence, Italy. Piscataway, NJ, pp. 1–5. IEEE (2015). https://doi.org/10.1109/ICL.2015.7318024

19. Zharkov, E., Kibzun, A.I., Martiushova, I.A.G., Mkhitaryan, GA.: The main interdisciplinary aspects of the development and software implementation of electronic textbooks in technical disciplines on the example of SDO MAI. Mod. Inf. Technol. IT Edu. **15.2**, 507–515 (2019). https://doi.org/10.25559/SITITO.15.201902.507-515

The Necessity of Forming Multilingual Competencies in the Educational Process Digitalization

Galina Ismagilova⬤, Elena Lysenko(✉)⬤, and Evgeniya Khokholeva⬤

Ural Federal University named after the first President of Russia B. N. Yeltsin,
620002 Yekaterinburg, Russian Federation
e.v.lysenko@urfu.ru

Abstract. The pandemic caused by the COVID-19 virus affected all spheres of society and had a significant impact on the education system, forcing an urgent transition to distance learning formats. As practice has shown, distance learning forms are especially difficult when working in a foreign language environment. The present article aims to highlight critically important to discuss multilingualism in combination with online learning the process of an emergency transition to remote work and distance learning through the eyes of teachers working in multilingual student classrooms and a multilingual educational environment where students are both taught in Russian as a foreign language and in English, which is not native to them. An empirical study using the authors' questionnaire, in which representatives of the "cadre core" took part, showed the professional and psychological readiness of teachers for the transition, on the one hand, and a number of problems associated with the level of administrative support, a decrease in the quality of education and technical problems of the transition, on the other. However, it should be emphasized that, despite some clear advantages of distance working format in a multilingual environment, the assessment of the organization of the transition and its effects should be the subject of close attention, because online learning promotes the development of multilingual skills and competencies and increases the requirements for teaching and strengthens the role of the teacher as their guide. Distance learning formats should be developed in a comprehensive manner, from technical and organizational aspects to content and distance learning technologies. The results of this work are part of a large-scale study aimed at implementing a multilingual approach to learning in new historical conditions caused by global transformations of the existing global world order.

Keywords: Multilingualism · Pandemic Covid-19 · Higher education · Distance learning system · Multilingual competencies · Remote form of work · Technologies of distance learning · Multilingual personality

1 Introduction

Global transformations of the existing world order, caused by social cataclysms in the last decade, have affected all spheres of human life. Education is one of the main areas

© The Author(s), under exclusive license to Springer Nature Switzerland AG 2023
D. Bylieva and A. Nordmann (Eds.): PCSF 2022, LNNS 636, pp. 246–267, 2023.
https://doi.org/10.1007/978-3-031-26783-3_22

that have undergone total functional changes associated with the necessity to work in a multilingual environment. These processes were additionally aggravated by the COVID-19 pandemic. Currently, society is facing a situation that forces professionals in the field of higher education to master not only foreign, but also technical languages, to form multilingual competence, including digital one. There is a lot of discussion about the quality of education, the form and tools used, as well as the advantages and disadvantages of the online format in general [1–3]. For instance, the experience of the Kazan Federal University shows that more than 63% of students experienced discomfort during online learning, and for teachers, the work became more difficult, requiring more thorough preparation [1, pp. 105–106]. Radina and others notes that in academic research it is proposed to carry out not a formal transition to remote technologies, but a real technological twist (using a wide range of IT resources) [2, p. 189]. Frolova and others in their study revealed that the key risk of switching to distance learning is associated with a drop in the quality of education due to the use of unadapt educational technologies by teachers and a decrease in student involvement in the educational process [3, p. 86]. The problem is so new that science is just beginning to accumulate information on it. The "Scientific Data Bank" is literally developing right before our eyes.

The problems of multilingualism are broad and affect both general scientific (Aronin [4], Shipunova et al. [5], Bylieva et al. [6]), and multidisciplinary issues так и мультидисциплинарного характера (Baranova et al. [7], Pokrovskaia et al. [8]). Now everyone has a unique opportunity to contribute to it, since at this stage any experiment or research regarding the problems in education associated with the pandemic has a huge impact in forming the multilingual competence base. In this regard, numerous scientific articles and studies on the topic of education during a pandemic are published in various information sources [2].

However, only few are devoted to the situation in education after the pandemic. The most interesting, in our opinion, are the works devoted to university practices of the introduction of multilingual technologies in the educational process in higher education. (Baranova et al. [9], Bylieva et al. [10], Aladyshkin et al. [11]).

Some of the works deal with issues of a strategic nature (Almazova et al. [12], Rubtsova et al. [13], Ipatov et al. [14]. Almazova et al. [15], Balyshev, P. [16]), other authors devote research to the analysis of key aspects of multilingual reality (Odinokaya et al. [17], Anosova&Dashkina [18]).

Of particular importance are the issues of the introduction of information technologies and services in the educational process (Almazova et al. [19], Bylieva et al. [20]), as well as an analysis of the problems that this creates (Bylieva et al. [21, 22], Pozdeeva et al. [23]). Thus, Bolgova and others predict rapid digitalization of most processes and the growing gap between "elite" and "mass" higher education in connection with the creation of digital universities and, finally, a change in the very understanding of teaching as a process and the teacher [24, pp. 25–26]. Kostina and Orlova believe that it is essential to find a balance between full-time and online learning formats [25, p. 53]. All of the above indicates that this topic is extremely popular in society and has a huge potential for development in the future, and, of course, various kinds of scientific research need to be conducted in its support.

1.1 Justification of the Problem

The digital age has enabled population to study in a multilingual environment, despite the covid restrictions in the country and the world. At the same time, in the whole situation, it is worth noting the speed in the transition to distance learning. Universities that fell under a number of restrictions related to the transfer of students to a remote format managed to mobilize their resources within 3–7 days and develop a unified platform for working in a remote format and open databases with online courses. Teachers and students, due to technological progress and digital literacy, quickly adapted to the distance learning process and the new format. This allowed in a short time and without unnecessary inconvenience to organize the educational process and form the necessary wide-range multilingual competence in teachers of higher education, e.g. language competencies, technical competences as well as digital ones.

The phenomenon of multilingualism is becoming more and more relevant at the turn of the XX–XXI centuries [26, p. 232] and is associated both with the processes of globalization, on the one hand, and with the change in the industrial structure—the transition to Industry 4.0, on the other. The European Commission defines multilingualism as the ability of "societies, organizations, groups and individuals to incorporate more than one language into their daily lives" [27, p. 6].

Vakhtin and Gorlovka [28, p. 41] note that multilingualism is "a powerful resource that allows people to interact more effectively". This means that the world is becoming multilingual at its core, uniting the use of both natural and artificial (symbolic, technical, digital, etc. languages), different types of communicative language practices and their combinations, the need to master multilingual competencies.

Multilingual reality poses many problems to tackle. The COVID-19 pandemic clearly manifested the complexity of this process.

On the other hand, new general problems of the educational process have been revealed. One of them is technical equipment. Quite often, one has to deal with the fact that there is only one computer in the family, which makes it difficult to master the educational program and complete tasks. Many people were simply not ready to reorganize their lives in new circumstances, and therefore did not have time to prepare the necessary equipment for the lack of opportunity or funds.

Another problem was that most of those who switched to distance learning had to face psychophysiological problems [29–31]. It would have been right to foresee the developing situation and prepare the entire education system for all possible scenarios, which could have greatly facilitated the process of adaptation and made it less difficult. Presumably, one should wait for a change in the results of the quality of education, since the format for obtaining knowledge is new and there have been no previous developments regarding this new format of education. Undoubtedly, this issue needs to be studied. The introduced mode has the prerequisites for the appearance of mental health problems in students, and with all this, one should not deny the connection between the form of education, preparation for exams and the expected exam results. U.S Zakharova U.S. and others in their studies emphasize that "replacing practical knowledge with the processing of empirical data… And video demonstration of real processes do not solve the problem of developing professional skills" in applied areas of training [32, p. 131].

Mastering technological and digital languages has caused certain difficulties. Abramova M.O. and others revealed that in addition to technical problems in the online training of foreign students, there are methodological problems: the system of online classes and the form of presentation of the material, which complicate the communication process in distance learning [33, p. 137].

The current situation expectedly drew the attention of the state to the education system in the country. The organization of the educational process and the adoption of restrictive measures only emphasizes the importance of issues related to education for the country. Undoubtedly, after the restrictions are lifted, the education system will no longer be looked at as before. Due to the circumstances, a lot of attention has been riveted to a new format both from society and from the state.

Here it is necessary to draw attention to the emergence of problems related to education itself with the announcement of a pandemic.

A) *Digital adaptation* in a multilingual environment has become the most anticipated problem for education during the pandemic. It was exactly this problem that most teachers and students had to face when switching to a distance format. Initially, it was difficult to get used to the new interface of the platforms and master the algorithm of working with them, but with the help of technical specialists and clear instructions, most users of distance learning systems quickly overcame the barrier in working with distance education platforms.

B) With the announcement of the pandemic, *the degree of participation in the educational process* has decreased. This is true especially for foreign students, for whom remote perception of information in a non-native language proved to be very difficult.

C) *The level of motivation for learning* has decreased among students. This phenomenon is undeniably massive, since the transition to a remote format has affected everyone (Lysenko&Tokareva, [34]). Large volumes of academic assignments, the opportunity to spend more time at home and go about their own business, a low level of control / lack of it on the part of parents and teachers reduced the level of motivation to study. The lack of multilingual digital competencies was the main demotivator.

D) One cannot but take into account such an important criterion for evaluating education as *the quality of knowledge*. It can be stated that, on the one hand, due to the above reasons, the quality of knowledge has decreased, and on the other hand, the challenge of the pandemic has been a powerful source of development for the students themselves and the educational process in a multilingual environment.

E) *Satisfaction with the organization of learning* has decreased significantly with the pandemic and transition to e-learning. One of the tools for the formation of satisfaction is the presence of multilingual competence in teachers and students, the immediate participants of the educational process.

2 Methodology

The Ural Federal University named after the first President of Russia B. N. Yeltsin (hereafter – UrFU) is the base of the present research. The present article aims to highlight critically important to discuss multilingualism in combination with online learning the process of an emergency transition to remote work and distance learning through the eyes of teachers working in multilingual student classrooms and a multilingual educational environment where students are both taught in Russian as a foreign language and in English, which is not native to them.

With the assignment of the status of a pandemic to a new virus, forces around the world had to be mobilized to urgently transfer the learning process to a remote online format. UrFU (Ekaterinburg, Russian Federation) was no exception.

Ural Federal University named after the first President of Russia B. N. Yeltsin is the largest university in the Urals, a leading scientific and educational center in the region and one of the largest universities in the Russian Federation.

The university employs about 5,000 teachers in various fields. The ratio of men to women is 47.3% to 52.7%. The average age of teachers increased from 49.1 years to 50.0 years. Data on the age structure of the UrFU shows that in 2017, 33.3% of the university teaching staff belonged to the age groups of 60 years and older. This group increased slightly for the researched period. A significant decrease in the share of staff under the age of 35 should be noted, namely from 23 to 17.8%, which suggests a further decrease in the share of older teaching staff in the future. Institutions of higher education in general and for UrFU in particular, have a certain gap in the number of people in two age groups—from 40 to 49 years old in 2012 and from 45 to 54 years old in 2018. One of the important indicators of the human resources potential in the university is the length of service of the teaching staff. Most of them are employees with a long total work experience (in 2012—62.5% of teachers, in 2017—64.1% have 20 or more years of total work experience). The share of people with teaching experience also increased from 42% to 48%. Thus, it can be assumed that the group of teachers consists of a certain permanent "cadre core" of experienced workers (the group of young people with up to 10 years of work experience is dramatically low—12.7% of the total number of teaching staff). It was the representatives of the "cadre core" who became the respondents of the present study.

The aim of the present research was to study the attitude of teachers of the UrFU to remote work and a new distance learning format introduced due to the Covid-19 pandemic.

The sample of respondents was formed in the amount of 10% of the general population, which is represented by the total number of so-called. "cadre core" - scientific and pedagogical workers (SPD) of the university (3,793 people), and amounted to 379 people. The quota sample is distributed in proportion to the number of faculty members of the 15 institutes that are part of the university and includes the respondents who differ in terms of qualifications (according to academic degrees and academic titles), gender, and age.

The instrument of the empirical research was the authors' questionnaire, which included 4 information blocks:

Block 1. Self-organization of activities, the changes in the nature of the work of the teaching staff and factors influencing them.
Block 2: Student academic performance and quality of education.
Block 3: Technical equipment and support of the teacher's work during the transition to remote operation.
Block 4: Evaluation of the transition to a distance format.

3 Results

Results of empirical research allow us to define follow trends by blocks.

Block 1. Self-organization of activities, the changes in the nature of the work of the teaching staff and factors influencing them

a) The vast majority of teachers—70% of the respondents to one degree or another planned their work, which is essential for the effectiveness of a teacher's work (Fig. 1).

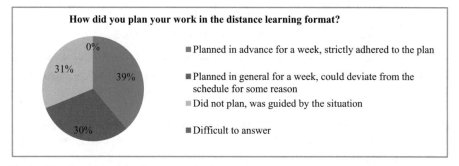

Fig. 1. Planning of work by the teachers in distance learning format (Source: own research, 2021)

b) The vast majority of teachers—69% of the respondents—prepared for their classes successfully (Fig. 2).

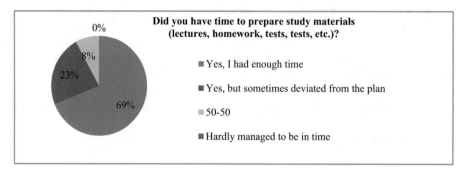

Fig. 2. The success of the preparation of teachers for their classes. (Source: own research, 2021)

c) Meanwhile, the majority of the respondents—77%—significantly increased the time spent on checking assignments given to students to complete at home (Fig. 3).

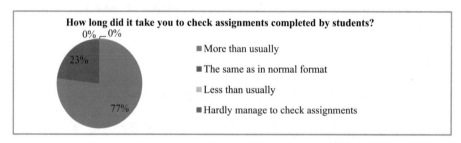

Fig. 3. Time spent by teachers on checking student assignments (Source: own research, 2021)

d) The difficulties arosed during the remote work format, which required restructuring of both quantitative (speed, volume, duration) and qualitative (structure) working formats (Fig. 4).

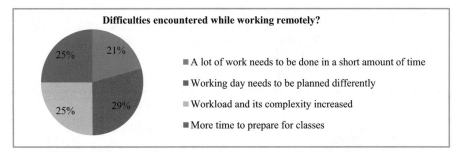

Fig. 4. Assessment of the difficulties of working remotely (multiple choice of answers) (Source: own research, 2021)

e) Among the priority reasons for difficulties in working remotely, the respondents note technical problems—equipment failures and incorrect operation of digital platforms (Fig. 5).

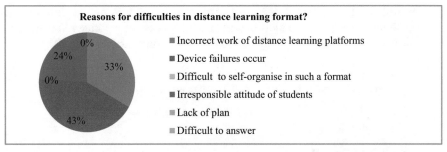

Fig. 5. Reasons for difficulties when working remotely (multiple choice) (Source: own research, 2021)

f) The majority of the respondents (more than 60%) note the presence of factors distracting from work (Fig. 6).

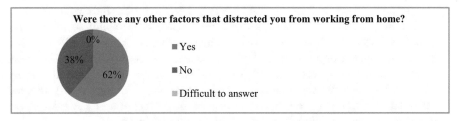

Fig. 6. The presence of distractions when working remotely (Source: own research, 2021)

g) Among the factors distracting from working remotely, respondents note both objective (calls/messages from work, insufficient equipment) and subjective (staying permanently at home, no opportunity to switch to other activities, spontaneous desire to be distracted, relatives) (Fig. 7).

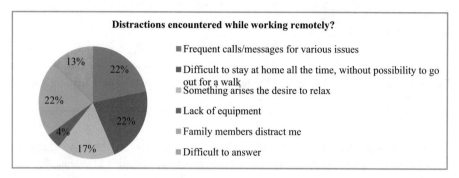

Fig. 7. The characteristic of distractions when working remotely (Source: own research, 2021)

h) The respondents attributed "personal responsibility" (69%) and "ability to self-organisation" (31%) to the tools that ensure the success of working remotely. This indicates a high level of professional and personal competencies (Fig. 8).

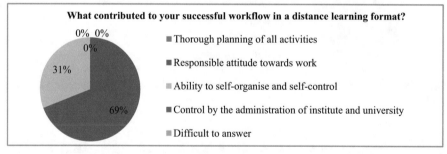

Fig. 8. Tools for the success of teachers in distance learning format (Source: own research, 2021)

i) The respondents noted factors based on personal preferences as the top factors influencing the success of distance working format, namely "the ability to adjust work schedules" (54%) and "the ability to work in a familiar environment" (38%). Other factors related to external control ("lack of management control"), on the one hand, and operational prospects ("the ability to prepare better for classes" and "the ability to work in a familiar environment"), on the other hand, the data was distributed in equal proportions (23%). It should also be noted that almost a quarter of the respondents found it difficult to answer this question (Fig. 9).

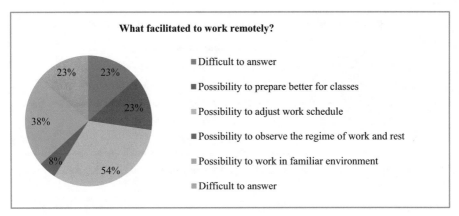

Fig. 9. Factors influencing the success of working distantly (multiple choice) (Source: own research, 2021)

j) The majority of the respondents (54%) noted the undoubted presence of advantages of remote work. However, the rest either reacted neutrally to the transition to distance learning format (23%), or rated this format more or less negatively (23%) (Fig. 10).

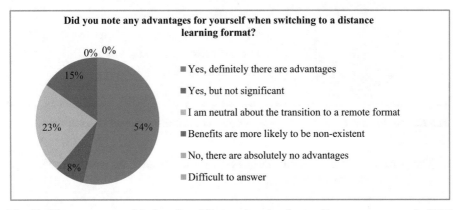

Fig. 10. The assessment of the benefits of distance learning format (Source: own research, 2021)

Block 2. Student academic performance and quality of education

a) More than half of the respondents (54%) assess student academic performance in distance learning format as average (50 × 50), only about 40% of students show higher results (Fig. 11).

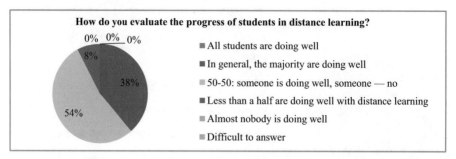

Fig. 11. The assessment of student academic performance in distance learning format (Source: own research, 2021)

b) As for assignments, a little less than half of the students handed in the completed assignments on time, and almost the same number handed assignments unstably—46%, respectively. (Fig. 12).

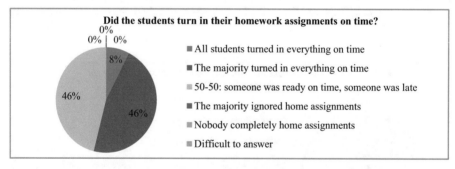

Fig. 12. The assessment of the academic assignment timeliness (Source: own research, 2021)

c) More than half of the students needed additional consultations with teachers when completing educational tasks (Fig. 13).

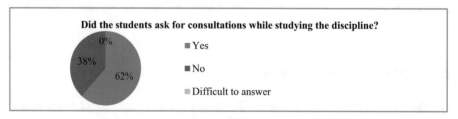

Fig. 13. The need of students for additional consultations (Source: own research, 2021)

d) Meanwhile, only a third of the students asked feedback from the teacher about the completed assignments, less than half (46%) asked for it sometimes (Fig. 14).

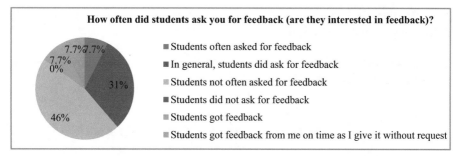

Fig. 14. The need of students for teacher feedback (Source: pwn research, 2021)

e) Slightly more than half of the teachers provide feedback constantly, the rest as needed depending on their preferences (Fig. 15).

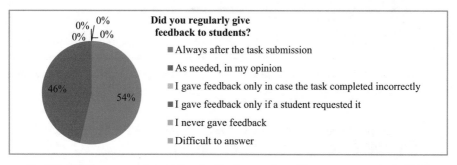

Fig. 15. The frequency of providing feedback to students (Source: own research, 2021)

f) Generally, the respondents state that when switching to a distance learning format, the quality of the acquired competencies improved significantly only in 15% of students, in a third the results remained at the same level, and in almost half slightly worsened (Fig. 16).

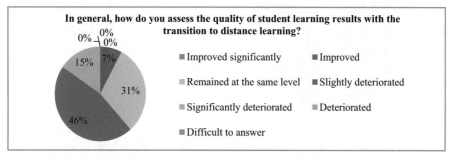

Fig. 16. The general assessment by teachers of the results of distance learning of students (Source: own research, 2021)

Block 3: Technical equipment and support of the teacher's work during the transition to remote operation

a) Teachers had practically no problems with communication means when giving lectures and other work (Fig. 17).

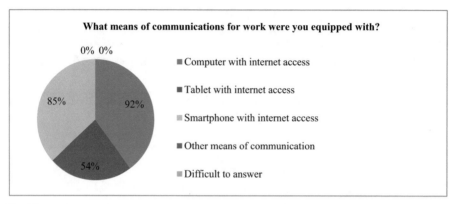

Fig. 17. Equipment of teachers with communication devices during the period of distance learning (multiple choice) (Source: own research, 2021)

b) The range of Internet resources used by teachers for work and learning is wide and sufficient for all types of communication (Fig. 18).

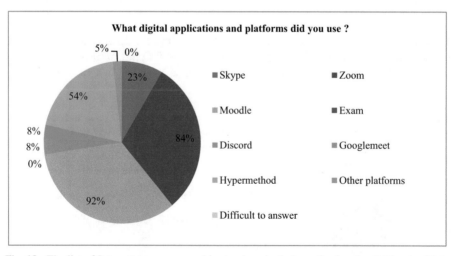

Fig. 18. The list of Internet resources used by teachers in their professional activities (multiple choice) (Source: own research, 2021)

c) More than half of the respondents (54%) indicated that distance learning platforms were unstable when they used them to give classes (Fig. 19).

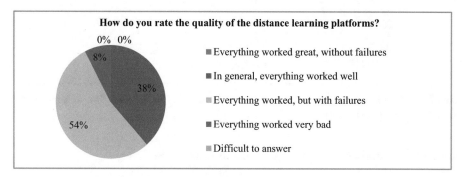

Fig. 19. Quality assessment of distance learning platforms (Source: own research, 2021)

d) In general, teachers adapted successfully to working on distance learning platforms. For a third of teachers, it was not difficult to adapt to working with distance learning platforms, and almost half of them had to make additional efforts (Fig. 20).

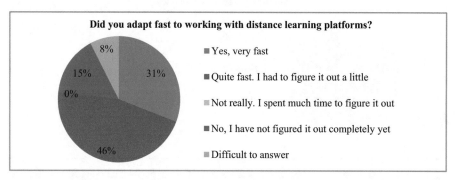

Fig. 20. The adaptation speed of teachers to working with distance learning platforms (Source: own research, 2021)

e) More than half of the teachers (54%) had enough digital competencies to switch to a remote learning and working format, but almost 40% experienced certain difficulties (Fig. 21).

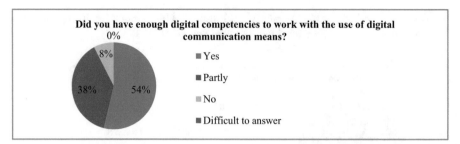

Fig. 21. The sufficiency of digital competencies of teachers in the transition to remote work (Source: own research, 2021)

Block 4: Evaluation of the transition to a distance format.

a) The assessment by teachers of the transition to distance learning during the period of pandemic restrictions is predominantly (62%) neutral. However, negative assignments prevail over positive ones by 8%. (Fig. 22).

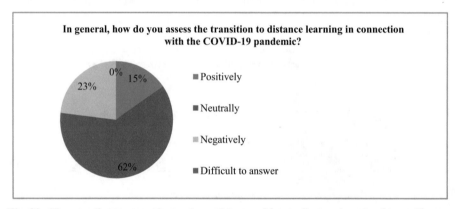

Fig. 22. The overall assessment by teachers of the transition to distance learning format (Source: own research, 2021)

b) More than half of the teachers positively assess the organisation of distance learning at the Ural Federal University, but the rest assess it as insufficient (50 × 50) (Fig. 23).

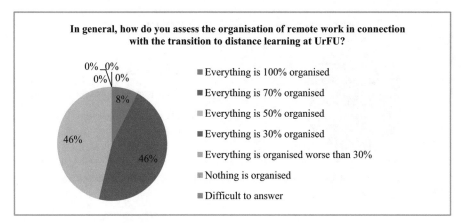

Fig. 23. The evaluation by teachers of the organisation of distance learning at the Ural Federal University (Source: own research, 2021)

c) More than half of teachers experienced positive emotions from working remotely, but almost 40% do not confirm this (Fig. 24).

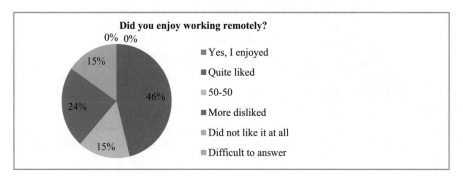

Fig. 24. Emotional evaluation of teachers when working remotely (Source: own research, 2021)

d) The overall assessment of the easiness to adapt to distance learning among teachers (Fig. 25).

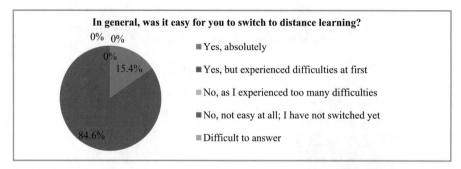

Fig. 25. The assessment of the ease of adaptation among teachers to distance learning format (Source: own research, 2021)

e) Regarding the speed and timelines of informing teachers about the necessary procedures and their adjustment by the university administration, the findings are quite positive: almost a quarter (23%) of the respondents are absolutely satisfied, the rest (77%) note slight delays in receiving necessary information (Fig. 26).

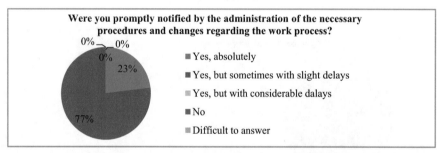

Fig. 26. The assessment of the timeliness of informing teachers by the university administration (Source: own research, 2021)

f) The assessment by the respondents of the conditions created by the university administration for distance learning format varies: a little more than a half (54%) consider the administration work sufficient, the rest (46%) are convinced that not all the necessary conditions have been created (Fig. 27).

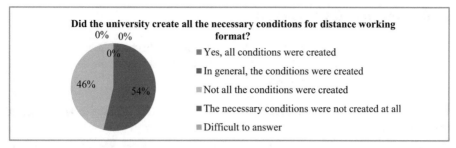

Fig. 27. Sufficiency assessment of the conditions for the transition to distance learning by administration (Source: own research, 2021)

g) The vast majority of teachers (more than 92%) indicated the need additional support for their work when implementing distance learning (Fig. 28).

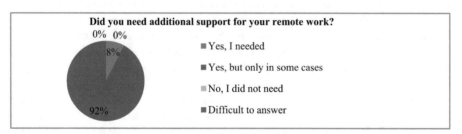

Fig. 28. The need for additional support for the work of teachers in the implementation of distance learning format (Source: own research, 2021)

h) The nature of the desired support of remote work expressed by teachers varies significantly, but technical support is the priority—92% (Fig. 29).

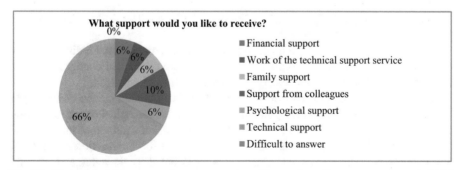

Fig. 29. The nature of remote work support expressed by teachers (Source: own research, 2021)

4 Discussion and Conclusion

The results of the study allow us to draw significant conclusions about the changes in the work nature during the COVID-19 pandemic, on the one hand, and the multilingual challenges of our time, on the other:

the complexity and intensity of teaching labour associated with working in a multilingual environment increased;
the need to strengthen contacts between participants of the educational process went up, as well as the necessity to search for tools for new types of communication, including technical and natural languages, in order to increase the effectiveness and quality of education;
the use of technical resources required for the implementation of the educational process soared, which required the formation of new multilingual competencies.. At the same time, the degree of complexity\\ease of mastering technological languages varied, which led to unstable learning results and necessitated constant technical support when mastering technological languages in a multilingual environment;
despite the relative ease of transition to distance learning and distance work, the adaptation of teachers can hardly be called quick. Factors limiting adaptation were organizational and technical, not directly dependent on employees.

Thus, the experience of developing a multilingual environment in UrFU during the COVID-19 pandemic proved to be generally positive. This situation encouraged the formation of multilingual competencies in the educational environment of the university. This affects the interests of all stakeholders: administrative and technical services, students, and faculty. However, it should be emphasized that, despite some clear advantages of distance working format in a multilingual environment, the assessment of the organization of the transition and its effects should be the subject of close attention, since online learning promotes the development of multilingual skills and competencies and increases the requirements for teaching and strengthens the role of the teacher as their guide to multilingual person.

We see the formation of a multilingual individual in a multilateral educational process as the main vector of development.

From the point of view of the multilingual approach, a multilingual individual is not just a "sum of several monolinguals", but a unique linguo-cognitive configuration, the study of which requires special complex research methods [35, p. 259].

The purpose of such studies is to combine ideas about the interaction of languages that an individual speaks. In the age of changes in the technological order, expanding the "existential" field of individuals (COVID-19 pandemic) and numerous international challenges, this is the primary task of sciences lying at the intersection of studying multilingualism, including not only natural languages, but also a wide range of artificial languages. The Dynamic Model of Multilingualism, considering the multilingual system of an individual as a single holistic formation, is the most interesting for integrating ideas about the nature of multilingualism in the modern world. Based on the system-theoretic and holistic principles, the authors of the concept F. Herdina and W. Jessner [36] start from the idea that language acquisition in a multilingual context is a complex non-linear process due to the interaction of a number of factors. This leads us to the thought that the current idea of multilingualism as a complex socio-, technical and psycholinguistic phenomenon requires the development of qualitatively new methods for studying this problem, especially on the basis of an interdisciplinary approach. The search and description of a multilingual personality and its characteristic features is the cornerstone of such aspirations. We support the opinion of Ostapenko T.C., who believes that the "ideal concept of multilingualism must meet the requirements of plasticity, adaptability and universality. The above listed requirements would enable to use this ideal concept to research different types of multilingualism, taking into account multiple factors: the number of languages the individual knows, the way they are mastered, the degree of their balance, etc." [26, pp. 237–238]. One way or another, the multilingual world and the existence of a multilingual personality in it is the closest vector for the development of human history and modern science.

References

1. Gafurov, I.R., Ibragimov, H.I., Kalimullin, A.M., Alishev, T.B.: Transformation of higher education during the pandemic: pain points. High. Educ. Russia **29**(10), 101–112 (2020). https://doi.org/10.31992/0869-3617-2020-29-10-101-112. (in Russian)
2. Radina, N.K., Balakina, Ju.V.: Challenges for education during the pandemic: an overview of literature. Educ. Stud. **1**, 178–194 (2021). https://doi.org/10.17323/1814-9545-2021-1-178-194. (in Russian)
3. Frolova, E.V., Rogach, O.V., Ryabova, T.M.: Benefits and risks of switching to distance learning in a pandemic. Perspect. Sci. Educ. **48**(6), 78–88 (2020). https://doi.org/10.32744/pse.2020.6.7. (in Russian)
4. Aronin, L.: Multilingualism in the age of technology. Technol. Lang. **1**, 6–11 (2020). https://doi.org/10.48417/technolang.2020.01.02
5. Shipunova, O.D., Evseeva, L., Pozdeeva, E., Evseev, V.V., Zhabenko, I.: Social and educational environment modeling in future vision: infosphere tools. In: E3S Web Conference, vol. 110, p. 02011 (2019). https://doi.org/10.1051/e3sconf/201911002011
6. Bylieva, D.S., Lobatyuk, V.V., Rubtsova, A.V.: Information and communication technologies as an active principle of social change. In: IOP Conference Series: Earth and Environmental Science, vol. 337, p. 012054 (2019). https://doi.org/10.1088/1755-1315/337/1/012054
7. Baranova, T., Khalyapina, L., Kobicheva, A., Tokareva, E.: Evaluation of students' engagement in integrated learning model in a blended environment. Educ. Sci. **9**, 138 (2019). https://doi.org/10.3390/educsci9020138
8. Pokrovskaia, N.N., Ababkova, M.Y., Fedorov, D.A.: Educational services for intellectual capital growth or transmission of culture for transfer of knowledge—consumer satisfaction at St. Petersburg Universities. Educ. Sci. **9**, 183 (2019). https://doi.org/10.3390/educsci9030183
9. Baranova, T., Kobicheva, A., Tokareva, E.: Total transition to online learning: students' and teachers' motivation and attitudes. In: Bylieva, D., Nordmann, A., Shipunova, O., Volkova, V. (eds.) PCSF/CSIS -2020. LNNS, vol. 184, pp. 301–310. Springer, Cham (2021). https://doi.org/10.1007/978-3-030-65857-1_26
10. Bylieva, D., Krasnoshchekov, V., Lobatyuk, V., Rubtsova, A., Wang, L.: Digital solutions to the problems of Chinese students in St. Petersburg multilingual space. Int. J. Emerg. Technol. Learn. **16**, 143–166 (2021). https://doi.org/10.3991/ijet.v16i22.25233
11. Aladyshkin, I.V., Kulik, S.V., Odinokaya, M.A., Safonova, A.S., Kalmykova, S.V.: Development of electronic information and educational environment of the university 4.0 and prospects

of integration of engineering education and humanities. In: Anikina, Z. (ed.) IEEHGIP 2022. LNNS, vol. 131, pp. 659–671. Springer, Cham (2020). https://doi.org/10.1007/978-3-030-47415-7_70

12. Almazova, N., Eremin, Y., Kats, N., Rubtsova, A.: Integrative multifunctional model of bilingual teacher education. In: IOP Conference Series Materials Science and Engineering, vol. 940, p. 012134 (2020). https://doi.org/10.1088/1757-899X/940/1/012134

13. Rubtsova, A.V., Almazova, N.I., Bylieva, D.S., Krylova, E.A.: Constructive model of multilingual education management in higher school. In: IOP Conference Series Materials Science and Engineering, vol. 940, p. 012132 (2020). https://doi.org/10.1088/1757-899X/940/1/012132

14. Ipatov, O., Barinova, D., Odinokaya, M., Rubtsova, A., Pyatnitsky, A.: The impact of digital transformation process of the Russian University. In: Katalinic, B. (ed.) Proceedings of the 31st DAAAM International Symposium, Vienna, Austria, pp. 0271–0275. DAAAM International (2020). https://doi.org/10.2507/31st.daaam.proceedings.037

15. Almazova, N., Rubtsova, A., Krylova, E., Barinova, D., Eremin, Y., Smolskaia, N.: Blended learning model in the innovative electronic basis of technical engineers training. In: Katalinic, B. (ed.) Proceedings of the 30th DAAAM International Symposium, Zadar, Croatia, pp. 0814–0825. DAAAM International (2019). https://doi.org/10.2507/30th.daaam.proceedings.113

16. Balyshev, P.: The stages of discourse-oriented virtual learning environment modeling. Technol. Lang. **3**, 88–105 (2022). https://doi.org/10.48417/technolang.2022.03.07

17. Odinokaya, M., Andreeva, A., Mikhailova, O., Petrov, M., Pyatnitsky, N.: Modern aspects of the implementation of interactive technologies in a multidisciplinary university. In: E3S Web Conference, vol. 164, p. 12011 (2020). https://doi.org/10.1051/e3sconf/202016412011

18. Anosova, N., Dashkina, A.: The teacher's role in organizing intercultural communication between russian and international students. In: Anikina, Z. (ed.) IEEHGIP 2022. LNNS, vol. 131, pp. 465–474. Springer, Cham (2020). https://doi.org/10.1007/978-3-030-47415-7_49

19. Almazova, N., Barinova, D., Ipatov, O.: Forming of information culture with tools of electronic didactic materials. In: Katalinic, B. (ed.) Annals of DAAAM and Proceedings of the International DAAAM Symposium, vol. 29, issue 1, pp. 0587–0593. Danube Adria Association for Automation and Manufacturing, DAAAM, Zadar, Croatia (2018). https://doi.org/10.2507/29th.daaam.proceedings.085

20. Bylieva, D., Zamorev, A., Lobatyuk, V., Anosova, N.: Ways of enriching MOOCs for higher education: a philosophy course. In: Bylieva, D., Nordmann, A., Shipunova, O., Volkova, V. (eds.) PCSF/CSIS -2020. LNNS, vol. 184, pp. 338–351. Springer, Cham (2021). https://doi.org/10.1007/978-3-030-65857-1_29

21. Bylieva, D., Lobatyuk, V., Tolpygin, S., Rubtsova, A.: Academic dishonesty prevention in e-learning university system. In: Rocha, Á., Adeli, H., Reis, L.P., Costanzo, S., Orovic, I., Moreira, F. (eds.) WorldCIST 2020. AISC, vol. 1161, pp. 225–234. Springer, Cham (2020). https://doi.org/10.1007/978-3-030-45697-9_22

22. Bylieva, D., Lobatyuk, V., Safonova, A., Rubtsova, A.: Correlation between the practical aspect of the course and the e-learning progress. Educ. Sci. **9**, 167 (2019). https://doi.org/10.3390/educsci9030167

23. Pozdeeva, E., et al.: Assessment of online environment and digital footprint functions in higher education analytics. Edu. Sci. **11**(6), 256 (2021). https://doi.org/10.3390/educsci11060256

24. Bolgova, V.V., Garanin, M.A., Krasnova, E.A., Khristoforova, L.V.: Post-pandemic education: falling or preparing for a jump? High. Educ. Russia **30**(7), 9–30 (2021). https://doi.org/10.31992/0869-3617-2021-30-7-9-30. (in Russian)

25. Kostina, E.Yu., Orlova, N.A.: «New Normal» of students' educational practices in the coronavirus pandemic reality. High. Educ. Russia **31**(4), 42–59 (2022). https://doi.org/10.31992/0869-3617-2022-31-4-42-59. (in Russian)

26. Ostapenko, T.S.: Multilingualism: problems of definition and main research trends in contemporary linguistics. Liberal Art. Rus. **7**(3), 232–240 (2018). (in Russian)

27. European Commission. Final report: High level group on multilingualism. European Communities, Luxembourg (2007)

28. Vahtin, N.B., Golovko, E.V.: Sociologistics and sociology of language. Humanitarian Academy, St. Petersburg (2004). (in Russian)

29. Novoselova, E.N.: The impact of the COVID-19 pandemic on the social practices of health care and the mental health of Russians. Moscow Univ. Bullet. Ser. 18. Sociol. Polit. Scien. **28**(1), 238–259 (2022). https://doi.org/10.24290/1029-3736-2022-28-1-238-259. (in Russian)

30. Baigushzin, P.A., Shibkova, D.Z., Aizman, R.I.: Factory influencing the psychophysiological and logical processes of information perception in the conditions of informatization of the educational environment. Sci. Educ. Today **9**(5), 48–62 (2019). https://doi.org/10.15293/2658-6762.1905.04. (in Russian)

31. Efimova, V.M., Makaricheva, A.A.: The influence of distance learning on the state of health of students: the pedagogical aspect of the problem. Probl. Moder. Teach. Educ. **73**(1) (2021). (in Russian)

32. Zakharova, U.S., Vilkova, K.A., Egorov, G.V.: It can't be taught online: applied sciences during the pandemic. Educ. Stud. **1**, 115–137 (2021). https://doi.org/10.17323/1814-9545-2021-1-115-137. (in Russian)

33. Abramova, M.O., Filkina, A.V., Sukhushina, E.V.: Challenges to internationalization in Russian higher education: the impact of the COVID-19 pandemic on the international student experience. Educ. Stud. **4**, 117–146 (2021). https://doi.org/10.17323/1814-9545-2021-4-117-146. (in Russian)

34. Lysenko, E., Tokareva, J.: Study of satisfaction with the organization of the studying process in remote mode during the COVID-19 pandemic: case UrFU named after the First President of Russia BN Yeltsin. In: SHS Web of Conference, vol. 92, p. 01026 (2021)

35. Grosjean, F.A.: Psycholinguistic approach to code-switching: the recognition of guest words by bilinguals. One Speaker, Two Languages: Cross-disciplinary Perspectives on Code-Switching. Cambridge University Press, Cambridge (1995)

36. Herdina, P., Jessner, U.: A Dynamic Model of Multilingualism: Perspectives of Change in Psycholoinguistics. Multilingual Matters Clevedon (2002)

Formation of Graduates Mathematical Culture in Multicultural Environment on the Example of Chinese Students in Russia

Victor Krasnoshchekov$^{(\boxtimes)}$ (ID) and Natalia Semenova (ID)

Peter the Great St. Petersburg Polytechnic University, St. Petersburg 195251, Russia
krasno_vv@spbstu.ru

Abstract. The article considers the problem of formation of mathematical culture of a graduate of a Russian university. The concept of mathematical culture is debatable due to the non-universality of the language of mathematics, as well as due to the contradictions between the supporters of "pure" and applied mathematics. The article analyzes the components of mathematical culture and shows the transformation of this concept over the past 100 years. The problem of the dependence of mathematical culture on the ethnicity of its creators and its bearers is debatable. This problem is relevant in connection with the development of international education. The article shows that the assessment of the level of mathematical culture is subjective. Chinese students numerically prevail in the contingent of foreign citizens in Russian universities. To objectify ideas about the level of mathematical culture of Chinese graduates, it is necessary to build the mathematics teaching process of them at the university with the involvement of Russian-Chinese bilingual textbooks. The authors have developed the manual on the course of Probability Theory and Mathematical Statistics with parallel texts in Russian and Chinese. The article presents the results of the experiment comparing the academic success of Chinese students in economics before and after the introduction of the manual. The results of the experiment demonstrate the increase in academic success. Testing the corresponding hypothesis by method of confidence intervals proves this conclusion.

Keywords: International education · Probability theory · Academic success

1 Introduction

Many authors, both in Russia [1] and abroad [2, 3], studied the formation of the mathematical culture of schoolchildren, students and university graduates. First of all, researchers ask themselves the question: what is mathematical culture? Based on the opinions of different researchers, this is an integral formation in the personality structure, which is the part of a general culture that embraces mathematical concepts and competencies. Y. K. Chernova and S. A. Krylova expanded the concept of mathematical culture, emphasizing the need for mathematical thinking and readiness for self-development [4]. Modern researchers have identified two more aspects of mathematical culture. First of all, we are talking about mathematical ethics [5]. Without mathematicians, it would be impossible to build financial pyramids that are beneficial for their creators, but they ruin gullible investors. It is mathematicians who are involved in the development of targeted advertising, including political advertising, which helps customers manipulate the minds of customers [6]. The second aspect is positive. It generalizes the idea of self-development, attributing the desire for creativity to the bearers of mathematical culture [7–9].

Discussions on the content of mathematical culture were relevant at the beginning of the 20th century in connection with the increasing role of applied mathematics. These discussions resumed in the middle of the 20th century, as the rapid development of computing technology filled the methods of computational mathematics with, in fact, a new meaning. Thus, the central problem of research in the last century in this aspect was the problem of attributing mathematical modeling, computational mathematical schemes and computer algorithms to the sphere of mathematical culture [10, 11]. Opponents of such inclusion were supporters of "pure" mathematics, whose voices began to be heard more and more quietly, and gradually fade away. Nevertheless, debatable issues remain in the problem of the content of mathematical culture, which the authors highlight in connection with the subject of this study. These questions touch upon cross-cultural aspects of mathematical culture and are of particular interest to authors who work in the field of international education.

2 Materials and Methods

Having determined the main goal of this work to study the cross-cultural components of mathematical culture, the authors turned to the study of sources on cross-cultural issues in the mathematical aspect, including both the philosophy of mathematics and the problems of mathematical training.

The dependence of mathematics on the cultural component stems from such a problem as the non-universality of mathematical culture in general, and the non-universality of mathematical language, in particular [12]. The problem of the non-universality of mathematical culture also has several dimensions and is the battlefield between "pure" mathematicians and their opponents, which the authors mentioned above.

The authors conditionally refer to the first dimension of non-universality a collision between the so-called "Mathematics with a capital letter" and "mathematics with a small letter" [13]. By the first of these, B. L. Yashin understands "pure", Platonic, ideal mathematics, which is taught in classical universities. Under the "second" mathematics, the same author means the so-called "everyday" mathematics, in which there is no place for linear spaces and "language $\varepsilon - \delta$" [14]. This "second" mathematics allows people to exist and function without feeling their "inferiority", including in the field related to the applications of mathematics. The authors begin their consideration of the problem with the fact that the language of mathematics is indeed difficult for both schoolchildren and students who have not graduated from a specialized school with in-depth study of mathematics [15]. A. Borovik talks about the imminent division of school training in mathematics into two streams - a larger one and a smaller one. In a larger stream, students will study only the basics of mathematics, since they do not intend to use mathematics in their future professional activities. On the contrary, students of a smaller stream will focus on a deep study of mathematics and natural sciences as the basis of fundamental education in their respective fields [16]. In a more "provocative" work, G. S. Aikenhead exacerbates the contradiction between the humanistic values of the majority of schoolchildren and the "anti-humanistic" models of teaching school mathematics in the Platonic spirit [17]. Accordingly, the mentioned author gives a chance for the existence of "pure" mathematics only within the framework of STEM education [18]. Softened echoes of this opposition are a significant number of studies that develop the problems of further diversification of mathematical courses - for engineers [19], for economists [20], etc.

The second discussion position on the topic of the non-universality of mathematics was expressed by O. A. G. Spengler back in 1918 in the first part of his famous book *Der Untergang des Abendlandes*. It consists in asserting the dependence of the forms of mathematics on the forms of the culture that gave rise to it [21]. This statement shook the foundations of "pure" mathematics and caused vortex-like indignations that are existing this day. Some followers of O. Splenler reinforce his idea with fresh examples, extending the idea of universality to more and more new ethnic groups [22]. Others generalize the position of the master, deriving it from the theory of scientific model pluralism, which is based on the idea that the forms of any science depend on the forms of the corresponding culture [23]. Still others refer to the lack of a generally accepted consistent concept of culture, which naturally leads to the idea that it is impossible to build a unified theory of mathematical culture [24]. R. Hersh gave a very simple and at the same time elegant substantiation of O. Splenler's position. Indeed, if mathematical knowledge is produced by a person, then it depends on a person, and a person belongs to a particular culture, i.e. mathematical knowledge is conditioned by culture [25].

From the non-universality of mathematical culture, one step remains to the non-universality in the teaching of mathematics. This step was taken by the Brazilian M. Ascher in 1991 [26]. Thus, ethno-mathematics was born. Of course, there have been impressive studies on this topic before, for example, [27, 28], but the term itself arose precisely in 1991. The topic remains relevant to this day. There are works that expand the base of cultures [29, 30]. Other studies develop theory of ethno-mathematics, for example, using the "language games" of L. Wittgenstein, which are based on the idea that the language of an ethnic group partially determines the development of mathematics in this ethnic group [31]. Of course, ethno-mathematics and its predecessor theories, which attributed differences in mathematical culture to different ethnic groups, immediately had opponents who defended the primacy of a "pure" and universal mathematical science [32, 33]. But their voices were weak, and now they are completely silent, since critics of ethno-mathematics can deserve accusations of intolerance, which will be followed by at least excommunication from the professional scientific environment.

The ideas that the authors expressed above formed the basis of the author's model of teaching mathematics to students in a foreign language environment. To confirm the provisions of this theory, a longitudinal pedagogical experiment was conducted with the participation of students from China. The authors processed the results of the experiment by methods of mathematical statistics, in particular, using the algorithm for testing hypotheses by the method of confidence intervals.

3 Results and Discussion

3.1 Overcoming the Limitations of Existing Theories of Cross-Cultural Learning

The model for the formation of the mathematical culture of a foreign graduate of a Russian university follows from the general theory of teaching in a non-native language by A.I. Surygin [34]. A. I. Surygin developed didactic principles that define both the goals and the process of learning and teaching. Of course, other researchers followed A. I. Surygin, developing his ideas, and some authors developed the subject of mathematical training for foreign students. Taking the theory of learning in a non-native language as a basis, the authors, however, show a number of limitations that currently do not make it possible to fully use the provisions of this theory.

Firstly, under teaching in a non-native language A. I. Surygin understands the education of foreign citizens at the preparatory faculty of a Russian university. At the same time, the authors, on the contrary, deliberately exclude students of the preparatory faculty from consideration. The fact is that Russian universities have created their online preparatory departments and remote preparatory departments abroad. In these departments, students receive training in the absence of a language environment and, very often, with very limited opportunities for communication with native Russian speakers. These students, of course, demonstrate a different degree of readiness for studying at the university in Russian than students who graduated from the preparatory faculty in Russia. The percentage of such students is constantly increasing. Experts explain this phenomenon not only by the impact of the COVID-19 pandemic, but also by economic reasons. Indeed, students spend less money studying in their own, and not a foreign country. A significant part of these "new" students are from Mainland China.

Secondly, each theory, unless, of course, it was created by a brilliant "creator for all times", bears the imprint of the time, place and conditions of its creation. Theory of A. I. Surygin meets the realities of teaching foreign students in Russia in the mid-1990s. First of all, the country was in the stage of moving away from the economic and political structure, which is commonly called socialist. The element of the free market has just begun to roll into the Soviet system of higher education, despite the fact that education is a conservative system by its very nature. Ideas about the high state mission of the preparatory faculty for foreign citizens, designed to forge cadres of friends of our country, remained in force. Accordingly, the curriculum for the student's hourly workload was very extensive. It amounted to more than 2000 classroom hours against 1008 standard hours in our time. At the same time, the workload norms for teachers of the preparatory faculty were small, and the salaries were high relative to the salaries of "ordinary" university teachers, since they increased with bonuses for the complexity of the work performed. On the other hand, the number of students in the academic groups of the preparatory faculty did not exceed 8 people. The total number of foreign students, even at one of the largest preparatory faculties - at Peter the Great St. Petersburg Polytechnic University (the authors give the current name of the university, SPbPU) was about 300 people. For comparison, in 2019, about 1,500 foreign students studied at the preparatory faculty of SPbPU. This means that in the mid-1990s, each student during the preparatory period received significantly more hours of Russian language and basic disciplines than now. Also, a foreign student received more attention, teachers had more opportunities for individualization of education due to the small size of academic groups. These were the starting points on which A. I. Surygin built his learning theory on a non-native language. It is clear that by now the situation has changed dramatically. The administration and teachers simply do not have the opportunity to implement all the recommendations on the organization of the educational process and the consistent application of teaching methods at the preparatory faculty, which follow from the theory of A. I. Surygin.

Thirdly, as the authors have already emphasized, A. I. Surygin created his theory more than twenty years ago. He relied on the results of research on the adaptation of foreign students, which foreign scientists, and after them, Russian ones received in the 1970s-1980s. The degree of adaptation to a foreign culture was considered, and sometimes is considered now, as one of the main factors in ensuring the goals of teaching in a non-native language. The researchers of that time considered the rapid mastery of the language of the country of temporary residence to be the main condition for the successful adaptation, which is supported by intensive communication with representatives of other cultures [35, 36]. Currently, the authors found the studies that link the success of teaching foreign students with the help of compatriots who communicate with them in their native language [37, 38]. This provision is true, of course, for students from large diasporas, which primarily include the citizens of Mainland China [39, 40].

In addition to the theory of teaching in a non-native language, the authors partially draw on the provisions of the modern theory of bilingual and multilingual education [41, 42]. It is impossible to fully rely on this theory in the case of teaching university students. According to the recent scientific tradition, researchers began to associate bilingual education with the education of ethnic minorities [43]. This leads to the following difficulties in scientific discussions. Firstly, the development of the theory and spread of the practice of bilingual education in minorities is resisted by a number of scientific schools that adhere to traditional views on adaptation, which the authors presented above [44]. Indeed, according to the traditional view, which partly bears the features of a stereotype, bilingualism in education slows down the process of adaptation. Secondly, since the issues of education of minorities are touched upon, the scientific discussion can acquire an ideological and political character [45], which, if possible, should be avoided in studies on educational topics. As an acceptable example, the authors can cite a study that is devoted to bilingualism in China [46]. Of course, this study also addresses the issues of education of the numerous ethnic minorities of Mainland China. However, in another part of the work, its authors touch upon the implementation of bilingual Anglo-Chinese educational programs. Since the study of Russian as a non-native language is somewhat similar to the study of Chinese as a non-native language, some of the ideas of Chinese colleagues turned out to be useful in terms of this study.

Thus, all the references that the authors made during the preliminary consideration of the problem turned out to be connected in one way or another with the teaching of Chinese students. Taking into account the large representation of Chinese citizens in the Peter the Great St. Petersburg Polytechnic University, the authors build their own model of the formation of the mathematical culture of graduates in a multicultural environment, based on the results of an experiment in which Chinese students participated.

3.2   Diagnosing the Level of Mathematical Culture

It is quite difficult to diagnose one or another level of mathematical culture of a schoolchild/student/graduate. First of all, researchers separately diagnose the components of the mathematical culture model [47]. For example, for a manager, two groups of components can be distinguished - structural (ideological, methodological, prognostic) and functional (value, cognitive, emotional, analytical) [48]. The easiest way is to check the level of formation of the methodological component, i.e. mathematical knowledge, skills, or competencies. The scale of levels in the available works has a heuristic character - "high-medium-low". The authors did not find quantitative indicators in the sources to measure the levels. Therefore, the authors propose to use data on the academic performance of students in mathematical disciplines, which some authors consider an acceptable assessment of academic success. Academic success, in turn, is an assessment of the formation of competencies [49].

To understand how a teacher assesses the level of a student's mathematical culture, one can use the "reference points" that are consciously or unconsciously used by many university mathematicians.

Example 1. A student clearly demonstrates a low level of mathematical culture if he does not change sign by expanding the brackets with a minus in front of them:

$$(2x + 3) - (5x - 8) \neq 2x + 3 - 5x - 8 = -3x - 5.$$

whereas the correct solution is:

$$(2x + 3) - (5x - 8) = 2x + 3 - 5x + 8 = -3x + 11.$$

Example 2. The actions of a student with a low level of mathematical culture when transforming irrational expressions are:

$$\sqrt{a^2 + b^2} \neq \sqrt{a^2} + \sqrt{b^2} = a + b.$$

Or in the numerical example:

$$\sqrt{4^2 + 2^2} \neq 4 + 2 = 6.$$

The correct solution is:

$$\sqrt{4^2 + 2^2} = \sqrt{16 + 4} = \sqrt{20}.$$

Example 3. When converting fractions, students usually know the rule:

$$\frac{a+b}{c+d} \neq \frac{a}{b} + \frac{c}{d}.$$

However, when making calculations, students with a low level of mathematical culture cannot apply this rule, for example:

$$\frac{6x^2 + 8x^3}{2x + 4x^2} \neq \frac{6x^2}{2x} + \frac{8x^3}{4x^2} = 3x + 2x = 5x.$$

While the correct answer is:

$$\frac{6x^2 + 8x^3}{2x + 4x^2} = \frac{2x^2(3 + 4x)}{2x(1 + 2x)} = \frac{x(3 + 4x)}{1 + 2x}.$$

The last answer is not as "beautiful" as the previous one. From the point of view of students with a low level of mathematical culture, the answers should be "beautiful" and short.

A variant of the actions of a student with a low level of mathematical culture in the latter case can be as follows:

$$\frac{6x^2 + 8x^3}{2x^2} \neq \frac{6x^2}{2x^2} + 8x^3 = 3 + 8x^3.$$

In this case, the student remembers the rule specified above, but cannot apply it correctly, interpreting it too broadly. Correct answer is:

$$\frac{6x^2 + 8x^3}{2x^2} = \frac{6x^2}{2x^2} + \frac{8x^3}{2x^2} = 3 + 4x.$$

Let's move on to examples from a university course in mathematics.

Example 4. When calculating derivatives, only students with an extremely low level of mathematical culture do this:

$$\left(e^{5x}\right)' \neq e^{5x}.$$

Basically, students remember the rules of differentiation correctly:

$$\left(e^{5x}\right)' = 5e^{5x}.$$

But already when solving the same task with a negative power, many students with a low level of mathematical culture choose the wrong path:

$$\left(e^{-5x}\right)' \neq 5e^{-5x}.$$

The correct solution is:

$$\left(e^{-5x}\right)' = -5e^{-5x}.$$

If we consider the problem of diagnosing the level as a whole, then errors in operations with fractions and negative values are the most reliable indicators of a low level of mathematical culture.

Example 5. The same error, but more often, students with a low level of mathematical culture do, integrating:

$$\int e^{5x} dx \neq e^{5x} + C.$$

The correct solution is:

$$\int e^{5x} dx = \frac{1}{5}e^{5x} + C.$$

Example 6. When solving a problem in probability theory, a student with a low level of mathematical culture can get the answer $p = 1.35$, or $p = -0.87$, since he cannot connect the axiom of probability theory:

$$0 \le p \le 1$$

with the solution of the specific problem.

Example 7. When solving a problem in probability theory or mathematical statistics, a student with a low level of mathematical culture can obtain a distribution of a discrete random variable in the form:

x_i	1	2	3	4
p_i	0.1	0.3	0.5	0.2

However, the completeness or normalization condition is violated here:

$$\sum_{i=1}^{n} p_i = 1.$$

In this case

$$\sum_{i=1}^{4} p_i = 0.1 + 0.3 + 0.5 + 0.2 = 1.1 > 1.$$

It is much more difficult to diagnose the differences between students of medium and high levels of mathematical culture.

Example 8. Consider again Example 2. A student with a medium level of mathematical culture will do this:

$$\sqrt{4^2 + 2^2} = \sqrt{16 + 4} = \sqrt{20}.$$

A student with a high level of mathematical culture can give the answer a finished look:

$$\sqrt{4^2 + 2^2} = \sqrt{16 + 4} = \sqrt{20} = \sqrt{4 \cdot 5} = 2\sqrt{5}.$$

Example 9. When transforming irrational expressions, a student with an average level of mathematical culture will stop at this form of the correct answer:

$$\frac{1}{\sqrt{3} - \sqrt{2}}.$$

A student with a high level of mathematical culture will carry out the transformations further, getting rid of the irrational expression in the denominator:

$$\frac{1}{\sqrt{3} - \sqrt{2}} = \frac{\sqrt{3} + \sqrt{2}}{\left(\sqrt{3} - \sqrt{2}\right)\left(\sqrt{3} - \sqrt{2}\right)} = \frac{\sqrt{3} + \sqrt{2}}{\left(\sqrt{3}\right)^2 - \left(\sqrt{2}\right)^2} = \frac{\sqrt{3} + \sqrt{2}}{3 - 2} = \sqrt{3} + \sqrt{2}.$$

Let's go to the examples from the university mathematics course.

Example 10. When solving a quadratic equation

$$x^2 - 2x + 10 = 0$$

student finds the roots:

$$x_{1,2} = \frac{2 \pm \sqrt{4 - 40}}{2} = \frac{2 \pm \sqrt{-36}}{2}.$$

A student with a medium level of mathematical culture concludes that there are no roots, because he knows from a secondary school mathematics course that it is impossible to extract roots from negative numbers. A student with a high level of mathematical culture knows about the existence of complex numbers and he writes the answer:

$$x_{1,2} = \frac{2 \pm \sqrt{4 - 40}}{2} = \frac{2 \pm \sqrt{-36}}{2} = \frac{2 \pm 6i}{2} = 1 \pm 3i.$$

It is possible that many colleagues will disagree with the authors, believing that the lack of knowledge about complex numbers automatically assigns a student a low level of mathematical culture. The authors allow the same doubts about the ambiguity of the definition of levels in tasks for finding limits, differentials and determining the convergence of series.

Example 11. When finding the limit

$$\lim_{x \to 2} \frac{x^2 - 3x + 2}{x^2 + 3x - 10}$$

A student with a low (medium) level of mathematical culture will remember the school rule: "You cannot divide by zero", and say that there are no solutions (there is no limit). At the same time, a student with a medium (high) level of culture will reveal an indefinite expression and find the correct answer $1/7$.

Example 12. In the task of finding the differential of the function

$$y = 3x^2 + 2x$$

at point 2, a student with a medium (low) level of mathematical culture can answer

$$dy = 14,$$

and a student with a high (medium) level of mathematical culture will write the correct answer

$$dy = 14dx.$$

Example 13. Examining the convergence of the series

$$\sum_{n=1}^{\infty} (-1)^n \frac{n+1}{n^3 + 1},$$

a student with a medium (low) level of mathematical culture can answer: "the series converges", while a student with a high (medium) level of mathematical culture will definitely clarify: "the series converges absolutely".

The considered examples show that diagnosing a low level of mathematical culture of students is quite simple: a student either does not know the rules of mathematics, or cannot apply them. Diagnostics of a medium and a high level of mathematical culture is more difficult. Does a mistake always indicate a low mathematical culture? Perhaps this error appeared by chance, for example, due to poor health or student stress. Does a medium level of mathematical culture always mean that a student can give correct answers? Accordingly, will a student with a high level of mathematical culture really choose the most rational ways of solving, will he receive an answer in a complete form? In addition, it is obvious that determining the level of a student's mathematical culture significantly depends on the diagnosing teacher, i.e. this process is subjective. To objectify the diagnostics of the level of mathematical culture, the community of mathematicians must develop criteria for differentiation and evaluation tools. Perhaps in the future, due to the increasing importance of mathematical training in the conditions of the Fourth Industrial Revolution, such a need will appear in society.

3.3   The Results of the Introduction of Russian-Chinese Manual on Probability Theory

In order to understand the problems of the formation of the mathematical culture of Chinese students at a Russian university, it is necessary to understand how the mathematical culture of Chinese schoolchildren is formed in Chinese secondary schools. There are several approaches both to identifying Chinese mathematical culture and to comparing Chinese and Western (Greek, Arabic) mathematical cultures. Some authors do not find specifics in the Chinese approach to the formation of mathematical culture at all [50]. However, most researchers derive the main characteristics of Chinese mathematical culture from the Confucian Heritage Culture [51]. First of all, this is a sense of deep responsibility of the student to society, a respectful attitude towards the teacher, the desire to please parents [52]. This, in turn, leads to the apparent passivity of students, the difficulties in involving Chinese students in class discussions that teachers face both in the West and in Russia. Accordingly, the question arises: why, with such "backward" methods of teaching and learning, do the Chinese demonstrate excellent results in international tests and math Olympiads [53]? Chinese authors give such an answer to the question about the differences between Western and Chinese approaches to the formation of mathematical culture. If the Western mathematical culture is based on the comprehension of the most complex abstract-logical constructions, then the Chinese paradigm focuses on the formation of practical skills of students [54]. In addition, if not in form, then in content, the system of mathematical training for Chinese schoolchildren copies the Soviet school system of the early 1950s [55]. This, on the one hand, guarantees a high level of mathematical culture, and, on the other hand, gives hope for the continuity of the formation of mathematical culture in the Chinese school and at the Russian university.

Among the studies that turned out to be close to the problems of this work, we can name the article by Kashyap et al. [56]. The authors of the cited article deduce the high level of mathematical culture of Chinese schoolchildren and students from the peculiarities of their thinking, namely, from the need to comprehend and memorize a large number of hieroglyphs from early childhood, as a basic element of Chinese culture.

It should be understood that hieroglyphic inscriptions-images are the natural environment of any Chinese. From this follows the most important conclusion of this study that it would be pointless to abandon hieroglyphic writing in the formation of the mathematical culture of Chinese students at a Russian university. The authors consider this a weighty argument in favor of the development of Russian-Chinese bilingual teaching manuals.

Turning to the analysis of practical tools for the formation of the mathematical culture of Chinese students, we must understand that a student who does not speak Russian well may simply not understand the task, especially if the text plays a big role in this task. Moreover, the student uses the knowledge and skills that he received in China. Therefore, based on the formula, he may misinterpret the task in accordance with the traditions of Chinese mathematical culture. Relying only on mathematical texts in Russian, teachers may underestimate the level of mathematical culture of Chinese students. This is especially true for a course in probability theory, in which text assignments are crucial. Accordingly, the authors consider the use of bilingual manuals to be the main component of the model for the formation of the mathematical culture of Chinese graduates of Russian universities.

In 2015, the authors prepared and published at SPbPU such the manual on the course of Probability Theory and Mathematical Statistics, in 2019–2022 they prepared and published the series of 5 manuals on the main sections of Calculus and Linear Algebra. The use of the manual on Probability Theory and Mathematical Statistics was taken as the basis of the pedagogical experiment to test a model for the formation of the mathematical culture of Chinese graduates of Russian universities. The experiment was carried out in face-to-face classes of 2nd year Chinese students of the Institute of Industrial Management, Economics and Trade of SPbPU from year 2014 to year 2018. The main results of the experiment are in Table 1.

Table 1. Academic success of Chinese students in Probability Theory1.

Academic year	Number of students	Number of successful students	Percentage of successful students	Confidence interval
2014/2015	69	48	69,5%	60,4% - 78,6%
2015/2016	73	61	83,6%	76,4% - 90,8%
2016/2017	72	65	90,3%	84,5% - 96,1%
2017/2018	84	75	89,3%	83,8% - 94,9%

Highlighting the first line of Table 1, the authors emphasize the fact that students of the 2014/15 academic year did not have the opportunity to use the bilingual aid. The authors note that the performance of successful students increased already in the first year of using the bilingual manual. Indeed, the percentage of 2015/16 academic year successful students is no longer included in the confidence interval of the percentage of 2014/15 academic year at a significance of 0.1 (with reliability 90%). The converse statement is also true, which means that the two-sided criterion works. The confidence intervals themselves intersect, which means that the criterion is fulfilled at the limit of reliability. This is explained by the fact that the authors used the manual for the first time, and Chinese students are conservative in relation to changes in the educational process that are made by foreigners [57].

Confidence intervals of academic success percentages, starting from 2016/17, no longer intersect with the 2014/15 percentage confidence interval, demonstrating a stable improvement in academic success after the introduction of the manual into the educational process. The reliability of these data is also 90%. Since the authors published the last of the series of Calculus tutorials only in 2022, it is necessary to continue the experiment to test the success of the entire series of tutorials.

4 Conclusion

As a starting point for building a model for the formation of the mathematical culture of Chinese graduates of Russian universities, the theory of learning in a non-native language is considered. Due to significant changes in the conditions of teaching foreign students at Russian universities, some provisions of this theory had to be revised. Also, elements of the theory of bilingual education were used to build the model, and the ideological references of this theory were not taken into account. Numerous examples of assessing the level of mathematical culture of students are given. It is shown that these estimates cannot be considered as reliable estimates in the case of teaching in a non-native language. Hieroglyphic writing is an essential component of the formation of mathematical culture of Chinese students, since they are surrounded by hieroglyphic images from childhood. The authors showed that it is difficult to assess the level of formation of the level of mathematical culture of Chinese students at a Russian university without the use of mathematical texts in Chinese. Therefore, the authors have developed a series of bilingual Russian-Chinese textbooks for the university course in mathematics. The results of an experiment on the introduction of a bilingual textbook on probability theory are presented. The method of confidence intervals is used to test the statistical hypothesis about the impact of the introduction of a bilingual manual on the academic success of students. Based on the analysis of the results of the experiment, it can be concluded that the use of a bilingual textbook really increases the academic success of Chinese students when they study Probability Theory and Mathematical Statistics. This means that the use of bilingual aids can be considered as one of the main tools for forming the mathematical culture of Chinese graduates of Russian universities.

References

1. Voronina, L.V., Moiseeva, L.V.: Mathematical culture of personality. Pedagogical Educ. Russia **3**, 37–45 (2012). https://pedobrazovanie.ru/archive/1/3/matematicheskaya-kul-tura-lichnosti
2. Löwe, B.: Philosophy or not? The study of cultures and practices of mathematics. In: Ju, S., Löwe, B., Müller, T., Xie, Y. (eds.) Cultures of Mathematics and Logic. THS, pp. 23–42. Springer, Cham (2016). https://doi.org/10.1007/978-3-319-31502-7_2
3. Löwe, B., Martin, U., Pease, A.: Enabling mathematical cultures: introduction. Synthese **198**(26), 6225–6231 (2021). https://doi.org/10.1007/s11229-020-02858-y

4. Chernova, Y.K., Krylova, S.A.: Mathematical culture and its components formation via education process. Toliatti Polytechnic University, Toliatti (2001). https://www.elibrary.ru/item.asp?id=23123046. Accessed 10 Sept 2022

5. Ernest, P.: The Ethics of mathematics: is mathematics harmful? In: Ernest, P. (ed.) The Philosophy of Mathematics Education Today, pp. 187–216. Springer, Cham (2018). https://doi.org/10.1007/978-3-319-77760-3_12

6. Chiodo, M., Clifton, T.: The importance of ethics in mathematics. EMS Newsl. **12**, 34–37 (2019). https://doi.org/10.4171/NEWS/114/9

7. Simonton, D.K.: Teaching creativity: current findings, trends, and controversies in the psychology of creativity. Teach. Psychol. **39**(3), 203–208 (2012). https://doi.org/10.1177/0098628312450444

8. Leikin, R., Pitta-Pantazi, D.: Creativity and mathematics education: the state of the art. ZDM – Int. J. Math. Educ. **45**(2), 159–166 (2013). https://doi.org/10.1007/S11858-012-0459-1

9. Arney, K.M., Blyman, K.K., Cepeda, J.D., Lynch, S.A., Prokos, M.J., Warnke, S.: Going beyond promoting: preparing students to creatively solve future problems. J. Hum. Math. **10**(2), 348–376 (2020). https://doi.org/10.5642/jhummath.202002.16

10. Pedersen, S.A.: Mathematics in engineering and science. In: Ju, S., Löwe, B., Müller, T., Xie, Y. (eds.) Cultures of Mathematics and Logic. THS, pp. 61–79. Springer, Cham (2016). https://doi.org/10.1007/978-3-319-31502-7_4

11. Borovik, A., Kocsis, Z., Kondratiev, V.: Mathematics and mathematics education in the 21st century. Preprint arXiv:2201.08364, pp. 1–22 (2022)

12. Krasnoshchekov, V., Semenova, N.: On the non-universality in mathematical language. Tech. Lang. **3**(3), 73–87 (2022). https://doi.org/10.48417/technolang.2022.03.06

13. Yashin, B.L.: Mathematics and culture: uniqueness and universality of mathematics. Teacher Century 21 **3**(1), 11–21 (2020). https://doi.org/10.31862/2073-9613-2020-3-11-21

14. Sinkevich, G.I.: To the history of epsilontics. Math. High. Edu. **10**, 149–166 (2012). https://www.spbgasu.ru/upload-files/vuz_v_licah/publish/sinkevich_gi/33.pdf

15. Bulaon, M.A.: Why is the language of mathematics confusing to students? Researchgate Publications **325968025** (2018). https://www.researchgate.net/publication/325968025_Why_is_the_language_of_mathematics_confusing_to_students. Accessed 10 Sept 2022

16. Borovik, A.: Mathematics for makers and mathematics for users. In: Sriraman, B. (ed.) Humanizing Mathematics and its Philosophy, pp. 309–327. Springer, Cham (2017). https://doi.org/10.1007/978-3-319-61231-7_22

17. Aikenhead, G.S.: Resolving conflicting subcultures within school mathematics: towards a humanistic school mathematics. Can. J. Sci. Math. Technol. Educ. **21**(2), 475–492 (2021). https://doi.org/10.1007/s42330-021-00152-8

18. Yildirim, B., Sidekli, S.: STEM applications in mathematics education: the effect of STEM applications on different dependent variables. J. Baltic Sci. Edu. **17**(2), 200–214 (2018). https://doi.org/10.33225/jbse/18.17.200

19. Rassokha, E.N., Antsiferova, L.M.: Mathematical culture of students of technical directions of training. Orenburg State Uni. Gerald **2**(220), 41–48 (2019). https://doi.org/10.25198/1814-6457-220-41

20. Melnikov, Y.B., Boyarskii, M.D., Lokshin, M.D.: Formation of mathematical culture of a graduate of an economic university as a means of increasing his professional competence. Mod. Edu. **1**, 99–111 (2017). https://doi.org/10.7256/2409-8736.2017.1.22616

21. Spengler, O.: The Decline of the West, vol. 1. Form and Actuality. George Allen & Unwin, London (1918)

22. McNaughton, D.: Cultural souls reflected in their mathematics: the spenglerian interpretation. Sci. Cult. **2**(1), 1–6 (2016). https://doi.org/10.5281/zenodo.34387

23. Veit, W.: Model pluralism. Philos. Soc. Sci. **50**(2) (2019). https://doi.org/10.1177/0048393119894897

24. Larvor, B.: What are mathematical cultures? In: Ju, S., Löwe, B., Müller, T., Xie, Y. (eds.) Cultures of Mathematics and Logic. THS, pp. 1–22. Springer, Cham (2016). https://doi.org/10.1007/978-3-319-31502-7_1

25. Hersh, R.: What is mathematics, really? Mitteilungen der Deutschen Mathematiker-Vereinigung **6**(2) (1998). https://doi.org/10.1515/dmvm-1998-0205

26. Ascher, M.: Ethnomathematics: A Multicultural View of Mathematical Ideas, 1st edn. 1991. 5th edn. Routledge, New York (2017). https://doi.org/10.1201/9780203756522

27. Menninger, K.: Number Words and Number Symbols: A Cultural History of Numbers. MIT Press, Cambridge (1969)

28. Stigler, J., Baranes, R.: Culture and mathematics learning. Rev. Res. Educ. **15**(1), 253–330 (1988). https://doi.org/10.3102/0091732X015001253

29. Rigney, L., Garrett, R., Curry, M., MacGill, B.: Culturally responsive pedagogy and mathematics through creative and body-based learning: urban aboriginal schooling. Educ. Urban Soc. **52**(8), 1159–1180 (2020). https://doi.org/10.1177/0013124519896861

30. Nur, A., Kartono, K., Zaenuri, Z., Waluya, S., Rochmad, R.: Ethnomathematics thought and its influence in mathematical learning. MaPan **8**(2), 205–223 (2020). https://doi.org/10.24252/mapan.2020v8n2a3

31. Yashin, B.L.: Ethnomathematics: socio-cultural and psycholinguistics aspects. Probl. Mod. Educ. **2**, 21–33 (2022). https://doi.org/10.31862/2218-8711-2022-2-21-33

32. Bishop, A.J.: The interactions of mathematics education with culture. Cult. Dyn. **1**(2), 145–157 (1988). https://doi.org/10.1177/092137408800100202

33. Rowlands, S., Carson, R.: Where would formal, academic mathematics stand in curriculum informed by Ethnomathematics? A critical review of Ethnomathematics. Educ. Stud. Math. **50**(1), 79–102 (2002). https://doi.org/10.1023/A:1020532926983

34. Surygin, A.I.: Fundamentals of Theory of Learning in a Non-native Language for Students, 1st edn, 1999, 2nd edn. Zlatoust, St. Petersburg (2000). https://elib.spbstu.ru/dl/1721.pdf/download/1721.pdf. Accessed 11 Sept 2022

35. Bochner, S.: Culture shock due to contact with unfamiliar cultures. Online Readings Psychol. Cult. **8**(1) (2003). https://doi.org/10.9707/2307-0919.1073

36. Bylieva, D., Lobatyuk, V.: Meanings and scripts in the linguistic landscape of Saint Petersburg. Open Linguist. **1**, 802–815. https://doi.org/10.1515/opli-2020-0180

37. Lobatyuk, V., Nam, T.: Everyday problems of international students in the Russian language environment. Tech. Lang **3**(3), 38–57 (2022). https://doi.org/10.48417/technolang.2022.03.04

38. Shipunova, O., Pozdeeva, E., Evseev, V., Romanenko, I., Gashkova, E.: University educational environment in the information exchange agents evaluations. In: Rocha, Á., Fajardo-Toro, C.H., Rodríguez, J.M.R. (eds.) Developments and Advances in Defense and Security. SIST, vol. 255, pp. 501–511. Springer, Singapore (2022). https://doi.org/10.1007/978-981-16-4884-7_42

39. Lin, C.: Culture shock and social support: an investigation of a Chinese student organization on a US campus. J. Intercult. Commun. Res. **35**(2), 117–137 (2006). https://doi.org/10.1080/17475750600909279

40. Bylieva, D., Krasnoshchekov, V., Lobatyuk, V., Rubtsova, A., Wang, L.: Digital solutions to the problems of Chinese students in St. Petersburg multilingual space. Int. J. Emerg. Technol. Learn. (iJET) **16**(22), 143–166 (2021). https://doi.org/10.3991/ijet.v16i22.25233

41. Schlabach, J., Hufeisen, B.: Plurilingual school and university curricula. Tech. Lang. **2**(2), 126–141 (2021). https://doi.org/10.48417/technolang.2021.02.12

42. Rubtsova, A.V., Almazova, N.I., Bylieva, D.S., Krylova, E.A.: Constructive model of multilingual education management in higher school. In: IOP Conference Series Materials Science and Engineering, vol. 940, p. 012132 (2020). https://doi.org/10.1088/1757-899X/940/1/012132

43. Csernicskó, I.: Bilingual education of minorities: always the best solution? In: Filipović, J., Vučo, J. (eds.) Minority Languages in Education and Language Learning: Challenges and New Perspectives, vol. 7, pp. 87–103. University of Belgrade, Belgrade (2017). https://doi.org/10.18485/fid.2017.7.ch5

44. May, S.: Bilingual education: what the research tells us. In: García, O., Lin, A.M.Y., May, S. (eds.) Bilingual and Multilingual Education. ELE, pp. 81–100. Springer, Cham (2017). https://doi.org/10.1007/978-3-319-02258-1_4

45. Wright, W.E., Baker, C.: Key concepts in bilingual education. In: García, O., Lin, A.M.Y., May, S. (eds.) Bilingual and Multilingual Education. ELE, pp. 65–79. Springer, Cham (2017). https://doi.org/10.1007/978-3-319-02258-1_2

46. Gao, X.A., Wang, W.: Bilingual education in the People's Republic of China. In: García, O., Lin, A.M.Y., May, S. (eds.) Bilingual and Multilingual Education. ELE, pp. 219–231. Springer, Cham (2017). https://doi.org/10.1007/978-3-319-02258-1_16

47. Yezhova, V.S.: Implementing and testing the model of efficient formation of the mathematical culture in future Math teachers. Sci. Search **2**(8), 16–19 (2013). http://sspu.ru/pdf/pages/journal/arhiv/2/2(8)%202013_v20210303165835.pdf

48. Kijkova, N.J.: The synergetic basis of mathematical culture model of future manager. Mod. Probl. Sci. Educ. **5** (2011). https://science-education.ru/ru/article/view?id=4812

49. Lyz', N.A., Lyz', A.E., Neshchadim, I.O.: The success of study at higher education institution: the engineering students' experience. Mod. Probl. Sci. Educ. **4** (2022). https://doi.org/10.17513/spno.31915

50. Stevenson, H.W., Hofer, B.K., Randel, B.: Mathematics achievement and attitudes about mathematics in China and the West. J. Psychol. Chinese Soc. **1**(1), 1–16 (2000). https://psycnet.apa.org/record/2001-11761-001

51. Wang, J.: How Chinese learn mathematics: perspectives from insiders. Front. Educ. China **10**, 495–498 (2015). https://doi.org/10.1007/BF03397082

52. Norton, S.J., Qinqiong Z.: Chinese students' engagement with mathematics learning. Int. J. Math. Teach. Learn. **14**(December 10), 1–24 (2013). https://www.cimt.org.uk/journal/norton2.pdf

53. Zheng, Y.: Mathematics education in china: from a cultural perspective. In: Leung, F.K.S., Graf, K.D., Lopez-Real, F.J. (eds.) Mathematics Education in Different Cultural Traditions-A Comparative Study of East Asia and the West, vol. 9, pp. 381–390. Springer, Boston (2006). https://doi.org/10.1007/0-387-29723-5_21

54. Martzloff, J.-C.: A History of Chinese Mathematics, 2nd edn. Springer, Heidelberg (2006)

55. Zhou, M., Gao, L.: A comparative study of chinese and western mathematical thoughts. In: Proceedings of the 2018 International Workshop on Education Reform and Social Sciences, Zhengzhou, China. Atlantis Press China (2019). https://doi.org/10.2991/erss-18.2019.120

56. Kashyap, R., Isaak, A., Kim, K.-H., Chia-Hsing, H.: The mathematics of symbols: a closer confrontation of the Chinese cultural command over calculations. SSRN Electron. J. (2017). https://doi.org/10.2139/ssrn.3060522

57. Antonova, Y.A.: National-psychological portrait of a Chinese student studying Russian language. Philol. Class **27**(2), 161–171 (2022). https://cyberleninka.ru/article/n/natsionalno-psihologicheskiy-portret-kitayskogo-studenta-izuchayuschego-russkiy-yazyk. Accessed 28 Sept 2022

Developing Digital Identity Management Skills Among University Students

Violetta V. Petrova(✉) ⓘD

Saint-Petersburg branch of the Financial University under the Government of the Russian Federation, Saint-Petersburg 197198, Russian Federation
violettap1@mail.ru

Abstract. Digital literacy is an umbrella term that covers a lot of competences or skills to be taught to those who are to function successfully in the multilingual digital world of the 21st century. Some competences such as information and data literacy, safety or problem solving have been paid a lot of attention to by educators. Other competences such as communication and collaboration have frequently been left outside the educational framework. The paper focuses on the analysis of formation of digital identity management skills as part of the communication and collaboration competence among university and college students. It introduces the concept of digital reputation, considers factors that improve or impede digital communication as compared to real life communication both in a monolingual and a multilingual society and elaborates on the idea that digital perception of communication partners by each other depends on their use of registration names, e-mail addresses and avatars. After analyzing students' use of those and carrying out experiments on the impact of names, addresses and avatars on the choice of communication partners, the author comes to the conclusion that identity management skills are not sufficiently developed and suggests teaching them in educational institutions in line with other digital literacy skills.

Keywords: Digital literacy · Digital identity management · Digital reputation · Digital communication

1 Introduction

Every educator is familiar with the concept of literacy - the ability to read and write as dictionaries define it [1]. The notion of functional literacy, accepted by many researchers, suggests that a literate person is able to engage in all the activities in which literacy is required for effective functioning of him/her or their group or community [2]. It is quite obvious that a person who is illiterate, who cannot read or write, will have difficulty in performing even ordinary daily tasks like paying for groceries or checking the bus timetable at the nearest bus stop. The chances to prosper in life will also be limited for such a person. It's impossible to get any professional training, fill in a job application form and find a high-paying job without the ability to read and write. It's difficult to get access to financial and banking services and set up and run one's own business. It's

D. Bylieva and A. Nordmann (Eds.): PCSF 2022, LNNS 636, pp. 283–294, 2023.
https://doi.org/10.1007/978-3-031-26783-3_24

difficult to participate in the community life or get social security or community benefits. It's impossible to get quality medical assistance as an illiterate person is unlikely to be able to apply for and obtain a medical insurance policy which poses a threat to a physical survival of an illiterate individual. UNESCO experts proposed defining literacy as the "ability to identify, understand, interpret, create, communicate and compute, using printed and written materials associated with varying contexts" [3, p. 13]. This concept of literacy reflected the meaning of literacy that tackled the problems of the 20th century.

The definition of literacy has evolved in the 21st century. One of the most important features of the period of the end of the 20th - the beginning of the 21st centuries is globalization which affects all the spheres of social life - telecommunication, technical and scientific cooperation, intercultural exchange, and the formation of the virtual environment of digital communication which provides opportunities for every person from any location in the world to participate in the international information exchange [4]. Technological progress, which is the main driver of globalization, has resulted in increasing the integration of various societies and communities across the globe. It is believed that very soon almost every adult or child will own a smart mobile device offering instant access to everything that has been digitized and is available on the World Wide Web [5]. Also, the possibilities of communicating with people speaking other languages and belonging to other cultures will become limitless [4]. These social, cultural and technological changes have transformed the world into a multilingual digital society - the society in which language contact occurs between two or more languages [6] and which has been formed as a result of adaptation and integration of advanced technologies into all spheres of social life and culture [7], - and extended the notion of literacy/illiteracy. Therefore, the new concept of literacy of the 21st century should integrate the multilingual domain and the digital domain which in many ways have become interdependent and interconnected. As a result, in the 21st century a person belonging to any type of linguistic community has to think about becoming digitally literate to be able to engage in various activities and to function successfully in a multilingual digital community.

The term "digital literacy" was invented in 1997 by an American writer and journalist Paul Gilster who believed that digital literacy was the ability to understand and use information presented in a large variety of formats and sources with the help of a computer. He thought that dealing with the Internet changes patterns of human behaviour, ways of searching information, ways of producing and sharing information, ways of communicating with other people [8]. In Gilster's opinion, digital literacy involves such skills as media literacy - the ability to access, critically evaluate, create and act using all forms of media, information literacy - the ability to search for the necessary information and the ability to use information instruments, communicative literacy - the ability to communicate with other Internet users, creative literacy - the ability to create information in different forms and formats.

Since 1997, as society has progressed in spreading technology into all spheres of life, turning digital transformation into a key factor of economic growth [9] and making technological achievements available to the majority of people, the concept of digital literacy (or digital competence) has been elaborated, clarified and refined to fit various contexts of the digital society [10–13]. Researchers point out that this concept is frequently mentioned in public discussions and there are many diverse definitions of it used in research reports and policy documents [14]. Many researchers agree that the concept of digital literacy is an umbrella concept that embraces a large number of elements - competences or skills. These elements are identified on the basis of the analysis of what is involved in being "digitally literate" and range from purely technical ones (being able to create a new folder or being able to identify the number of bytes in a megabyte) to sociocultural ones [15].

In our paper we would adhere to the definition proposed by UNESCO in A Global Framework of Reference on Digital Literacy Skills for Indicator 4.4.2: "Digital literacy is the ability to access, manage, understand, integrate, communicate, evaluate and create information safely and appropriately through digital technologies for employment, decent jobs and entrepreneurship. It includes competences that are variously referred to as computer literacy, ICT literacy, information literacy and media literacy" [16, p. 6]. Competences identified by UNESCO as elements of digital literacy are presented in the Table 1 below.

Table 1. Digital literacy competence framework (by DigComp 2.0, UNESCO)

Competence area	Competences
1. Information and data literacy	1.1 Browsing, searching and filtering data, information and digital content 1.2 Evaluating data, information and digital content 1.3 Managing data, information and digital content
2. Communication and collaboration	2.1 Interacting through digital technologies 2.2 Sharing through digital technologies 2.3 Engaging in citizenship through digital technologies 2.4 Collaborating through digital technologies 2.5 Netiquette 2.6 Managing digital identity
3. Digital content creation	3.1 Developing digital content 3.2 Integrating and re-elaborating digital content 3.3 Copyright and licenses 3.4 Programming
4. Safety	4.1 Protecting devices 4.2 Protecting personal data and privacy 4.3 Protecting health and well-being 4.4 Protecting the environment
5. Problem solving	5.1 Solving technical problems 5.2 Identifying needs and technological responses 5.3 Creatively using digital technologies 5.4 Identifying digital competence gaps

The importance of teaching digital literacy skills has been recognized all over the world. In 2013 UNESCO developed the Global Media and Information Literacy (MIL) Assessment Framework that provides a conceptual and theoretical framework for MIL and introduces the rationale and methodology for conducting an assessment of country readiness and existing competences on MIL at the national level. This Framework also includes practical steps for adaptation of its recommendations at national level. Special emphasis is given to the MIL competences assessment of teachers in service and their training [17, 18].

The Government of the Russian Federation adopted "The main directions in development of activities of the Government of the Russian Federation for the period until 2024" on 29 September 2018 which emphasize the importance of education as the main resource of innovative technological and socio-economic development. In August 2017 the programme "Digital economy" was adopted. The Government has also developed a document, "The Professional Standard" [19], which sets up professional responsibilities of employees of educational institutions. This document makes it compulsory for employees of educational institutions to use digital educational resources and digital equipment, to develop methods and materials for distance and e-learning.

According to research data [20–23], in Russia educators are more advanced in using digital devices than the majority of the population (88% against 52%). But their positive attitude to using technological innovations is lower than their proficiency level (78%). The research report by the National Agency for Financial Research (NAFI) of the Russian Federation, which carried out a series of research on the level of digital literacy of the population based on the methodology of DigComp, showed that only about 27% of the Russians have a high level of digital literacy. The index of digital literacy of the Russians in the first quarter of 2020 was estimated to be 58%. According to the report, about a quarter of people (24%) understand the significance of digital competences for their careers and consider constant updating of digital skills important for the purpose of keeping their jobs [24].

Research also shows that development of some competences within Digital Literacy Competence Framework is paid more attention to. Both in Russia and all over the world the development of competences related to information and data literacy, safety or problem solving are focused on [21, 25–27], but communication and collaboration skills are frequently left outside the educational framework [28, 29]. As a result, there is no understanding of appropriate digital behavior and factors that may affect digital communication and collaboration both in a monolingual and a multilingual community.

Real life communication can be affected by a number of factors. Some factors influencing communication are related to the complexity of the communication process itself. Psychologists believe that communication involves many aspects such as information exchange, interaction between people and perception - the way people see and understand each other. Needless to say, social perception affects other communication aspects and can improve or impede them. Communication can be verbal and non-verbal. Researchers noted that the total impact of a message is about 7% verbal, about 38% vocal and 55% non-verbal; a similar study showed that the verbal component of a face-to-face conversation is less than 35% and over 65% of communication is done non-verbally [30, 31]. Researchers agree that the verbal channel of communication is used for conveying information, while a non-verbal channel is used for negotiating interpersonal attitudes.

Communication can also be affected by external factors - factors not depending on the communicator. One of the most important ones is the personality and the image of a communication partner [32, 33]. If the communicator appreciates the appearance and other personal features of their communication partner, the communication process is more likely to be successful. Very often a communication partner is judged by their reputation which is defined as somebody's opinion about a person, their advantages and disadvantages, their lifestyle and behaviour. This opinion is based on the past experience of both the communicator and the communication partner and adds to the expectations of the communication partner's behaviour for the future [34].

We believe that digital communication involves the same aspects - information exchange, interaction and perception. Consequently, information exchange and interaction can be affected by perception - by the way people see and understand each other in the digital world. While we may not be present in the digital world in person, our digital image will always be there. Opinion about us will be formed on the basis of everything we put online.

Digital or online reputation is the opinion about a person, their advantages and disadvantages, their lifestyle and behaviour based on the search results. It has been noticed that people are placing more and more trust in search results regardless of what exists in the real world [35]. According to the Pew Research Center, 91% of people trust what they see and read in search engine results, and the Edelman Trust Barometer reports that 65% of people trust these results more than any other source.

2 Methods and Materials

Nowadays, people leave a lot of footprints in the digital world - images and comments in the social networks, stories in blogs, dialogues in chatrooms etc. People can make up their names to be used in the digital world - names under which they register their e-mail accounts and e-mail addresses for their e-mail accounts. People can choose the way they are seen by their digital communication partners - they choose pictures for their avatars which reflect their inner selves at a given moment or their attitudes to the communication process [34]. Those can be formal or informal pictures of communicators or other images not related to communicators' appearances. All these footprints affect the digital perception of communicators by their communication partners and form their digital reputation. Since the rules of digital behavior haven't been developed, the ideas of how one can be seen in a digital world and how this can affect digital communication may not have come to minds of the majority of digital communicators.

We aim to find out how well students' digital identity management skills are developed. For that, we analyze the students' e-mail addresses from which we regularly receive class and home assignments, names under which students register their email accounts and students' avatars. In our study we also make an attempt to describe whether the students understand the concept of a digital reputation - how they are seen in a digital world by other people, what are the constituent parts of their digital reputation, what impression these constituent parts make on the on-lookers and what digital skills can be taught to students to improve their digital reputation.

2.1   Research Methods

We employed various research methods at the Saint-Petersburg branch of the Financial University under the Government of the Russian Federation and the Department of secondary professional training of the Saint-Petersburg branch of the Financial University under the Government of the Russian Federation (college). In 2019–2022 we used the existing data research method to collect, classify and analyse e-mail addresses, registration names and avatars used by the students in communication with the University administrative and teaching bodies.

We also used the method of survey to get information about the students' e-mail accounts, about the origin and purpose of establishing those e-mail accounts, about the students' opinions about various types of e-mail addresses, various pieces of writing and images. The method of survey was also used to get information about the students' awareness of factors affecting their digital reputation and communication.

In addition, we used the method of experiment to find out whether students have any knowledge about the difference between various types of writing and images suitable for various target audiences.

2.2   Research Questions

How much do students know about the concept of digital reputation?
How much do students know about the factors that may affect their digital communication?
Do students judge their communication partners on the basis of their registration names and e-mail addresses?
How do images chosen for avatars affect the way students see their communication partners?

3   Results

In this part of the paper, first we present the descriptive statistics of the origin of the students' e-mail addresses and of our analysis of the students' registration names, e-mail addresses and avatars. We also present students' explanations on the choice of their registration names, e-mail addresses and avatars. Second, we present the results of the survey carried out among the students about the concept of digital reputation and other the factors that may affect their digital communication. Third, we present the results of surveys and experiments carried out among the students on their judgements of possible communication partners by their registration names, e-mail addresses and avatars.

3.1   Descriptive Statistics of the Students' Registration Names, E-mail Accounts and Avatars

The Origin of E-mail Addresses

Students are free to choose an e-mail account from which they send their class and home assignments to the teachers or letters to various university bodies. We found out that 95%

of students have, at least, 1 e-mail account. 5% of students admitted using somebody else's accounts. These e-mail accounts were set up for the purpose of communicating with friends (15%), when joining a social network (22%), when getting loyalty cards from some shops (39%), when joining a club or a society (17%) or when subscribing for a service (7%).

27% of students have more than 1 e-mail account. 33,5% of students, who have more than 1 e-mail account, set up an account when they entered the University or college and found out that some of their communication with the university administrative bodies and teachers was to be carried out by e-mail. Other students believed that they didn't need another e-mail account to communicate with the University bodies. Besides these individual e-mail accounts, students have to set up a group e-mail account to communicate with the University administrative bodies and teachers (information about the changes in their timetables is sent to a group's e-mail account by the dean's office, homework and reminders about deadlines for projects and course papers can be sent by teachers). Registration names, e-mail addresses and avatars for group e-mail accounts are chosen by all group members.

Descriptive Statistics of the Students' E-mail Addresses
E-mail addresses range from a combination of a student's first and last name (i.e. pavlovaalexandra@) (24,5%), names (either first names or last names) combined with other symbols (i.e. serezhka73judo@) (43%) to a set of symbols not related to either first or second names (i.e. trampampuz@, boxathome@, admininfield@, ?????????@, finyashka18@) (32,5%).

E-mail addresses for the accounts created after their owners entered the University or the College were mostly a combination of a student's first and last name (90%), e-mail addresses created before their owners entered the University or the College were mostly names combined with other symbols or sets of symbols not related to their owners' names.

Descriptive Statistics of the Names Under Which the Students Register Their E-mail Accounts
75% of students register all their e-mail accounts using their own first and last names. About 18% of students use only one of their names, often their deminutive names to register their accounts (i.e. Lizzavetta instead of Elizaveta Nefedova; Am Am instead of Amalunga Menkeeva). About 7% of students use names, phrases or sets of symbols not related to their own names (i.e. Don't disturb!; Paavy vya vay vya; Vezunchik Vezunchikov).

Descriptive Statistics of Students' Avatars
Only about 50% of students use pictures with images for their e-mail accounts. Of those pictures about 70% are images of students and 30% are images not related to students' appearances. Images of students can be classified as formal (35%) and informal (65%). Images not related to students' appearances are either cartoon characters, animals or pictures of other people.

Explanations on the Choice of Registration Names, E-mail Addresses and Avatars
Students who registered their e-mail accounts under their own names and whose e-mail addresses correspond to their names explained their choice by the necessity to be

recognized by their addressees and by their wish to look more serious, professional and reliable before their addressees. Students who registered their e-mail accounts under other names and whose e-mail addresses do not correspond to their names or partly correspond to their names explained their choice by the desire to look cool or funny (about 55%), by the desire to shock the others (about 16%), by not caring about the meaning of their addresses or names at all (10%), by doing something their parents or friends suggested (10%) or by borrowing someone else's e-mail account and not being able to change anything (9%).

Students who don't use avatars with their e-mail accounts explained it by the fact they haven't found a proper picture (43%), they never pay attention to the others' avatar and don't expect others' to pay attention to theirs (31%), not wanting to be seen (15%), not caring about avatars at all (11%).

Students who have formal pictures for their avatars explained their choice by the desire to look more serious, professional and reliable for all their addressees. Students who have informal pictures for their avatars explained their choice by the fact that they often change pictures and the chosen picture has a particular meaning for a moment in question (56%), that the picture on their avatar is their favorite picture (22%), that they didn't have a better picture (15%) or they didn't care what picture to use (7%).

Students who chose pictures not related to their appearance explained their choice by the fact that the picture showed their true appearance or feelings (34%), the picture reflected their current attitude to the world (27%), they picture portrayed someone or something they love (25%), they wanted others to feel confused (8%) and they didn't care about what the others might think (6%).

3.2 Students' Awareness of Their Digital Reputation

We found out that students are not familiar with the concept of digital reputation although they all know what reputation is and how important it is in the real world.

We also found out that about 70% of students are unaware of the fact that something may affect the way they are seen in the digital world. They find it highly unlikely that anyone would ever care to analyse everything they have ever put on the Internet except for the addressees of their posts. They understand that the addressees of their posts may react differently to what they read or see but, if misunderstanding occurs, the situation can be rectified in person.

Students also admit that they rarely pay attention to e-mail addresses from which they receive e-mail letters. However, they pay attention to the registration names (100%) and avatars (80%). They are happy to get letters from their family, friends or any other people they know from any e-mail address, registered under any name and accompanied by any avatar. But they would not be so happy to get such e-mails from strangers (people or organizations).

Students also believe that they can never be punished for inappropriate digital behaviour. Inappropriate digital behaviour is understood as writing inappropriate letters to addressees - the content or the style is not appropriate for a particular addressee (75%), sending inappropriate images to addressees - the content or the style is not appropriate for a particular addressee (60%), uploading inappropriate content on social networks (65%).

About 20% of students understand that their posts can be seen by anyone and action can be taken on the basis of their posts as they have heard of cases when people were punished (fired) for inappropriate posts. But, as they believe they are of no importance to anyone who can or has reasons to be engaged in reviewing and analysing their posts, their digital footprints (everything they put on the Internet) are highly unlikely to be scrutinized.

Only about 10% of students are careful about anything they put on the Internet as they believe that their digital behaviour affects their image in just the same way their real behaviour does.

Students' Judgements of Possible Communication Partners by Their Registration Names and E-mail Addresses

When exposed to a set of registration names and e-mail addresses, students were able to classify them as being serious, reliable and professional (the addresses that correspond to the owners' names) and silly, funny, strange, stupid (the addresses that do not correspond to the owners' names). It should also be noted that if the students were familiar with the owners of strange e-mail addresses or registration names, their responses were rather positive than negative (funny, silly, crazy). If the students were not familiar with the owners of strange e-mail addresses, their responses were mostly negative and much harsher (stupid, idiotic, insane).

The Impact of Images on the Digital Reputation

We found out that the majority of students understand the importance of images published in social networks. There is a clear division between formal photographs and informal photographs. Moreover, students believe that people who have a formal status, like University officials and teachers or politicians, are supposed to publish only formal or neutral photographs in social networks. Students regularly check accounts of their teachers, view all the photographs published there and may complain about inappropriate (in their opinions) photographs of teachers they find to other teachers.

As students attribute an informal status to themselves, they believe they can publish any photographs and pictures. They also use Photoshop techniques to modify their pictures and photographs. In a photograph with a teacher, they can modify their own images but leave their teacher's image untouched (this is true, at least, for photographs that can be accessed by the teachers as well as by the other students).

The Impact of Avatars on the Choice of Communication Partners

The last series of our experiment was a simulation: students were exposed to a set of specially made letters sent from certain e-mail addresses registered under different names and accompanied by various avatars. The avatars chosen were blank pictures, formal pictures, informal pictures and images not related to human appearance. The students were requested to choose a) candidates for a job interview and b) volunteers for a community project.

When choosing candidates for a job interview, 85% of students opted for letters sent from accounts with formal pictures on their avatars and 15% of students opted for letters sent from accounts with informal pictures on their avatars. When choosing volunteers for a community project, about 53% of students opted for letters sent from accounts with formal pictures on their avatars, 39% of students opted for letters sent from accounts with informal pictures on their avatars, 8% of students opted for letters sent from accounts with blank avatars. Such choices were explained by the necessity to see people the students will have to work with. A more formal situation (a job interview) suggested more formal pictures, a less formal situation (a volunteer project) suggested a lower level of formality.

4   Conclusions

Our studies show that students are not as well prepared to act in the digital environment as in the real life environment. They do not consider the digital environment to be part of their real life environment that is why the majority of them (90%) are not careful about what they upload on the Internet and do not believe that their digital behaviour can affect their real life image or have real life consequences. The rules of behaviour in the digital environment have not been formalized and codified as compared to the rules of behaviour in the real life environment. So, they are difficult to express and to apply for the majority of digital users. Therefore, some students do not follow any rules while other students borrow some rules of behaviour from the real life environment and transfer them to the digital environment. However, these rules exist in the passive form - the majority of students believe that they are exempt from following them and apply them only when judging other people.

Students are also unaware of the concept of digital reputation and digital footprints and its difference from the real life reputation. Therefore, they lack digital identity management skills.

5   Discussions

Students arrive at a university campus and have to act both in a real life university environment and in a digital university environment - they have to join digital platforms to do some courses, they may have to send their homework to the teachers and get feedback from them, they may have to request information from the dean's office, send applications for the participation in various educational and extracurriculum activities or even job applications for some vacancies available on campus. Sometimes, students may want to send applications for jobs or participation in events outside the campus. They may also be invited to participate in community or educational projects organized by domestic or international commercial or non-commercial organizations or compete for grants provided by them. Nowadays, all these things are done through digital technologies. So, engaging in on-campus and off-campus activities requires one to be digitally literate - to have a high level of communication and collaboration competence in general and digital identity management skills in particular. However, students lack knowledge and skills needed to communicate and collaborate successfully with other people in the digital world. Implicit knowledge of how things should be done is not sufficient for proper skill development.

We suggest that communication and collaboration skills and identity management skills should not be left outside the educational framework. Educational institutions should look for ways of developing communication and collaboration skills among students - formalizing and codifying rules of digital behaviour, developing courses on building digital reputation of a person/a company, developing courses on successful digital communication etc. At the same time we want to note that all the digital literacy skills outlined in this paper have to be taught simultaneously. More research should be done on assessment frameworks design for these skills and on the methodology for developing them.

References

1. Literacy. In: The Dictionary by Merriam-Webster. https://www.merriam-webster.com/dictionary/literate#h1. Accessed 1 Oct 2022
2. Keefe, E.B., Copeland, S.R.: What is literacy? The power of a definition. Res. Pract. Pers. Sev. Disabil. **36**(3–4), 92–99 (2011). https://doi.org/10.2511/027494811800824507
3. UNESCO Education Sector Position Paper: The Plurality of Literacy and Its Implications for Policies and Programs. UNESCO Education Sector Position Paper, Paris (2004). https://unesdoc.unesco.org/ark:/48223/pf0000136246. Accessed 10 Oct 2022
4. Drobot, G.A.: Globalization: the term, the phases, the contradictions and the opinions. Soc.-Hum. Knowl. **2**, 105–127 (2008)
5. Wolf, M.: Shaping globalization. Finance Dev. **51**(3), 22–25 (2014)
6. Valdes, G.: Multilingualism. Linguistic Society of America (2021). https://www.linguisticsociety.org/resource/multilingualism. Accessed 1 Oct 2022
7. Paul, P.K., Aithal, S.: Digital society: it's foundation and towards an interdisciplinary field. In: National Conference on Developments in Information Technology, Management, Social Sciences and Education, pp. 1–6 (2018). https://www.researchgate.net/publication/330025821. Accessed 10 Oct 2022
8. Pool, C.R.: A new digital literacy: a conversation with Paul Gilster. Educ. Leadersh. **55**(3), 6–11 (1997)
9. Rassadnev, E.S., Osipenko, A.A., Lubeyankov, A.S.: Digital literacy of the population as a factor in the development of the digital economy in Russia. Perm Univ. Bull. Math. Mech. Comput. Sci. **1**(52), 75–80 (2021). https://doi.org/10.17072/1993-0550-2021-1-75-80
10. Pozdeeva, E., et al.: Assessment of online environment and digital footprint functions in higher education analytics. Educ. Sci. **11**(6), 256 (2021). https://doi.org/10.3390/educsci11060256
11. Aurora, S.: Natural language as a technological tool. Technol. Lang. **2**(2), 86–95 (2021). https://doi.org/10.48417/technolang.2021.02.08
12. Pozdeeva, E., Shipunova, O., Evseeva, L., Kulsariyeva, A.: Systems analysis of the digital agent's role in hybrid social interaction forms. In: Vasiliev, Y.S., Pankratova, N.D., Volkova, V.N., Shipunova, O.D., Lyabakh, N.N. (eds.) System Analysis in Engineering and Control. SAEC 2021. LNNS, pp. 153–165. Springer, Cham (2022). https://doi.org/10.1007/978-3-030-98832-6_14
13. von Xylander, C.: Quipping equipment: apropos of robots and Kantian Chatbots. Technol. Lang. **3**(1), 82–103 (2022). https://doi.org/10.48417/technolang.2022.01.09
14. Spante, M., Hashemi, S.S., Lundin, M., Algers, A.: Digital competence and digital literacy in higher education research: systematic review of concept use. Cogent Educ. **5**(1), 1519143 (2018). https://doi.org/10.1080/2331186X.2018.1519143
15. Lankshear, C., Knobel, M.: Digital literacy and digital literacies: policy, pedagogy and research considerations for education. Nord. J. Digit. Lit. **1**(1), 8–20 (2006). https://doi.org/10.18261/ISSN1891-943X-2006-01-03

16. Law, N., Woo, D., Torre, J., Wong, G.: A global framework of reference on digital literacy skills for indicator 4.4.2. Information Paper, 51. UNESCO Institute for Statistics, Quebec (2018). https://uis.unesco.org/sites/default/files/documents/ip51-global-framework-reference-digital-literacy-skills-2018-en.pdf. Accessed 17 Oct 2022
17. UNESCO CIS: Global Media and Information Literacy Assessment Framework: Country Readiness and Competencies. UNESCO, Paris (2013). https://unesdoc.unesco.org/ark:/48223/pf0000224655. Accessed 19 Sept 2022
18. UNESCO UNAOC: Core Teacher Competencies. Media Information Literacy for Teachers. http://www.unesco.mil-for-teachers.unaoc.org. Accessed 19 Sept 2022
19. Ministry of Labour and Social Security: Professional Standard "Pedagogue of Professional Training, Professional Education and Additional Professional Education" (2015). http://fgosvo.ru/01.004.pdf. Accessed 8 Oct 2022
20. Aimaletdinov, T.A., Baimuratova, L.R., Zaitzeva, O.A., Imamaeva, G.R., Spiridonova, L.V.: Digital Literacy of Russian Pedagogues. Readiness to Use Digital Technologies in Teaching. NAFI, Moscow (2019)
21. Gaisina, S.V.: Digital Literacy and Digital Educational Environment in School. Karo, Saint-Petersburg (2018)
22. Bylieva, D., Moccozet, L.: Messengers and chats – technologies of learning. Technol. Lang. **2**, 75–88 (2021). https://doi.org/10.48417/technolang.2021.03.06
23. Evseeva, L.I., Shipunova, O.D., Pozdeeva, E.G., Trostinskaya, I.R., Evseev, V.V.: Digital learning as a factor of professional competitive growth. In: Antipova, T., Rocha, Á. (eds.) DSIC 2019. AISC, vol. 1114, pp. 241–251. Springer, Cham (2020). https://doi.org/10.1007/978-3-030-37737-3_22
24. NAFI: Digital literacy of the Russian: 2020 research. https://nafi.ru/analytics/tsifrovaya-gramotnost-rossiyan-issledovanie-2020/?ysclid=laqh5y89f667005177. Assessed 11 Oct 2022
25. Hobbs, R.: Digital and Media Literacy: A Plan of Action. The Aspen Institute, Washington (2010)
26. Luque, S.G., Becerra, T.D., Abengozar, A.E., Simon, I.M.V.: MIL competences: from theory to practice. Measuring citizens' competences on media and information literacy. In: eLearning Papers 38. Barcelona (2014) http://wwwopeneducationeuropa.eu/en/elearning_papers. Accessed 9 Sept 2022
27. Perez-Escoda, A., Rodriguez-Conde, J.: Digital literacy and digital competences in the educational evaluation: USA and IEA contexts. In: TEEM 2015 Conference, Oporto (2015). https://doi.org/10.1145/2808580.2808633. Accessed 7 Sept 2022
28. Bylieva, D., Hong, J.-C., Lobatyuk, V., Nam, T.: Self-regulation in e-learning environment. Educ. Sci. **11**, 785 (2021). https://doi.org/10.3390/educsci11120785
29. Krylov, E., Khalyapina, L., Nordmann, A.: Teaching English as a language for mechanical engineering. Tech. Lang. **2**(4), 126–143 (2021). https://doi.org/10.48417/technolang.2021.04.08
30. Pease, A.: The Body Language. Sheldon Press, London (1981)
31. Pease, A., Pease, B.: The Definitive Book of Body Language. Pease International, Buderim (2006)
32. Nemov, R.S.: Psychology, vol. 1. Vlados, Moscow (2004)
33. Ilyin, E.P.: Psychology of Communication and Interpersonal Relationship. Piter, Saint-Petersburg (2013)
34. Voiskunskiy, A.E.: Social perception in social networks. Psychology **3**, 90–104 (2014)
35. Donnelly, T.: Why your digital reputation matters and how to influence it. Forbes Agency Council (2018). https://www.forbes.com/sites/forbesagencycouncil/2018/05/07/why-your-digital-reputation-matters-and-how-to-influence-it/?sh=d4fc4d449a59. Accessed 20 Sept 2022

Foreign Language Anxiety in Multilingual Students Before and After Corona-Lockdown: A Comparative Analysis

Fatima Valieva[(⊠)] [iD]

Peter the Great St. Petersburg Polytechnic University, Polytechnicheskaya, 29, St. Petersburg 195251, Russia
jf.fairways@mail.ru

Abstract. The study is devoted to linguistic anxiety of multilingual students in the conditions of adaptation to foreign-language educational environment. The work was carried out in two stages covering the period of study before COVID and after COVID. The main goal was to reveal structural and contents differences of the phenomenon of linguistic anxiety depending on changes in educational conditions and general psychological atmosphere in Lockdown conditions. A total of 120 foreign students studying humanities were interviewed. A comparative analysis of the data obtained showed that the samples did not differ significantly in terms of the overall index of linguistic anxiety and resilience. However, there were differences in the basic constructs. Thus, bilingual students before Corona-Lockdown had the highest test anxiety score. Anxiety about communication apprehension was lower. The situation changed after Corona-Lockdown. Multilingual respondents revealed the maximum anxiety concerning live communication and the minimum concerning testing. Foreign language anxiety and resilience appeared to correlate. Factor analysis of the data for both samples revealed differences in the number of components and their compounds. In the first sample, the components with a communicative orientation prevailed, while in the second sample, the components with an evaluation-test orientation dominated. The study showed the variability of the components that make up the basic subconstructs of linguistic anxiety. The influence of individual cognitive characteristics and styles on linguistic anxiety was suggested.

Keywords: Foreign-language anxiety · FLCAS · Multilinguals · Resilience

1 Introduction

The changes that have occurred in the education system of most countries due to the spread of coronavirus infection are extensively discussed in public speeches, articles, and monographs by education researchers and specialists. This is primarily due to significant changes both in the structure and content of education and in teaching methods at all levels of the pedagogical system.

Against the background of actively debated methodological aspects, of particular interest are the issues related to the impact of the period of distance learning during the pandemic on the individual-psychological characteristics of the students. Such a long forced distant format, as the results of the surveys reveal, caused an ambiguous reaction from all participants in the educational process. Many people associate teaching and learning during the COVID period with various kinds of difficulties and problems, which resulted in stressful states, decreased stamina and attention, reduced emotions and communication skills [1, 2].

Of particular interest against the background of post-COVID changes is a complex and contradictory psychological formation - anxiety. There are different approaches to the interpretation of the causes, nature, function, structural components of the phenomenon of anxiety.

Causes of anxiety can be both objective and subjective. In various sources, anxiety is usually attributed an adaptive function [3], warning and preparing the person for certain difficulties and stressors. Anxiety tends to be associated with the anticipation of difficulties, even threats, it promotes the appearance of helplessness, fear, significantly reduces a person's cognitive and emotional resources.

In recent years, the concept of foreign language anxiety (FLA) has attracted much attention from many scholars and theorists in the fields of psycholinguistics, foreign language studies, and communication studies [4]. The FLA concept is believed to be related to the individual's affective filter, developed in the works of Krashen [5, 6]. The scholar believed that comprehensible input material and the state of the affective filter that regulates it lead to the second language (L2) assimilation. Krashen believed that this filter was related to a number of affective indicators, such as motivation, attitude, self-confidence, and anxiety.

In contrast to the above, MacIntyre (1998) defined FLA as the individual emotional experience of feeling negatively as a reaction to the process of learning a foreign language [7]. Emphasizing specific learning contexts, Horwitz et al. [8] described FLA as an individual set of feelings, perceptions, beliefs, and behaviors that emerge at different stages of foreign language learning and use. In our opinion, the question about the differences in the linguistic anxiety experiencing among foreign students who belong to multilingual and bilingual groups is also interesting. Practical experience proves that knowledge of several languages and their use can influence the manifestation of foreign language anxiety in students. Such knowledge together with certain individual-psychological characteristics can both contribute to FLA and reduce it.

2   Literature Review

Linguistic anxiety is a special complex of beliefs, feelings, self-esteem and way of action that manifests itself when learning L2 in a learning situation and is related to the specificity of this process. The structural and content aspect of linguistic anxiety has been studied by many specialists in the field of linguistics, linguodidactics, psycholinguistics. Noteworthy is the work of Park (2014), which presented an in-depth analysis of research on the main factors and components of linguistic anxiety [4]. In different years scientists distinguished different components of the basic constructs of the mentioned phenomenon: fear of communication, fear of negative evaluation and test anxiety. The components in different ratios and combinations were analyzed in the works of Aida

(1994), Matsuda and Gobel (2004), Liu and Jackson (2008), Yashima et al. (2009), Mack (2011), and others [9–13].

As a result of empirical research, some specialists [9, 14, 15] have outlined that language anxiety is a form of situational anxiety and should not be considered as a character trait or a condition that causes emotional tension only in a certain moment. Proponents of the second point of view believe that emotionally unstable individuals tend to experience linguistic anxiety. This condition can arise as a reaction to a wide range of emotional factors, in particular when learning a foreign language. It is the predisposition of the subject to experience strong psychological tension that explains the stupor that occurs as a result of linguistic anxiety in the process of mastering a foreign language, even in gifted and capable students.

Some experts identify prerequisites of foreign language anxiety in connection with the personality features of the subject of the learning process: low self-esteem, intro-version, bashfulness, exaggerated level of competitiveness, fear of possible loss of the sense of their own uniqueness, high level of fear of making mistakes.

Other authors attribute the emergence of anxiety in the process of L2 acquisition to insufficiently formed language competence in L2: lack of lexical, grammatical and intercultural knowledge, problems with the production of spontaneous speech [16].

The third group of researchers point out that the learning process is often fixed on the fight against mistakes, which makes students even with high L2 proficiency perceive any mistake as a failure.

A new approach can be traced in the works of D.B. Nikulicheva (2009), who asso-ciates language anxiety with negative cognitive programs in students: generalization - active generalization on the basis of single negative facts; polarization - an extremely negative or positive interpretation of events; dramatization - when a situation is projected in gloomy colors; perfectionism; projection - inventing the attitude of others toward one-self; emotional confirmation of one's negative qualities in any problem situation, etc. [17].

Summarizing the publications on the topic of the study, it should be noted that mostly the authors drew attention to the negative aspects of the influence of FLA on the processes of learning L2. However, it should be recognized that to a greater or lesser degree anxiety is inherent in every person. A non-substantial level of anxiety acts as a mobilizer and promotes goal attainment. The current trend in the study of FLA reflects the interest of specialists in recognizing the positive influence of emotions on L2 learning. It has been argued that positive thinking psychology can be an important positive factor in successful learning [18] as it aims to activate students' strengths and self-regulate learning [19, 20].

Studies of FLA have acquired a massive experimental character. Thus, the following have been studied: the active influence of this phenomenon on learners' cognitive abilities [21]; the negative correlation between language anxiety and L2 proficiency [22]; the influence of language anxiety on various language skills, such as reading [23]; listening [24]; speaking [25].

Research on L2 learning in a distance format notes the presence of both negative and positive aspects. Among the positive phenomena of distance learning it is possible to note the use of metacognitive learning strategies by students to a greater extent in comparison with full-time students. The use of metacognitive strategies is associated with the emergence of foreign language anxiety [26]. Many modern works are devoted to the changes in the organization of the educational process, which appeared in the post-college period. Specificity of this stage consists in certain adaptation difficulties of students, first, those who spent the most part of the study time during the pandemic in the distance abroad.

Considering all of the above, we posed the following questions as a part of this study: *Does linguistic anxiety in multilingual students before the Coronavirus-Lockdown coincide in its structure and content with what they experienced after they came out of the lockdown?*

Is there a significant correlation between foreign language anxiety and resilience? Which anxiety construct is more closely correlated with the resilience phenomenon?

3 Research Method

The study was conducted in two steps: the first step affected the period of 2019–2020 academic year, the second step came in 2022. A little over 70 foreign students (mostly from China) studying in the humanities at Peter the Great St. Petersburg Polytechnic University were interviewed before Corona-Lockdown period. The students belonged to the bilingual group. There were also 39 multilingual foreign students as respondents in the post COVID period. The average age of the respondents was 19–23 years old. The survey was administered using google forms. Students were sent a link with a brief description of the purpose of the survey and relevant instructions.

Two volumetric scales were used as research tools: the Foreign Language Classroom Anxiety Scale (FLCAS), consisting of 33 items and the Resilience Scale, an adapted version of the author's scale, including 23 items [27]. All respondents were asked to fill in the questionnaire data relevant to the study: age, gender, country of origin, native language, the number of languages spoken by the interviewee.

The Foreign Language Classroom Anxiety Scale (FLCAS) (with 33 items) is a system developed by foreign scientists (Horwitz et al., 1986) to measure language anxiety in the form of a test consisting of 33 points on a 5-point Likert scale ranging from 1 = "Strongly Disagree," 2 = "Disagree," 3 = "Not Sure," 4 = "Agree," to 5 = "Strongly Agree". This scale includes 3 constructs responsible for the communicative aspect, the assessment process, and testing. The resilience scale includes several subscales: emotional, social, and cognitive flexibility; successful self-identification at different contextual levels; value system; and others [28]. The affirmation was also assessed on a 5-point scale.

The main methods of research in the work were: the study of contemporary materials on the topic, the survey, mathematical and statistical processing of the obtained data, which included: correlation and regression types of analysis, determination of the coefficient of Alpha-Cronbach, exploratory and confirmatory factor analysis of the results of the survey.

First, we were interested in differences in the FLA aspect between samples made before the pandemic and after leaving the long distance learning. As the research practice shows, often exploratory factor analysis made on samples with differences in culture and other characteristics showed significant differences in the distribution of FLA constituents by components. Thus, nowadays, more than 3 possible models of linguistic anxiety have already been identified. Moreover, all of them have different factors in number and composition.

4   Results

4.1   Descriptive Statistics

As a reminder, the first instrument to measure foreign language anxiety was the FLCAS: a 33-item, 5-point Likert Foreign Language Classroom Anxiety Scale (FLCAS). This scale has been the basis of numerous studies to determine the level of foreign language anxiety in a multilingual learning environment.

According to the results obtained, there are differences between the first sample (before COVID-19) and the second sample (after COVID-19), both at the statistical mean level and according to the indicators of factor analysis.

The group of respondents who filled out the pre-lockdown questionnaires was mainly (more than 85%) bilingual, that is, in addition to their native language (Chinese), they knew, at the level of understanding and speaking skills at an average level, Russian. Most of the sample, 57%, were girls; 43% were guys. According to the age characteristics, the students were distributed as follows: 20–21 years old - 44% of the respondents, 22 years old and older - 29%, 18–19 years old - 27%.

The second group of 40 people differed significantly in language characteristics. All respondents were multilingual, that is, they spoke three or more languages. In fact, 88% of the respondents knew Chinese, English and Russian at the required level. The remaining 12% were students who knew Persian, Russian, and English. About 35% knew other languages Spanish, Korean, Arabic, Uzbek, and Turkmen. Most of this group began learning foreign languages between the ages of 7 and 12, with some even earlier. About 57% of the second sample were girls and 43% were boys. In terms of age characteristics, these were students aged 20–21 years (63%) and 22–23 years or more (37%).

In general, the level of linguistic anxiety in the statistical mean in representatives of the two samples did not reveal any significant differences. In accordance with descriptive statistics: FLA = 3,08, Resilience = 3,51 (before COVID-19); FLA = 3,09, Resilience = 3,48.

The diagram below (Fig. 1) clearly indicates that, compared to students surveyed before Corona-Lockdown, multilingual students, after completing two years of distance learning, have higher rates of fear of communicating with others and of being negatively evaluated.

We found no meaningful differences between the two samples at the level of descriptive statistics. However, comparing scores at the level of mean values, the first sample (before COVID19) had a clear superiority in scores for the components related to emotional reactions and feelings about the communicative aspects of the FLA. According to the second sample, the aspects related directly to testing scored the highest.

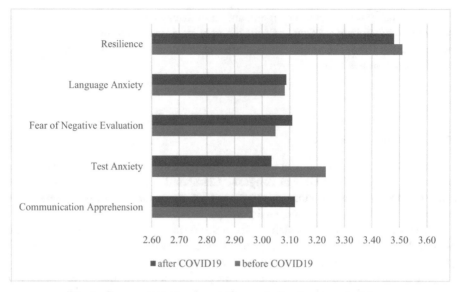

Fig. 1. Comparative data chart: before and after the corona lockdown

4.2 Correlation and Factor Analysis

The Pearson correlation coefficient for variables with normal distribution was taken as the main one. The alpha Cronbach coefficient was used to test the reliability of the research tools, which showed on average a high level of reliability with $\alpha = 0.893$ for the first group of respondents (before COVID-19) and with $\alpha = 0.924$ for the second group of respondents (after COVID-19). The study examined correlation relations at $p \leq 0.05$ and $p \leq 0.01$.

One of the most important steps in our study was an exploratory analysis of the results of the FLCAS survey in two samples. Nine factors were obtained from the first cutoff data, which were reduced to three as a result of additional processing. Factors with very low scores and factors with two or fewer components were removed from the analysis. Exploratory analysis of the first, pre-sample data highlighted the following components: 1 - Emotional and Cognitive component (16 indicators); 2 - Communicative Projection component (3 indicators); 3 - Academic Communication component (3 indicators).

The exploratory factor analysis of the indicators obtained from the post-pandemic survey of multilingual students differed in no small measure from our first attempt. The data processing system allowed us to identify eight factors, which, as in the first case, were subjected to additional processing. At the final stage we obtained four factors, but their composition differed significantly: 1 - Fear of Negative Evaluation (14 indicators); 2 - Test Anxiety (5 indicators); 3 - Communication Apprehension (3 indicators); 4 - Speech Anxiety (3 indicators).

Correlation analysis was performed using the indicators of the second stream of multilingual respondents. To determine the Pearson coefficient, the most important constructs of the anxiety and resilience phenomena under study were involved. The results are presented in Table 1.

Table 1. Correlations: FLA and resilience

		Communication apprehension	Fear of negative evaluation	Test anxiety	Language anxiety	Resilience
Communication apprehension	r-Pearson	1				
Fear of negative evaluation	r-Pearson	,803**	1			
		<,001				
Test anxiety	r-Pearson	,712**	,787**	1		
		<,001	<,001			
Language anxiety	r-Pearson	,950**	,924**	,857**	1	
		<,001	<,001	<,001		
Resilience	r-Pearson	,328*	,348*	,399*	,384*	1
		,041	,030	,012	,017	

As can be seen from the correlation data between the FLA and Resilience constructs there is a significant positive relationship of different levels, expressed in the range between r = .328 (Resilience and Communication Apprehension) to r = .950 (Language Anxiety and Communication Apprehension). Most of the correlations have a very high level of significance - < .001.

The fact that Resilience and Test Anxiety have the highest interdependence is of particular interest. We decided to look at the relationships between the constructs presented in the table through regression analysis.

Regression linear analysis of the following data: Communication Apprehension, Fear of Negative Evaluation, Test Anxiety, Language Anxiety, Resilience showed the highest correlation between Resilience and Test Anxiety. ANOVA has regained its influence (with p-value below 2%).

Confirmatory factor analysis did not reveal any significant changes. The total number of factors did not exceed 3.

5 Discussions

As can be seen from the name and the number of components, the factors obtained do not coincide with the version recommended by the authors of the scale. The factors we obtained are more related to the communicative aspect, which most likely caused the greatest anxiety in students.

Such researchers as Aida (1994), Liu and Jackson (2008) [9, 11] pointed out about the movable nature of constructs. In some cases it led to an increase in the number of constructs with heterogeneous composition, including test anxiety, fear of communication, and fear of negative evaluation. In our case, the results of the first ascertaining experiment pointed to the priority of communication. To improve the communicative skill, first, the trainees' efforts were directed. For international students, the communicative skill is one of the most important for successful adaptation in a foreign language environment. Mathematical and statistical processing of the data of the second stream of respondents showed the priority of test anxiety. We are inclined to explain it by the attitude to solve the immediate task and certain difficulties of the post-COVID adaptation period for foreigners in a foreign-language environment.

The results of exploratory factor analysis enable us to suggest the feasibility of identifying a new model of linguistic anxiety, the main system-forming indicator of which may be anxiety prior to assessment and passing through the evaluation procedure. Here we can assume the influence of students' cognitive features and programs, which could be expressed in perfectionism, an overly emotional representation of upcoming events in the academic process, etc. The role of students' cognitive features and cognitive processes in general has been mentioned in other studies more than once [29]. A significant difference in the emotional experience of students belonging to different cultures (e.g., Eastern and Western cultures) on the upcoming performance in front of a group, live communication with the teacher has been repeatedly noted.

One pattern that has been observed in studies of foreign language anxiety, both in the classroom and in multilingual interactions in everyday life, is that FLA less affects participants who know more languages in different languages [30]. It has been suggested that multilinguals are more skilled communicators, able to overcome communicative difficulties, and that their increased communicative confidence leads to lower levels of FLA [31]. In our study, marked linguistic anxiety in multilingual students can be explained by the enforced prolonged lack of live communication both in the educational environment and in everyday life.

Indicators from the questionnaire data, including age, gender, culture did not show meaningful correlations. The component that got our interest was students' multilingualism. At a certain stage of the study, this characteristic played a significant role.

We believe that the study of the dependence of cognitive programs and specific styles with features of the experience of language anxiety can play a more significant role than it might seem at first view. One must not forget that FLA is a type of psychological anxiety, which is composed of personal and situational components.

We assume that multilingualism played an important role for the second sample. Otherwise, their scores on resilience and FLA would have been even lower. Prolonged exposure to enforced distance does have some negative effects on students' psychological resilience. The experience of communicating in different languages, accompanied by new knowledge and communicative experiences, helps to overcome unexpected challenges and stresses.

6　Conclusions

Summarizing all of the above, it is reasonable to summarize the research, which in total took several years to complete. Language anxiety is not a stable entity, even if it concerns only one person. Various global and smaller-scale complexities can lead to significant changes in the structure of the phenomenon of language anxiety. Much depends on the available sociocultural and communicative experience of the individual. In addition, the specifics of the FLA experience depend on individual cognitive scripts.

Resilience, as a unique characteristic that allows a quick recovery in the most challenging situations, has significant correlations with linguistic anxiety, especially in moments that have a situational nature, for example, experiences before a test, a survey.

The results obtained, we believe, require further investigation using more in-depth psychological research tools.

As a limitation of this research work, we can point out the composition of the representative sample. The fact that most of the interviewees were of Chinese culture somewhat limits the audience for the dissemination of the findings of this paper.

References

1. Valieva, F., Fomina, S., Nilova, I.: Distance learning during the corona-lockdown: some psychological and pedagogical aspects. In: Bylieva, D., Nordmann, A., Shipunova, O., Volkova, V. (eds.) PCSF/CSIS-2020. LNNS, vol. 184, pp. 289–300. Springer, Cham (2021). https://doi.org/10.1007/978-3-030-65857-1_25
2. Kruse, I., Lutskovskaia, L., Stepanova, V.V.: Advantages and disadvantages of distance teaching in foreign language education during COVID-19. Front. Educ. **7**, 964135 (2022). https://doi.org/10.3389/feduc.2022.964135
3. Hjell, L., Ziegler, D.: Theories of Personality. Peter, St. Petersburg (2013)
4. Park, G.P.: Factor analysis of the foreign language classroom anxiety scale in Korean learners of English as a foreign language. Psychol. Rep. **115**(1), 261–275 (2014). https://doi.org/10.2466/28.11.PR0.115c10z2
5. Krashen, S.: Principles and Second Language Acquisition. Pergamon, Oxford (1982)
6. Krashen, S.: The Input Hypothesis: Issues and Implications. Addison-Wesley Longman, Boston (1985)
7. MacIntyre, P.D.: A review of the research for language teachers. In: Young, D.J. (ed.) Affect in Foreign Language and Second Language Learning, pp. 24–45. McGraw-Hill, Boston (1998)
8. Horwitz, M.B., Horwitz, E.K., Cope, J.: Foreign language classroom anxiety. In: Horwitz, E.K., Young, D.J. (eds.) Language Anxiety: From Theory and Research to Classroom Implications, pp. 27–39. Prentice Hall, Englewood Cliffs (1991)
9. Aida, Y.: Examination of Horwitz, Horwitz, and Cope's construct of foreign language anxiety: the case of students of Japanese. Mod. Lang. J. **78**(2), 155–168 (1994). https://doi.org/10.1111/j.1540-4781.1994.tb02026.x
10. Matsuda, S., Gobel, P.: Anxiety and predictors of performance in the foreign language classroom. System **32**(1), 21–36 (2004). https://doi.org/10.1016/j.system.2003.08.002
11. Liu, M., Jackson, J.: An exploration of Chinese EFL learners' unwillingness to communicate and foreign language anxiety. Mod. Lang. J. **92**(1), 71–86 (2008). https://doi.org/10.1111/j.1540-4781.2008.00687.x
12. Yashima, T., Kimberly, A.N., Shizuka, T., Takeuchi, O., Yamane, S., Yoshizawa, K.: The interplay of classroom anxiety, intrinsic motivation, gender in the Japanese EFL context. Foreign Lang. Educ. Stud. **17**, 41–64 (2009)
13. Mak, B.: An exploration of speaking-in-class anxiety with Chinese ESL learners. System **39**, 202–214 (2011). https://doi.org/10.1016/j.system.2011.04.002
14. MacIntyre, P.D.: How does anxiety affect second language learning? A reply to Sparks and Gangschow. Mod. Lang. J. **79**, 90–99 (1995)
15. Wood, D.: Fundamentals of Formulaic Language: An Introduction. Bloomsbury Academic, London (2015)

16. MacIntyre, P.D., Gardner, R.C.: Anxiety and second language learning: toward a theoretical classification. Lang. Learn. **39**, 251 (1989)
17. Nikulicheva, D.B.: How to Find Your Way to Foreign Languages. Linguistic and Psychological Strategies of Polyglots. Flint Science, Moscow (2009)
18. Tugade, M.M., Fredrickson, B.L.: Resilient individuals use positive emotions to bounce back from negative emotional experiences. J. Pers. Soc. Psychol. **86**(2), 320–333 (2004)
19. Dewaele, J.M., MacIntyre, P.D., Boudreau, C., Dewaele, L.: Do girls have all the fun? Anxiety and enjoyment in the foreign language classroom. Theory Pract. Second Lang. Acquis. **2**(1), 41–63 (2016)
20. Mercer, S., MacIntyre, P., Gregersen, T., Talbot, K.: Positive language education: combining positive education and language education. Theory Pract. Second Lang. Acquis. **4**(2), 11–31 (2018)
21. Horwitz, E.: Language anxiety and achievement. Ann. Rev. Appl. Linguist. **21**, 112–126 (2001)
22. Kitano, K.: Anxiety in the college Japanese language classroom. Mod. Lang. J. **85**(4), 549–566 (2001)
23. Argaman, O., Abu-Rabia, S.: The influence of language anxiety on English reading and writing tasks among native Hebrew speakers. Lang. Cult. Curric. **15**(2), 143–160 (2002)
24. Kim, J.S.: Korean EFL learners' listening anxiety, listening strategy use, and listening proficiency. Engl. Lang. Lit. Teach. **17**(1), 101–124 (2011)
25. Young, D.J.: An investigation of students' perspectives on anxiety and speaking. Foreign Lang. Ann. **23**, 539–553 (1990)
26. White, C.: Language learning strategies in independent language learning: an overview. In: Language Learning Strategies in Independent Settings. Multilingual Matters, Clevedon (2008)
27. Valieva, F.: The higher school community resilience as a predictor of efficient system of education. IOP Conf. Ser.: Mater. Sci. Eng. **940**, 012146 (2020). https://doi.org/10.1088/1757-899X/940/1/012146
28. Horwitz, E.K., Horwitz, M.B., Cope, J.: Foreign language classroom anxiety. Mod. Lang. J. **70**(2), 125–132 (1986)
29. MacWhinnie, S.G.B., Mitchell, C.: English classroom reforms in Japan: a study of Japanese university EFL student anxiety and motivation. Asian-Pac. J. Second Foreign Lang. Educ. **2**(1), 1–13 (2017). https://doi.org/10.1186/s40862-017-0030-2
30. Dewaele, J.-M.: Multilingualism and affordances: variation in self-perceived communicative competence and communicative anxiety in French L1, L2, L3 and L4. Int. Rev. Appl. Ling. **48**, 105–129 (2010)
31. Dewaele, J.-M., Petrides, K.V., Furnham, A.: The effects of trait emotional intelligence and sociobiographical variables on communicative anxiety and foreign language anxiety among adult multilinguals: a review and empirical investigation. Lang. Learn. **58**(4), 911–960 (2008)

Self-assessment of Psychological Issues of Bangladeshi and Russian Students on the Online Learning During the Pandemic

Marianna Yu. Ababkova[1]([⊠]) [ID], Khandakar K. Hasan[2] [ID], Debarshi Mukherjee[2] [ID], S. K. Mamun Mostofa[3] [ID], and Roslina Othman[3] [ID]

[1] Saint Petersburg Electrotechnical University LETI, ul. Professora Popova 5, 197022 Saint Petersburg, Russia
miuababkova@etu.ru

[2] Tripura University (A Central University), Suryamaninagar 799 022, Tripura (W), India

[3] International Islamic University, Jalan Gombak, 53100 Kuala Lumpur, Selangor, Malaysia

Abstract. Currently the educational systems of many countries have a growing influence of the changing educational paradigm due to global changes in global politics, economics' transition to the information society and the challenges of pandemics. COVID-19 has significantly affected tertiary learning systems in Bangladesh, Russia and all around the world, and has forced curricula to be transformed into an online format. The research was conducted at the University of Dhaka and the University of Rajshahi in Bangladesh, as well as at the Peter the Great St. Petersburg Polytechnic University in Russia. The analysis of the participants' answers helped to identify the following psychological issues on learning continuity by the university students, such as: attention, motivation, emotion and anxiety of students. According to the research, students from Bangladesh got tired faster, felt loneliness more acutely, and experienced health problems than their Russian colleagues during online learning. Russian students took notes of their lectures more often, claimed to use various memorization techniques, spent more time online while studying, and were more demanding of handout materials and presentations. Bangladeshi students faced more difficulties with concentration and focusing their attention during online training. The motivation and interest of Bangladeshi students in online learning were significantly lower than Russian students' learning motivation. Also, the level of stress and anxiety among Bangladeshi students was higher. Thus, psychological issues on learning continuity based on learner satisfaction are vital to minimize the negative impact of rapid changes in the educational process and ensure effective online education during digital transformation.

Keywords: Learning continuity · Online learning · Learner satisfaction · Students' psychology · COVID-19

1   Introduction

The pandemic of COVID-19 forced educational institutions, including universities, to move quickly toward distance learning and online learning, according to Almaiah, Al-Khasawneh and Althunibat [1]. UNESCO has established that roughly 1,524,648,768 students belonging to various ethnic groups all over the world were excluded from the regular training practice throughout the COVID-19 era [2]. Almarashdeh concluded that a lot of online educational and instructional tools, including the LMS Moodle, were carried out on numerous distance education platforms [3].

The pandemic in Bangladesh and Russia showed the lack of technical equipment, the disability, and often the unwillingness of instructors and students to join the new technology. The success of the online class depends on the capacity, willingness and acceptability of students to that system, as Almaiah and Alismaiel described [4], still the lack of the adequate regulations, online toolkit and research data on students' reactions and issues prevents the advantages from being achieved. Research is still in its infancy on this topic according to Tarhini, Masa'deh, Al-Busaidi, Mohammed and Maqableh [5], as students' viewpoints are not investigated thoroughly, Alamri, Almaiah and Al-Rahmi report [6]. El-Masri and Tarhini believe that studying the implementation of an online system can allow institutions to understand their students' demands better and eventually lead to a successful Internet class system [7] and Alksasbeh, Abuhelaleh, Almaiah, AL-jaafreh and Abu karaka consider that of the most important factors for the successful development of online learning system [8]. The certain qualities of an online educational platform and the diverse learning material should be given more significant consideration, as Hasan, Mukherjee and Saha stated [9]. The student's desire to study online seems to focus on several different variables such as student's characteristics, learning motivation and skills, as well as courses and online learning platforms. The scope of research also is enhanced by the additional critical issues of online learning, such as teachers' and students' joint attention [10], mixed digital communication tools of social media, messengers, and video conferences [11] and communication management in the multilingual groups of the students bogged down by non-contact learning process [12]. The last issue is to be explored more thoroughly in the context of multinational student groups [13], as the mixed teams encounter certain issues related to understanding and processing online content.

The study aims to detect issues and challenges of the students in Bangladesh (University of Dhaka and Rajshahi University) and Russia (Peter the Great Saint Petersburg Polytechnic University) after launching online courses during the pandemic.

The research questions were the following:

– What are the main psychological issues of Russian and Bangladeshi students during online learning?
– If there is any dependence between LMS quality, teacher-students and student-student interaction on psychological issues of the learners?

The following hypotheses were formulated:

Attention, motivation, emotion, stress and anxiety during online learning has a significant influence on psychological issues.

LMS quality, teacher-student interaction and student-student interaction during online learning has a significant relation to psychological issues in learner satisfaction.

The universities of Russia and Bangladesh were chosen out of the authors' research interests. The author from Russia is interested in cognitive research for enhancing students' skills and comparative analysis of international and Russian students' behavior, as well as the implementation of multilingual technologies into international students' activities. The authors from Bangladesh, India, and Malaysia share the interest in the blended education issues for the Asian region in general and Bangladesh in particular.

The authors set a task to define the range of coincidences within the frame of the psychological issues of Russian and foreign students during the pandemic and the forced transition to online education. The sample from Russian and Bangladeshi universities was chosen out of the researchers' convenience.

The limitation of the study is that a wider sample of students from Russia, Europe, Asia, North, and Latin America is needed for a more accurate assessment of the common and various psychological challenges of online learning.

2 Literature Review

Learners' psychological and personal qualities can be cited as constraint to the perception and assimilation of the online material of the webinars and e-lectures, and, as Agung, Surtikanti and Quinones conclude, online learning require more friendly LMS platforms to increase students' participation [14]. Abbasi, Ayoob, Malik and Memon insist, that students did not prefer online teaching to face-to-face teaching during the pandemic [15]. Accountability, self-management, interactivity with a teacher and other students, and inner locus of control proved to be another side of distance learning - online classes require significantly more motivation, constraint and attention. Currently, there is a need for systematic research of the psychological aspects of online learning, particularly educational interaction features in the digital environment, according to Pokrovskaia, Ababkova and Fedorov [16].

Wijayanti, Nikmah and Pujiastuti specify that a teacher plays a crucial role in strengthening students' attention and recognize the relevance of learning for students during and after classes [17]. The teacher can also provide students with an understanding by clarifying the training material. According to one recent research based on biofeedback, the presence of a teacher in offline learning requires more effort and mental resources from students than self-study activities via LMS. At the same time, students' confidence in judgments decreases, and stress increases due to the inability to check one's progress and performance instantly, as during the same classroom activities, as Ababkova, Leontieva and Pokrovskaia found out in their study based on biofeedback technique [18].

The motivation was found as also pivotal in the world of education, as Chai and Lim accentuated, and a key condition to the implementation of constructivist-oriented teaching and learning practices [19]. Previous research has established a link between learner motivation and a range of critical learning outcomes, including persistence, as Vallerand and Blssonnette surmised [20]; retention and motivational benefits of minutely designed educational activities, according to Lepper and Cordova [21]; achievement, and course satisfaction, as Wigfield in his expectancy-value model proposed [22]. Motivation should be addressed seriously in online learning environment, according to this research. Improved performance and satisfied emerging learning demands were the key factors encouraging the students to continue their online learning, as Li, Chen, Zhu and Zhang found out in their research on students' anxiety about H1N1 influenza [23]. In addition, the relation between student motivation and attitudes has been established during online training, which can influence their desire to take more online courses or utilize specific online learning platforms, as Kim, Lee, Leite and Huggins-Manley determined in their exploration of student and teacher dependent factors during the usage of online learning platform [24].

It is typical to feel stressed, anxious, or upset, amongst other emotional reactions, in unpredictable situations like COVID-19, and as Tarman emphasized, the uncertainty, the urgency and the boost of online learning made it more fearful [25]. Physical and psychological pressure also observed among medical staff, children, patients with probable illness and quarantine family members, Duan and Zhu pointed [26].

Depression and anxiety are frequent mental diseases, and the fourth-largest cause of morbidity is the incidence of 10–44% in developing nations, according to Azad, Shahid, Abbas, Shaheen, and Munir [27]. Students at university are at high risk for symptoms of depression and anxiety, as Zivin, Eisenberg, Gollust and Golberstein thought [28], and exposed to many stressors and irritants unique to this age and level of development, as the research of Beiter, Nash, McCrady, Rhoades, Linscomb, Clarahan and Sammut proved [29]. Some research during SARS and H1N1 in China found that university students showed apparent anxiety and stress, and proposed coping measures, as Jia, Fan and Lu reported [30]. Moreover, COVID-19 can significantly affect the emotional well-being of learners, leading to lower levels of academic engagement and to a greater risk of behavioral problems, and, as Ismaili ascertained, difficulties of e-learning increased personal health problems [31]. Therefore, educators are more concerned with students' emotional development in this unique period and to give priority to students' physical and mental health, though the issues and challenges of social and psychological adaptation of students aren't unique, as Hasan and Leontieva investigated, though significantly enhanced [32].

The "teacher-student" connection can be seen as one of the main variables to increase student happiness and participation in online courses, so Lohmann, Boothe, A., Hathcote and Turpin stated that diversity of classroom encouraged universities to meet diverse learning needs [33], and online learning environments, as well as asynchronous classes, require quality interaction between learners and instructors, according to Nandi, Hamilton and Harland [34]. The engagement methods to encourage interaction between instructor and student were more appreciated by students than the methods used to foster interactions in "student-students" and "student-learning content" context, Bolliger, and Martin argued [35]. Holzweiss, Joyner, Fuller, Henderson and Young found out, that some students assigned crucial role to the engagement with their professors in delivering the best knowledge, expertise, and training experience during online learning, with instructor accessibility being critical to their overall happiness [36]. The relationship between teachers and students on the Internet can also positively influence students' results. For student-instructors engagement to be promoted, regular and consistent interaction should also occur, as Gaytan and McEwen stated [37]. Additionally, these contracts should be transparent, according to Garrison, Anderson and Archer [38], timely (it is mandatory condition to enhance student engagement in online courses, as Robinson and Hullinger stressed in their study [39]), and multimodal, including emails, phone and video conferences, messages, and announcements, to incorporate multiple design modes to fit in multiliterate, multiskilled students' needs, as Iyer and Luke specified in their research [40]. The latter requirement of multimodality especially urgent in the context of multilingual international students' groups during the online learning, when the educational content and instructions need to be tailored to the learners' different levels of language command.

Teachers play an essential role in the students' online participation through courses to promote communication, involvement, and interaction amongst students. Students express more satisfaction with the course, particularly early in the learning, when there are opportunities to connect with other participants online. In particular, at the beginning of a semester, researchers discovered that students saw icebreaker talks as the primary engagement method online. Research on the difference in students' satisfaction of two countries (Germany and Sweden) with the Quality Characteristics of the Digital Learning Systems based on gender and frequency of interaction between students-students and students-teachers was done by Waheed and Leišytė to reveal a statistical difference in the German students' satisfaction [41].

The literature analysis suggests that most studies are conducted to ascertain students' happiness and attitude toward learning continuity and that study examining the psychological aspects of online learning during the COVID-19 epidemic are scarce and infrequent. According to some research, social support can help alleviate psychological discomfort during epidemics. It shows that practical and substantial social support is needed during a pandemic. More research demonstrates that social support is adversely associated, along with prior findings, with student attitude, motivation, anxiety, readiness, and interaction between teacher and students, as Thompson, McBride, Hosford and Halaas found out [42].

3   Materials and Methods

A total of 161 respondents have participated in this study. Of this number, 90 participants (55.9%) were from Bangladesh and 71 participants (44.1%) were from Russia. Among them, 63 respondents (39.1%) were male and 98 (60.9%) of them were female. More than half of the respondents (102 participants, 63.4%) constituted the age group of 17–21 years. One third of them (52 participants, 32.3%) were from the age group 22–26 years, The smallest group (7 participants, 4.1%) was aged 26–30. 143 respondents (88.8%) were bachelor students, 11 participants (6.8%) represented masters students, the rest 7 (4.4%) were Ph.D. students.

A research model demonstrates the relationships between the target variables in a study. A research model to study the factors affecting the learner satisfaction was proposed during the research. It has six independent variables – attention; motivation; emotion, stress and anxiety; LMS, teacher-student and student-student interaction.

Google forms were employed to build the questionnaire, and the link was sent to students' email groups and placed on WhatsApp groups of the students through communication with their instructors. Students in the tertiary institutions in Bangladesh and Russia were requested to discuss their experiences in a digital setting as they had just moved from conventional learning to online one. The participants engaged in the study voluntarily and informed consent was acquired before the study began.

The first section of the questionnaire profiled the demographic and labor identity, and general tendencies of the respondents' behavior on the Internet. The second half measured the psychological questions of digital learning during pandemic, such as: the attention during online learning; motivating online learning; emotional and stressful factors during learning; readiness to learn online; interaction between teacher and student, as well as student and student interaction. The Digital Learning Program compiled the second portion. Finally, the researchers examined the issues facing the respondents and requested interventions for online education quality. The choice of a 7-point Likert scale was based on the study that anticipated that participants of Asian ethnicity tended to choose the middle score or be non-partisan in their responses since this phenomenon was viewed as creating a non-attractive research outcome, according to Jebb, Ng and Tay [43]. The random sampling methodology was applied to acquire responses from the respondents in the study with a sample population for the quantitative data analysis. The survey was open to students from undergraduate to postgraduate level.

The reliability of the questionnaire has been tested by the Statistical Package for Social Sciences (SPSS) program. The reliability coefficient of the questionnaire Cronbach's Alpha test shows alpha is 0.922 which can be considered as reliable as alpha p $< .5$ is good to run the next series of tests.

Bartlett's test of Sphericity has been used along with Kaiser-Meyer-Olkin (KMO) statistics. Here, the KMO value is .827, which is between 0.5 and 1.0 and the result is excellent because the KMO value is close to 1. The approximate chi-square statistic is 7980.696 with 1540 degrees of freedom, which is significant ($p < 0.05$). Thus, factor analysis could be taken as an adequate technique for analyzing these data.

To ensure the validity of the constructs, the measurement items and variables were developed from prior studies and divided into 7 groups ("Perception", "Attention", "Motivation and interest", "Emotions, Stress and Anxiety during online learning", "LMS quality", 'Teacher-Student interaction", "Student-Student interaction"), see the Tables 1, 2, 3 and 4.

4   Results

Among the 56 statements of the psychological issues in online learning, the 31 statements were found after factor analysis premised on the variables having a factor loading of above 0.6 and the communality value above 0.5. The value of KMO statistic (0.851) is also large for these factors and shows good reliability values ($\alpha > 0.5$).

4.1   Perception During Online Learning

The differences in Russian and Bangladeshi students' perception were following. The statement "I get quickly tired during online learning" and "I suffer stress and frustration during digital learning" occupied the highest mean of 4.49 for Bangladeshi students, while for Russian students this statement scored 3.82 both. The statement "I feel loneliness and isolation" for Bangladeshi and Russian students scored 4.37 and 3.93 respectively. The statement "I do not feel confident and mentally relaxed during online classes" showed the mean score of 4.37 (for Russian students 3.58 respectively) followed by "Constant feedback from a teacher and my classmate inspires me to study more" (4.23), the same scores were for Russian students.

The statement "I was able to interact with the teacher during the online course discussion" scored 4.22 for Bangladeshi students, for Russian students it was almost the same – 4.19. Another high scored statement for Bangladeshi students was "I have health problems after online learning" (4.20), while for Russian students it was 3.86. The statement "I haven't any great wish to learn more in online than in offline" scored 4.19 by Bangladeshi students and 3.96 by Russian students.

Bangladeshi students scored the statement "I cannot relax physically during online learning" as 4.17, while Russian students marked it 3.86. Bangladeshi students pointed, that "The teacher encouraged me to become actively involved in the online course discussion" (4.14), while Russian students estimated this statement lower (3.84) than expected. "The teacher provided me feedback on my work with comments" for Bangladeshi students achieved the mean score of 4.06 (for Russian students 3.99), "I was able to share online learning experiences with other students" for Bangladeshi students was 3.97 (4.09 for Russian students respectively).

Russian students pointed out that the statement "I can integrate and organize myself during online classes" occupied the highest mean score 4.34 for Russian students (for Bangladeshi students 3.91 respectively), "I can manage the pressure and intensity during my online classes" occupied the 2nd highest mean score 4.27 (for Bangladeshi students 3.29 respectively).

The statements that got the lowest scores both by Bangladeshi and Russian students are "The quality of the online course is high in comparison to other offline courses" (2.72 and 3.92 respectively), and "I was satisfied with this online course" (2.93 and 3.75 respectively).

The main differences between Bangladeshi and Russian students in perception of online courses were that Bangladeshi students grew quickly tired and frustrated during online learning, more acutely felt loneliness and isolation, had health problems after online learning more often than Russian colleagues, according to their self-assessment. The students from both countries accentuated their need for constant feedback from the teacher and their classmates.

4.2   Memory During Online Learning

A report on how the students from two countries use their memory and some memorizing aspects during online learning is presented in the table below (Table 1).

Table 1. The comparison of Russian and Bangladeshi students' memory aspects during the online learning.

Statements	Average scores, Bangladeshi students	Average scores, Russian students
I think that online classes and digital presentations have to be more captive and emotional to help students remember them	3.44	4.47
I spend more time to learn online material than offline one	3.36	4.04
I memorize written content better than video lectures	3.63	3.96
I can easily memorize, produce and apply online material	2.69	4.22
I think modern student doesn't have to memorize the educational content; all can be found online anytime	3.33	3.95
I know how to combine different techniques to memorize online materials	3.07	4.45
I reread the lectures' notes to prepare for my exams	3.82	4.74
I reread the lectures' notes to prepare for my exams	3.82	4.74
I think online classes requires that teacher has to be a good storyteller to engage students into learning process	4.19	4.65
I agree that the university has to prepare more informative and thorough learning material to enable students to learn online effectively	4.3	4.81

The statements that seriously differ in the answers of Bangladeshi and Russian students are the following:

"I reread the lectures' notes to prepare for my exams" scored 4.74 and 3.82 (Russian and Bangladeshi students comparatively);

"I think that online classes and digital presentations have to be more captive and emotional to help students remember them" scored 4.47 and 3.44 (Russian and Bangladeshi students comparatively);

"I know how to combine different techniques to memorize online materials" scored 4.45 and 3.07 (Russian and Bangladeshi students comparatively);

"I can easily memorize, produce and apply online material" scored 4.22 and 2.69 (Russian and Bangladeshi students comparatively);

"I spend more time to learn online material than offline one" scored 4.04 and 3.36 (Russian and Bangladeshi students comparatively).

Thus, we can conclude that Russian students use lectures' notes more frequently, than their Bangladeshi colleagues, harness some memorizing techniques and have more practice in memorizing online material and spend more time learning online. Also, Russian students are more critical and demanding to the online material and visual presentations.

4.3 Attention During Online Learning

Some issues of attention during online learning reported by Russian and Bangladeshi students are presented in the table below (Table 2).

Table 2. The comparison of Russian and Bangladeshi students' attention aspects during online learning.

Statements	Average scores, Bangladeshi students	Average scores, Russian students
I can easily focus my attention constantly and concentrate during online learning	2.13	3.83
I get easily distracted by different factors during online classes	3.23	3.96
I can manage the pressure and intensity during my online classes	2.67	4.26
I mobilize all my efforts and thinking capacity to succeed in digital learning	3.13	4.14
I need constant reminders, stimuli, and feedback from my teacher and classmates and teachers to prevent me from falling out from online learning	3.43	3.89
I can integrate and organize myself during online classes	3.06	4.31

The issues with attention reported by Bangladeshi students were as follows:

"I can easily focus my attention constantly and concentrate during online learning" (2.13 to 3,83);

"I can manage the pressure and intensity during my online classes" (average 2.67 in comparison to Russian students' 4.26).

This part of questionnaire showed that Bangladeshi students have more problems in concentrating and mobilizing their attention than their Russian colleagues.

4.4   Motivation and Interest During Online Learning

Some aspects of motivation and interest during online learning reported by Russian and Bangladeshi students are presented in the table below (Table 3).

Table 3. The comparison of Russian and Bangladeshi students' motivation and interest during online learning.

Statements	Average scores, Bangladeshi students	Average scores, Russian students
I think it is just a temporarily measure to move in online during Pandemic, I don't see any other reasons for digital learning when everything is OK	3.76	4.14
Online class is really a waste of time	3.06	3.7
I feel online learning enable participants to vary and enhance their academic skills	3.17	4.19
I think the use of ICT increases students' involvement to online learning	3.82	4.07
It is not important for me to learn in online	2.56	4.19
The online learning meets my learning needs not only during Pandemic	3.04	4.04
I haven't any great wish to learn more in online than in offline	3.37	3.95
Online learning is a great source of inspiration to me	3	4.01
I feel highly motivated during online learning	2.6	3.89

(*continued*)

Table 3. (*continued*)

Statements	Average scores, Bangladeshi students	Average scores, Russian students
Constant feedback with a teacher and my classmate inspires me to learn more	3.31	4.22
I don't have any stimulus to move on during online learning	3.24	3.82
Online learning helps me to improve my communications skills	3.356	3.87
The online learning meets my learning needs	3.01	3.93

The main differences in motivation between Russian and Bangladeshi students were the following: "Constant feedback with a teacher and my classmate inspires me to learn more" (4.22 against 3.31 comparatively); "It is not important for me to learn in online" (4.19 against 2.56 comparatively). Bangladeshi students were more sceptic about the statement "Online learning enable participants to vary and enhance their academic skills", than their Russian mates (3.17 against 4.19 comparatively) and didn't admit that they feel highly motivated during online learning in comparison with Russian students (2.6 against 3.89 comparatively).

The results on questions concerning attitude during online learning reported by Russian and Bangladeshi students showed that Russian students feel more enthusiastic about online learning in comparison with their Bangladeshi mates, who show their ambiguous attitude toward online learning and don't trust online learning completely (2.59 to 3.63 respectively), but Bangladeshi students show fuller disagreement with the statement "It is not important for me to learn in online", than their colleagues from Russia (2.86 to 3.82 respectively).

4.5 Emotions, Stress and Anxiety During Online Learning

The evaluation of the statements on "Emotions, Stress and Anxiety during online learning" by Bangladeshi and Russian students is presented on Fig. 1.

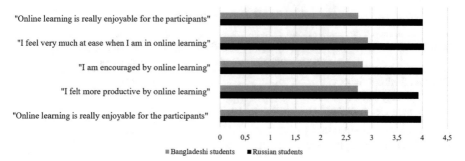

Fig. 1. The comparison of the statements' evaluation on Emotions, Stress and Anxiety during online learning by Bangladeshi and Russian students.

Bangladeshi students don't consider online learning as really enjoyable for the participants, feel less productive, not very much at ease and encouraged by online learning in comparison with their Russian mates.

4.6 LMS Quality During Online Learning

Evaluation of LMS quality is presented in Table 4.

Table 4. The comparison of LMS quality in Russian and Bangladeshi universities during online learning.

Statements	Average scores, Russian students	Average scores, Bangladeshi students
The quality of the online course compared favorably to other offline courses	3.90	2.1
I was very satisfied with this online course	3.73	2.2
I would gladly take another course via this mode	4.08	2.37
I was very interested in the subject of this online course	4.05	2.38
This online course served my needs well	4.04	2.42
I would recommend this course to another students	3.9	2.58

As can be seen from the Table 4, Bangladeshi students weren't satisfied at all with the quality of the LMS offered by their institution in comparison with Russian students.

4.7 Teacher-Student and Student-Student Interaction Quality During Online Learning

The students' evaluation of Teacher-Student Interaction quality from Bangladeshi and Russian students doesn't differ a lot, only the interaction with a teacher during the online course discussion gained the higher score from Russian students (4.12). The total average scores evaluating Teacher-Student Interaction are 3.89 and 3.2 (for Russian and Bangladeshi, respectively). That means that the system Teacher-Student Interaction need a serious revision and upgrading according to the opinion of the students of both countries.

The evaluation of Student–Student Interaction from Bangladeshi and Russian students differ, it has an average score 4.3 from Russian students, while the interaction between Bangladeshi students has total average score 3.04, that is better score for Russian Student-Student Interaction than Russian Teacher-Student Interaction.

Linear Regression analysis was also generated to describe the statistical relationship between dependent (Learner satisfaction) and independent variables, such as attention, motivation, emotion; LMS, Teacher - Student, and Student - Student Interaction. Preliminary analysis was applied to ensure that there is no violation of the assumptions of normality and multicollinearity. Figure 2 indicates that the normal probability plot of standardized residuals as well as the scatter plot indicated that the assumptions of normality and multicollinearity were met. The PP-Plot also pointed out that in our multiple linear regressions' analysis, there is no tendency in error terms.

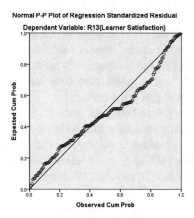

Fig. 2. The PP-Plot.

The regression summary and overall fit statistics were found, and R^2 of the model is .493 that means the regression explains 49.3% of the data variance. Adjusted R^2 change is .493. This change was statistically significant; p less than .005 (F change 4.220; Df1 = 30; Df2 = 130; Sig. F change .000, Durbin-Watson 2.016). The Durbin Watson d = 2.016, which is between $1.5 < d < 2.5$ and therefore, there is no first-order linear auto-correlation in our multiple linear regression data (Table 5).

Table 5. Model fit summary.

Model	R	R^2	Adjusted R^2	Std. error of the estimate	Change statistics						Durbin-Watson
					R^2 change	F change	Df1	Df2	Sig. F change		
1	.702[a]	.493	.376	1.364	.493	4.220	30	130	.000		2.016

[a]Dependent Variable: Learner Satisfaction
[b]Predictors: (Constant) Attention, Motivation, Emotion; LMS, Teacher-Student, Student-Student

The result of the hypotheses testing showed that there are no significant relationships among attention, emotion, stress and anxiety and motivation during online learning, and all aforementioned psychological processes have no significant influence with learner satisfaction.

5 Discussion

The research objectives were achieved. The main psychological issues troubling students in online education during the pandemic (the first research question) are as follows:

- Concerning the peculiarities of perception for the students from both countries, Bangladeshi students pointed that they grew quickly tired and frustrated during online learning, more desperately felt loneliness and isolation, and had health problems after online learning more often than Russian mates. A need for constant feedback from the teacher and their classmates is crucial for the students from both countries.
- As to memory during online learning, Russian students prefer lectures notes more than their Bangladeshi colleagues, more expertise in memorizing techniques and dealing with online material. They also spend more time learning online and have certain demands for online material and presentations.
- Bangladeshi students have pointed out such problems in their focusing attention, such as attention span and attention mobilization during online learning.
- Dealing with the questions concerning motivation and interest during online learning, Bangladeshi students didn't fully agree that online learning enable participants to vary and enhance their academic skills, and didn't feel highly motivated in comparison with Russian students.
- The attitude of Bangladeshi students toward online learning isn't as enthusiastic as Russians, and they don't trust online learning as their Russian colleagues.
- Stress and anxiety among Bangladeshi students are higher, and they don't think online learning is enjoyable for the participants, don't feel very much at ease, and encouraged by online learning in comparison with their Russian mates.

This research found some influencing factors which were considered as psychological issues in online learning. Bangladeshi students prefer smartphones and Russian students prefer laptops for the Internet access for educational purposes. The study found that most of Bangladeshi students use Zoom for joining online classes and Google Meet is their second priority. On the other hand, Russian students' first priority is Microsoft Teams, and Zoom was the second choice for Russian students.

Bangladeshi students pointed out that their Internet using depends on the situation. The largest number of Russian students replied that they used the internet 5 to 7 h a day. About one third of the respondents replied that they used the internet more than 8 h a day. The present study also showed that poor Internet connection is the major challenge for Bangladeshi students, and they indicated that electricity failure is another difficulty for joining online class. Russian students replied that poor Internet connection is a major problem for them. It revealed that the students of both countries face the same problems during the online class.

Concerning the second research question, the hypotheses test result showed no significant relationship among attention, emotion, stress and anxiety and motivation during online learning with learner satisfaction.

The consequence for online education is that teachers need to be aware that students should not just dichotomize psychology. Students have several motives for taking part in the class in online learning. It may encompass internal factors such as self-fulfillment, interest and joy. Students may also have extrinsic causes, such as fear of becoming outmoded, coerced by governments for better pay, or forced to examine, as Jang showed [44]. The students' unique motivations were shown by the longitudinal study by Otis, Grouzet and Pelletier, and it could have a continuing effect on their class attitudes and behaviors [45]. Online instructors should spend time learning about their students' intentions and choose strategies for students' retention, as Muljana and Luo proposed [46], as well as give personalized guidance to lessen students' uncertainty and anxiety, and to encourage them to become more reliable and autonomous and to begin to enjoy their learning online. Under the following circumstances, the curriculum can be efficiently transformed into an online format during the COVID-19 in Bangladesh and Russia:

- Assistance in overcoming psychological difficulties associated with conducting online instructional activities;
- Development of material and technical base, including both hardware (computing, high-speed internet, etc.) and software (LMS, educational and electronic textbooks, diagnostic, control, etc.);
- Organizational and methodological assistance, including ideas on teaching in online learning environment for multilingual students' groups;
- Professional training programs for teachers to combine digital training, multilingual technologies and scaffolding for international students' groups [47];
- Establishing a university teacher's academic workload when working online based on university regulation.

6   Conclusion

Higher education is currently in high uncertainty in Bangladesh, Russia and around the world and it is necessary to strengthen the readiness of teachers of universities for online education and training in an intentional manner. In Bangladesh and Russia, the higher education system was confronted by significant issues that demand conventional education to be restructured psychologically, technologically and methodologically. Innovative educational activities based on collaboration between teachers and students and efficient use of current technology are required to organize the educational process in a digital education environment. Thus, during the COVID-19 epidemic university professors must be ready, via remote implementation of LMS, ICT tools and so on, to organize and conduct student educational and scientific research activities, to control the education and evaluation of students online, and to carry out online conferences, webinars, etc., given the specific characteristics of virtual interaction.

This study contains several limitations such as sample, data and time limitations. Initially, the factors generated from the one-time management of the survey instrument during the semester should be noted. The stability of the satisfaction criteria has thus not been confirmed throughout a whole semester. Though the investigators have collected crucial demographic information from the responding students, many of the student characteristics which could have influenced the outcomes have not been controlled. Finally, inclusion of some other countries would have facilitated more generalization of the findings, but the present study could not do so because of time limitations. Therefore, this research will prompt further research on different aspects of the psychological issues at different universities in Bangladesh and Russia.

References

1. Almaiah, M.A., Al-Khasawneh, A., Althunibat, A.: Exploring the critical challenges and factors influencing the E-learning system usage during COVID-19 pandemic. Educ. Inf. Technol. **25**(1), 5261–5280 (2020). https://doi.org/10.1007/s10639-020-10219-y
2. Hoq, M.Z.: E-learning during the period of pandemic (COVID-19) in the kingdom of Saudi Arabia: an empirical study. Am. J. Educ. Res. **8**(7), 457–464 (2020)
3. Almarashdeh, I.: Sharing instructors experience of learning management system: a technology perspective of user satisfaction in distance learning course. Comput. Hum. Behav. **63**, 249–255 (2016). https://doi.org/10.1016/j.chb.2016.05.013
4. Almaiah, M.A., Alismaiel, O.A.: Examination of factors influencing the use of mobile learning system: an empirical study. Educ. Inf. Technol. **24**(1), 885–909 (2018). https://doi.org/10.1007/s10639-018-9810-7
5. Tarhini, A., Masa'deh, R., Al-Busaidi, K.A., Mohammed, A.B., Maqableh, M.: Factors influencing students' adoption of e-learning: a structural equation modeling approach. J. Int. Educ. Bus. **10**(2), 164–182 (2017). https://doi.org/10.1108/JIEB-09-2016-0032
6. Alamri, M.M., Almaiah, M.A., Al-Rahmi, W.M.: Social media applications affecting Students' academic performance: a model developed for sustainability in higher education. Sustainability **12**(16), 6471 (2020). https://doi.org/10.3390/su12166471
7. El-Masri, M., Tarhini, A.: Factors affecting the adoption of e-learning systems in Qatar and USA: extending the unified theory of acceptance and use of technology 2 (UTAUT2). Educ. Technol. Res. Dev. **65**, 743–763 (2017). https://doi.org/10.1007/s11423-016-9508-8
8. Alksasbeh, M., Abuhelaleh, M., Almaiah, M.A., AL-jaafreh, M., Abu karaka, A.: Towards a model of quality features for mobile social networks apps in learning environments: an extended information system success model. Int. J. Interact. Mob. Technol. (iJIM) **13**(05), 75–93 (2019). https://doi.org/10.3991/ijim.v13i05.9791

9. Hasan, K.K., Mukherjee, D., Saha, M.: Learning continuity during COVID-19 pandemic using the virtual classroom – a cross-border experimental multi case approach. J. Educ. Cult. Soc. **12**(1), 335–354 (2021). https://doi.org/10.15503/jecs2021.1.335.345

10. Shi, L., Stickler, U.: Eyetracking a meeting of minds: teachers' and students' joint attention during synchronous online language tutorials. J. China Comput.-Assist. Lang. Learn. **1**(1), 145–169 (2021). https://doi.org/10.1515/jccall-2021-2006

11. Pokrovskaia, N.N., Leontyeva, V.L., Ababkova, M.Y., Cappelli, L., D'Ascenzo, F.: Digital communication tools and knowledge creation processes for enriched intellectual outcome - experience of short-term e-learning courses during pandemic. Future Internet **13**, 43 (2021). https://doi.org/10.3390/fi13020043

12. Ababkova, M.Y., Leontieva, V.L.: Metaphor-based research for studying Russian and Chinese students' perception of the university. In: Shipunova, O.D., Bylieva, D.S. (eds.) Professional Culture of the Specialist of the Future & Communicative Strategies of Information Society, Volume 98, European Proceedings of Social and Behavioural Sciences, pp. 89–98. European Publisher (2020). https://doi.org/10.15405/epsbs.2020.12.03.9

13. Gao, L.X., Zhang, L.J.: Teacher learning in difficult times: examining foreign language teachers' cognitions about online teaching to tide over COVID-19. Front. Psychol. **11**(549653), 1–14 (2020). https://doi.org/10.3389/fpsyg.2020.549653

14. Agung, A.S.N., Surtikanti, M.W., Quinones, C.A.: Students' perception of online learning during COVID-19 pandemic: a case study on the English students of STKIP Pamane Talino. SOSHUM: Jurnal Sosial Dan Humaniora **10**(2) 225–235 (2020). https://doi.org/10.31940/soshum.v10i2.1316

15. Abbasi, S., Ayoob, T., Malik, A., Memon, S.I.: Perceptions of students regarding E-learning during Covid-19 at a private medical college. Pak. J. Med. Sci. **36**(COVID19-S4) 57–61 (2020). https://doi.org/10.12669/pjms.36.COVID19-S4.2766

16. Pokrovskaia, N.N., Ababkova, M.Y., Fedorov, D.A.: Educational services for intellectual capital growth or transmission of culture for transfer of knowledge - consumer satisfaction at St. Petersburg universities. Educ. Sci. **9**(3), 183 (2019). https://doi.org/10.3390/educsci9030183

17. Wijayanti, K., Nikmah, A., Pujiastuti, E.: Problem solving ability of seventh grade students viewed from geometric thinking levels in search solve create share learning model. Unnes J. Math. Educ. **7**(1), 8–16 (2018). https://doi.org/10.15294/ujme.v7i1.21251

18. Ababkova, M.Yu., Leontieva, V.L., Pokrovskaia, N.N.: S prepodavatelem ili bez? Metod biologicheskoj obratnoj svyazi pri vy'bore formy' organizacii uchebnogo processa = Does a teacher matter? Biofeedback for modelling educational process. ZhivayaPsikhologia **7**(1), 8–20 (2020) (in Russia)

19. Chai, C.S., Lim, C.P.: The Internet and teacher education: traversing between the digitized world and schools. Internet High. Educ. **14**(1), 3–9 (2011). https://doi.org/10.1016/j.iheduc.2010.04.003

20. Vallerand, R.J., Blssonnette, R.: Intrinsic, extrinsic, and motivational styles as predictors of behavior: a prospective study. J. Pers. **60**(3), 599–620 (1992)

21. Lepper, M.R., Cordova, D.I.: A desire to be taught: instructional consequences of intrinsic motivation. Motiv. Emot. **16**(3), 187–208 (1992). https://doi.org/10.1007/BF00991651

22. Wigfield, A.: Expectancy-value theory of achievement motivation: a developmental perspective. Educ. Psychol. Rev. **6**, 49–78 (1994). https://doi.org/10.1007/BF02209024

23. Li, J., Chen, S., Zhu, L., Zhang, W.: Investigation of knowledge of prevention and control and psychological anxiety about H1N1 influenza among college students in a university, Zhengzhou. Mod. Prev. Med. **38**, 3036–3042 (2011)

24. Kim, D., Lee, Y., Leite, W.L., Huggins-Manley, A.C.: Exploring student and teacher usage patterns associated with student attrition in an open educational resource-supported online learning platform. Comput. Educ. **156**, 103961 (2020)

25. Tarman, B.: Reflecting in the shade of pandemic. Res. Soc. Sci. Technol. **5**(2), 1–4 (2020). https://doi.org/10.46303/ressat.05.02.ed

26. Duan, L., Zhu, G.: Psychological interventions for people affected by the COVID-19 epidemic. The Lancet Psychiatry **7**(4), 300–302 (2020). https://doi.org/10.1016/S2215-0366(20)300 73-0

27. Azad, N., Shahid, A., Abbas, N., Shaheen, A., Munir, N.: Anxiety and depression in medical students of a private medical college. J. Ayub Med. Coll. Abbottabad **29**(1), 123–127 (2017)

28. Zivin, K., Eisenberg, D., Gollust, S.E., Golberstein, E.: Persistence of mental health problems and needs in a college student population. J. Affect. Disord. **117**(3), 180–185 (2009). https://doi.org/10.1016/j.jad.2009.01.001

29. Beiter, R., et al.: The prevalence and correlates of depression, anxiety, and stress in a sample of college students. J. Affect. Disord. **173**, 90–96 (2015). https://doi.org/10.1016/j.jad.2014.10.054

30. Jia, N., Fan, N., Lu, Z.: A survey of the undergraduate anxiety in the SARS infected areas. J. Hebei Normal Univ. Educ. Sci. Ed. **5**, 57–60 (2003)

31. Ismaili, Y.: Evaluation of students' attitude toward distance learning during the pandemic (Covid-19): a case study of ELTE university. On the Horizon **29**(1), 17–30 (2021). https://doi.org/10.1108/OTH-09-2020-0032

32. Hasan, K.K., Leontieva, V.L.: Social'no-psihologicheskie trudnost I iosobennosti adaptacii studentov Yugo-vostochnoj Azii k obucheniyu v Rossii = Socio-psychological difficulties and peculiarities of adaptation of Southeast Asian students to study in Russia. XXIX Nedelya nauki SPbGTU. Peter the Great Polythecnical University **6**, 1–12 (2001). (in Russia)

33. Lohmann, M.J., Boothe, K.A., Hathcote, A.R., Turpin, A.: Engaging graduate students in the online learning environment: a universal design for learning (UDL) approach to teacher preparation. Netw. Online J. Teach. Res. **20**(2), 2–21 (2018). https://doi.org/10.4148/2470-6353.1264

34. Nandi, D., Hamilton, M., Harland, J.: Evaluating the quality of interaction in asynchronous discussion forums in fully online courses. Distance Educ. **33**(1), 5–30 (2012). https://doi.org/10.1080/01587919.2012.667957

35. Bolliger, D.U., Martin, F.: Instructor and student perceptions of online student engagement strategies. Distance Educ. **39**(4), 568–583 (2018). https://doi.org/10.1080/01587919.2018.1520041

36. Holzweiss, P.C., Joyner, S.A., Fuller, M.B., Henderson, S., Young, R.: Online graduate students' perceptions of best learning experiences. Distance Educ. **35**(3), 311–323 (2014). https://doi.org/10.1080/01587919.2015.955262

37. Gaytan, J., McEwen, B.C.: Effective online instructional and assessment strategies. Am. J. Distance Educ. **21**(3), 117–132 (2007). https://doi.org/10.1080/08923640701341653

38. Garrison, D.R., Anderson, T., Archer, W.: The first decade of the community of inquiry framework: a retrospective. Internet High. Educ. **13**(1–2), 5–9 (2010). https://doi.org/10.1016/j.iheduc.2009.10.003

39. Robinson, C.C., Hullinger, H.: New benchmarks in higher education: student engagement in online learning. J. Educ. Bus. **84**(2), 101–109 (2009). https://doi.org/10.3200/JOEB.84.2.101-109

40. Iyer, R., Luke, C.: Multimodal, multiliteracies: texts and literacies for the 21st century. In: Pullen, D., Cole, D. (eds.) Multiliteracies and Technology Enhanced Education: Social Practice and the Global Classroom, pp. 18–34. IGI Global (2010). https://doi.org/10.4018/978-1-60566-673-0.ch002

41. Waheed, M., Leišytė, L.: German and Swedish students going digital: do gender and interaction matter in quality evaluation of digital learning systems? Interact. Learn. Environ. (2021). https://doi.org/10.1080/10494820.2021.1965626

42. Thompson, G., McBride, R.B., Hosford, C.C., Halaas, G.: Resilience among medical students: the role of coping style and social support. Teach. Learn. Med. **28**(2), 174–182 (2016). https://doi.org/10.1080/10401334.2016.1146611

43. Jebb, A.T., Ng, V., Tay, L.: A review of key Likert scale development advances: 1995–2019. Front. Psychol. **12**, 1590 (2021). https://doi.org/10.3389/fpsyg.2021.637547

44. Jang, H.R.: Teachers' intrinsic vs. extrinsic instructional goals predict their classroom motivating styles. Learn. Instruction **60**, 286–300 (2019). https://doi.org/10.1016/j.learninstruc.2017.11.001

45. Otis, N., Grouzet, F.M., Pelletier, L.G.: Latent motivational change in an academic setting: a 3-year longitudinal study. J. Educ. Psychol. **97**(2), 170–183 (2005). https://doi.org/10.1037/0022-0663.97.2.170

46. Muljana, P.S., Luo, T.: Factors contributing to student retention in online learning and recommended strategies for improvement: a systematic literature review. J. Inf. Technol. Educ. Res. **18**, 019–057 (2019). https://doi.org/10.28945/4182

47. Dubinina, G., Konnova, L., Stepanyan, I.: Technologies for teaching mathematics in a multilingual digital environment. Educ. Sci. **12**(9), 590 (2022). https://doi.org/10.3390/educsci12090590

Psychological Risks of a Successful Pupil

Sofya Tarasova(✉) [iD]

Psychological Institute of Russian Academy of Education, Moscow, Russia
syurarasov@yandex.ru

Abstract. This study examines risk factors of a successful pupil's psychological health. Introduction of new technologies and multilingualism in educational technologies influence emotional life of an adolescent. The purpose is to discover correlations between trait anxiety and aggressiveness among adolescents in prestige educational institutions. The study explains the differentiation between adaptive and maladaptive perfectionism. The overall excellent performance of a pupil may hide risks of adaptation disruption. The analysis of psychological profiles of adolescents from risk group for maladaptive perfectionism was performed using qualitative and quantitative research methods. The total of two hundred pupils of the 8th grade from three prestige schools took part in the research. According to the research, the markers of maladaptive behavior of a successful pupil are trait anxiety, maladaptive perfectionism, hostility, and anger. About 20% of the participants fall into the risk group for anxiety. Objectively, their performance is high, but teenagers have doubts about their success, demonstrate tendency for delaying, the inability to start doing something. In this case the anxiety is caused by fear of not conforming to self-imposed high standards. Self-esteem anxiety is related to the components of Self-Image: intelligence, situation at school, communication, self-confidence. Non-conformity to self-imposed standards and procrastination are related to anger. As qualitative research methods showed psychological risks may reach the level of destructive personality tendencies. Anxiety markers of a successful pupil, including psychosomatic manifestations were described. They can be useful for teachers. Prevention work in schools has to be complex and interdisciplinary. It is a teacher-psychologist tandem that is needed.

Keywords: Successfulness · Anxiety · Perfectionism · Destructive personality tendencies · Risk factors

1 Introduction

In recent years the number of destructive manifestations among adolescents has been rising. Sometimes these destructive manifestations, aggression and auto-aggression are found in prestige schools, high-rated schools, language schools, and gymnasiums. This happens all of a sudden when teachers, classroom teachers and administration don't expect. Is it possible to identify this "sleeping" destruction in advance using psychological methods? What a teacher can do in this respect?

The thesis which was carried out on the premises of a high-rated Moscow school was focused on anxiety [1]. One of them is a school with a focus on learning English in

Moscow (located at Kutuzovsky Prospekt). Below is a detailed description of this school and its language environment that places higher standard demands on its students. In a longitudinal research we studied anxiety of children and adolescents on behavioral, psychological and physiological levels. We discovered a risk group for anxiety described by A.M. Prikhozhan as well-being, inadequately anxious children. While their position in pupil's community, appearance, their parents' social status and performance are good, those children demonstrate constant spread anxiety. They often lack self-confidence and have strong fear of making mistakes. On the one hand, a conflicted sense of self-esteem makes them strive for success. On the other hand, it causes constant doubts like "what if I haven't achieved enough?" They want to be successful, and teachers sometimes even add fuel to this desire. High performance is important for teachers, but the price for high scores may be psychological health. In elementary school the Phillips Questionnaire factor *A Fear of Knowledge Check Situation* is the most informative for estimation of anxiety level. Sometimes children and adolescents quite frankly state in a self-report: "I fear low grades. I fear that I won't have friends. I'm afraid of losing the class leader status". The real "self" never approaches the level of the ideal "self", unwinding the vicious circle of self-improvement. In teenage period the psychological marker is a high value of maladaptive perfectionism factor in the Slaney model [2]. During the whole educational period the risk group constantly includes about one third of the pupils. The same pupils according to the results of both participant and consultancy observation are prone to auto-aggressive behavior. We called this group "maladaptive perfectionists", and it is the main subject of this conversation.

Anxiety as a stable personality trait is related to defense of usual self-perception. Self-esteem anxiety is related to self-criticism, personal standards, attitude towards success and failure. Self-esteem anxiety is an important regulatory instrument for a person's behavior and depends on relationships with others, especially meaningful ones, relatives, and friends. High level of self-esteem anxiety may become a need as it allows a person to avoid making a choice: "I'm a loser, there is no point in trying". Self-esteem anxiety may be considered as a defense mechanism and analyzed as a result of communication between a personality and people around. Self-esteem anxiety is connected with maladaptive perfectionism [3–11].

The idea of differentiating between positive and negative perfectionism is reflected in existing terminology: normal and neurotic, self-oriented and socially prescribed, destructive narcissism. In this study we base on the R. Slaney model of adaptive and maladaptive perfectionism [12]. "Almost Perfect Scale (APS) by Slaney includes important aspects of perfectionism that no other questionnaire has. Like Frost, Slaney assumes that perfectionism can have not only negative but also positive aspects. Positive aspects include high performance standards and need for order. Negative aspects include feeling of non-conformity to standards and performance, anxiety, procrastination, inability to start doing something and difficulties in relationships" [13, p. 240]. In the Slaney model the key factor is *Non-conformity* – fear of not comforting to self-imposed high standards [12]. Besides, our choice of the Slaney model was based on successful though still little experience of application of this questionnaire in prestige schools with high pupils' performance. This study is a contribution to its further validation.

Let's analyze the social aspect of self-esteem anxiety from the very beginning, i.e. from the first years of life of a baby. We refer to psychoanalysis as it considers adaptation to environment as a basis of human development. Born in Russian psychology cultural-historical approach by Lev Vygotsky does not contradict psychoanalysis. Since birth both humans and animals learn to identify signs of potential threats in the outside world. Starting from early days a child experience anxiety and stress when physiological needs are not satisfied or there are threats to safety. In a preschool period, a child may feel uncomfortable because of separation from mother or when basic needs are not satisfied. Stress experienced by a mother transfers to her baby. According to the author of the interpersonal approach in psychiatry Harry Stack Sullivan, this is "anxiety theorem #1". Only thanks to mothering care the world "retracts its claws": the cruel biological law according to which the survival of one living being requires killing of other living beings softens. In a normal course of development basic trust to the world and full emotional contact with reality form. In a "neurotized" family anxiety appears in the mother-child diad in the first place being further cemented in other important relationships. From the psychoanalysis point of view neurotization almost always takes place to some extent. Thus, it is reasonable to talk about anxiety as a function of interpersonal relationships. Self-esteem anxiety appears in a situation of conditional acceptance: "I will love you if you... get high scores for Algebra", "You will be good for me if you...speak English well, are admitted to a prestige university, earn a lot in your adulthood". Conditional acceptance is a variant of pedagogical practice, family scenario that cultivates maladaptive perfectionism.

Introduction of new technologies and high pressure for acquisition of foreign languages and computer (robot) languages change human affective life [14]. Digital technologies bring forth new problems. "Bernard Stiegler defines digital technologies as φάρμακον – the term used in philosophy and critical theory that combines two - medicine and poison – or even three meanings (Stiegler, 2019). It refers to inherent tension of digital world and technologies which promise to provide more opportunities for human culture, on the one hand, and have destructive power, on the other hand" [15, p. 171]. A successful and much in-demand man of today speaks different languages and masters new technologies. Modern society requires a person to be multilingual even in everyday life. Along with international English language a person traditionally loves his or her native language. For example, Germans are great lovers of German language. Thus, to be a much-needed specialist and a successful member of community one has to speak more than one foreign language. Therefore, parents put such pressure on their children in relation to foreign language acquisition ("You have to learn English, otherwise it will be a great shame").

By teenage defined, enrooted standards internalize. In a normal variant adaptive perfectionism forms – high standards of performance and strive for order. Children of neurotized parents develop maladaptive perfectionism – feeling of not comforting to standards and performance, anxiety, procrastination as inability to make a decision and start doing something, relationship problems. Inability to constantly meet one's own standards generates and cements high level of self-esteem anxiety.

Family, school, society that demands to be successful – everything matters. The first edition of "The Neurotic Personality of Our Time" by Karen Horney came out in 1937:

"The quest for power, possession can, in our culture, serve as a protection against help-lessness and insignificance or humiliation as wealth gives both power and prestige... The specific fear against which possession is a protection is that of impoverishment, destitution, dependence on others – the fear of impoverishment may be a whip driving a person to work incessantly and never miss a chance of earning money" [16, p. 135]. Has the world changed a lot since then? New communication technologies and social networks have made the space dwindle. A shooter from a US school was seen by adolescents around the world. Information pressure to be supercool, successful has also risen. Information pressure overlaps the inherent emotional immaturity of teenagers. Let's take as an example a heading of the video in the web: "My lips are as big as my success'.

Meta-analysis proves that perfectionism can be related to different types of destructions – from self-esteem problems to eating disorders and depression [17–19]. Several experts think that perfectionism is a transdiagnostic factor of anxiety, eating disorder and depression [20, 21]. That's why it is important to timely identify and work on perfectionism before symptoms of a disease appear. According to Blatt, a perfectionist is so self-focused that interpersonal communication suffers [22, 23]. Here are other words of the classic: "It is obvious that the tendency to deprive or exploit, like all the other hostile tendencies we have discussed, not only arises from impaired personal relations but results in further impairment" [16, p. 143]. In addition, there are data proving the correlation between perfectionism and anger among adolescents [24].

The other side of fear and anxiety is aggression. A. Guggenbühl thinks that fear and aggression are "anthropological constants". In the history of society only forms of fear and aggression change while meaning remains the same. There is no consensus about the definition of aggression. Guggenbühl refers to K. Lorenz and defines it as follows: "...aggression is not reaction to external factors but real instinct that serves self-preservation purposes in animal world. In human world this instinct was hypertrophied, derailed and became grotesque" [25, p. 11]. The same opinion is shared by some other German-speaking authors [26]. We preferred the following definition of aggression: it is behavior directed at doing physical or psychological harm or damage contradicting to the rules of human existence and bringing negative emotions, fear and depression [27]. Violence threat is considered violence too. Aggressiveness is a fairly stable personality trait that depends on many factors. The aggressiveness level is related to adoption of social code, penalty fear, adequate strategies of overcoming stress in the arsenal, etc. Hostility is a negative setting towards another person or a group of people. Aggressiveness and hostility are factors of proneness to aggressive behavior.

We suggested that adolescents from the risk group for maladaptive perfectionism have personality traits manifested as destructive tendencies. These destructive manifestations are not on the surface and are not obvious for a school teacher. Moreover, a teacher often doesn't even expect any problems from such a successful pupil. Despite this, destructive tendencies of a maladaptive perfectionist are revealed in his or her behavior.

The purpose of this study is to discover correlations between trait anxiety and aggressiveness among adolescents in prestige educational institutions. At the same time, we are interested in manifestations of maladaptive perfectionism. In addition, we made an attempt to describe behavioral signs of anxiety in a successful pupil for school teachers.

2 Materials and Methods

The method by Prikhozhan for studying the Self-Image of teens in the age group 12–17 years. Factors: *Behavior*: self-assessment of one's behavior against demands of adults. *Intelligence and position at school*: self-assessment of one's intelligence and school performance. It is compared with the real situation in the class. *Situation at school*: self-assessment of a school situation, whether it brings about positive emotions or anxiety. *Appearance*: self-assessment of physical attraction to peers. *Anxiety*: self-assessment of the level of anxiety. *Communication*: self-assessment of communication skills. *Happiness and satisfaction*: satisfaction or dissatisfaction with the present life situation (e.g. How fortunate am I right now?). *Family position*: self-assessment of the family position. *Self-confidence*: self-assessment of self-confidence.

The Buss-Perry Aggression Questionnaire as adapted by S.N. Yenikolopov. Scales: *Physical aggression* – self-report on one's behavioral tendency for aggression; *Anger* – self-report on one's tendency for irritation; *Hostility* – self-report on one's readiness for aggressive behavior.

The APS-R Perfectionism Scale as adapted by S.N. Yenikolopov. The scales of adaptive perfectionism are: *Standards* – pursuit of high personal standards. *Order* – tendency to maintain order, be accurate, «be on a friendly note» with time. *Maladaptive perfectionism*: *Non-conformity* – the feeling of being unable to adhere to self-established high standards. *Relationships* – difficulties in interpersonal relationships as a consequence of constant distress. *Procrastination/Anxiety*, where procrastination is the propensity for delaying, the inability to start doing something, and anxiety is caused by the inability or fear of not conforming to self-imposed high standards. This factor reveals such personal traits as emotional instability, dependency on opinions of others, anxiety, and the propensity to delay things.

Incomplete sentences. This method was used to identify the participants' personal problems.

The Hand Test. An interpretative projective technique used to interpret the meaning of hand poses for those surveyed. This was used to identify meaningful needs, motives, and personality conflicts. Also, the test is aimed at detecting *expectation of aggression* from the immediate surroundings.

Pathopsychological test. This method was used for examination of the pupils' adaptation potential.

Extended clinical conversation. Gathering of psychological histories. They were used to examine the participants' development social situation.

Torrance figurative subtests. They are focused on creativity. In this work they were analyzed in the same way as a projective drawing technique.

Sociometry. This method was used in the given social group (class).

In this study quantitative surveys were combined with qualitative semi-projective methods. This allows to mutually validate the results.

Methods of Data Analysis
Methods of descriptive statistics were used.

Participants of the Research. The research was carried out on the premises of three prestige schools:

– School 1 – a gymnasium in Dmitrov (Moscow region)
– School 2 – a school with a focus on learning English in Moscow (located at Kutuzovsky Prospekt that is considered a prestige district in Moscow)
– School 3 – a general secondary school in Moscow

The schools are called prestige because they have the selection principle of admission. A great focus is placed on foreign languages (English and German), Russian language and Literature in these schools. For example, in the gymnasium in Dmitrov the following subjects are taught: Literature and Cinematography, Native Russian Literature, Native Russian Language, Russian Language Arts. The total of 200 adolescents – 110 boys and 90 girls – took part in the research. Their participation was voluntary.

Language Aspect in Schools

There are six first foreign language lessons (English) per week. Starting from the 5th grade the second foreign language is added (German, French, or Chinese).
Here is the timetable of the 8th grade in the gymnasium in Dmitrov:

№	Monday	Tuesday	Wednesday	Thursday	Friday
1	Geometry	Geography	Language arts	Russian language	Social studies
2	Physics	Geometry	Biology	Geography	Chemistry
3	Geometry	Algebra	History	Chemistry	English language
4	Algebra	History	English language	Physics	Russian language
5	Choreography	Literature	English language	English language	Physical education
6	English Language	Economics	Algebra	Russian language	Physical education
7	Literature	Handicraft	Russian language	English language	Literature
8	Physical education	Biology	Safety management in emergencies	German language	Algebra
9		Music	Classroom hour		

The Dmitrov school has the following departments: Foreign Languages, Sciences, Mathematics, Russian Philology, Elementary School, Social and Humanitarian Sciences, Technology and Computer Science, Physical Education, Classroom Teachers.
Below are the departments' annual reports for 2021–2022.

The Foreign Languages Department. The teachers completed the following continuing education courses: Development of reading skills in pupils using PISA test; Effective communication in foreign and Russian languages in professional activity; Management of Educational Content in the context of functional literacy development among pupils at all educational levels. Also, below are the examples of demo lessons:

Arrangement and execution of All-Russian Testing on English language in the 8th grade.
"Critical thinking development technology. Yes-or-no questions and problem questions".
Robert Burns' date of birth
Extraordinary lifestyle. Infinitive/Gerund.
"Bad habits"
"English- speaking countries. Quest game" (Open house day for parents).
"Reading literacy: privilege for elites or a useful habit of a man of today?"
How to create catchy slogans/Compound nouns.
Battle as a group project and a training before final examination essay in English.
Preparing for the final exams: a letter about ecology problems. The 8th grade.
Development of reading literacy in English classes.
Buy! Commercial tricks.

Russian Philology Department. Examples of demo lesson topics:

Working with talented children.
Business game "Imagine that…" (Business style. Execution of documents: CV, explanatory letter).
Development of a programme on off-hour activity "Literature and Cinematography".
Voyage to Russian Language Country.
Anton Chekhov. Life and Work.
Living Classics Competition.

Extracts from the Classroom Journal. English Language. What kind of knowledge and skills was acquired during the lesson?

Everyday English. Buying formal clothes (dialogue).
Helping people by means of arts. Monologue (share impressions about being the main character's assistant).
Conjunctive.
Arts festival. Events. Verbs.
News and Media. Information channels. Newspapers.
Metadisciplinarity: media courses. IT project "How holography can change your life?"
Reviews, summaries. Review of an adventure movie (compliance test).
Adjectives, adverbs. Antonyms. Short article about the favourite movie.
Advanced level tests on Passive voice.
Living Statue Life. A Personal Letter.
Advanced level tests on Conditionals.
English in Practice. Phrasal verbs with the *off*-preposition.
Science (Vocabulary). Indirect speech.
Indirect speech questions.
Vocabulary activization. Science and health.
Science (Vocabulary). Indirect speech.

Indirect speech questions.
Everyday English. Persuasion. "Visiting a Science Museum".
Exploring the unknown. Causative form.
Way to success. Personal traits. Countable and uncountable nouns.
Metadisciplinarity. Science. "One or two brains?" IT project "Left or right hemisphere?"
Quantitative words. Countable/uncountable nouns
"Eureka! A Museum Tour". What do you know about Hermitage?
English in practice. Phrasal verbs with the *down*-preposition. IT project "Hermitage Exhibited Items". Lost cities. IT project "Important archeological finding in Russia". Culture corner. "What do you know about Dickens?" IT project "A Prominent Russian writer from the past". Final test analysis. Revision and summing-up.

Let's introduce a school with a focus on learning English at Kutuzovsky prospect. Official information:

The school sets its goal to provide high-quality general and linguistic education that will prepare a student for professional activity, will provide solid background for speaking English fluently and reading literature in the original. High school pupils have Business English and Original English literature classes in their timetable. The aim of the educational programme is to arrange profound studying process with a special focus on English language as means of developing communicative and social competency of a student. In the 10–11th grades subject-oriented teaching of Anglo-American literature and Business English is introduced.

It is interesting to note comments about the school in the Internet. They are made by the parents who had attended this school and now had their children studying there. "In this school we had the Theatre and Children Week (non-existent for now), field trips, school fests, visits to museums (though not often). There was an exchange programme with the US school BB&N. I think it is a usual programme with the remainder of the previous high-class English education It's easier to tell what has changed in this school: lower volume of home reading of English classics, no theatre week (including in English), English history and literature (of course, in English and backed by English materials). As for typing in English I don't know if it has any sense nowadays. Once this school was the best not only in Moscow, but also in the USSR". Other people describe this school as a good language school of today. In the context of perfectionism demands we need to add that this school had closed the question of suicide in Soviet times.

3   Results

Below is interpretation of the results. We estimated the risk group for trait anxiety in each school in percentage terms. For this purpose, we counted the number of adolescents from the risk group according to the *Anxiety, self-assessment of the anxiety level* factor from the Prikhozhan method. The risk group included pupils with the score of > 8 according to this factor. In the gymnasium in Dmitrov no emotionally well-being pupils were found but only highly anxious ones. In two other schools the number of adolescents from the risk group varies from 16% to 20%. The results were as such though we tried to exclude situations of anxiety growth among pupils, i.e. we carried out our research when there were no tests and exams. May was excluded from the research period as well as it precedes the final round of exams before summer holidays. It should be noted that the gymnasium takes an active part in academic competitions. The share of those graduating with honors in the gymnasium accounts for up to 35% of all graduates. More than 80% of the gymnasium graduates enter non-profit educational organizations every year.

Anxiety correlates with aggressiveness. The values of procrastination/anxiety according to the Slaney model positively correlate with anger and hostility in adolescents (from $r = 0.28$ to $r = 0.52$ where $p < 0.05$). According to the sociometry method, the most hostile pupils are more isolated in the class. This study also proves difficulties in communication linked to inadequately high self-imposed standards.

Let's look at statistics more closely.

The schools revealed statistically meaningful connections between the Aggression Questionnaire values and the Perfectionism Scale values (Tables 1 and 2).

Table 1. The results of the correlation analysis of the aggression questionnaire values and perfectionism scale values in the school 1.

Method scales	Spearman's rank correlation coefficient		
	Physical aggression	Anger	Hostility
	r	r	r
Non-conformity	.31*	.49***	.63***
Relationship	n.s.	.41**	.55***
Procrastination/anxiety	n.s.	.51***	n.s.

*Symbols: n.s. – a non-significant value; symbols indicate the significance level: * –p < 0.05; ** –p < 0.01; *** –p < 0.001.*

Table 2. The results of the correlation analysis of the scales values from the aggression questionnaire and the perfectionism scale in the school 3.

Method scales	Spearman's rank correlation coefficient		
	Physical aggression	Anger	Hostility
	r	r	r
Non-conformity	.28*	.31**	.46***
Relationship	n.s.	.36**	.51***
Procrastination/anxiety	.27*	.28*	.52***

The school 2 demonstrated similar results. The pictures of anxiety and aggressiveness in three schools have slight differences, but they look similar. And how the results of adaptive and maladaptive perfectionism correlate with the Self-Image components? (Table 3).

The values of maladaptive perfectionism are connected with the components of Self-Image (Table 3).

Table 3. The results of the correlation analysis of the Self-Image components and maladaptive perfectionism in the gymnasium.

Method scales	Spearman's rank correlation coefficient		
	Order (adaptive perfectionism)	Interpersonal relationships (maladaptive perfectionism)	Procrastination/anxiety (maladaptive perfectionism)
	r	r	r
Behavior	.64**	−.62*	−.70*
Intelligence	.54*	−.72*	−.74***
Situation at school	.62**	−.74***	−76***
Appearance	.53*	n.s.	n.s.
Anxiety	−.70**	.73***	.76***
Communication	.66**	−.53*	−.69*
Happiness and satisfaction	.61**	−.58*	−.62*
Family situation	n.s.	−.59*	n.s.
Self-confidence	.66**	n.s.	−.48*

In two other schools the pictures of self-esteem anxiety are similar. The most important Self-Image components are: situation at school, self-assessment of intelligence, communication, happiness and satisfaction.

Earlier we identified the risk group of pupils for anxiety. These adolescents according to the pathopsychological test, the hand test, the Torrance figurative subtest, and the participant observation data demonstrate destructive personality tendencies.

We want to note the efficiency of the Torrance figurative test as a projective technique. First, nowadays many techniques from the psychologist's arsenal are widespread and well-known. That's why techniques that are new to the participant work better. Standard instruction encourages to give non-standard answers, to create unusual plots as much as possible, driving the participant to unveil the whole range of personality traits. In this way we "deceive" the participant. Second, the technique is interesting for talented children as it allows them to demonstrate maximum creativity. If we analyze the Torrance figurative subtests like projective tests, we can see destructive personality tendencies.

To sum up, adolescents from the risk group for maladaptive perfectionism possess personality traits manifested in the form of destructive tendencies. High level of self-esteem anxiety is difficult to identify without professional psychological diagnostics. Procrastination as inability to make a choice, start doing something because of anxiety is manifested in the teenager's behavior. Strive for perfection brings about problems where they don't have to appear otherwise.

In the course of a more comprehensive research we also studied non-prestige schools. Using the same battery of tests, we painted a different picture of anxiety and maladaptation.

4 Discussion

The Role of a Teacher. What Characteristic Should a Teacher Have?

Like a teacher of a talented pupil, a teacher of a successful pupil has to possess special personality traits. It is desirable to develop and maintain programmes on popularization of psychological health in the teaching circle to diminish negative effects of difficult work. It is necessary to promote the value of caring for psychological health. It is a preventive measure against professional burnout.

The analysis of the psychotherapeutic alliance "doctor - patient" showed that the more perfectionistic is the patient the less realistic is his or her perception of the psychotherapist (Miller, 2017). In the primary analysis of such alliance performed by Blatt (1996) he examined the connection between perfectionistic settings of the patient before treatment and perception of Rogerian traits of the therapist by the patient (empathetic understanding, unconditional respect). Even in the beginning of the therapy patients with higher level of maladaptive perfectionism think of his or her therapist as less empathetic and more critical than do other patients. We think that these findings can be carried on to the alliance "teacher - pupil". A teacher has to be keen on teaching, to see a personality in the pupil, not performance. Such teachers communicate with pupils like with interesting people. They discuss topics not directly related to school subjects more often, show respect towards their interlocutors, defer to their opinion. Empathy and only empathy. A robot should not replace a teacher! Or, it has to be an empathetic robot. Whether development of new technologies makes it possible, we'll see in the future. In fantasy novels empathetic nature of robots is often dual: robots make human life useful as long as this doesn't damage them. The robot pilot Calder in the story "The Inquest" about Pirx the Pilot by Stanisław Lem is a great example. So far, we are only seeking Man.

The above-mentioned important traits of a teacher can be developed during trainings. Such a training has three purposes. First, reflection on oneself, one's own character and understanding of other people. Second, development of skills to create individual programmes for children. Third, improvement of personality sensitivity and communication and hearing skills. Many successful pupils experience difficulties primarily because teachers and school administration are not able not hear them.

Anxiety Signs:

Let's outline the markers of anxiety of a successful pupil. Its main marker is maladaptive perfectionism. Destructive tendencies of a maladaptive perfectionist are manifested in behavior. Maladaptive perfectionists have some popularity in the class, but they also have strong fear of mistakes, materialized more often in public situations. Such children and adolescents fear to lose their status in the social group. They can refer to a teacher, a classroom teacher or a school psychologist: "Can you help me to restore my position in the class?" In private conversations they complain about their doubts in success and that achievements don' t bring much (or any!) joy.

Suffering from non-compliance with the ideal may materialize at body level. By teenage female participants from this risk group may have eating disorders related to dissatisfaction with their appearance. They like to cook, to offer food to their relatives and friends, to enthusiastically talk about food, but at the same time they eat very little. Thus, psychosomatic manifestations up to eating disorders form the second marker of anxiety.

The third marker of a successful pupil from the risk group is procrastination. Inability to start doing something is more than visible to an empathetic teacher.

Aggression is the most obvious sign of anxiety. According to sociometric studies, maladaptive perfectionists are usually not stars, but not outcasts too. They are very sensitive to their position in the class among their classmates and friends. If someone in the class tries to change the established situation, "rejection and betrayal drive them wild".

Aggression may be redirected inside personality and transform into auto-aggression. Teenagers often write somber poems in the Silver Age style. If they hide them and show them sneakily and rarely only to a selected teacher they trust – this is a worrying sign, a cry for help. Or if children get interested why someone who is well-being commits a suicide and decide to carry out a sociological survey on this topic. High interest to the plot of "Anna Karenina" by Lev Tolstoy is also, in our opinion, a sign of destructive personality tendencies. This means there is work for psychologist.

5   Conclusions

1. Language schools that admit pupils on the basis of a selection principle were called prestige. The risk group for anxiety accounts for about 20% of the research participants. Objectively, their performance is good, but the adolescents have doubts about their success, demonstrate tendency for delaying and inability to start doing something. In this case anxiety is caused by fear not to comply with self-imposed high standards.
2. Markers of maladaptive state of a successful pupil are trait anxiety, maladaptive perfectionism, hostility, and anger. Self-esteem anxiety is related to the components of Self-Image: intelligence, situation at school, communication, self-confidence.
3. In earlier works we identified highly anxious pupils. According to the results of the qualitative research methods, psychological risks for these pupils may reach the level of destructive personality tendencies. The psychological test battery that allows a timely identification of destructive personality tendencies of successful pupils was developed and tested.

References

1. Tarasova, SJu., Osnitsky, A.K.: Physiological and behavioral indicators of school anxiety. Psychol. Sci. Educ. **20**(1), 59–68 (2015). https://doi.org/10.17759/pse.2015200107

2. Tarasova, S.: A picture of trait anxiety and aggressiveness among adolescents from different types of educational institutions. Psychol. Russ. State Art **14**(4), 130–148 (2021). https://doi.org/10.11621/pir.2021.0409

3. Egan, S., Wade, T., Shafran, R.: Perfectionism as a transdiagnostic process: a clinical review. Clin. Psychol. Rev. **31**(2), 203–212 (2011). https://doi.org/10.1016/j.cpr.2010.04.009. Epub 5 May 2010

4. Stoeber, J., Harvey, L.N., Almeida, I., Lyons, E.: Multidimensional sexual perfectionism. Arch. Sex. Behav. **42**(8), 1593–1604 (2013). https://doi.org/10.1007/s10508-013-0135-8

5. Lloyd, S., Schmidt, U., Khondoker, M., Tchanturia, K.: Can psychological interventions reduce perfectionism? A systematic review and meta-analysis. Behav. Cogn. Psychother. **43**(6), 705–731 (2015). https://doi.org/10.1017/S1352465814000162

6. Halldorsson, B., Creswell, C.: Social anxiety in pre-adolescent children: what do we know about maintenance? Behav. Res. Ther. **99**, 19–36 (2017). https://doi.org/10.1016/j.brat.2017.08.013. Epub 1 Sep 2017

7. Dobos, B., Pikó, B.: The role of perfectionism, social phobia, self-efficacy and life satisfaction in the background of trait anxiety. Psychiatr. Hung. **33**(4), 347–358 (2018)

8. Reichenberger, Y., Blechert, J.: Malaise with praise: a narrative review of 10 years of research on the concept of fear of positive evaluation in social anxiety. Depression Anxiety **35**(12), 1228–1238 (2018). https://doi.org/10.1002/da.22808

9. Abdollahi, A.: The association of rumination and perfectionism to social anxiety. Psychiatry **82**(4), 345–353 (2019). https://doi.org/10.1080/00332747.2019.1608783. Epub 21 May 2019

10. Pietrabissa, G., et al.: Measuring perfectionism, impulsivity, self-esteem and social anxiety: cross-national study in emerging adults from eight countries. Body Image **35**, 265–278 (2020). https://doi.org/10.1016/j.bodyim.2020.09.012

11. Doyle, I., Catling, J.: The influence of perfectionism, self-esteem and resilience on young people's. Ment. Health J. Psychol. **156**(3), 224–240 (2022). https://doi.org/10.1080/00223980.2022.2027854. Epub 24 Feb 2022

12. Slaney, R.B., Rice, K.G., Mobley, M., Trippi, J., Ashby, J.S.: The revised almost perfect scale. Measur. Eval. Couns. Dev. **34**(3), 130–145 (2001)

13. Shchipitsyna, A.S.: Perfectionism measuring techniques. PSHPU Newsl. Psychol. Pedagogical Sci. **1**, 235–245 (2014)

14. Pezzica, L.: On talkwithability. Communicative affordances and robotic deception. Technol. Lang. **3**(1), 104–110 (2022). https://doi.org/10.48417/technolang.2022.01.10

15. Saltanovich, I.: Global-local cultural interactions in a hyperconnected world. Technol. Lang. **3**(2), 162–178 (2022). https://doi.org/10.48417/technolang.2022.02.10

16. Horney, K.: The Neurotic Personality of Our Time. Self-Analysis. Progress, Moscow (1993)

17. Thomas, M., Bigatti, S.: Perfectionism, impostor phenomenon, and mental health in medicine: a literature review. Int. J. Med. Educ. **11**, 201–213 (2020). https://doi.org/10.5116/ijme.5f54.c8f8

18. Robinson, K., Wade, T.: Perfectionism interventions targeting disordered eating: a systematic review and meta-analysis. Int. J. Eat. Disord. **54**(4), 473–487 (2021). https://doi.org/10.1002/eat.23483. Epub 17 Feb 2021

19. Morgan-Lowes, K., et al.: The relationships between perfectionism, anxiety and depression across time in paediatric eating disorders. Eat. Behav. **32**, 101305 (2019). https://doi.org/10.1016/j.eatbeh.2019.101305

20. Bardone-Cone, A., Lin, S., Butler, R.: Perfectionism and contingent self-worth in relation to disordered eating and anxiety. Behav. Ther. **48**(3), 380–390 (2017). https://doi.org/10.1016/j.beth.2016.05.006. Epub 7 June 2016

21. Drieberg, H., McEvoy, P., Hoiles, K., Shu, C., Egan, S.: An examination of direct, indirect and reciprocal relationships between perfectionism, eating disorder symptoms, anxiety, and

depression in children and adolescents with eating disorders. Eat. Behav. **32**, 53–59 (2019). https://doi.org/10.1016/j.eatbeh.2018.12.002. Epub 19 Dec 2018

22. Blatt, S.J.: Level of object representation in anaclitic and introjective depression. Psychoanal. Study Child **29**, 107–157 (1974)
23. Miller, R., Hilsenroth, M., Hewitt, P.: Perfectionism and therapeutic alliance: a review of the clinical research. Res. Psychother. **20**(1), 264 (2017). https://doi.org/10.4081/ripppo.201 7.264
24. Ćorluka Čerkez, V., Vukojević, M.: The relationship between perfectionism and anger in adolescents. Psychiatr. Danubina **33**(Suppl 4), 778–785 (2021)
25. Guggenbühl, A.: Incredible Fascination of Violence. Prevention of Child Aggressiveness and Cruelty. Cogito Center, Moscow (2006)
26. Walter, G., Nau, J., Oud, N.: Aggression und Aggressions-Management. Huber, Bern (2012)
27. Tarasova, S.Ju., Osnitsky, A.K., Enikolopov, S.N.: Social-psychological aspects of bullying: interconnection of aggressiveness and school anxiety. Psychol. Sci. Educ. **8**(4), 102–116 (2016). https://doi.org/10.17759/psyedu.2016080411

Psychological Readiness for Intercultural Communication in Students Learning German as Second Foreign Language

Valentina Dolgova[(⊠)] [ID] and Ivan Scorobrenko [ID]

South Ural State University for the Humanities and Education, Chelyabinsk, Russia
23a12@list.ru

Abstract. Globalization processes erase political, ideological and cultural borders, hence promoting constant intercultural cooperation. However, there is currently a practical problem to deal with as on the one hand, there is a clear awareness of an unsatisfactory level of readiness for intercultural communication in university graduates, and on the other hand, higher education has proven to be rather ineffective in providing such training. Thus, the present study was initiated to address the issue of ability for intercultural interaction in students learning German as their second foreign language. The study implemented theoretical methods in the form of an analysis of literature on psychology and education, synthesis, compilation of findings and definition of objectives, and empirical methods in the form of testing: a Psychological Readiness for Intercultural Communication questionnaire by V.I. Dolgova, E.A. Vasilenko and A.S. Baronenko, and a Self-Evaluation of Mental Conditions test by G.Yu. Ayzenk. Statistical methods were also applied. The study was conducted at the South Ural State Humanitarian Pedagogical University in Chelyabinsk. In total, 27 fourth-year full-time students of the German and Teaching of German Department of the Foreign Language Faculty participated in the study. The ascertaining experiment showed a rather low level of readiness for intercultural communication in students learning German as their second foreign language.

Keywords: Psychological readiness · Intercultural communication · Students · Mental state · Mental condition · Self-reflection

1 Introduction

Many Russian and foreign researchers addressed the issue of readiness for intercultural communication in their studies [1–6]. Literature review showed that although there are few practical applications of these studies, the issue of readiness for intercultural communication is currently being rather actively researched theoretically [7–11].

The number of works published in the last 5 years proves a high academic interest to the issue [12]. A large number of publications on psychological readiness to intercultural communication also shows a high scientific relevance of the topic. Many academics from all over the world constantly highlight the importance of a formed psychological readiness for intercultural communication. Therefore, all of the above proves the relevance of this study in the current conditions of a multicultural world associated with globalization, widening cultural arenas and growing communication processes between countries.

For many people, the most difficult thing is to understand the cultural mindset of other nations and to comprehend their ways of thinking, living and behaving [13]. This necessitates new methods, forms and techniques of psychological and pedagogical work that can be realized through a programme of psychological and pedagogical support in such a process [14, 15].

Contemporary conditions of today's life define a modern society's need for a multicultural individual who is ready and eager to efficiently cooperate in the global environment incorporating many different cultures.

The very understanding that there are real differences between cultures is already a key to successful international business relations today [16]. This statement truly helps realize the relevance of the issue of the building of psychological readiness for intercultural communication and cooperation in the modern multicultural world.

Experts often highlight the importance of building psychological readiness for intercultural communication in a foreign language [17]. This idea is supported by R. R. Sharma as well; the scholar believes that the quality of relationships mediates the connection between cultural intelligence and success in addressing issues arising due to differences in the communication environment [18].

Intercultural cooperation defines the desire of cultures for integration and synthesis, which is a process of cooperation between various elements with the end result of a new cultural phenomenon, model of a social and cultural system, style or trend, therefore it is qualitatively different from a simple sum of these elements [19].

These two mechanisms become especially crucial in the process of communication within a multicultural environment as they lead to a better acceptance (or tolerance) and understanding of other cultures [20].

Such scholars as T.O. Smoleva and S.N. Morozyuk discussed the significance of self-reflection in the communication. When addressing the role of self-reflection in overcoming uncertainties during intercultural communication, they conclude that in order to secure successful communication, it is necessary to identify a type of uncertainty in the context of intercultural communication and then adjust the self-reflection competence as a part of intercultural communication competences accordingly [21].

Many scholars believe that the building of psychological readiness for intercultural communication must be conducted with the help of programmes of psychological and pedagogical support. For instance, I. Gol and O. Erkin propose to introduce such courses into education curricula that would improve cultural sensitivity and cultural intelligence of students in order to help them overcome cultural differences and choose culturally appropriate behavioural patterns [22]. A. Presbitero and A. Hooman suggest a peer education model for the building of psychological readiness for intercultural communication as it includes structured training activities that support students' studying [23].

Students must learn to integrate their knowledge, skills and competences of both the linguistic and the psychological nature when addressing various challenges in their studying. They must get acquainted with psychological mechanisms of empathy and self-reflection and understand their high psychological potential in ensuring efficient and conflict-free intercultural communication. Such introduction must be done both theoretically and practically until it's completely automatic.

The whole process of the building of psychological readiness for intercultural communication in students learning German as their second foreign language must duplicate real situation of communication. Training sessions must be organized based on the combination of competency-based and communication approaches in order to monitor the degree of psychological readiness for intercultural communication.

So, a strong academic interest to the issue of psychological readiness both in Russia and abroad and an increasing number of publications on the issue demonstrate its high relevance and academic potential in the current multicultural world.

Review of literature on psychology and education showed that the building of psychological readiness for intercultural communication in students learning German as their second language is seen as clearly necessary; however, there are a few peculiarities. It is crucial that students learning German as their second language build psychological readiness for intercultural communication as people tend to feel uncomfortable in the process of intercultural communication due to their perception of a foreign culture and its specific characteristics, a fear of being misunderstood, underestimation of their own knowledge, skills and competences in the context of their second foreign language proficiency and underestimation of their own readiness for intercultural communication, which results in a higher anxiety.

Students learning a second foreign language feel such discomfort due to their lack of knowledge and skills regarding intercultural communication and cooperation. It is clear that many people suffer from their own shortage of such knowledge and skills, even though intercultural competences are seen as a must in the current process of globalization.

Research Questions: To conduct a theoretical analysis of psychological and pedagogical studies. To study special characteristics in students learning German as their second foreign language. To define research stages, choose methods and techniques and describe them. To describe the selection of subjects and analyse the results of an ascertaining experiment.

The purpose of the study was to address the issue of psychological readiness for intercultural interaction in students learning German as their second foreign language.

2 Materials and Methods

The following theoretical methods were applied in the study: review of literature, analysis, synthesis, and definition of objectives. The following psychodiagnostic techniques were used to analyse psychological readiness for intercultural communication in students learning German as their second foreign language: a questionnaire on the level of psychological readiness for intercultural communication [24–26], and a Self-Evaluation of Mental Conditions test by G.Yu. Ayzenk [27, 28]. The determined stages, methods and psychodiagnostic techniques fully comply with the purpose and objectives of the study.

27 fourth-year full-time students of the Foreign Language Faculty participated in the study. On average, their academic performance can be described as above the median.

Teachers describe them as disciplined. Their main interests revolve around academic and professional activities, hence some students participate in scientific and art events of various levels. The students know and understand each other rather well and get along well as they are of the same age and share many interests. The general psychological environment in their group is positive.

3   Findings and Discussion

Figure 1 demonstrates findings obtained with a Psychological Readiness for Intercultural Communication questionnaire by V.I. Dolgova, E.A. Vasilenko and A.S. Baronenko, which is aimed at analysing four components of psychological readiness for intercultural communication.

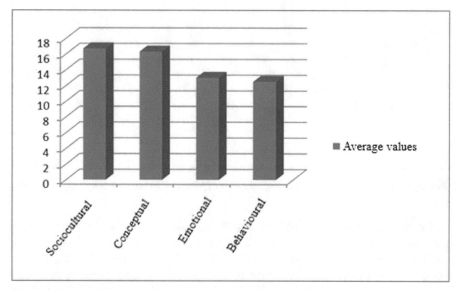

Fig. 1. Findings obtained with a psychological readiness for intercultural communication questionnaire by V.I. Dolgova, E.A. Vasilenko and A.S. Baronenko.

The findings show that the most pronounced components in students are sociocultural and conceptual ones with average values of 16.7 and 16.3 respectively. Thus, students learning German as their second language rather quickly acquire knowledge on specific characteristics of the country and its culture. In the foreign language learning process, facts about the culture, customs and popular sights of the target language country are the first to learn and the easiest to remember.

Emotional and behavioural components showed rather low results with average values of 13 and 12.5 respectively. It demonstrates that students face difficulties in understanding and accepting the mindset of other nations and their ways of thinking, living and behaving. Low values of the emotional component exhibit the fact that students do not possess a fully formed psychological readiness for the emotional perception of intercultural communication. Therefore, it should be crucial to develop and improve emotional intelligence in them in the context of intercultural communication through intercultural training sessions.

Figure 2 shows findings obtained with a Self-Evaluation of Mental Conditions test by G.Yu. Ayzenk.

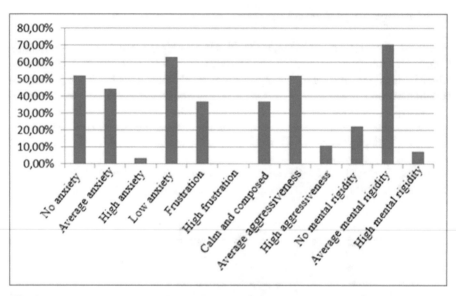

Fig. 2. Findings obtained with a Self-Evaluation of Mental Conditions test by G.Yu. Ayzenk.

The findings above obtained with a Self-Evaluation of Mental Conditions test by G.Yu. Ayzenk show that most students have average values on all scales. 44.4% of the students demonstrate an average level of anxiety, which is acceptable. 37% have an average level of frustration, 51.9% have an average level of aggressiveness, and 70.4% have an average level of mental rigidity. Therefore, the findings above demonstrate that students learning German as their second foreign language in general possess readiness for intercultural interaction. However, the question arises as to what extent and how efficiently they can apply various psychological techniques to achieve successful intercultural communication.

A successful intercultural communication is possible provided there is no stress and each participant of communication is able to identify with foreign parties and ready to apply various psychological techniques to relieve communication stress. In order to avoid conflicts in intercultural communication, it is necessary to possess knowledge on culture-specific behavioural patterns of a foreigner and the context of communication, which might be of a partial or a foundational nature in finding common ground between participants of intercultural communication [29–31]. This statement is shared by other academics as well; in their works, they analysed the content of textbooks for learning English as a foreign language, concluded that the emphasis is placed on grammatical functions of the English language rather than its communicative functions and proposed to change the content of language textbooks and make it communication-oriented and contextualized [32].

Therefore, in the process of language learning, besides studying lexical, grammatical and syntactic features of a foreign language - in other words, its linguistic features - students are also required to learn an adequate psychological reaction to a native speaker speaking, how to use facial expressions and gestures, to apply the rules of speech etiquette and to acquire correct cultural and historical knowledge on the target country, communication contexts and taboo words. It is expected that in the process of building readiness for intercultural interaction, people learning German as their second foreign language would treat their communication partners from other countries with tolerance and change for the better themselves. However, having analysed second foreign language teaching processes in higher education from the psychology and pedagogy perspective, it can be concluded that teachers often prioritise the development of speaking and linguistic skills in students rather than their communication skills associated with readiness to initiate communication with native speakers.

The building of psychological readiness for intercultural communication begins once a person learns intercultural communication; this process should consider cultural and mindset-specific differences between communication participants. Therefore, the whole phenomenon of readiness for intercultural communication is viewed as a requirement for a successful dialogue between cultures. We agree with other academics [33, 34] in their beliefs that negative stereotypes perceived weaken the connection with the dominant culture and enhance the connection with the ethnic culture, therefore social resources associated with any culture are useful for acculturation and personal improvement of intercultural communication participants.

Students need specific knowledge, skills and competences that help students under-stand foreigners and choose a speech behaviour pattern in accordance with objectives, situation and context. Such understanding is deemed impossible if a person does not possess cultural intelligence or does not improve it, as it is cultural intelligence that is believed to be a precursor of speech behaviour that defines the main facilitative mech-anism of communication and qualitatively influences relationships between communi-cation partners coming from different cultures [34]. Therefore, besides acquiring main linguistic concepts, students are also required to have rather developed communication skills, or verbal communication competences, for various communication situations depending on objectives or a recipient. We believe that intercultural training sessions can prove to be highly efficient in this regard.

4   Conclusion

The testing undertaken in the present study showed a rather low level of readiness for intercultural communication in students learning German as their second foreign language.

It is crucial that every participant of communication acknowledges and accepts on a psychological level a possibility that some problems might arise in the process of intercultural communication between people from different cultures as well as under-stands that other cultures have their own traditions, customs, norms, values and rules of behaviour in order to achieve successful intercultural communication.

References

1. Toybazarova, N.A., Nazarova, G.: The modernization of education in Kazakhstan: trends, perspective and problems. Bull. Natl. Acad. Sci. Republ. Kazakhstan **6**(376), 104–114 (2018)
2. Kassymova, G.K., et al.: Personal self-development in the context of global education: the transformation of values and identity. Bull. Natl. Acad. Sci. Republ. Kazakhstan **6**(376), 195–207 (2019)
3. Karabalina, A., Yesengulova, M., Kulbayeva, B., Sarkulov, M., Hayrusheva, A., Summers, S.A.B.: Self-education: model, process and competency. Bull. Natl. Acad. Sci. Republ. Kazakhstan **2**(372), 48–56 (2018)
4. Dmitrieva, L.G., Politika, O.I., Nagumanova, E.R.: Factors of psychological resilience among adolescends. Bull. Natl. Acad. Sci. Republ. Kazakhstan **6**(376), 209–219 (2018)
5. Almurzayeva, B., Summers, D., Karabalina, A., Yesengulova, M., Zholdassova, M., Summers, B.: Teaching in a modern school on base of cognitive-constructive theory. Bull. Natl. Acad. Sci. Republ. Kazakhstan **2**(372), 57–62 (2018)
6. Summers, D., et al.: Psychological and pedagogical aspects of moral education in Kazakhstan. Bull. Natl. Acad. Sci. Republ. Kazakhstan **3**(373), 121–129 (2018)
7. Agnoli, S., Runco, M.A., Kirsch, C., Corazza, G.E.: The role of motivation in the prediction of creative achievement inside and outside of school environment. Thinking Skills Creativity **28**, 167–176 (2018). https://doi.org/10.1016/j.tsc.2018.05.005
8. Munk, K.: Motive orientations at work. Learn. Cult. Soc. Interact. **27**, 100225 (2018). https://doi.org/10.1016/j.lcsi.2018.04.005
9. Kleptsova, E.Y., Shubnitsyna, T.V., Kleptsov, N.N., Mishutinskaya, E.A., Tsvetkova, N.V.: Psychological structure of humane interpersonal relations among the subjects of educational activity. Espacios **39**(49), 2 (2018). https://elibrary.ru/item.asp?id=38679318
10. Mazilkina, E.I., Panichkina, G.G.: Adaptation in the team. IP Air Media, Saratov (2012)
11. Tabolina, A.V., Gulk, E.B.: Gaming technologies as a mean of development of motivation of students. In: 18th PCSF 2018 - Professional Culture of the Specialist of the Future. European Proceedings of Social and Behavioural Sciences. EpSBS, vol. LI, pp. 1672–1678 (2018). https://doi.org/10.15405/epsbs.2018.12.02.179
12. Shipunova, O.D., Berezovskaya, I.P., Mureiko, L.V., Evseev, V.V., Evseeva, L.I.: Personal intellectual potential in the e-culture conditions. Espacios **39**(40), 15 (2018). http://www.rev istaespacios.com/a18v39n40/18394015.html
13. Zhang, X., Zhou, M.: Interventions to promote learners' intercultural competence: a meta-analysis. Int. J. Intercultural. Relat. **71**, 31–47 (2019). https://doi.org/10.1016/j.ijintrel.2019.04.006
14. Dolgova, V., Kondratieva, O., Sencheva, L., Rokitskaya, Ju.: Psychocorrection of conflictual behaviour of college students. In: Advances in Economics, Business and Management Research, vol. 90, pp. 17–20 (2019). https://www.atlantis-press.com/proceedings/ispcbc-19/125914459
15. Dolgova, V.I., Rokitskaya, Y.A., Kondrateva, O.A., Arkaeva, N.I., Kryzhanovskaya, N.V.: The interactions of individual religiosity and axiological orientations of students. Eur. J. Sci. Theol. **13**(4), 47–53 (2017). http://www.ejst.tuiasi.ro/Files/65/Contents%2013_4_2017.pdf
16. Khoteeva, M.S.: Ethics and psychology of business communications in an intercultural context. Historical, philosophical, political and legal sciences, cultural studies and art history. Quest. Theor. Pract. **3–1**(77), 174–177 (2017). https://elibrary.ru/item.asp?id=28417090
17. Ana, A.D., Starcevic, J., Zorana, J.M.: Can ability emotional intelligence help explain intercultural effectiveness? Incremental validity and mediation effects of emotional vocabulary in predicting intercultural judgment. Int. J. Intercultural Relat. **69**, 102–109 (2019). https://doi.org/10.1016/j.ijintrel.2019.01.005

18. Sharma, R.R.: Cultural intelligence and institutional success: the mediating role of relationship quality. J. Int. Manag. **25**, 100665 (2019). https://doi.org/10.1016/j.intman.2019.01.002
19. Bystrai, E.B., Belova, L.A., Vlasenko, O.N., Zasedateleva, M.G., Shtykova, T.V.: Development of second-language communicative competence of prospective teachers based on the CLIL technology (from the experience of a pedagogic project at a department of history). Espacios **39**(52), 12 (2018). https://elibrary.ru/item.asp?id=38588081
20. Troitskaya, E.A.: Features of empathy in a multicultural environment. Bull. Moscow Stat. Ling. Uni. Educ. Pedag. Sci. **6**(814), 217–229 (2018). https://elibrary.ru/item.asp?id=38499986
21. Smoleva, T.O., Morozyuk, S.N.: Reflection status in a situation of overcoming the uncertainty of intercultural communication. Soc. Human. Technol. **1**(1), 139 (2016). https://elibrary.ru/item.asp?id=29161072
22. Gol, I., Erkin, O.: Association between cultural intelligence and cultural sensitivity in nursing students: a cross-sectional descriptive study. Collegian **26**, 485–491 (2019). https://doi.org/10.1016/j.colegn.2018.12.007
23. Presbitero, A., Hooman, A.: Intercultural communication effectiveness, cultural intelligence and knowledge sharing: extending anxiety-uncertainty management theory. Int. J. Intercultural Relat. **67**, 35–43 (2018). https://doi.org/10.1016/j.ijintrel.2018.08.004
24. Gretsov, A.G., Azbel, A.A.: Get to know yourself. Psychological tests for adolescents. Peter, St. Petersburg (2016)
25. Zagvyazinskiy, V.I.: Technique and Methods of Psycho-pedagogical Research: Study Guide. Academy, Moscow (2008)
26. Dolgova, V., Scorobrenko, I., Bogachev, A., Golieva, G., Kondratieva, O.: Addressing difficulties in intercultural communication with communication training sessions. In: European Proceedings of Social and Behavioural Sciences, vol. 112, no. 3, pp. 20–28 (2021). https://doi.org/10.15405/epsbs.2021.06.04.3
27. Dolgova, V., Grasmik, E., Nurtdinova, A., Gamova, E.: Correlation between self-esteem and state anxiety in professional self-determination formation. Bull. Natl. Acad. Sci. Republ. Kazakhstan **2**(390), 274–280 (2021). http://www.bulletin-science.kz/index.php/en/
28. Milosevic, I.T.: Skidding on common ground: a socio-cognitive approach to problems in intercultural communicative situations. J. Pragmatics **151**, 118–127 (2019). https://doi.org/10.1016/j.pragma.2019.05.024
29. Bylieva, D., Krasnoshchekov, V., Lobatyuk, V., Rubtsova, A., Wang, L.: Digital solutions to the problems of Chinese students in St. Petersburg multilingual space. Int. J. Emerg. Technol. Learn. **16**, 143–166 (2021). https://doi.org/10.3991/ijet.v16i22.25233
30. Lobatyuk, V., Nam, T.: Everyday problems of international students in the Russian language environment. Tech. Lang. **3**(3), 38–57 (2022). https://doi.org/10.48417/technolang.2022.03.04
31. Huang, P.: Textbook interaction: a study of the language and cultural contextualisation of English learning textbooks. Learn. Cult. Soc. Interact. **21**, 87–99 (2019). https://doi.org/10.1016/j.lcsi.2019.02.006
32. Fons, J.R.V., Pekerti, A., Miriam, M.T., Okimoto, G.: Intercultural contacts and acculturation resources among International students in Australia: a mixed-methods study. Int. J. Intercultural Relat. **75**, 56–81 (2020). https://doi.org/10.1016/j.ijintrel.2019.12.004
33. Tukaev, S.V., Vasheka, T.V., Dolgova, O.M.: The relationships between emotional burnout and motivational, semantic and communicative features of psychology students. Proc. Soc. Behav. Sci. **82**, 553–556 (2013). https://doi.org/10.1016/j.sbspro.2013.06.308
34. Afsar, B., Shahjehan, A., Shah, S.I., Wajid, A.: The mediating role of transformational leadership in the relationship between cultural intelligence and employee voice behavior: a case of hotel employees. Int. J. Intercultural Relat. **69**, 66–75 (2019). https://doi.org/10.1016/j.ijintrel.2019.01.001

Issues of Training Multilingual Professionals in Higher Education

Elena Vdovina[✉] and Liudmila Khalyapina

Peter the Great St. Petersburg Polytechnic University, Polytechnicheskaya, 29,
195251 St. Petersburg, Russia
evdovina@spbstu.ru

Abstract. The paper considers the results of a preliminary study conducted among the first-year undergraduate students (N = 64) who entered English-medium international educational programs at Peter the Great St. Petersburg Polytechnic University. The study was focused on three goals. The first goal was to find out what language related difficulties students encountered during the first academic year. The second goal was to find out what methods and techniques Russian lecturers used to help students to overcome difficulties in studying non-linguistic disciplines. And the third goal was to find out respondents' opinion of a graphic organizer used in the micro- and macroeconomics courses as an instrument of deeper comprehension of the theoretical content. The data was collected from students' answers to the questionnaire. The study revealed a contradiction between the respondents' belief in an adequate level of English proficiency and the significant language difficulties they encountered in their studies. Though some students observed improvements in academic skills, others continued experiencing serious problems in the use of English for academic studies. According to the respondents, the instructors helped them better understand disciplinary content. However, the comments demonstrated a limited range of techniques used by lecturers to support them in overcoming language and comprehension problems. A highly positive responses pertaining to the use of a graphic organizer in the economic theory stresses the importance of more diversified and carefully planned instructional support to ensure not only students' higher academic achievements but also their acquisition of multilingual competences.

Keywords: English as a Medium of Instruction (EMI) · English as a lingua franca · Multilingual competence · CLIL (Content and Language Integrating Learning) · Scaffolding · Graphic organizer · Concept Flowchart (CFC) · Higher education

1 Introduction

In contemporary education, the development of a specialist's professional culture is expected to involve multilingualism, by which, in the context of this study, we mean professionals' awareness of the importance of practical ability to use more than one language both at the stage of education and in their professional sphere [1]. Despite an obvious trend towards an increase in the number of specialists who, even at the level of secondary and tertiary education, gained experience in using two or more languages in education, it is in tertiary education that the main non-native language of instruction continues to be English. According to the studies, the English used as a language of instruction deviates from ordinary English as a native language. The academic discourse in which meanings are negotiated has an effect on how interlocutors achieve mutual comprehensibility within multi-participant groups [2].

Traditionally, university language instructors place a heavy focus on form. In light of increasing empirical evidence, it is more obvious now that students benefit to a greater extent from getting training in academic language as social practice in the multilingual world, English obtaining essentially multilingual nature [3]. Even those graduates who have been trained in English are quite likely to work in their professional area using a language other than English or a native language. The question of how this state of affairs relates to the dominance of the English language in education is not only an existential question but also a matter of perspective on the problems of methodology in education and multilingualism, in general.

One of the approaches which may reconcile professionals with the prevalence of English in tertiary education though not necessarily in graduates' professional life stems on the idea that an individual who has gained significant experience in using English as a lingua franca in vocational training develops a multilingual competence due to his/her engagement in the language contacts in which multicultural awareness increases. In these circumstances, professionals are more likely to be able to master, if necessary, an additional language of professional communication than someone who received professional training in a monolingual framework. In other words, the procedural nature of the use of a non-native language in obtaining professional knowledge and multilingual competence in the classroom may become a determining factor of the further development of human potential, an ability to process and systematize professionally significant information even if their linguistic experience and knowledge are initially limited. To generalize, learning to use one foreign language through academic instruction and in the framework of multicultural dialogue has a potential to help a person flexibly adapt to changing conditions later, including, if necessary, the transition to a different non-native language in their professional field.

This initial assumption binds a lot, since what conditions are created for a person to develop his or her abilities to critically and creatively work with large blocks of information in a non-native language is likely to determine the success of the development of personal multilingual and professionally significant competences. The advances in the research methods brought into light an idea of superiority of guided instruction based on knowledge of 'human cognitive architecture' and 'expert – novice differences' [4]. A wide adoption of modern technologies in education has decisively increased the choice between types of instruction diversifying methods and media, widening Web-based environment, and facilitating the adaptation of the instructional design to the new challenges in tertiary education [5–7]. Hence the fundamental importance of which educational methods and instructional techniques to choose to ensure that the professional training in a foreign language is a success as a process and as a result.

Furthermore, this choice should take into account the students' needs in a particular educational environment since in different national contexts, both university minimum requirements for initial English language proficiency level and the students' expectations of how they are going to study in a foreign language vary significantly. One of the large-scale, multi-faceted studies which involved 5,000 undergraduates from all 26 departments at the English-medium university in Hong Kong identified students' language problems which hinder their progress in learning disciplinary content with the emphasis on receptive and productive vocabulary, academic speaking, and academic reading [8]. A later longitudinal study at the same university examined the language-related challenges which first-year students face when using English as the language of instruction. As a result of this study, four particular problems were indicated during the first year at university: understanding technical vocabulary, comprehending lectures, achieving an appropriate academic style and meeting disciplinary and institutional requirements [9].

The importance of supporting students after they switch from a 'comfortable zone' of their secondary schools to a new challenging educational environment attracts the attention of the university instructors. An ever-growing cohort of researchers who investigate English-medium university practices unanimously support an idea of integrating two components in the search for efficient support of students' learning efforts. The improvement of students' academic achievements lies nor only with the linguistic component but with cognitive component as well. Linguistic proficiency should not underestimate the role of learning strategies through the adjustment of instruction to the learners' needs [5]. Thus, the instructional approaches with a strong emphasis on the guidance of the students through the learning process should prevail [4]. Researchers' and practitioners' attention is drawn to the inseparability of the two components in the instructional support: academic language proficiency and knowledge of academic content [10].

Interestingly, this duality of focus can be found in an innovative educational approach known as CLIL (Content and Language Integrating Learning) in which a dual focus on the target language and discipline-specific content constitutes the central distinguishing feature of CLIL. Studies of how CLIL evolves at any educational level prioritize the development of methods and tools to support learners, in other words, scaffolding tools without which the dual focus may not ensure a positive outcome [11].

Scaffolding, initially defined by Wood et al. (1976) as a process "that enables a child or novice to solve a task or achieve a goal that would be beyond his unassisted efforts" [12 p. 90], became a key notion in the theory and practice within the framework of CLIL at any educational level, including higher education. Based on Vygotsky's idea of a Zone of Proximal Development, scaffolding facilitates the learning process in case learners face difficulties in the compression of new concepts and need an expert's or instructor's support to advance in their learning. Gradually, scaffolding can be removed or changed when learners can progress independently [13].

The types of scaffolding are multiple, and the choice depends on the learners' needs, the instructors' priorities and competences, the available resources, and the types of learning activities in which students are engaged. For example, there exists a strong correlation between reading in a foreign language and vocabulary knowledge. Thus, in the EMI context, students can benefit from such a scaffolding tool as domain-specific glossary of terms organized by subject topic or unit. Evidence confirms a current trend towards the provision of the on-line platforms created at universities to support university students in particular areas of study [14]. This is an indispensable linguistic support of reading comprehension development as researches identified a strong correlation between the size of the vocabulary and the development of reading skills [15].

On the other hand, academic reading is more than a language process; it is predominantly a cognitive activity which involves verbal and non-verbal components of the written input and which can be facilitated and intensified due to guidance, support and methods of instruction. Researches indicate that visualization as a process of creating non-verbal images to communicate a message enhances cognitive processes including reading comprehension, especially referring to their findings pertaining to visual literacy which is exhibited by 'millennial learners' also known as 'digital natives' [16]. This ability justifies the reliance on visually-oriented instructional techniques. And more than that, visual types of scaffolding support both language skills development and cognition as they have a positive effect on learners' thinking skills.

2   A Graphic Organizer as a Conceptualizing Tool

A graphic organizer is recognized by the professionals as an effective tool to visualize the hierarchy of the links of terms and the logical sequence of theoretical concepts within a particular theory by 'making things visible' thus extending and deepening students' thinking and learning [17].

One type of a graphic organizer called a CFC (Concept Flowchart) was successfully used at the World Economy Department as one of the scaffolding tools to support students' learning efforts in a number of CLIL disciplines designed and developed in the framework of a longitudinal research conducted at the department in 2010–2017 by the authors of this paper. The CLIL disciplines had dual focus: on the economics knowledge acquisition and on the academic language development in one learning process. One of the main purposes of using CLIL approach was for students with different levels of English language proficiency to acquire appropriate academic language and cognitive skills in order to be able to participate in the international mobility. As initial language level varied widely, it was crucial to create the methods and techniques supporting the students in both goals, disciplinary and linguistic, one of the scaffolding tools being a particular type of graphic organizer like CFC [18].

In a CFC, the new key terms are laid out in a hierarchical structure which ensures the logical sequencing of the concepts so that a theory under consideration unfolds from top till bottom. The links between the terms which represents categories and the subcategories, and the relationships between concepts are shown by arrows forming a tree diagram which unfolds downwards representing the logical development of a theory presented in the text. Appropriate graphs and equations are located below the concepts whose relationships they demonstrate. The theoretical topic which may cover from five to ten pages of the course book text should be presented on one A4 piece of paper, with all definitions of new terms and concepts being written on the back side of the CFC. This requirement ensures students' developing the skills of optimization and generalization of the abstract considerations. One example of a CFC is shown in Fig. 1.

The longitudinal research provided the ample evidence of the benefits of using a graphic organizer in order to systematize and conceptualize written academic input using newly-obtained domain-specific vocabulary. It was constantly observed that the graphic visualization of predominantly abstract knowledge obtained from academic written texts encourages active processing of new knowledge, one of the prerequisites being an active communication of the participants (the students and the instructor) in the analysis of the topical texts through questioning [19].

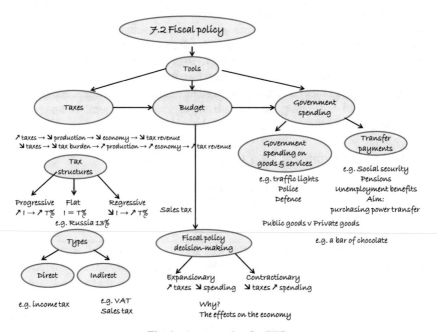

Fig. 1. An example of a CFC.

This type of a graphic organizer as a scaffolding tool was transferred by the authors to the EMI context as it proved efficient in supporting students who had no experience in obtaining theoretical disciplinary knowledge in the English language. And the stakes were high as the Russian-native speakers who enrolled in International Educational Programs taught in English were required to do the things they had never done before: to extensively read authentic course book texts, to listen to two-hour lectures, and to analyze disciplinary content orally during tutorials. The Concept Flowchart was used by the instructor to support students' conceptualization of the economic theory which also required the use of the new domain-specific language and academic language in general.

Students were familiarized with the purposes of CFC and practical steps how to design a layout of the concept. Discussions about the structure of each CFC were aimed at the optimization of the students' cognitive and language skills.

3   Methodology and Results of the Study

The study was conducted among Russian first-year students who chose to study in the English language under the international educational programs "International Business", "International Trade" and "Digital Enterprise" at Peter the Great St. Petersburg Polytechnic University (Russia) at the end of the academic year 2021–2022.

The study had three main goals. The first goal was to find out students' perceptions of whether their initial language proficiency level was sufficient for successful studies and what language related difficulties students encountered and whether their learning motivation changed by the end of the first academic year. The second goal was to find out what methods and techniques Russian lecturers used to help students to overcome difficulties in studying non-linguistic disciplines. And the third goal was aimed at their opinion about the possible benefits of a graphic organizer used only in the study of the economic theory as an instrument facilitating the conceptualization of theoretical content.

For the purposes of the study, a questionnaire of 8 items was developed:

1. What was your experience of studying in English before entering university?
2. Do you rate your initial level of English as adequate for learning in this language?
3. Have you met your expectations regarding the difficulty of learning in English?
4. Has your motivation for learning in English changed?
5. What was difficult for you: listening to lectures, reading authentic textbooks, or speaking in class?
6. In what academic language skills do you see improvement?
7. How did your lecturers help you overcome the difficulties in understanding the disciplinary content?
8. Do you consider the use of a graphic organizer useful in the study of theoretical disciplines? And if so, what are its benefits.

The last question of the questionnaire concerned the respondents' experience in using a graphic organizer which was used by the instructor who taught Microeconomics in the autumn semester and Macroeconomics in the spring semester.

64 respondents answered the questionnaire in writing. The study was of a qualitative nature, and statistical data in percentages are rounded to the nearest integer in the four following figures.

According to Fig. 2, only 6 out of 64 respondents (9%) had a short-term experience of learning English abroad (from one week to one month). Answering question 2 about the adequacy of the language level for learning, all respondents (100%) rated their initial level as sufficient. However, all of them expected that learning in a foreign language would be difficult. Interestingly, more than half (62%) adapted to learning in a non-native language during the academic year; the rest of the respondents were still experiencing significant difficulties. Nevertheless, all respondents, with the exception of three students (about 4% of the total), showed unanimity, stating that the motivation for learning in English remained high.

The data of answers to the fifth and sixth questions are presented in Fig. 3. Answers to the fifth question about the areas of problems during the training period showed the

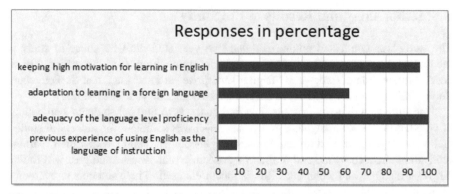

Fig. 2. The responses to the first four questions concerning students' previous experience, adequacy of language level, adaptation and level of motivation in percentage.

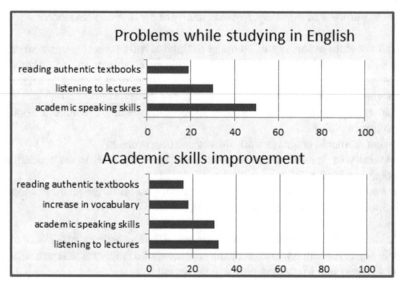

Fig. 3. Students' problems while studying in English and their progress by the end of the first year.

following: every fifth respondent (19%) experienced difficulties in reading authentic textbook materials; 30% of respondents indicated problems with listening to lectures; half of the respondents (50%) faced the problem related to speaking in the academic context and answering instructors' questions during the tutorials.

Also, Fig. 3 demonstrates the percentage of the respondents who reported about the improvement in particular language skills. According to the data, an improvement in reading academic texts and an increase in vocabulary, in addition to new disciplinary terms, were reported by 16% and 18% respondents, respectively. The improvement in academic speaking skills and lecture listening skills were reported by 32% and 30% respondents, respectively.

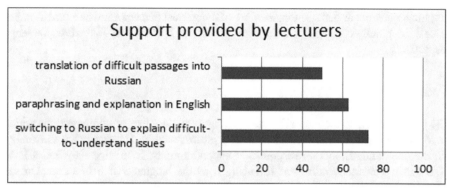

Fig. 4. Types of support provided by instructors to enhance students' understanding of disciplinary content.

Figure 4 summarizes the distribution of techniques used by instructors to support students' understanding of English-language materials during tutorials and lectures. According to the data, the respondents remembered the following methods which their instructor used during tutorials and lectures: the oral translation of difficult paragraphs in textbooks into Russian during the tutorial (50% of respondents); paraphrasing the text and explaining points in English using simpler linguistic constructions (63% of respondents); switching to the Russian language to explain difficult-to-understand issues (73% of respondents).

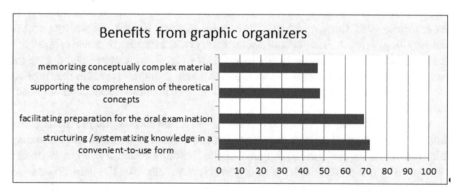

Fig. 5. Students' opinions about the benefits of a graphic organizer.

Students' opinions about the benefits of a graphic organizer are summarized in Fig. 5. Answering the question related to the CFC, all respondents, without exception, noted the usefulness of this form of support for learning the theoretical material.

The responses are grouped according to the frequency of reference to a particular factor. Most often, respondents gave the following justification for their opinion: graphic organizers help to structure / systematize knowledge in a form that is convenient for perception (72%). Almost the same number of students (69%) emphasized the positive role of this scaffolding tool in preparing for the oral examination. Such advantages of graphic organizers as better understanding of theoretical concepts and as a tool to memorize conceptually complex material were mentioned by 48% and 46% of respondents, respectively.

4 Discussion

The interpretation of the research data shed light on a number of contradictions which undoubtedly affect the success of obtaining professional knowledge at the university. On the one hand, almost all respondents unconditionally stated that they had a level of English proficiency sufficient for using it as the language of instruction. On the other hand, students readily identify problems with such important skills central for gaining knowledge as reading texts, listening to lectures, and the ability to use discipline-specific academic language for speaking during tutorials. Actually, the emergence of these problems is quite predictable, given the lack of experience in using a foreign language as a language of instruction in the EMI context (Fig. 1).

It can be assumed that the respondents' confidence in having an adequate level of English language proficiency, which they unanimously demonstrated in their answers, is directly related to the results of the written language test, which applicants take when entering international educational programs. The results of the tests are considered acceptable at the B2 level (CEFR), which actually indicates a level of language adequate for the needs of everyday communication, but not for academic purposes.

Moreover, the answers of respondents to other questions confirm the seriousness of the problem concerning the level of language proficiency. As we can see in Fig. 1, only 62% of the respondents adapted to learning in the EMI environment during the first academic year. This implies that the practice of using English constantly during a relatively long period of time does not automatically ensure the improvement of academic language skills. The data support this assertion. In Fig. 3, we can see that during the year, only about a third of respondents reported about an improvement in the listening to lectures skill and in academic speaking, and only 16% of the respondents improved reading skills.

Another contradiction is observed between the seriousness of the problems which students face in the EMI environment and a limited number of methods and techniques which instructors regularly use to support students' learning efforts, mostly by switching to the native language (Russian), translating some paragraphs from English into Russian, and, finally, paraphrasing or explaining difficult issues in other words (Fig. 4). Moreover, switching to the native language in order to facilitate the understanding of disciplinary material or translating difficult passages in the text into the native language reinforces two circumstances which may have negative consequences for students in the long run. Firstly, the use of the native language in the classroom reduces students' accumulation of the necessary experience of successful use of the new academic English, and thereby pushes the achievement of a professional level of language proficiency to a more distant perspective. In this case, the psychological dependence on the native language use for the purposes of learning increases. As a result, students' self-confidence they need to overcome the language barrier does not grow, which, in turn, hinders their move to a new comfort zone in their university life.

Unlike in the above methods, interpreting and paraphrasing texts, or giving explanations in other words using simpler language, techniques reported by almost two-thirds of the respondents (Fig. 4), can be considered valuable didactic tools which facilitate understanding. They do not conflict with the goal of developing foreign language academic competence. In addition, they reduce the level of difficulty which they experience in their academic activities, which is of great importance for maintaining a positive attitude and removing the psychological barriers which were mentioned by some of the respondents in their comments, such as the fear of speaking during the tutorials, the fear of not understanding the instructor's questions, etc.

Techniques such as 1) drawing schemes when explaining, 2) using examples, and 3) providing lists of new terms with definitions were mentioned by one or two respondents. It indicates that these forms of support played a minor role in the learning process, although they are extremely valuable methods of optimizing students' cognitive processes.

The answers to the last question about the benefits of graphic organizers indicate that this form of support resonated with respondents not only in relation to the economic theory where this scaffolding tool was used. Respondents submitted a wide variety of responses to this question listing from one to three positive effects. No one presented negative opinions or sentiments. In the comments, we can see some statements which generalize the effects and refer to other disciplines, and to the university studies in general. Some of the comments follow: "this is an ideal technique for studying any discipline", "the most interesting part of CFC was the process of conceptualization of theoretical material on one piece of paper described on ten pages of a textbook", "the scheme allows you to connect many concepts in one logical concept", "creating CFC requires repeated reference to the text and thus encourages to read carefully", "CFC help to cover a large amount of material in a limited period of time; it takes a lot of time, but when you learn to systematize the material, then everything becomes clear".

Indeed, the use of graphic organizers allows students to visualize a multifaceted theoretical topic in a systematic way, to generalize the new theoretical input in a compact and easy-to-store way, which the students can benefit from for the purpose of delayed revision, for example before the examination. Such instruments as graphic organizers and mind maps enhance more accurate in-depth understanding of a whole theory, and have a positive effect on the further development of the students' multilingual competence as they have to collaborate and communicate in the creative conceptualization process using the target language. And an extra benefit of graphic organizers is related to the instructors as they optimize instructors' control of students' comprehension of the theoretical input and timely intervention for possible corrections and explanations, if necessary.

The study has its own limitations such as a relatively small number of respondents, the general reliance on their opinions and self-evaluation, and the focus on only one specially designed scaffolding tool like CFC. As the challenges of both teaching and learning in a foreign language are fundamental for the high quality of higher education, more studies are needed to find out which types of support and guidance provided by the university instructors can be optimal in what particular disciplines. This requires closer cooperation of the university departments in designing such instruments and in providing special in-service training for the lectures and instructors with little or no experience of teaching in a foreign language so that it is not only a disciplinary content that the students acquire but also their multicultural competence is taken into account.

5 Conclusion

The results received from the study allow us to draw some preliminary conclusions. Firstly, it is obvious that the transition from learning in a native language to learning in the target language is associated with a number of difficulties, which are due to the fact that successful acquisition of the language of everyday communication does not mean that the learner can successfully use this language as the language of instruction at the tertiary level. At the same time, the data obtained indicate that students are highly motivated to study in international educational programs and that they maintain high level of learning motivation despite the difficulties they experience.

However, students feel the need for guidance and support in their learning efforts to bridge the gap between what they can already do in a foreign language and what they need to do in their professional university training. Due to the lack of such experience, they face a deficit of such central academic skills as listening to lectures, reading specialized literature, and academic speaking skills to express complex theoretical points.

University instructors understand the difficulties that their students experience when studying disciplines in a foreign language and are willing to help, as the respondents wrote in their comments. Nevertheless, as we have found out in the study, in this process, preference is often given to simple forms of problem solving, such as switching to the students' native language. In fact, this is not a solution to the problem, but its avoidance. For successful professional training, students need such scaffolding methods which can contribute to a real improvement in academic foreign language skills through the activation of cognitive activity and the reduction of the level of difficulty when learning new disciplinary content. One of the multiple methods includes any form of visualization and conceptualization of the disciplinary input, a graphic organizer being one of them. Among the tools to facilitate the process of learning, there may be lists of key terms and their clear definitions for each topic written by the lecturers, on-line glossaries and digital mind map sites, other computer-assisted techniques such as data-driven learning, and the use of various instructional Internet resources.

The further research can focus on the procedures used to stimulate students' cognitive processes by involving them in dialogical communication while considering disciplinary content. These procedures require specially designed support. Also, students should have a saying in the search for more beneficial learning experiences.

To sum up, high quality international education is not only about improved instruction. It is about the creation of beneficial learning environment, which needs a systemic approach to changing 'class dynamics' [20 p. 8]. In the digital era students who study in the EMI context can benefit from digitalized leaning environment which the ever-growing number of universities creates for them. What needs specific attention is the provision of more instructional support including English language and discipline-specific conceptual scaffolding incorporated in the Distant Learning university platforms [21].

International educational programs provide students with the most valuable experience in widening cultural and linguistic horizons in the process of professional training thus contributing to the development of multilingual competence. Adding an additional language of instruction leads to the development of cognitive abilities and makes the process of building the scientific picture of the world more exciting due to the arising dialogue of languages and cultures, thus building the road to multilingualism.

References

1. Hackett-Jones, A.V.: From bilingualism to polylingualism: concepts of multilingualism in new educational reality. Intern. Sci. Res. J. **3–4**(45), 104–106 (2016). https://doi.org/10.18454/IRJ. 2016.45.042

2. Mauranen, A.: Features of English as a lingua franca in academia. Helsinki Eng. Stud. E-J. Depart. Eng. Uni. Helsinki **6**, 6–28. http://blogs.helsinki.fi/hes-eng/files/2010/12/Mauranen_HES_Vol6.pdf. Accessed 28 July 2022

3. Jenkins, J.: Repositioning English and multilingualism in English as a lingua franca. Englishes Pract. **2**(3), 49–85 (2015). https://doi.org/10.1515/eip-2015-0003

4. Kirschner, P.A., Sweller, J., Clark, R.E.: Why minimal guidance during instruction does not work: an analysis of the failure of constructivist, discovery, problem-based, experiential, and inquiry-based teaching. Educ. Psychol. **41**(2), 75–86 (2006). https://doi.org/10.1207/s15326 985ep4102_1

5. Hunt, J.G.: The transformation of instruction: a critical analysis of trends in English education in twenty-first century Korea. Korea TESOL J. **13**(2), 3–24 (2017). https://koreatesol.org/con tent/korea-tesol-journal-13-2. Accessed 28 July 2022

6. Shostak, E., Khalyapina, L., Khodunov, I.: LMS Moodle interactive exercises sequence for developing linguistic competence. In: Alexandrov, D.A., Boukhanovsky, A.V., Chugunov, A.V., Kabanov, Y., Koltsova, O., Musabirov, I. (eds.) DTGS 2019. CCIS, vol. 1038, pp. 517–529. Springer, Cham (2019). https://doi.org/10.1007/978-3-030-37858-5_44

7. Liaw, S.S.: Considerations for developing constructivist web-based learning. Int. J. Instruct. Media **31**(3) (2004). https://www.learntechlib.org/p/63314/. Accessed 28 July 2022

8. Evans, S.: Why EAP is necessary: a survey of Hong Kong tertiary students. J. Eng. Acad. Purp. **6**(1), 3–17 (2007). https://doi.org/10.1016/j.jeap.2006.11.005

9. Evans, S., Morrison, B.: Meeting the challenges of English-medium higher education: the first-year experience in Hong Kong. Eng. Specif. Purp. **30**(3), 198–208 (2011). https://doi.org/10.1016/j.esp.2011.01.001

10. Krashen, S.: Academic proficiency (language and content) and the role of strategies. TESOL J. **2**, 381–393. https://doi.org/10.5054/tj.2011.274624

11. Coyle, D., Hood, P., Marsh, D.: CLIL: Content and Language Integrated Learning. Cambridge University Press, Cambridge (2010)

12. Wood, D., Bruner, J., Ross, G.: The role of tutoring in problem solving. J. Child Psych. Child Psychiat. **17**, 89–100 (1976)

13. Vygotsky, L.S.: Mind in Society: The Development of Higher Psychological Processes. Harvard University Press, Cambridge (1978)

14. Barrios, E., López-Gutiérrez, A., Lechuga, C.: Facing challenges in English Medium Instruction through engaging in an innovation project. Procedia-Soc. Behav. Sci. **228**, 209–214 (2016). https://doi.org/10.1016/j.sbspro.2016.07.031

15. Hacking, J.F., Tschirner, E.: The contribution of vocabulary knowledge to reading proficiency: the case of college Russian. Foreign Lang. Ann. **50**(3), 500–518 (2017). https://doi.org/10.1111/flan.12282

16. Brumberger, E.: Visual literacy and the digital native: an examination of the millennial learner. J. Vis. Liter. **30**(1), 19–46 (2011). https://doi.org/10.1080/23796529.2011.11674683

17. Perkins, D.: Making thinking visible. Harvard Graduate School of Education (2003). http://www.visiblethinkingpz.org/VisibleThinking_html_files/VisibleThinking1.html.   Accessed 28 July 2022

18. Vdovina, E.: University CLIL: multimodal conceptualization of the academic content in English. In: Mirola, T. (ed.) Update 2017 on Higher Education, pp. 80–89. Saimaa University of Applied Sciences, Lappeenranta (2017). http://www.theseus.fi/handle/10024/133680. Accessed 28 July 2022

19. Vdovina, E., Popova, N., Kogan, M.: Questioning as a cognitive instrument and a language issue in content and language integrated learning. In: Gómez Chova, L., López Martínez, A., Candel Torres, I. (eds.) ICERI 2019 Proceedings 12th International Conference of Education, Research and Innovation, pp. 8431–8440. IATED Academy, Seville (2019). https://doi.org/10.21125/ICERI.2019.2013

20. Bamond, V., Strotmann, B.: Book overview and Q&A with David Marsh, Victor Pavón-Vázquez, and María Jesús Frigols-Martín: review of the book 'the higher education language landscape: ensuring quality in English language degree programmes'. High. Learn. Res. Commun. 5(1), 4–10 (2015). https://doi.org/10.18870/hlrc.v5i1.241

21. Khalyapina, L., Shostak, E., Koltsova, S., Vdovina, E.: Examining plurilingual training principles in teaching foreign languages to engineering students. In: Anikina, Z. (ed.) IEEHGIP 2022. LNNS, vol. 499, pp. 233–242. Springer, Cham (2022). https://doi.org/10.1007/978-3-031-11435-9_25

The Experience of Forming Multilingual Competences in Universities in Russia and Turkey

Natalia V. Goncharova[1] , Liudmila V. Daineko[1(✉)] , Viola Larionova[1] ,
and Çağlar Demir[2]

[1] Ural Federal University, 19 Mira Street, 620002 Ekaterinburg, Russia
{n.v.goncharova,l.v.daineko,v.a.larionova}@urfu.ru
[2] Balıkesir Üniversitesi Çağış Yerleşkesi, 10145 Balıkesir, Turkey

Abstract. The Russian and Turkish systems of higher education are similar in structure, both have three main levels: bachelor's, master's, and postgraduate. The duration of education is the same: 4–6 years. The higher education system in Russia and Turkey includes universities, academies, and high schools. Russian and Turkish universities are very popular among foreign students. However, Russia and Turkey are traditionally characterized by a rather poor level of foreign language skills. Accordingly, the insufficient level of foreign language proficiency may be one of the obstacles to inclusion in the multilingual new reality. The aim of this article is to analyze the capabilities of universities in Turkey and Russia to form multilingual competences of students.

Based on the analysis of the experience in the formation of multilingual competences of students in two universities - in Russia and Turkey - the authors have concluded that it is necessary to replicate the best practices of the two universities in the formation of multilingual competences of students and teachers.

Keywords: Learning · University · Students · Multilingual competencies · Multilingual educational environment · Higher education

1 Introduction

In today's digital international world, borders between countries are disappearing, and knowledge of foreign languages is becoming one of the necessary competencies of a highly qualified specialist. Knowledge of languages allows you to search and exchange multilingual information, establish international relations with foreign partners, solve professional tasks in an international environment. Specialists with knowledge of foreign languages at a high level are required in almost every sphere of economics. As a consequence, specialists with knowledge of a foreign language are more competitive in the labor market, their work is better paid because of the shortage of such personnel.

As a rule, at universities a foreign language is studied in non-linguistic specialties only in the first year of the bachelor's degree and during one semester in the master's program. Consequently, the level of language training of graduates often corresponds only to the A2-B1 level. For a highly qualified specialist who is in demand on the labor market it is not enough to have a basic vocabulary, it is necessary to know professional terminology in a foreign language, to be able to communicate with international partners, to understand special documentation, etc.

Universities can become a driver of growth of multilingual competences in graduates of all specialties due to the availability of highly qualified personnel potential and multilingual student environment (academic mobility, mixed groups with foreign students).

2 Problem Statement

The steady increase in the number of people and communities using three or more languages has led many researchers to study multilingualism and multilingual mastery [1–3]. Researchers have explored the linguistic challenges faced by international students and how to mobilize their multilingual resources [4–6], investigated the impact of global English on motivation to learn other languages [7–9], methods of building teacher professional competence in multilingual education [10], exploring the promotion of disciplinary and linguistic learning for multilingual students developed a guiding document, The Fair Learning Framework, aimed at supporting learning based on a system of language standards, assessments, and professional learning [11]. Pirhonen notes that foreign language teaching in higher education faces the challenge of preparing students due to students' lack of understanding of the meaning of language learning [12]. In turn, Baranova et al., evaluating the multilingual approach in the context of international educational programs, note the effectiveness and progressiveness of the multilingual approach, including through increasing students' interest in the educational process [13]. Jiang, Gu & Fang, looking at the interaction of native English teachers note that a lot of research is based on communication with non-native speaker teachers, although multimodality can be used as a bridge to create a translingual space [14]. Rubtsova et al., looking at a constructive model for managing multilingual education in a university, note that the management mechanisms of educational systems require improvement and contemporary approaches to the process of modeling management strategies and tactics [15]. Paksoy & Çelik, examining the Turkish educational system's readiness for multiligual education, note that providing equal opportunities to learners and minimizing cultural conflicts can allow for improved academic performance [16]. Kuppers & Bozdag, exploring diversity in education, noted that it could be achieved using multilingualism, media, and mobility [17]. Prokhorova substantiates the role and place of multilingual communicative competence among other professional competences for Russian technical graduates [18]. Thompson explores the motivation of Turkish students to learn foreign languages [19]. However, there is not enough research on the impact of multilingual educational environment of the university on the formation of multilingual competences of graduates.

3 Materials and Methods

In order to solve the problem of formation and development of language competences of university graduates, the authors conducted a comparative analysis of the activities carried out in two major universities in Turkey and Russia, aimed at the formation of multilingual competences of teachers and students. The hypothesis of the study is the assumption that the multilingual educational environment of the university contributes to the formation of multilingual competences.

4 Results and Discussion

4.1 The Ural Federal University Experience

Ural Federal University is one of the largest universities in the Russian Federation; over the 100 years of its existence, it has educated more than 350,000 graduates. There are currently more than 36,000 students studying at the university, including more than 4,600 international students and interns and 172 international postgraduate students from 110 countries. The university employs 3,782 research and teaching staff, including 24 international faculty members from 16 countries. The list of the UrFU partners includes more than 500 foreign universities from 72 countries. The University has 12 research laboratories headed by leading international scientists, 28 international joint study programs, 15 master's degree programs and 47 postgraduate programs in English.

About 100 foreign delegations representing foreign universities, international organizations and companies, diplomatic missions, etc. visit the university each year. About 200 university students, teachers and researchers annually participate in international academic mobility programs.

In order to build multilingual competencies of teachers and students, the following activities are implemented at UrFU:

1. Teaching foreign languages to students of all specialties and forms of study.
2. UrFU has a department of foreign languages, whose professors teach foreign languages to students of all specialties for two semesters of the first year of study. Since September 2020, the university began to approve a mechanism for implementing adaptive learning [20] in a foreign language course, which involves entrance testing to determine the level of entrance knowledge and skills of first-year students, affecting the formation of level groups and the appointment of teachers.
3. Professional training of bachelors and masters in the linguistic direction. The basis of this direction is mastering foreign languages at the professional level, using serious theoretical training to master the modern communicative culture [21]. Varieties of digital tools are actively used in the educational process [22]. During their studies, students complete internships at prestigious European and Asian universities, as well as at various international organizations. After graduation, graduates work as translators, specialists in international relations and foreign economic activities, teachers, language experts and project managers.
4. Preparation of UrFU students and staff for foreign language examinations. In 2013, the University established the first Cambridge Center in Russia - a licensed examination center of Cambridge Assessment English Association, which has the right to take international examinations PET, CET, FCE, CPE, CAE, BEC Preliminary, BEC Vantage, and BEC Higher.

5. Preparation of UrFU teachers for Cambridge qualification examinations in foreign language teaching methodology. Since 2013, the Cambridge Center has received the right to take the CELTA exams from Cambridge English Department of the University of Cambridge.

6. Development and improvement of linguistic and communicative competences of foreign languages (English, Spanish, Italian, French, Russian (as a foreign language), Arabic, Chinese, Turkish) through conversational practice with native speakers. In 2012 the University established UrFU Foreign Languages Club, which provides an opportunity to learn several languages at once within the university, practicing them with native speakers. Various methods are used in the classes: interactive games, film screenings, debates and others. Instructors - Ural Federal University students, fluent in a particular foreign language or international students - native speakers.

7. In 2008 the first Confucius Institute in the Ural Federal District and the sixteenth in Russia was established at the university. At the beginning of 2009, the Confucius Institute began its educational activities. Three groups of Chinese language students were selected: two are beginners and one is a continuation. Currently, there are 5–8 basic groups and 10–15 advanced groups on a permanent basis. In December 2019, the Confucius Institute of UrFU was awarded the title "Best Confucius Institute 2019". At the Confucius Institute of UrFU, Chinese is taught both by professional Russian teachers and by native speakers sent to Russia by the Confucius Institutes Headquarters. Eight Russian teachers, most of them graduates from the Departments of Oriental Studies and Linguistics of Ural Federal University, and 18 volunteer Chinese teachers work at the Confucius Institute today. The Institute introduces students and residents of the region to Chinese culture through events, seminars and clubs: Chinese Conversation, Chinese Dance, Tai Chi chuan, and Chinese Song. In addition, the Confucius Institute has a library of Chinese literature, which can be used as a reading room or borrow books from home. Students can take HSK and HSKK qualification tests on average 2–3 times a year, and also take BCT (Business Chinese) and YCT for high school students.

8. Providing service support to specialized institutions in creating and promoting internationally oriented educational programs in foreign languages in the world scientific and educational market. Ural Federal University has a Department of International Educational Programs, which assists specialized institutions in modernizing existing higher and additional education programs in English and searching for foreign partners for programs in a foreign language, including network and experimental forms of interaction.

4.2 The Experience of Ege University

Ege University (EGE) is one of the largest universities in Turkey, which started its activities in 1955. Ege University is one of the top 20 universities in Turkey and one of the top 5% of the most famous and prestigious educational institutions in the world. Ege University is a large educational institution with more than 65,000 students and 3,165 research and teaching staff. The university has about 4% foreign students. Today, Ege University has 17 faculties, 9 graduate schools, 4 four-year schools, 6 institutes, 10 two-year vocational training schools and 37 research and application centers.

The university's main language of instruction is Turkish, with the exception of some individual faculties and departments, so the Faculty of Chemical Engineering teaches all subjects in English. At the Faculty of Economics and Administrative Sciences, two-thirds of the curriculum is offered in English. In other faculties and departments, the curricula include courses in English and German. Ege University collaborates with foreign universities for educational and research purposes to help raise the quality of education to international standards.

International and European cooperation plays an important role in the strategic development of Ege University. Internationalization encompasses all university activities such as education, research, knowledge transfer and administration. In line with its mission to become an international university, Ege University has entered into 82 academic cooperation protocols with many foreign institutions around the world to promote enrichment of teaching and research programs.

In 2011, Ege University was awarded the European Commission's prestigious ECTS Label Award. In April 2005, 2009 and 2013, the European Commission awarded Ege University the Diploma Supplement label.

In order to build the multilingual competence of faculty and students, the following activities are implemented at Ege University:

1. Teaching foreign languages to students in all faculties. The School of Foreign Languages employs 180 faculty members, 2 of whom are native English speakers, other staff and nearly 8,000 students. The school offers preparatory English classes for new students whose departments require mandatory preparatory language education. In addition, new students whose departments do not require prerequisite preparatory language instruction can also enroll in the school and receive a year's worth of English language instruction. Each classroom at the school is equipped with projectors, computers and audio equipment. The Curriculum Development Department, Professional Development Department, Testing Department, Accreditation Department, Project Department, Private Courses Department, Information Technology Department, Joint Compulsory English Courses Department, Materials Development Department, Procurement and Cataloging Department and Student Support Department are critical components of the school. The school's Project Center organizes faculty and student participation in projects and programs sponsored by the European Union. In November 2013, the Preparatory English class program at the School of Foreign Languages received the Pearson Assured quality label. In an effort to continuously improve the quality of education, all aspects of the school system are regularly reviewed; measures and actions are taken to ensure quality standards within the school's established quality management system. Upon successful completion of the school year, students will receive a Pearson Assured certificate.

2. In 2015, the School of Foreign Languages developed the Blended Learning Project in collaboration with the Department of Computer Education and Instructional Technology. In addition, the "BAP (Research Project)" provided two "Book Vending Machines" and students can buy English stories of different levels in them. Through the coordination of the Private Courses Division, the school offers the following courses: General English Course, YDS Preparation, TOEFL Preparation, English Conversation Club, Translation Course, German Course, French Course, Japanese

Course, Italian Course, and Spanish Course. These courses are open to university academic staff, graduate students, doctoral students and those who want to learn a foreign language for a specific purpose. The school also cooperates with EGESEM and holds such courses as Business English, Kids Club for kids 4–6 years old, and Junior Club for kids 8–11 years old.

3. The Doctoral Program in Translation Studies was opened in the Department of Translation and Interpretation during the 2019–20 academic year at the Institute of Social Sciences and began with the admission of students that same year. Students who successfully complete this program and successfully defend their dissertation are awarded the title of Doctor of Philosophy (Ph.D. Doctor of Philosophy) at the end of the Ph.

4. The Center for the Study and Application of European Languages and Cultures (ADIKAM) analyzes the language, literature, and cultures of various European countries in the context of academic research. In terms of intellectual disciplines, it is firmly rooted in the humanities with strong literary and linguistic components. The inclusion of the word "language" in the Center's name indicates our commitment to "translinguistic and transcultural competence, that is, the ability to work with the multilingual reality that is Europe. The objectives of the Center are the development of communication, the promotion of mutual research and study, cooperation through translation, and the organization of scientific and cultural encounters. The main scope of ADIKAM's activities focuses on:

   – Conducting basic and applied research in the field of European languages and cultures and shaping national and international cooperation,
   – The promotion of scientific projects related to culture and language,
   – Cooperation with Ministries such as the Ministries of National Education, Tourism and Culture,
   – Organizing national and international scientific events such as debates, seminars, symposia, conferences and congresses related to research in the field of languages and cultures.

   The Center plays an important role in developing our understanding of the complex social transformations of European languages and cultures. The Center's research and seminars consistently address broader themes of regional integration and globalization with a focus on cultural diversity.

5. Ege University has an Institute for Turkish Studies. The goals of the Institute for Turkish World Studies are:

   – To bring together qualified scholars and have a large library, auditoriums, conference rooms, etc. for research based on Turkology;
   – To develop the relationship between the Turkish people and the world through its institutional capacity;
   – To lead scientific research on the linguistic, cultural, literary, historical, artistic, political, social and economic structures of Turkish republics and Turkish peoples, and to develop Turkology.

The objectives of the Institute for the Study of the Turkish World are;

- To publish archaic and modern Turkish literature in Turkish;
- To investigate the structures and development of historical, linguistic, literary, artistic, cultural, political, social and economic themes of the Turkish world and to publish these studies as scholarly works;
- Collect qualified publications on the Turkish world and establish a modern archive and library;
- Participate in exchange programs and collaborate with scientific institutions and academic foundations of the Turkish World;
- Conduct research projects, conferences, seminars, and publications with scholars of the Turkish World.

6. International and European cooperation plays an important role in the strategic development of Ege University. Internationalization includes all university activities such as education, research, knowledge transfer and administration. Ege university strives to improve the quality of learning, research and teaching by enhancing international education and cooperation.
7. Ege University stands out among Turkish universities for its large number of exchange students and staff through the Erasmus Program. The Erasmus Program contributes to the international strategy of Ege University. The Erasmus Program contributes greatly to the international recognition and preference of Ege University.

A comparative analysis of the multilingual competency opportunities in two major universities in Russia and Turkey is shown in Table 1.

Table 1. Opportunities implemented by universities for students to master multilingual competencies.

Indicators	Ural Federal University	Ege University
Country, region	Russia, Ural Federal District	Turkey
Number of students, pers	36 000	65 000
Number of foreign students, pers	4 772	2 400
Number of teachers, pers	3 782	3 165
Teaching foreign languages to students of non-linguistic specifications	2 semesters in the 1st year of the bachelor's degree; 1 semester in the master's program	Preparatory Class Program -2 semesters in the 1st year of the bachelor's degree
The use of level-education in foreign languages	Yes	Yes

(continued)

Table 1. (*continued*)

Indicators	Ural Federal University	Ege University
The presence of a special linguistic area of study	Bachelor's and Master's degrees	As of 2010-Pearson Assured
Availability of educational programs implemented in a foreign language	Bachelor's and Master's degrees	Bachelor's and Master's degrees
Possible to learn foreign languages	English, Spanish, Italian, French, Russian (as a foreign language), Arabic, Chinese, Japanese, Turkish	English, German, French, Japanese, Italian, Spanish
Ability to learn foreign languages with native speakers	Yes	Yes
Preparing for Foreign Language Exams	PET, CET, FCE, CPE, CAE, BEC Preliminary, BEC Vantage, BEC Higher, CELTA, HSK, HSKK, BCT, YCT	YDS, TOEFL
The presence of a structure that supports international mobility	Yes	Yes

According to the results of a comparison of the practice of formation of multilingual competencies of two different universities from different countries, we can conclude that both universities use similar approaches to the implementation of universal foreign language acquisition. All students at the beginning of training are given the opportunity to master a foreign language at the basic level, the universities have the opportunity to pass exams in foreign languages, there are international exchange programs, programs are implemented in a foreign language, there is an opportunity to communicate with native speakers, etc. The universities even have a similar "mix" of possible foreign languages, mostly European. However, at Ural Federal University, in addition to European languages, there is an opportunity to study Chinese because of the large number of students from China studying at the university. It should be noted that there is a significant difference in the proportion of foreign students at the universities, which is 13.26% at Ural Federal University and 3.69% at Ege University. The low proportion of native speakers among university students may have a negative impact on the level of their language competency acquisition. So, according to the EF EPI 2021 (Education First English Proficiency Index), the world's largest English language proficiency rating, which is based on the testing results of 2 million adults in 112 countries, Turkey is in 70th place of 112 countries and in 34th place of 35 European countries, EF EPI 478, English proficiency is low. Russia ranks 51st out of 112 countries and 32nd out of 35 European countries, EF EPI Index: 511, English language proficiency - average. According to the EF EPI authors the best command of English is among people aged 26–30 years old, that is people who have already graduated from higher educational institutions and have work experience, and university graduates (age 21–25 years) are in the second place for the level of language skills. Consequently, language practice in the workplace and perhaps additional language study can improve language skills in the early stages of a career.

A positive thing about teaching foreign languages to university students is the level-based instruction of students in separate groups. In universities, at the beginning of the first year of study, foreign language proficiency is determined and then students are allocated to level groups. According to the authors, one of the possible ways to increase the efficiency of students' acquisition of multilingual competences is to use the mechanism of adaptive foreign language teaching [23], in level groups of students. The use of adaptive learning techniques can allow students to increase their motivation to master a foreign language, so necessary in a multilingual world [24]. Understanding that adaptive learning allows each student to build their own trajectory of foreign language acquisition by using their existing knowledge and learning style can significantly increase the effectiveness of the educational process in foreign language learning, and therefore increase the level of multilingual competences of students.

5   Conclusion

The experience of forming multilingual competences of teachers and students at large universities in Russia and Turkey shows that universities often choose standard ways of forming multilingual competences of students. However, the university multilingual environment itself and the individual adaptive approach to each student can increase motivation in learning foreign languages, and consequently increase the level of their multilingual competences. Expansion of the multilingual educational environment of the university by attracting native speaker teachers, increasing the proportion of international students, development of informal activities - foreign language clubs, communication with native speakers, development of group research activities can act as a driver of increasing the level of foreign language proficiency of university students. Understanding the importance of foreign language proficiency not only as a competitive advantage of graduates in the labor market, but also as an acceleration of communication between people contributes to the increase in the level of its proficiency. The use of the best successful practices of large universities to improve the efficiency of the process of forming multilingual competences can be useful for many educational institutions. The experience of implementing adaptive learning at Ural Federal University can be replicated to improve multilingual competencies of higher education students.

References

1. Kang, E.Y.: Multilingual competence. Stud. Appl. Ling. TESOL **13**(2) (2013). https://doi.org/10.7916/salt.v13i2.1334
2. Hufeisen, B., Nordmann, A., Liu, A.W.: Two perspectives on the multilingual condition - linguistics meets philosophy of technology. Technol. Lang. **3**(3), 11–21 (2022). https://doi.org/10.48417/technolang.2022.03.02

3. Bylieva, D., Nordmann, A.: Technologies in a multilingual world. Technol. Lang. **3**(3), 1–10 (2022). https://doi.org/10.48417/technolang.2022.03.01

4. Li, J., Xie, P., Ai, B., Li, L.: Multilingual communication experiences of international students during the COVID-19 Pandemic. Multilingua **39**(5), 529–539 (2020). https://doi.org/10.1515/multi-2020-0116

5. Lobatyuk, V., Nam, T.: Everyday problems of international students in the Russian language environment. Technol. Lang. **3**(3), 38–57 (2022). https://doi.org/10.48417/technolang.2022.03.04

6. Bylieva, D., Krasnoshchekov, V., Lobatyuk, V., Rubtsova, A., Wang, L.: Digital solutions to the problems of chinese students in St. Petersburg multilingual space. Int. J. Emerg. Technol. Learn. **16**, 143–166 (2021). https://doi.org/10.3991/ijet.v16i22.25233

7. Ushioda, E.: The impact of global English on motivation to learn other languages: toward an ideal multilingual self. Modern Lang. J. **101**(3), 469–482 (2017). https://doi.org/10.1111/modl.12413

8. Cheung Matthew Sung, C.: Learning English as an L2 in the global context: Changing English, changing motivation. Chang. Eng. **20** (4), 377–387. (2013). https://doi.org/10.1080/1358684X.2013.855564

9. Lasagabaster, D.: Language learning motivation and language attitudes in multilingual Spain from an international perspective. Modern Lang. J. **101**(3), 583–596 (2017). https://doi.org/10.1111/modl.12414

10. Pigovayeva, N., Kumar, T.: The formation professional competencies of a future teacher based on multingual education. Mir Nauki Obrazovaniya **1**(86), 280–282. (2021). (in Russian). https://doi.org/10.24412/1991-5497-2021-186-280-282

11. Molle, D., Wilfrid, J.: Promoting multilingual students' disciplinary and language learning through the WIDA framework for equitable instruction. Educ. Res. **50**(9), 585–594 (2021). https://doi.org/10.3102/0013189X211024592

12. Pirhonen, H.: Towards multilingual competence: examining beliefs and agency in first year university students' language learner biographies. Lang. Learn. J. **5**(50), 1–14 (2021). https://doi.org/10.1080/09571736.2020.1858146

13. Baranova, T., Mokhorov, D., Kobicheva, A., Tokareva, E.: The assessment of a multilingual approach in the context of international educational programs in English. In: Anikina, Z. (ed.) IEEHGIP 2022. LNNS, vol. 499, pp. 157–167. Springer, Cham (2022). https://doi.org/10.1007/978-3-031-11435-9_17

14. Jiang, L., Gu, M., Fang, F.: Multimodal or multilingual? Native English teachers' engagement with translanguaging in Hong Kong TESOL classrooms. Appl. Ling. Rev. (2022). https://doi.org/10.1515/applirev-2022-0062

15. Rubtsova, A., Almazova, N., Bylieva, D., Krylova, E.: Constructive model of multilingual education management in higher school. IOP Conf. Ser.: Mater. Sci. Eng. **940**(1), 012132 (2020). https://doi.org/10.1088/1757-899X/940/1/012132

16. Paksoy, E., Çelik, S.: Readiness of Turkish education system for multicultural education. Educ. Res. Rev. **14**(8), 274–281 (2019). https://doi.org/10.5897/ERR2017.3171

17. Kuppers, A., Bozdag, Ç.: Doing diversity in education through multilingualism, media and mobility. Instanbul Policy Center, Instanbul (2015)

18. Prokhorova, A.: Multilingual approach to power engineering students' language teaching. Eur. J. Natur. Hist. **4**, 41–45 (2019). https://doi.org/10.1007/978-3-030-47415-7_2

19. Thompson, A., Erdil-Moody, Z.: Operationalizing multilingualism: language learning motivation in Turkey Intern. J. Biling. Educ. Biling. **19**(3), 314–331 (2016). https://doi.org/10.1080/13670050.2014.985631

20. Goncharova, N., Daineko, L., Larionova, V.: Development of an adaptive learning model based on digital technologies and the work of the teacher. In: 15th International Technology,

Education and Development Conference, INTED2021, INTED2021 Proceedings, pp. 6549–6558. IATED, Valencia (2021). https://doi.org/10.21125/inted.2021.1307

21. Zaitseva, E., Zapariy, V.: Role of the university corporate culture in university management. In: 10th International Days of Statistics and Economics, pp. 2089–2095. Melandrium, Prague (2016)

22. Reshetnikova, O.: Effective learning tools in e-learning. In: Busch C., Steinicke M., Frieß R., Wendler T. (eds.) Proceedings of the 20th European Conference on e-Learning 2021 (ECEL), pp. 387–393. Academic Conferences and Publishing International Limited, Berlin (2021). https://doi.org/10.34190/EEL.21.091

23. Daineko, L., Goncharova, N., Larionova, V.: Creating an adaptive learning model based on the learner's digital footprint. In: EDULEARN21: Proceedings of 13th International Conference on Education and New Learning Technologies, pp. 4230–4237. IATED (2021). https://doi.org/10.21125/edulearn.2021.0896

24. Demir, Ç.: English teachers' role in boosting English learner motivation. In: 2nd International Conference on New Trends in Education and Their Implications, pp. 1189–1200. Iconte, Antalya (2011). https://www.researchgate.net/publication/331023751_English_teachers'_role_in_boosting_English_learners'_motivation

Assessment of Student Satisfaction with Technology and Organization of the Learning Process at the University

Ekaterina V. Zaitseva(⊠) 🆔 and Natalia V. Goncharova🆔

Ural Federal University, 19 Mira Street, 620002 Ekaterinburg, Russia
{e.v.zaitceva,n.v.goncharova}@urfu.ru

Abstract. A university graduate must possess a set of necessary competencies that meet modern requirements. The authors identify three groups of competencies, in addition to professional knowledge. The first group includes language competences. They allow you to know the skills of business communication in a foreign language, know the international professional terminology, and understand technical documentation that has not been translated. The second group includes social competences. The language of socio-cultural communications allows university graduates to fully interact in a multicultural and multi-ethnic environment. The third group includes IT competences of the future - knowledge of different programming languages, digital services, application in practice of modern technologies, information management methods, and ways to create information systems. Today, students are focused on the needs of employers. The labor market values both professional and additional competences of university graduates. This, in turn, increases students' requirements to the content, process, learning technologies and forms of learning activities. The article deals with the results of a survey to identify the level of student satisfaction with various aspects of the educational process. Regular assessment of student satisfaction with technologies and organization of the learning process allows: 1) identify problems and improve the quality of educational services provided; 2) shift the focus from purely professional competencies to extended multilingual: language, social, IT competencies. The results obtained can be used in the organization of the educational process in higher education institutions.

Keywords: Learning · Multilingualism · Educational technology · University · Students · Extended multilingual competencies

1 Introduction

Until recently, the term multilingualism was defined as the ability to speak several foreign languages. In today's world of global digitalization, multilingual competences are interpreted in a much broader way and presuppose the presence of social competences and digital literacy in addition to language competences. To increase competitiveness and demand in the labor market, a university graduate should have not only professional but also extended multilingual competences, i.e. linguistic, social and digital competences.

D. Bylieva and A. Nordmann (Eds.): PCSF 2022, LNNS 636, pp. 370–384, 2023.
https://doi.org/10.1007/978-3-031-26783-3_31

The university is a complex socio-economic system, constantly undergoing organizational and technological change. The leaders of transformations in Russia are federal and national universities.

Due to the change in the statuses of universities, the tasks facing them have changed. Previously, the university, or rather the activity of its educators, was aimed at transferring knowledge, skills and abilities. This approach was affected by the crisis of higher education, especially postgraduate and doctoral studies [1]. Today universities implement innovative technologies of knowledge transfer. They are aimed at forming various competences in students, including foreign students [2]. A modern university is a project university, i.e. it implements project-based learning technologies [3], which allows bringing students closer to the production environment and developing the professional competences demanded in the labor market [4]. Today the universities of the country actively attract for training foreign citizens of both post-Soviet and non-CIS countries. Language competences are formed in the process of direct educational activities. Students from different countries also work together during joint educational and project team work. Universities also face the task of developing social competences in their graduates. They are necessary for the implementation of subsequent employment in a multicultural and multi-ethnic domestic and professional environment. The third area of implementation of educational technologies in universities is the formation of competencies for generating and implementing multilingual digital solutions through various programming languages.

2 Problem Statement

A large number of works are devoted to the study of multilingual training implementation in the educational process of higher education institution. Prokhorova, Bezukladnikov suggest that multilingual competence will facilitate adaptation to new professional conditions, allow competing in the labor market with various specialists, including foreign ones [5]. Egorova defines multilingual competences as important qualities of future employees with disabilities. Having multilingual competences will allow them to act as linguistic and cultural mediators, mediators in professional activities [6]. Rubtsova et al. note the need to develop a constructive model for managing multilingual higher education to form mechanisms to assess the quality of educational services and their demand in the global market [7].

In the professional community, under the conditions of multinational, multiconfessional, multicultural society, there are expectations regarding the future graduate of the university in active interpersonal interaction [8]. Studenthood is the age and period of human life, when the main qualities of the future professional, his self-awareness and worldview are laid and formed. The correct pedagogical approaches will allow the student to be prepared for interactions in a multicultural environment [9]. In the future it will serve as a basis for interethnic and interconfessional understanding in the society and the future labor collective. Therefore, the discourse of multicultural learning is becoming popular. Multicultural learning will allow, along with the acquisition of professional competences, to acquire social experience in the unity of professional, social, cultural, national characteristics of the country and region [10]. Prokhorova explains the presence of communicative competences the need for graduates to exist in a multicultural environment in the future [11]. Anum et al. also note that intercultural communication is central to human relations in a community, organization, social and political environment [12]. Therefore, socio-communicative skills, i.e. competencies that facilitate social interaction, will be a complementary component for a future professional. Jiménez-Bucarey et al. to develop a quality assurance strategy for the digital transformation of online learning as a result of the constraints born by COVID-19 investigate the elements that affect student satisfaction and propose a model that measures student satisfaction, considering three dimensions: quality of teachers, quality of technical services and quality of services [13]. Many researchers analyze the problems that university professors and students faced in the transition to online learning during the COVID-19 pandemic [14], assess the level of student satisfaction with online learning services in higher education [15].

Researchers study student satisfaction with various components of online programs: course design, course organization, learning environment, and preference for teaching methods [16], analyze student satisfaction levels and the relationship between student satisfaction and service quality [17], compare student satisfaction with the teaching process, teaching technology, and educational infrastructure depending on gender, institution of higher education, and academic year of students [18].

Researchers study graduate students' preferences for developing their digital competencies. They believe that this will prepare them to work with digital technologies in the real sector [19]. Pesha and Shramko conduct a theoretical and empirical analysis of the demand for graduate communication competencies in the digital age [20].

Barinova and al. Analyze the problem areas of assessing students' learning achievements and propose criteria for evaluating general professional competencies at different stages of their formation in the educational process [21].

Other authors study students' and graduates' opinions on the self-assessment of their competences and the correspondence of their training to the requirements of the modern labor market [22, 23]. The training of qualified "specialists of the future" for the new digital economy, in the international environment of Russian urbanized systems, is aimed at forming the ability to effectively solve professional problems for the needs of the country and regions, improve business processes based on data processing and analysis, significantly increase the profitability of various types of production activities [24, 25]. Training in the "bachelor's/specialty, master's degree" system is carried out within the framework of the Strategy for Information Society Development in the Russian Federation for 2017–2030. Not for nothing the scope of the strategy is "information and communication technologies aimed at the development of information society, the formation of the national digital economy" [26].

Recruiting and HR specialists have formed the top 10 professions that will be irrelevant in 8–10 years. Despite the growing processes of globalization and the formation of a multicultural and multi-ethnic world, the profession of interpreter is one of them [27]. Of course, they don't connect this with the fact that everyone will acquire multilingual competences and will be able to communicate freely in several foreign languages, but there will appear such software products which will make it possible to carry out simultaneous and asynchronous translation by means of technical devices. Therefore, in our opinion, the main competences of future graduates will not be linguistic - knowledge of several foreign languages - but digital competences. Accordingly, in the context of an international environment and the development of Internet technologies [28], one of the priorities of modern education, is the formation of specialists with various IT-competences in the context of technology development in a multilingual environment - programming languages. Slugina and Trofimov point out that in the coming years professions related to website design, creation and maintenance will be in demand [29].

3 Materials and Methods

The modern student understands that the era of postgraduate employment is over. Today employers apply more effective methods of personnel selection and recruitment. Employers need a graduate with a wide range of professional and other competences, work experience, practice of project activities implementation [30]. A modern graduate wants to have not just an education diploma, but a full set of professional and extended multilingual competencies that allow him/her to be competitive and in demand in the labor market.

Therefore, today's "specialist of the future" wants to be more than a passive consumer of educational services. He or she is an actor who not only participates in various forms of educational activities, but also determines the trajectory of his or her individual learning. This leads to the formation of students' requirements for both the process and content of learning, as well as teaching technologies.

It is possible to assess the degree to which the modern university environment meets these requirements by studying the level of satisfaction with the educational process and its individual components [31].

However, there are not enough studies devoted to assessing student satisfaction with technology and the organization of the learning process in the context of their acquisition of professional and additional competencies.

The object of the empirical study is students of Ural Federal University (UFU) at all levels of education; Bachelor's, Specialist's and Master's degrees. The aim of the study is to assess student satisfaction with the technologies and organization of the learning process in the context of their acquisition of necessary competences. For this purpose, UrFU conducted a questionnaire survey of students. In total, more than 5,700 people were surveyed in 2020, 5,670 questionnaires were selected in the process of primary processing and checking the quality and. Completeness of the answers. In 2021, more than 7,100 people were surveyed and 7,040 questionnaires were selected during the initial processing and quality and. Completeness of responses. Based on the hypothesis that additional competences can be divided into 3 groups (communication in a foreign language (language), social, IT-competences) and they are acquired in the process of learning, we divided the questionnaire of student satisfaction into several blocks. Students assessed their satisfaction with the educational process in the following blocks: R&D, quality of teaching, additional education, practice, conditions for mastering the educational program, employment. The questionnaire contained a total of 37 questions, which students rated on a 5-point scale (where 1 – minimum score, 5 – maximum score). These blocks were chosen to assess student satisfaction, as professional competencies are formed in the process of current educational activities, and additional - in the process of practice, research, professional development, etc. In shaping the study design, we were focused on the SEEQ questionnaire. It has elements of effective learning: learning, examinations, organization, individual rapport, breadth, group interaction, assignments, overall rating, enthusiasm [32].

4   Results and Discussion

Ural Federal University is one of the largest universities in the Russian Federation. For 100 years the university has trained more than 350 thousand graduates. At present there are over 36,000 students studying at the university, including over 4,600 international students and trainees, 172 international postgraduate students from 110 countries. The university employs 3,782 scientific and pedagogical staff, including 24 foreign lecturers from 16 countries. Each year the University is visited by about 100 foreign delegations representing foreign universities, international organizations and companies, diplomatic missions, etc.

To study the level of satisfaction with various aspects of the educational process, UrFU conducted a large-scale survey of students at all levels of study in 2020–2021: over 5,700 people in 2020 and over 7,100 people in 2021. The distribution of the number of surveyed students by levels of education is presented in Table 1.

Table 1.   Education level of the students surveyed.

Level of education	Number of respondents in 2020	% of the number of respondents in 2020	Number of respondents in 2021	% of the number of respondents in 2021
Bachelor's degree	4080	71,96	5412	76,88
Specialty	659	11,62	1142	16,22
Master's degree	931	16,42	486	6,90
Total	5670	100,00	7040	100,00

Students evaluated their satisfaction with the educational process (Fig. 5, Table 2). The following blocks: Research work, quality of teaching, additional education, practice, conditions of the educational program, employment. The average values of satisfaction on the issues of the block Research work of students for 2021–2022 academic years are shown in Fig. 1.

The results of different studies are difficult to compare because of the variability in the selected aspects of the indicators. However, studies by other authors show an interest in the problem of teaching quality. So Dericks G. et al. write about the importance of indicators of satisfaction of graduate students with the support of supervisors, the support of the department and the university. Conducting research on graduate students

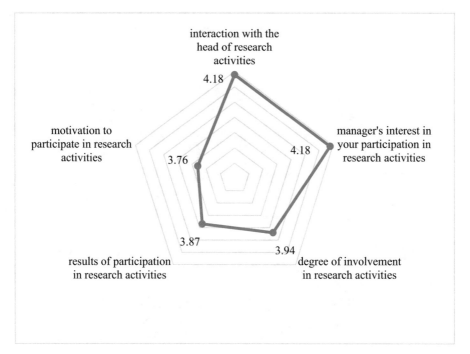

Fig. 1. The level of satisfaction with the block Research work of students.

in natural sciences, social sciences and humanities in 63 universities from 20 countries (n = 409) indicate the impact of managerial satisfaction with the educational and research activities of graduate students on the results of their learning [33].

Students' research activity is one of the important outcomes of students' preparation for the profession. As the results of the survey show, students rate most highly "interaction with the research activity supervisor" (4.18 points) and "interest of the research activity supervisor in participation in research activities" (4.18 points). This demonstrates their mutual readiness to develop professional and social competences of the student (ability to work with special literature and with various techniques; to apply the acquired knowledge in practice; to develop analytical thinking, to be able to form their own position; to solve complex problems). The lowest scores for students are "motivation to participate in research activities" (3.76) and "results of participation in research activities" (3.87).

This suggests that students do not see a correlation between active scientific activity and successful employment in the future. University training is based on knowledge and skills, which form the competence of a graduate. Today, the student is not a passive absorber of information; the result - the quality of his or her training - depends on the activity of his or her position. Therefore, it seems quite important how satisfied students are with the quality of teaching. The level of satisfaction with the quality of teaching and learning for 2020–2021 and 2021–2022 academic years are shown in Fig. 2.

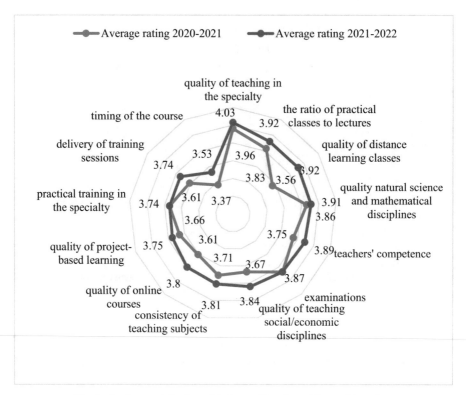

Fig. 2. Student satisfaction with the quality of teaching and learning.

As we can see from the graph, students rate the quality of teaching the disciplines in the specialty the highest (3.96 and 4.03 points in 2020–21 and 2021–2022 academic years, respectively). This shows the focus of students on obtaining professional competencies. However, the indicator of sufficiency of practical classes of disciplines in the specialty does not allow students to acquire applied professional competencies. In the future it leads to uncertainty of future graduates, refusal of employment in the profession (for example, only 8% of metallurgical graduates remain in the profession). However, we will focus on the problems with employment a little later.

The assessment of the quality of distance learning classes has changed significantly. It increased from 3.56 in 2020–21 to 3.92 in 2021–2022 academic years. The indicator we obtained correlates with the data obtained by Fatani T. on student satisfaction synchronous educational activities during the COVID-19 pandemic in Saudi Arabia. He points out that the students' satisfaction score (n = 662) with online learning in general was 4.1 [34].

The second highest level of satisfaction with the quality of teaching are natural sciences and mathematics disciplines (3.86 and 3.91 respectively), the third is humanities and socio-economic disciplines (3.67 and 3.71 respectively). One of the lowest indicators is the correspondence between the importance of the subject and the amount of time allocated to it. This raises some concerns and questions to the developers of educational programs of the university. Let us assume that students focused on expanding their own competencies, or rather their redistribution from purely professional to multilingual: language, social, IT competencies. Therefore, they are not satisfied with the number of hours allocated to "non-professional" disciplines and forms of work, which is reflected in the level of their satisfaction.

Perhaps the solution to these problems is additional professional education, implemented by the university, consider its components. It helps students to improve and expand their competences, to adapt them to modern economic and social conditions, to conduct professional activities. The average values of satisfaction in the block of additional education for 2020–2021 and 2021–2022 academic years are shown in Table 2.

Table 2. Average values of satisfaction on the issues of the block of additional education for 2020–2021 and 2021–2022 academic years.

№	Question	Average score 2020–2021	Average score 2021–2022
1	The content of additional education	3,70	4,14
2	The cost of obtaining additional education	3,30	4,05
3	The effectiveness of additional education	3,64	4,03
	Average for the whole block	3,55	4,07

Satisfaction rates are quite high, especially for the academic year 2021–22 (4.07), moreover, it increased from 3.55 in 2020/21. Experiencing a lack of formation of the three types of multilingual competences, including IT-competences in the educational process, students extinguish deficiencies through programs of additional education [35, 36]. It can be noted the work of the university to reduce the cost of educational programs of additional education and increase their availability, the indicator of satisfaction with the cost of programs has increased and in 2021–2022 academic years was 4.05 points.

Industrial and pre-graduation internships are an important part of the educational process and contribute to the development of professional and multilingual competencies. During the internship students learn how to independently apply the obtained knowledge in practice; to study technology and equipment used within a particular production; to develop teamwork skills; to perform self-monitoring. Average values of satisfaction with internship implementation at the university for 2020–2021 and 2021–2022 academic years are shown in Table 3.

Table 3. Satisfaction of students with the implementation of internships in production.

№	Question	Average score 2020–2021	Average score 2021–2022
1	Assistance of the head of practice from the company during the practice	-	4,01
2	Assistance of the head of the internship at the university in organizing and carrying out internships	-	3,92
3	Organization of internships at enterprises and organizations	3,15	3,87
4	University assistance in selecting internships	3,13	3,82
	Average for the whole block	3,14	3,91

Students rate the "assistance of the head of practice from the enterprise in organizing and passing the practice" (4.01) and "assistance of the head of practice at the university in organizing and passing the practice" (3.92) the highest. Sufficiently high scores of satisfaction on the questions of the block "practice" can be explained by the fact that since 2021 UrFU introduced the information service "Practice" in the Personal Office of the partner, which is designed to manage the interaction with partners-employers as part of the practice of students in enterprises. The service allows you to coordinate the agreements according to the standard template of UrFU, as well as the partner form, download scans of signed agreements, view the register of practice partners, view scans of concluded agreements. Allows you to apply to employers for internships for students with a choice of the necessary training direction. The service informs the partner about the status of enrollment on its application, about students enrolled, etc. Despite the introduction of the "Practice" service, students rate the "organization of internships in general the worst: 3.15 in 2020 and 3.87 in 2021 and the university's assistance in selecting places of practice: 3.13 in 2020 and 3.82 in 2021. To increase student satisfaction, the university is recommended to expand the functionality of the service for students and the list of partners-employers for the practical training of students.

The average values of satisfaction on the questions of the block Conditions of mastering the educational program in 2021–2022 academic years are shown in Table 4.

Low satisfaction of students with the conditions of the implementation of the educational program is due to the low values of scores on the question: The ability to choose an individual learning path (3.58 of 5). This assessment can be explained by the fact that every year the demand from students for individualization of educational trajectories increases. This is necessary to minimize the competence gaps in the transformation of the requirements of the labor market. It is well known that the labor market reaction outpaces the university reaction to such changes. In order to increase student satisfaction according to this indicator, the university needs to expand educational opportunities for students, increase flexibility, variability of educational trajectories according to students' choice. On the other hand, we should not forget that for the last 2 years universities mostly worked in a distance format. Of course, this form of work, the lack of opportunity to go to practice at the enterprise, of course, affected the decrease in the satisfaction index.

Table 4. Student satisfaction with the conditions of the educational program.

№	Question	Average score 2021–2022
1	Class Schedule	3,90
2	Ability to choose an individual learning path	3,58
3	Transparency and clarity of assessment criteria for current and interim evaluations	3,52
4	The level of encouragement of students for results in studies, science, extracurricular achievements, etc.	3,52
	Average for the whole block	3,63

Let's consider satisfaction with postgraduate employment opportunities for 2020–2021 and 2021–2022 academic years, they are shown in Table 5.

Table 5. Satisfaction with postgraduate employment opportunities for students.

№	Question	Average score 2020–2021	Average score 2021–2022
1	The University's assistance in finding a job for a student (by specialty) during the period of study	3,21	3,08
2	The University's assistance in finding employment after graduation	3,31	3,04
3	Average for the whole block	3,26	3,06

Low student satisfaction in the block "employment" with negative dynamics from 3.26 in 2020 and 3.06 in 2021/22 academic year is caused by the annual decrease in the share of university graduates who plan to work in their specialty. In 2021, 6,645 people graduated from Ural Federal University, including 5,181 full-time students. In 2021, 6,645 people graduated from Ural Federal University, including 5,181 full-time students. As of December 30, 2021, 90.3% of the 2021 full-time Ural Federal University graduates were employed. Of these, 56% were employed, 34% were continuing their studies, 0.5% were drafted into the RAs, and 10% were unemployed. The "employed" includes employed graduates, graduates who continued their studies, and graduates drafted into the ranks of the Russian Army. However, despite the rather high percentage of employed, the decrease in satisfaction shows that the university does not work enough in this direction. The matter is also that not all graduates, especially bachelors, are employed according to the specialty/direction of training, they are set to obtain broader competencies, which can be useful not only in production, but also in everyday life.

For example, de Oliveira Silva, J.H., et al. actualize the topic of graduate employability. They point out that students' skills and behavior depend on the quality of educational services, by which they mean academic quality, consistency of subjects, practice, efficiency of course coordination. This constitutes student satisfaction with higher vocational education institutions. The results of their study showed (n = 505) that if students positively evaluate employment opportunities, their satisfaction with education increases. The authors believe that the university should focus on developing job-based curricula that focus on the managerial and technical skills needed in the job market [37].

In our case, for to increase student satisfaction with this indicator, the university needs to implement career guidance and counseling of students, expand cooperation with potential employers.

5   Conclusion

Students study to acquire competencies that will allow them to compete after graduation in the municipal, regional, country and international labor market. All competencies are acquired in the process of learning at the university.

There was a slight decrease in satisfaction with the educational process on the block "conditions of mastering the educational program". In order to ensure a smooth educational process during the COVID-19 pandemic, Russian universities urgently transferred the educational process to a distance learning format, which led to difficulties in the transition period for both students and teachers [38]. The decrease in the indicators can be explained by technical and psychological problems of students in the transition to distance learning. It is recommended to analyze how the transition to the distance form occurred and to take the necessary measures.

Despite an increase in student satisfaction with the work of teachers compared to the previous academic year, the average level of satisfaction, reflecting different aspects of the educational process is not high enough, is 3.75 on a 5-point system.

The level of student satisfaction with scientific research work and research support is quite high, the score is 3.99 points. However, students do not see positive results in employment from active engagement in research and development.

Analysis of student satisfaction with the educational process at the university showed that the lowest level of satisfaction students show in relation to employment procedures. It decreased from 3.26 points in 2020/21 to 3.06 in 2021/22. This correlation can be explained by the fact that more than half of the respondents are in their first and second years of undergraduate studies. That is, most respondents have just begun their studies and are concerned about their future, concerned about what they have not yet explored in their experience. To increase satisfaction with this indicator it is necessary to transform the educational process in the direction of practice-oriented, to implement project-based learning together with the enterprises of the real economy, to hold career fairs, which allows to bring students closer to the production environment and develop the necessary competences in the labor market.

The data show that the lowest satisfaction of students with those aspects of the educational process that cause the greatest concern (the passage of practices, employment). Also decreased compared to last year, indicators related to practice, which can be explained by the transition to distance learning and practical training in the laboratories of the university, rather than in the workplace. Students are concerned about the lack of opportunity to acquire not theoretical but practical professional and extended multilingual competencies.

There are high marks and a significant increase in the level of satisfaction in relation to additional education. We can see that additional education is becoming more and more in demand, as it helps students to build individual educational trajectories, acquire extended multilingual competencies necessary for the modern labor market. The university responds most quickly to the changing needs of students to acquire the necessary competencies by providing Supplementary Education. Ural Federal University has established 19 centers for Supplementary Education at its institutes, the number of students trained in Supplementary Education programs is growing annually, reaching 10,927 in 2020 and 11,950 graduates in 2021. This explains the high satisfaction of students with the educational process in the block of additional education. Student satisfaction shows the quality of educational services and acquired competencies.

Responding to the requirements of society, universities have to transform the education system in accordance with the needs of the digital economy and the needs of students. This requires identifying problems and improving the quality of educational services, as well as shifting the focus from the formation of only professional competences to extended multilingual competences: language, social, IT-competences, so demanded in the constantly changing labor market.

References

1. Zaitseva, E., Zapariy, V., Asryan, G.: The results of reforms in the field of training highly qualified personnel in the late XX– early XXI century: scientific personnel. Hist. Mod. Perspect. **4** (2), 65–75 (2022). (In Rus). https://doi.org/10.33693/2658-4654-2022-4-2-65-75

2. Daineko, L., Larionova, V., Yurasova, I., Davy, Y., Karavaeva, N.: Educational process digitalization in Ural Federal University. In: Busch, C., Steinicke, M., Wendler, T. (eds.) 19th European Conference on e-Learning, ECEL 2020, vol. 2020, pp. 146–153. Academic Conferences International Limited, London (2020). https://doi.org/10.34190/EEL.20.029

3. Daineko, L.V., Yurasova, I.I., Karavaeva, N.M.: Creative projects as a link between theory and practice. In: Bylieva, D., Nordmann, A. (eds.) PCSF 2021. LNNS, vol. 345, pp. 558–572. Springer, Cham (2022). https://doi.org/10.1007/978-3-030-89708-6_47

4. Daineko, L., Reshetnikova, O.E.: Project method - an effective instrument for developing competencies of future professionals. In: Shipunova, O.D., Bylieva D. S (eds.) Professional culture of the Specialist of the Future & Communicative Strategies of Information Society. European Proceedings of Social and Behavioural Sciences, vol. 98, pp. 221–230. European Publisher (2020). https://doi.org/10.15405/epsbs.2020.12.03.23

5. Prokhorova, A., Bezukladnikov, K.: Multilingual teaching of technical university students: rational arguments. Lang. cult. **52**, 215–231 (2020). (In Rus). https://doi.org/10.17223/199 96195/52/14

6. Egorova, P., Prokhorova, A., Sorokoumova, S.: Application of multilingual approach in inclusive education in higher education. Lang. cult. **54**, 152–166 (2021). (In Rus). https://doi.org/10.17223/19996195/54/9

7. Rubtsova, A., Almazova, N., Bylieva, D., Krylova, E.: Constructive model of multilingual education management in higher school. IOP Conf. Ser. Mater. Sci. Eng. **940**, 012132 (2020). https://doi.org/10.1088/1757-899X/940/1/012132

8. Sinenko, T., Zalipaeva, O.: Formation of inter-ethnic tolerance by means of learning situation at the stage of students' adaptation to the educational environment of the university. Rus. J. Educ. Psych. **13**(2), 48–61 (2022). (In Rus). https://doi.org/10.12731/2658-4034-2022-13-2-48-61

9. Belousova, O.: Formation of tolerance in students in the conditions of a departmental university (on the example of the Kuzbass Institute of the Federal Penitentiary Service of Russia). Theor. Pract. Socio-Hum. Sci. **4**(12), 4–7 (2020). (In Russian)

10. Aldoshina, M., Komarova, E., Bakleneva, S., Fedorov, V.: A subject-centered approach to the development of multicultural education at university. Lang. Cult. **56**, 146–163 (2021). (In Russian). https://doi.org/10.17223/19996195/56/9

11. Prokhorova, A.: Formation of socio-cultural literacy of future multilingual specialists of the new generation. In: Innovative Scientific Research in Humanities, Natural, Technical and Social Sciences. Methodology, Theory, Practice, pp. 200–202. KultInformPress, St. Petersburg (2014). (In Rus)

12. Anum, V., Shaibu, G., Okeme, P.: Intercultural communication as a tool for relationship building among groups in a multi-ethnic system. Sau J. Manag. Soc. Sci. **6**(1), 220–228 (2021)

13. Jiménez-Bucarey, C., Acevedo-Duque, Á., Müller-Pérez, S., Aguilar-Gallardo, L., Mora-Moscoso, M., Vargas, E.C.: Student's satisfaction of the quality of online learning in higher education: an empirical study. Sustainability **13**(21), 11960 (2021). https://doi.org/10.3390/su132111960

14. Almazova, N., Krylova, E., Rubtsova, A., Odinokaya, M.: Challenges and opportunities for Russian higher education amid COVID-19: teachers' perspective. Educ. Sci. **10**, 368 (2020). https://doi.org/10.3390/educsci10120368

15. Surahman, E., Sulthoni: Student satisfaction toward quality of online learning in indonesian higher education during the Covid-19 pandemic. In: 2020 6th International Conference on Education and Technology (ICET). IEEE, pp. 120–125 (2020). https://doi.org/10.1109/ICET51153.2020.9276630

16. Yawson, D.E., Yamoah, F.A.: Understanding satisfaction essentials of E-learning in higher education: a multi-generational cohort perspective. Heliyon **6**(11), e05519 (2020). https://doi.org/10.1016/j.heliyon.2020.e05519

17. Pedro, E., Mendes, L., Lourenço, L.: Perceived service quality and students' satisfaction in higher education: the influence of teaching methods. Int. J. Qual. Res. **12**(1), 165 (2018)

18. Herrera Torres, L., Souza-Soares de Quadros, M.R., Sánchez-Sánchez, L.C., Ramiro-Sánchez, T.: Satisfaction with self and external regulation of learning in higher education students in Brazil. Int. J. Environ. Res. Public Health, **18**(11), 5914 (2021). https://doi.org/10.3390/ijerph18115914

19. Cham, K., Edwards, M.L., Kruesi, L., Celeste, T., Hennessey, T.: Digital preferences and perceptions of students in health professional courses at a leading Australian university: a baseline for improving digital skills and competencies in health graduates. Aust. J. Educ. Tech. **38**(1), 69–86 (2022). https://doi.org/10.14742/ajet.6622

20. Pesha, A., Shramko, N.: The importance of developing communicative competencies of future specialists in the digital age. In: 2nd International Scientific and Practical Conference "Modern Management Trends and the Digital Economy: From Regional Development to Global Economic Growth" (MTDE 2020), pp. 886–892. Atlantis Press (2020). https://doi.org/10.2991/aebmr.k.200502.145

21. Barinova, D., Ipatov, O., Odinokaya, M., Zhigadlo, V.: Pedagogical assessment of general professional competencies of technical engineers training. In: Katalinic, B. (ed.) Proceedings of the 30th DAAAM International Symposium, pp. 0508–0512. DAAAM International, Vienna (2019). https://doi.org/10.2507/30th.daaam.proceedings.068

22. Gawrycka, M., Kujawska, J., Tomczak, M.: Self-assessment of competencies of students and graduates participating in didactic projects–case study. Int. Rev. Econ. Educ. **36**, 100204 (2021). https://doi.org/10.1016/j.iree.2020.100204

23. Ardeleanu, M., Popescu, D.: The impact of the applied teaching activities in the training of technical and digital competencies and abilities. In: 2021 International Conference on Applied and Theoretical Electricity (ICATE), pp. 1–5. IEEE, Craiova (2021). https://doi.org/10.1109/ICATE49685.2021.9465014

24. Antonova, A., Aksyonov, K., Ziomkovskaia, P.: Development of method and information technology for decision-making, modeling, and processes scheduling. In: Simos, T., Tsitouras, Ch. (eds.) International Conference on Numerical Analysis and Applied Mathematics, ICNAAM 2020, vol. 2425(1), p. 130003. AIP Conference Proceedings, AIP Publishing LLC (2020). https://doi.org/10.1063/5.0082100

25. Khalyasmaa, A.I., Zinovieva, E.L., Eroshenko, S.A.: Performance analysis of scientific and technical solutions evaluation based on machine learning. In: Proceedings - 2020 Ural Symposium on Biomedical Engineering, Radioelectronics and Information Technology, USBEREIT 2020, pp. 475–478. IEEE, Ekaterinburg (2020) https://doi.org/10.1109/USBEREIT48449.2020.9117774

26. Strategy for the Development of Information Society in the Russian Federation for 2017–2030. http://pravo.gov.ru/proxy/ips/?docbody=&nd=102431687. Accessed 20 Aug 2022

27. Professions that will disappear by 2030. https://zaochnik.ru/blog/professii-kotorye-ischeznut-k-2030-godu/. Accessed 20 Aug 2022

28. Reshetnikova, O.: Effective learning tools in e-learning. In: Busch, B.C., Steinicke, M., Frieß, R., Wendler T. (eds.) Proceedings of the 20th European Conference on e-Learning, ECEL 2021, pp. 387–393. Academic Conferences and Publishing International Limited (2021). https://doi.org/10.34190/EEL.21.091

29. Slugina, N., Trofimov, M.: Raising the level of training of specialists in web-programming with regard to the needs of the labor market. Mod. Probl. Scien. Educ. **3**, 224 (2013). (In Rus)

30. Daineko, L., Davy, Y., Larionova, V., Yurasova, I.: Experience of using project-based learning in the URFU hypermethod e-learning system. In Orngreen, B R., Buhl, M., Meyer B. (eds.) Proceedings of the 18th European Conference on e-Learning, ECEL 2019, pp. 145–150. Academic Conferences and Publishing International Limited (2019). https://doi.org/10.34190/EEL.19.066

31. Zaitseva, E., Zapariy, V.: Role of the university corporate culture in university management. In: 10th International Days of Statistics and Economics, pp. 2089–2095. Melandrium, Prague (2016)

32. Marsh, H.W.: A longitudinal perspective of students' evaluations of university teaching: ratings of the same teachers over a 13-year period. In: Annual Meeting of the American Educational Research Association, ED353282, pp. 1–18. ERIC Document, San Francisco (1992). https://files.eric.ed.gov/fulltext/ED353282.pdf

33. Dericks, G., Thompson, E., Roberts, M., Phua, F.: Determinants of PhD student satisfaction: the roles of supervisor, department, and peer qualities. Assess. Eval. High. Educ. **44**(7), 1053–1068 (2019). https://doi.org/10.1080/02602938.2019.1570484

34. Fatani, T.: Student satisfaction with videoconferencing teaching quality during the COVID-19 pandemic. BMC Med. Educ. **2**(1), 1–8 (2020). https://doi.org/10.1186/s12909-020-02310-2

35. Fadeeva, O.: Development of teacher's ICT-competence within the framework of teacher-centered professional learning in the system of professional education. Open Educ. **2**(4), 34–41 (2018). (In Rus). https://doi.org/10.21686/1818-4243-2018-4-34-41

36. Fadeeva, O., Simonova, A.: Deficits of ICT-competence of teachers in Krasnoyarsk region. Bull. Krasnoyarsk State Pedagog. Univ. Named After V. Astafyev. **4**(42), pp. 89–99 (2017). (In Rus). https://doi.org/10.25146/1995-0861-2017-42-4-24

37. de Oliveira Silva, J.H., de Sousa Mendes, G.H., Ganga, G.M.D., Mergulhão, R.C., Lizarelli, F.L.: Antecedents and consequents of student satisfaction in higher technical-vocational education: evidence from Brazil. Int. J. Educ. Vocat. Guid. **20**(2), 351–373 (2019). https://doi.org/10.1007/s10775-019-09407-1

38. Goncharova, N., Zaitseva, E.: Responses of Russian universities to the challenges of Covid-19 pandemic. In: Busch, C., Steinicke, M., Wendler, T. (eds.) 19th European Conference on e-Learning, ECEL 2020, vol. 2020, pp. 221–228. Academic Conferences International Limited, London (2020). https://doi.org/10.34190/EEL.20.140

Features of the Transformation of Russian Universities into Digital Universities

Natalia V. Goncharova$^{(\boxtimes)}$ (ID) and Liudmila V. Daineko (ID)

Ural Federal University, 19 Mira Street, 620002 Ekaterinburg, Russia
{n.v.goncharova,l.v.daineko}@urfu.ru

Abstract. Traditionally, specialists with multilingual competences have been more highly valued in the labor market in any field of activity. However, the term "multilingualism" is losing original meaning, which implies mastery of several foreign languages in modern universal digital transformation's world. Today, multilingual competences imply not only a high level of foreign language proficiency, but also the ability to quickly perceive digital languages, i.e. to intuitively use various digital services, software packages and applications, understanding the principles of algorithm and computer program development. The global digitalization of the economy, accelerated by pandemic restrictions, has accelerated the digitalization of higher education institutions. Digitalization has forced students, faculty, and university administrators to rapidly adopt new digital services. The article examines the features of the digital transformation of universities that won the competition of the Ministry of Science and Higher Education held within the framework of the national program "Digital Economy in the Russian Federation" to develop a digital university model. The analysis revealed general trends in the digital transformation of universities, showing that the development of a digital university model is a complex task aimed at re-engineering business processes with the capabilities of modern IT technologies by introducing digital services in all areas of university activities: administrative, educational, scientific and innovative. The authors conclude that it is necessary to intensify the replication of the developed model of the digital university and the possibility of disseminating the best practices of digitalization of business processes by universities around the world.

Keywords: Digital university · Higher education · Transformation of universities · Graduate competencies

1 Introduction

The annual acceleration of the growth rate of technological development leads to a huge rate of obsolescence of knowledge and the emergence of new technologies, which requires a review of the entire system of higher education. Traditional planning of educational programs, which have been implemented unchanged for many years, is irrevocably outdated, as it is difficult for universities to foresee what kind of knowledge, skills and competencies today's applicants will need in 4–5 years after graduation. To improve the employability of tomorrow's university graduates, universities are actively introducing digital technologies into the educational process and digital literacy courses into their curricula. Digital technologies definitely imply a new modern digital language - the language of technology, which needs to be mastered, and this is often a difficult task for users of digital services. Digital literacy implies not only the knowledge of programming languages, designed for writing computer programs and representing a set of rules that allow the computer to perform any calculations or actions or organize the management of various objects, but also the understanding of the basic principles of information systems, electronic devices, social networks, applications, e-mail and other digital services. In addition, digital literacy involves knowledge of information security and digital ethics, and the ability to use information technology to make decisions, model and plan processes [1, 2].

The formation of an individual who perceives the diversity of the surrounding multilingual world, who takes an active lifestyle, who is able to interact with representatives of different cultures, who can competently use the language of technology is an important task of universities. In addition, global digitalization is changing the educational space and requires the development of multilingual communicative competence in the educational environment through communication in scientific and educational online communities, on professional websites, through online training, and etc. [3].

The pandemic has shown that universities differ significantly in the level of digitalization, and more advanced ones are ready to share their resources and best practices [4]. Another important aspect is that the digital transformation of universities varies quite a lot in orientation, speed and results due to differences in priorities, level of financial condition, available facilities and resources, educational policy, as well as implemented educational programs. In order to maintain and accelerate the digitalization of higher education, the Ministry of Science and Higher Education selected universities on a competitive basis to disseminate the best international practices of training, retraining and internship of advanced digital economy personnel, as well as to develop a digital university model.

It should be noted that the pandemic, on the one hand, exacerbated the problems associated with the unwillingness of universities to use digital technologies in the educational process; and on the other hand, it accelerated the processes of digital transformation of universities [5]. In a fairly short period of time, all universities were forced to switch to distance learning, but in most cases it was the transfer of the traditional training to the format of webinars without using the extensive opportunities provided by the platforms of mass open online courses, virtual communication platforms and digital feedback collection services [6]. In this regard, the issue of ensuring the proper quality of education and competitiveness of graduates has become acute. To solve this problem, the Ministry of Science and Higher Education, within the framework of the national program "Digital Economy in the Russian Federation", set the task of digital transformation of Russian universities, allocated funding and selected on a competitive basis five universities, where a network of centers for the development of a "Digital University" model was created for further broadcasting of best practices to other institutions of higher education. The competition allowed to mobilize advanced universities in a short time to form the concept of the model "Digital University".

2   Problem Statement

As long as ten years ago, Davies wondered whether universities could survive in the era of digitalization [7] and came to the conclusion that traditional universities would feel more and more pressure from digital technologies, in particular, through free digital educational platforms. And more and more researchers are inclined to believe that it is vital for universities to move into the digital environment, realizing the need to rethink interaction with the surrounding multilingual world [8]. The relevance of this transition was clearly demonstrated in 2020, when the whole world, and in particular, universities, had to urgently transfer all work to a remote format [3, 9–17]. Researchers, considering the issues of the general concept of a digital university [18], describe a set of necessary transformations of a classical university into a "digital" one [19], analyze the formation of a "digital university" model and the key elements of this model [20] consider the needs of stakeholders in the development of a digital university [21], analyze leading Russian and foreign universities in the fields of vocational guidance of applicants, the formation of student competencies and the support of graduates' careers [22], describe their national experience of digital transformation of universities [23–29]. Thus, Rozhkova, Rozhkova & Blinova, studying the prospects and problems of digital universities in Russia [30] identified eight criteria by which it is possible to assess the current digitalization of universities. According to the researchers, the main difficulties arise with the development and selection of an information system to support the individual profile of a student's competencies. Habib et al., considering the process of transformation of a university into a digital one, also revealed that digital opportunities are not fully used by the universities [31], including due to regulatory requirements. Kasatkin, Kovalchuk & Stepnov, assessing the role of universities in the formation of the digital economy, came to the conclusion that the importance of the influence of universities on the formation of the upward wave of Kondratiev cycles is confirmed in a long period [32]. Coccoli et al., considering the concept of a smart university [33], urge not to confuse a digital university with a smart university, explaining that digital technologies are a means, but not an end. However, there are not enough studies devoted to the analysis of the features of the digital transformation of universities.

3   Materials and Methods

The Ministry of Science and Higher Education provided substantial support to the process of digitalization of the educational process within the framework of the national program "Digital Economy in the Russian Federation", defining a list of universities for the development of a digital university model. The goal of the Digital Economy in the Russian Federation program is to train competent personnel for the digital economy, capable of quickly mastering the language of technology. Education is the sphere in which digitalization is happening right now. It is in educational institutions that most young specialists acquire their competencies. Understanding of this has directed the vector of attention of the state to the features of digitalization of universities. The competition of the Ministry of Education and Science was aimed at identifying the universities that are more advanced towards digitalization, that have accumulated methodological approaches to the implementation of digital practices, that have the necessary infrastructure and digital technologies to organize all aspects of university activities in a digital environment. During the competition, five universities were selected out of 43 that participated in the competition, on the basis of which a network of centers was created for the development of the "Digital University" model (DU model): HSE University, Ural Federal University, ITMO University, Tomsk State University and I.M.Sechenov First Moscow State Medical University. Three universities-developers of the "Digital University" model: HSE University, Ural Federal University and Tomsk State University were also selected by the Ministry of Higher Education and Science of the Russian Federation to disseminate the best international practices for training advanced digital economy personnel in the field of mathematics, computer science and technology through the creation of five international scientific methodological centers (Center P and PP). To analyze the components of a digital university, the authors analyzed the official websites of five universities selected for participation in the program "Digital Economy in the Russian Federation", in order to identify common approaches to the concept of "Digital University".

4 Results and Discussion

To develop a model of a "Digital University", the winners of the selection of the Ministry of Science and Higher Education propose to implement the following practices.

4.1 Higher School of Economics

At the Higher School of Economics, the model of a "Digital University" includes four large units:

- Education. The unit includes a single accounting content management system ACAB 2.0, which provides a high level of integration with the university information system through convenient and intuitive interfaces of users' personal accounts, a new generation SmartLMS training system equipped with a proctoring service that allows users to organize and conduct knowledge control or exam in a remote format while maintaining the reliability of the results obtained. It should be noted that the HSE is actively developing online education (125 online university courses are located on the national open education platform), offering the opportunity to obtain high-quality knowledge for everyone - both students and schoolchildren who are going to enroll, and just listeners interested in obtaining certain knowledge from one of the leading universities in the country.
- Working environment is a unit that optimizes business processes to ensure the effective operation of the university. This unit of the digital university is implemented using the SmartPoint multiservice "single window", which allows all users to access administrative services, to contact technical support to solve problems. Users can independently configure the elements of their personal account through the navigation panel. Login to the system is carried out through a single authorization system and the choice of a role model (student, employee, external user). The service provides access to 16

information systems, including the ability to receive notifications on submitted applications and answers to questions from support. The information in the service can be displayed in Russian and English (work is underway to connect Chinese, French and German). To sign the necessary documents, the service implements the possibility of using an electronic digital signature. The university also has a SmartBOSS platform that combines the programs and services of the university's back office on personnel, financial and accounting issues, in procurement and legal support of activities. The possibility of using the working environment through mobile applications is also implemented.

- Partnership. The unit includes the possibility of using the university's educational platforms for studying courses, passing exams and obtaining certificates by students and trainees around the world, including through the conclusion of network agreements for the implementation of educational programs with other universities. Another possibility of the unit is the transfer of non-core areas for the Higher School of Economics to partners.
- SmartData Infrastructure. This unit of the digital University of the Higher School of Economics is a reliable foundation for the digital transformation of education, implementing the Data Driven approach (decisions are made based on data analysis, not intuition or personal experience). The unit includes a comprehensive information security program Smart Security, which includes protection against spam and phishing, anti-virus protection, educating users based on the developed Information Security Policy and Regulations on confidential information. The University has implemented a digital support service "Hotline 55555" for the operation of all information systems and services. Another area of the unit's work is the organization of infrastructure as IaaS services, in which equipment (servers, data warehouses, etc.) are not purchased by the organization, but are used as a service of third-party providers.

The work of the HSE International Centre for Research and Teaching[1], established in 2019 for the implementation of the federal project "Personnel for the Digital Economy", is aimed at research and teaching staff and graduate students of universities who want to improve their competencies in teaching mathematics, computer science and digital technologies, as well as in conducting research in these areas of knowledge. Advanced training courses at the HSE are organized in three modules:

- Computer science;
- Mathematical sciences;
- Engineering sciences.

Another important project of the HSE ICRT is an aggregator platform that allows conducting research in the field of data analysis and machine learning. With the help of this platform, teachers can support the scientific work of their students. The key partners of the HSE ICRT are Yandex, the Steklov Mathematical Institute and the Center for Pedagogical Excellence.

---

[1] https://mnmc.hse.ru/.

4.2 Ural Federal University

The digital model program of the Ural Federal University involves the implementation and improvement of the following areas:

- Digital educational technologies is the direction that involves the organization of networking between universities, including organizational and technical support for online learning, including support for remote test and control measures and intermediate certification in an online format with identification, as well as the creation of a designer of educational programs and adaptive online courses by means of using the repository of the online technology center;
- Individual educational trajectories (IET) based on project-based learning - the direction that involves the management of the educational space, including the workload of teachers and the individual schedule of students. An important part of the direction is the Digital Tutor service [34], which provides decision-making support (recommendations) on the selection of online courses and the formation of students' IET, assistance to heads of educational programs to assess the quality of online courses, including using digital profiles of work programs of disciplines. The University has developed a unique digital platform for interaction with partners and implementation of students' project activities;
- Digital Economy Competencies (DECs), a tool for partner universities - is the direction that involves the construction of ontologies of demanded DECs using the designer of the dynamic model of the DECs based on interaction with the market in the areas of training and market segments. Collection and analysis of monitoring data of the DECs expert community and analysis of the assessment of the dynamics of the demand for DECs. Assessment of teachers' DECs, formation of additional professional retraining programs for specific DECs;
- The data-based management system aimed at outsourcing and integration includes the services of the applicant's, student's and employee's personal account, including a data collection system for an effective employee contract, a B2B meta-service for interaction with partners, a project management system, a portfolio management service for educational programs, a researcher's console and a management system for scientific equipment.

The International Scientific and Methodological Center for the Transfer of Competencies of the Digital Economy of UrFU is designed to disseminate the best international practices of training, retraining and internship of advanced personnel in the fields of mathematics, computer science and technology. UrFU ISMC implements two types of free programs for university teachers and researchers:

- Educational: advanced training programs and professional retraining programs;
- Internships: scientific and innovative;

Areas of study in the UrFU ISMC:

- Mathematics;
- Computer science;
- Digital economy technologies.

4.3 Tomsk State University

Tomsk State University considers the development of a digital university as an opportunity to transform learning and teaching. This involves updating and creating original technologies, content and formats based on fundamental research in the field of education and the development of new educational technologies. TSU adheres to the model TPACK of the University of Michigan [35], which emphasizes the need to combine the transformation of educational content and the transformation of pedagogical approaches. To implement the digital transformation of the university on the basis of research works in the field of new educational technologies, the university is focused on the development of the following projects:

- The Digital Mentor service is a digital assistant that accompanies students on academic, psychological, career and information issues. The service either provides the necessary information or redirects to the necessary services of the university, reducing the number of deductions, academic debts, increasing motivation and satisfaction of students;
- Leadership Academy service, which includes a whole set of micro-online courses dedicated to the development of soft-skills and self-skills by students in the mode of an individual educational trajectory. Soft-skills and self-skills competencies are universal and in demand in the modern world, regardless of the specialty being mastered;
- A training program to improve the digital skills of teachers to stimulate the translation of standard educational content into a digital and mixed educational environment.
- A unit of modular programs of additional education for different categories of students, provided with a set of electronic teaching materials and consulting support of teachers;
- Learning Spaces service for designing smart learning environments based on the needs of students, different styles and teaching methods, learning psychology and the best world practices. Such learning environments make it possible to implement learning in a new format using the latest achievements in the field of didactics, pedagogy and digital learning technologies;
- Together with partners, projects are being implemented to develop an adaptive mathematics learning system (with ENBISYS IT company), an educational environment for learning English (with Skyeng online school), to create a VR/AR Technology Laboratory (with Rubius IT company), to create a large-scale Virtual University 4.0 platform, and others.
- Service for the management of projects, which are implemented TSU, provides university teams with tools to work on their projects in the system "1C-Bitrix24". The service operates on a new model of organization of business processes and is adapted to project activities with support for electronic document management. Project passports have been completely digitalized, which reduces the time for creating and approving documents.

The university believes that technology, taking over the routine part of affairs, helping, accompanying, providing resources, leaves space for interpersonal communication.

The International Scientific and Methodological Center of Tomsk State University offers free training, retraining and internship programs for digital economy personnel in the fields of mathematics, computer science, digital technologies, including technologies from the field of artificial intelligence. In 2019, the Center launched three retraining programs in educational design and IT development in the digital economy for employees of Russian universities.

4.4 ITMO University

ITMO University is conducting a sequential transformation into a digital university. Digitalization of auxiliary processes is planned at the first stage:

- Communications;
- Reference books;
- Primary analytics;
- Administrative processes.

The result of the first stage is the development of algorithms for further concentration on the main processes. The University has put into operation data-based models of interaction with users and digital tools to improve the effectiveness of management decisions. The information management system, being the core of the corporate information ecosystem and a platform for the digital transformation of the university, includes more than 150 systems and services, providing work opportunities for more than 15 thousand participants.

At the second stage of transformation, it is planned to reorganize the main processes under the "digital" model of work, through:

- introduction of models of flexible construction of the educational process;
- working out and ordering, preventing the digitalization of chaos.

At the third stage of transformation (currently being implemented), two main processes of the university are being digitalized – educational and research:

- digitization of information for processes: databases on educational programs, information about students (achievements, interests, extracurricular activities, soft skills);
- digitization of new models of the educational process.

The University is implementing the ITMO Avatar program – a tool for adapting the digital environment of the university to the request of a specific participant: a student or a teacher. Further expansion of users is planned, including applicants, employees, partners and graduates of the university. The implemented platform will include a number of services that perform specific actions, for example, recommend optimal trajectories, support decision-making, provide personal recommendations based on previous requests, behavioral patterns and personal achievements. The Avatar is trained by regularly tracking feedback from the project participants, analyzing the participants' actions. The Avatar platform is based on a dynamic "digital lake", replenished with new data about project participants, for example, exams results, published scientific articles, completed tasks in LMS systems, attended events, etc. The system can also remove a project participant from a number of undesirable activities. ITMO Avatar also provides participants with personal news services, selection of events by interests, assistance in choosing a supervisor, etc. In the future, the capabilities of the ITMO Avatar will increase and it will be able to build an individual educational trajectory, recommend disciplines, universities for academic mobility, will be able to help in finding co-authors of scientific papers, will be able to register a platform user for various events, make appointments and much more. It is planned that the ITMO Avatar platform will become a convenient tool for communication and time management among students and university staff. The system of all ITMO Avatars will be a digital university developing both inside and outside the system. To use the platform, the user must pass an entrance voluntary testing to identify personality traits and assess existing competencies.

The ITMO.EXPERT service, an open platform for exchanging experiences on teaching and learning organization issues, operates at the university. This is a non-classical professional development program with open access to the best practices of ITMO teachers. Here you can also find materials on the organization of classes in a distance format from the expert community of ITMO University.

The upcoming fourth stage will include the introduction of deep analytics systems and the development of fundamentally new models of the main processes. It is planned to:

- use AI, BI, expert systems;
- adapt third-party practices;
- develop an individual educational tracks based on the information received.

4.5   I.M.Sechenov First Moscow State Medical University

The I.M.Sechenov First Moscow State Medical University, being the leading medical university in Russia, is actively implementing its own model of a digital university, replicating the best practices in cooperation with medical universities in the country. The digital transformation program involves providing students, staff and specialists with access to the university's services, resources and infrastructure through a single digital platform, as well as providing high-quality medical services using new technologies. The Sechenov University has identified four areas of work of the digital platform:

- The open digital portal of medical education Sechenov.online, being an interuniversity platform of electronic medical education, contains more than 60 online courses in Russian and English. The purpose of creating the portal was to provide all interested parties with relevant and high-quality professional information;
- Information system "University-Student" contains announcements, class schedules, samples of documents required by students, as well as a large number of scientific

publications. The unified educational portal for students is located in the system dl.sechenov.ru (Moodle);

- The digital platform of additional professional education is separately allocated on do.sechenov.ru, which contains training modules, knowledge testing and certification modules;
- The unique module of the digital platform is the international recruiting platform of Sechenov University ir-sechenov.ru containing a large number of current vacancies, information about various international competitions, etc. This platform helps potential candidates from Russia and foreign countries to find a job and build a career at the university, providing not only information support, but also the opportunity to submit the necessary documents (resumes, documents for the competition, etc.) in electronic form.

It should be noted that it is not easy to find information about the digital transformation of a university on every official websites of universities, for example, on the ITMO University website, whereas a whole section is devoted to the digital university on the websites of UrFU and HSE.

4.6 Comparative Analysis

The results of the analysis of the constituent elements of the "digital university" are presented in Table 1.

Table 1. Digital transformation of universities.

University	Contest winner	Digitalization of the educational process	Digitalization of the research process	Digitalization of administrative processes	Digitalization of relationships with counterparties
Higher School of Economics[a]	DU model, Center P and PP	Services: Teaching and learning materials for educational programs, Tools for organizing student learning, Tools for the teacher, Working with data	Reference Resources: Researcher's Handbook, Competitions and Grants, Resource Base, etc.	Services: SmartBOSS, SmartPoint Multi-Service Single Window, Electronic Signature, Mobile Applications	Services: Networking with universities, Educational platforms, Technological partnership
Ural Federal University[b]	DU model, Center P and PP	IET building service, project management system, educational programs portfolio management service, educational programs portfolio management service	Management service scientific equipment, a model for the collection of data collection and promotion of the results of scientific research activities	University property management service	Services:Partner's Personal Office, networking between universities

(continued)

Table 1. (*continued*)

University	Contest winner	Digitalization of the educational process	Digitalization of the research process	Digitalization of administrative processes	Digitalization of relationships with counterparties
Tomsk State University[c]	DU model, Center P and PP	Digital Mentor Service, Leadership Academy Service, Learning Spaces Service	Help section on the site	Service for project management and creation of project documentation	Help section for partners on the website
ITMO University[d]	DU model	Services: ITMO Avatar Platform, ITMO.Students Platform, ITMO.EXPERT	Help section on the site, support for the ITMO avatar service	Support for the ITMO Avatar service	Help section for partners on the website
The I.M. Sechenov First Moscow State Medical University[e]	DU model	Services: Educational portal Sechenov.online, Information system "University-student", portal DE do.sechenov.ru	Help section on the site	Help section on the site	Service: international recruiting platform ir-sechenov.ru

[a] https://digital.hse.ru/

[b] https://urfu.ru/ru/about/digital/

[c] https://alt.ihde.tsu.ru/

[d] https://itmo.ru/ru/

[e] https://www.sechenov.ru/univers/tsifrovoy-universitet.php

Despite the different initial level of digitalization for building a digital university Model, universities actively used the introduction of various digital services in all areas of the university's activities: administrative, educational, scientific, relationships with counterparties.

Almost all universities started the implementation of the Digital University Model with the digitalization of the educational process. Digitalization of the educational process usually includes various services for applicants, students and teachers. Almost all digital universities have implemented the applicant's personal account service, which includes the ability to access various information systems of the admission campaign, allowing to automate the process of submitting documents by applicants. Students and teachers in all digital universities have access to an electronic schedule of classes, curricula of academic disciplines, electronic statements and electronic test books, access to online courses and an electronic library, to a point-rating system. HSE, UrFU, TSU and ITMO implemented services for building Individual educational trajectories of students.

In all universities, a lot of attention is paid to digitalization of administrative activities. The digital services pool includes an employee's personal account, which provides access to the electronic document management system, HR services, financial services, training services, etc. Universities implement the possibility of rapid communication between university staff, management and students, provide the possibility of requests to technical support, administrative issues and other functionality. It should be noted that the employee's personal account provides access to digital educational and scientific services, and, if necessary, with the university's counterparties.

Digitalization of the scientific process provides access to electronic library resources, including a list of scientific achievements of each university employee (scientific publications, participation in conferences, participation in scientific grants, participation in meetings of dissertation councils, membership in editorial boards of scientific journals, etc.), as well as the ability to use systems to check for plagiarism, request examination of publications, etc. The digitalization of the scientific process at universities is usually at a prospective stage of development; most universities have only implemented a reference section on the website with access to various databases. However, UrFU has already implemented the Scientific Equipment Management Service and the Model of Data Collection and Promotion of Research Results.

Digitalization of relationships with counterparties is not implemented in all universities. For example, the Ural Federal University has created a special partner's personal account service that provides an opportunity to communicate with employers to organize student practices and participate in project activities [36]. HSE provides partners with the opportunity to use the educational services of the university, as well as the transfer of non-core areas to partners.

Users of digital university services are usually applicants (admission campaign), students (IT management, schedule, access to online courses, educational and scientific achievements), employees (management of the academic process, calculation of teaching load, anti-plagiarism, scientific services, etc.), administration (budget management, monitoring and control of business processes, formation of consolidated reports, etc.), government (interaction with the Ministry of Science and Higher Education, state bodies, etc.).

Having studied the approaches to the formation of a digital university model by universities selected by the Ministry of Higher Education and Science, the authors identified clear trends in the digital university Model presented (see Fig. 1).

The model of the digital university is not limited to one university; it involves communication not only with students, staff and university administration, but also with applicants, various counterparties (former employees and students, employer partners, other universities and educational institutions, government agencies, etc.). The pool of digital services includes tools for the digitalization of educational, administrative and scientific processes and various interactions with third-party contractors. In terms of increasing the accessibility of digital services, it is desirable to have a single access point to various services (for example, through a personal account) and intuitive services to facilitate the improvement of multilingual competences in the interaction with various digital platforms.

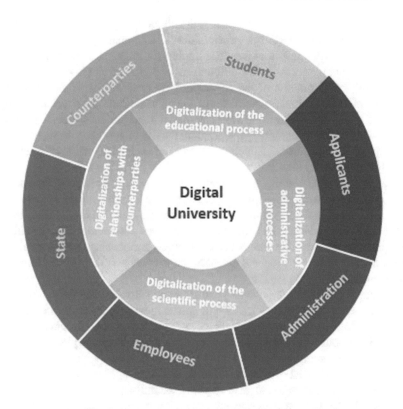

Fig. 1. Generalized model of a digital university.

5 Conclusion

Global digitalization inevitably leads to the blurring of spatial boundaries of communication between people through international online stores, social networks, various application packages, smartphone applications, etc., which in turn increases the importance of multilingual competences and intensifies attempts to use different ways of communication in society.

The digital transformation of universities and the introduction into the educational process of all specialties of new educational modules for the development of digital competences require that teachers and students have a high level of proficiency in technical languages, information technology, programming languages designed to write computer programs used in the new multilingual world.

The analysis of the experience of using information technologies in the winner universities of the competition organized by the Ministry of Science and Higher Education within the framework of the national program "Digital Economy in the Russian Federation" showed that the digital transformation of universities is a complex task, which is aimed at reengineering business processes taking into account the capabilities of modern IT technologies by introducing digital services in all areas of the university: administrative, educational, scientific and relations with contractors.

Due to the lack of a unified approach to the digital transformation of education and science on the part of the Ministry, universities have implemented their own transformation strategies. As a result, universities implement information systems at different times, on different platforms with a different set of digital services. The main objective of the digital transformation of the university is to ensure transparency, convenience and optimization of educational activities of students and staff, the development of new feedback channels and the transition to a data-based management system. Digitalization of universities is carried out taking into account the use of new and promising technologies: adaptive learning, artificial intelligence and Big Data, blockchain, cloud platforms, UX design, augmented reality, etc.

In order to spread the digital university model among Russian universities, it is necessary to organize an exchange of experience in the implementation and use of information services of the selected universities to develop a unified digital transformation strategy that will standardize the basic services provided, implement the compatibility of digital products developed by universities, establish data exchange between universities with the Ministry of Education and Science, and provide access to this information for external contractors.

The spread of the digital university model among Russian and foreign universities will accelerate the digital transformation of universities and activate the introduction of digital technologies into the educational process, as well as provide access to educational and scientific resources of any university for an external audience. By 2024, at least 50% of the universities subordinate to the Ministry of Education and Science of Russia should implement a digital university model. The federal project "Personnel for the Digital Economy" provides for each student to acquire a second digital profession, which cannot be done within the framework of one university. For this reason, it is necessary to create a single digital university for the whole country, adapted to the requirements of the new multilingual reality. A single digital university will allow universities not to "reinvent the wheel", but to save time and money on the introduction of digital technologies and the search for the most effective technological solutions by replicating the best practices of digital universities. Today's multilingual world is undergoing an era of radical change and requires universities to train specialists with digital literacy. Russian universities are quite strong when it comes to digitalization. The digital transformation of universities is a complex task, universities around the world face similar problems in implementing digital technology in all areas of their activities. Despite the diversity of solutions, the Russian model of the digital university can be used as a basis for digital transformation by universities around the world.

References

1. Antonova, A., Aksyonov, K., Ziomkovskaia, P.: Development of method and information technology for decision-making, modeling, and processes scheduling. In: Simos, T., Tsitouras, Ch. (eds.) International Conference on Numerical Analysis and Applied Mathematics, ICNAAM 2020, vol. 2425(1), p. 130003. AIP Publishing LLC, Melville (2020). https://doi.org/10.1063/5.0082100
2. Khalyasmaa, A.I., Zinovieva, E.L., Eroshenko, S.A.: Performance analysis of scientific and technical solutions evaluation based on machine learning. In: Proceedings - 2020 Ural Symposium on Biomedical Engineering, Radioelectronics and Information Technology, USBEREIT

2020, pp. 475–478. IEEE, Ekaterinburg (2020). https://doi.org/10.1109/USBEREIT48449.2020.9117774

3. Bylieva, D., Lobatiuk, V.: The image of society's digital future through the prism of the pandemic. Philosophical Thought. **2**, 11–23 (2021). https://doi.org/10.25136/2409-8728.2021.2.35169. (In Russian)

4. Daineko, L., Larionova, V., Yurasova, I., Davy, Y., Karavaeva, N.: Educational process digitalization in Ural Federal University. In: Busch, C., Steinicke, M., Wendler, T. (eds.) 19th European Conference on e-Learning, ECEL 2020, vol. 2020, pp. 146–153. Academic Conferences International Limited, London (2020). https://doi.org/10.34190/EEL.20.029

5. Agarkov, G., Sandler, D., Sushchenko, A.: The year after the COVID-19 outbreak: potential students' perceptions of higher education quality in the context of digitalisation and blended learning. Integr. Educ. **25**(4(105)), 646–660 (2021). https://doi.org/10.15507/1991-9468.105.025.202104.646-660. (In Russian)

6. Daineko, L., Yurasova, I., Larionova, V., Karavaeva, N.: Reflecting on the experience of forced transition to distance learning during the Covid-19 pandemic. In: Busch, C., Steinicke, M., Frieß, R., Wendler, T. (eds.) 20th European Conference on e-Learning, ECEL 2021, pp. 119–129. Academic Conferences International Limited, London (2021). https://doi.org/10.34190/EEL.21.052

7. Davies, M.: Can universities survive the digital revolution? Quadrant **56**(12), 58–66 (2012). https://search.informit.org/doi/10.3316/aeipt.197288

8. Rubtsova, A., Almazova, N., Bylieva, D., Krylova, E. A.: Constructive model of multilingual education management in higher school. IOP Conf. Ser. Mater. Sci. Eng. **940** (1), 012132 (2020). https://doi.org/10.1088/1757-899X/940/1/012132

9. Goncharova, N., Zaitseva, E.: Responses of Russian universities to the challenges of Covid-19 pandemic. In: Busch, C., Steinicke, M., Wendler, T. (eds.) 19th European Conference on e-Learning, ECEL 2020, vol. 2020, pp. 221–228. Academic Conferences International Limited, London (2020). https://doi.org/10.34190/EEL.20.140

10. Bylieva, D., Zamorev, A., Lobatyuk, V., Anosova, N.: Ways of enriching MOOCs for higher education: a philosophy course. In: Bylieva, D., Nordmann, A., Shipunova, O., Volkova, V. (eds.) PCSF/CSIS -2020. LNNS, vol. 184, pp. 338–351. Springer, Cham (2021). https://doi.org/10.1007/978-3-030-65857-1_29

11. Reshetnikova, O.: Effective learning tools in e-learning. In: Busch, C., Steinicke, M., Frieß, R., Wendler, T. (eds.) 20th European Conference on e-Learning, ECEL 2021, pp.387–393. Academic Conferences International Limited, London (2021). https://doi.org/10.34190/EEL.21.091

12. Pazos, A.J.B., Ruiz, B.C., Pérez, B.M.: Digital transformation of university teaching in communication during the COVID-19 emergency in Spain: an approach from students' perspective. Rev. Lat. Comun. Soc. **78**, 265–287 (2020). https://doi.org/10.4185/RLCS-2020-1477

13. García-Peñalvo, F.J.: Digital transformation in the universities: implications of the COVID-19 pandemic. Educ. Knowl. Soc. **22**, 1–6 (2020). https://doi.org/10.14201/eks.25465

14. Bogdandy, B., Tamas, J., Toth, Z.: Digital transformation in education during covid-19: a case study. In: 11th IEEE International Conference on Cognitive Infocommunications (CogInfoCom), pp. 173–178. IEEE, Budapest (2020). https://doi.org/10.1109/CogInfoCom50765.2020.9237840

15. Abdulrahim, H., Mabrouk, F.: COVID-19 and the digital transformation of Saudi higher education. Asian J. Distan. Educ. **15**(1), 291–306 (2020). https://doi.org/10.5281/zenodo.3895768

16. Mhlanga, D., Moloi, T.: COVID-19 and the digital transformation of education: what are we learning on 4IR in South Africa? Educ. Sci. **10**(7), 180–192 (2020). https://doi.org/10.3390/educsci10070180

17. Kutnjak, A.: Covid-19 accelerates digital transformation in industries: challenges, issues, barriers and problems in transformation. IEEE Access. **9**, 79373–79388 (2021). https://doi.org/10.1109/ACCESS.2021.3084801

18. Larionova, V., Karasik, A.: Digital transformation of universities: notes on the global conference on technology in education Edcrunch Ural. Uni. Manag. Pract. Anal. **23**(3), 130–135 (2019). (In Russian)

19. Kuzina, G.: The concept of digital transformation of a classical university into a "digital university. E-management **3**(2), 89–96 (2020). https://doi.org/10.26425/2658-3445-2020-2-89-96

20. Golyshkova, I.: Analysis of the key components of the digital university model. E-management. **3**(3), 53–61 (2020). https://doi.org/10.26425/2658-3445-2020-3-3-53-61

21. Neborsky, E., Boguslavsky, M., Ladyzhets, N., Naumova, T.: The digital university: rethinking the frame model within the framework of stakeholder theory. World Sci. Pedag. Psychol. **8**(6) (2020). https://doi.org/10.15862/22PDMN620

22. Brodovskaya, E., Dombrovskaya, A., Petrova, T., Pyrma, R., Azarov, A.: Digital environment of leading universities of the world and Russia: results of comparative analysis of website data. High. Educ. Russia **12**, 9–22 (2019). https://doi.org/10.31992/0869-3617-2019-28-12-9-22

23. Faria, J.A., Nóvoa, H.: Digital transformation at the University of Porto. In: Za, S., Drăgoicea, M., Cavallari, M. (eds.) IESS 2017. LNBIP, vol. 279, pp. 295–308. Springer, Cham (2017). https://doi.org/10.1007/978-3-319-56925-3_24

24. Kerroum, K., Khiat, A., Bahnasse, A., Aoula, E.S.: The proposal of an agile model for the digital transformation of the University Hassan II of Casablanca 4.0. Proc. Comput. Sci. **175**, 403–410 (2020). https://doi.org/10.1016/j.procs.2020.07.057

25. Kaminsky, O., Yereshko, Y., Kyrychenko, S.: Digital transformation of university education in Ukraine: trajectories of development in the conditions of new technological and economic order. Inf. Technol. Teach. Tools. **64**(2), 128–137 (2018). https://doi.org/10.33407/itlt.v64i2.2083

26. Zulfikar, M., bin Hashim, A., bin Ahmad Umri, H., Dahlan, A.: A business case for digital transformation of a Malaysian-Based University. In: 2018 International Conference on Information and Communication Technology for the Muslim World (ICT4M), pp. 106–109. IEEE, Kuala Lumpur (2018). https://doi.org/10.1109/ICT4M.2018.00028

27. Aditya, B., Ferdiana, R., Kusumawardani, S.: Identifying and prioritizing barriers to digital transformation in higher education: a case study in Indonesia. Intern. J. Innov. Sci. **14**(3/4), 445–460 (2022). https://doi.org/10.1108/IJIS-11-2020-0262

28. Daineko, L., Yurasova, I., Karavaeva, N.: University as an educational ecosystem. In: Chova, L., Martinez, A., Torres, I. (eds.) 14th annual International Conference of Education, Research and Innovation, ICERI2021, pp. 5455–5464. IATED Academy, Valencia (2021). https://dx.doi.org/10.21125/iceri.2021.1238

29. Goncharova, N.V., Pelymskaya, I.S., Zaitseva, E.V., Mezentsev, P.V.: Green universities in an orange economy: new campus policy. In: Bylieva, D., Nordmann, A. (eds.) PCSF 2021. LNNS, vol. 345, pp. 270–284. Springer, Cham (2022). https://doi.org/10.1007/978-3-030-89708-6_23

30. Rozhkova, D., Rozhkova, N., Blinova, U.: Digital universities in Russia: prospects and problems. In: Antipova, T., Rocha, Á. (eds.) DSIC 2019. AISC, vol. 1114, pp. 252–262. Springer, Cham (2020). https://doi.org/10.1007/978-3-030-37737-3_23

31. Habib, M.N., Jamal, W., Khalil, U., Khan, Z.: Transforming universities in interactive digital platform: case of city university of science and information technology. Educ. Inf. Technol. **26**(1), 517–541 (2020). https://doi.org/10.1007/s10639-020-10237-w

32. Kasatkin, P., Kovalchuk, J., Stepnov, I.: The modern universities role in the formation of the digital wave of Kondratiev's long cycles. Voprosy Ekon. **12**, 123–140 (2019). https://doi.org/10.32609/0042-8736-2019-12-123-140. (In Russian)

33. Koehler, M., Mishra, P., Cain, W.: What is technological pedagogical content knowledge (TPACK)? J. Educ. **193**(3), 13–19 (2013). https://doi.org/10.1177/002205741319300303

34. Brown, K., Khalfin, A., Larionova, V., Sandler, D., Sinitsyn, E., Tolmachev, A.: Sistem of digital services for supporting the individualized learning process. In: Chova, L., Martinez, A., Torres, I. (eds.) 13th annual International Conference of Education, Research and Innovation, pp. 8598–8607. IATED Academy, Valencia (2020). https://dx.doi.org/10.21125/iceri.2020.1911

35. Coccoli, M., Guercio, A., Maresca, P., Stanganelli, L.: Smarter universities: a vision for the fast changing digital era. J. Visual Lang. Comp. **25**(6), 1003–1011 (2014). https://doi.org/10.1016/j.jvlc.2014.09.007

36. Daineko, L., Goncharova, N., Larionova, V., Ovchinnikova, V.: Experience of introducing project-based learning into university programmes. In: Chova, L., Martinez, A., Torres, I. (eds.) 13th International Conference on Education and New Learning Technologies, pp. 8038–8043. IATED Academy, Valencia (2021) https://dx.doi.org/10.21125/edulearn.2021.1632

Technologies for Higher Education Digitalization

Natalia Kopylova[(✉)] [iD]

National Research University "Moscow Power Engineering Institute", Moscow 111250, Russia
nakopylova@yandex.ru

Abstract. The article discusses the questions of higher education digitalization technologies nowadays. The important definitions of this research like digitalization and digitalization in education are considered in the article. The main idea of the educational digital transformation is the movement towards the educational process personalization based on the digital technologies' usage. The properties and the main elements of digitalization are given in the article. The modern E-technologies, online boards, multifunctional online generators, timeline in education, gamification, their pros and cons, and their possibilities for the transition to digital education are considered. Massive online education is developing at a rapid pace nowadays. Massive Open Online Courses (MOOCs) are online courses with free or partially free access that an unlimited number of people can watch. The advantages and disadvantages of MOOCs are given. A diagram of the various people groups' attitude to the idea of obtaining online education, respondent attitude to the quality of MOOC, to the quality of the educational process in MOOC, to the variety of educational interaction forms in MOOC is presented in the article.

Keywords: Digitalization · Digitalization in education · E-Learning · Digital transformation · Educational technologies · Massive Open Online Courses (MOOCs)

1 Introduction

The use of the educational digitalization technologies in education is connected to the need for a new level of education development [1].

Digitalization is the process of modernizing human civilization through introducing digital technologies in all spheres of society. Digitization has the following properties:

- Complexity. Digitalization is a complex phenomenon, that takes place immediately in all spheres of society. Socio-economic, political, legal, educational, spiritual and cultural institutions of mankind are being transformed simultaneously at the local, regional, national and global levels.
- Globality. Digitalization is a global phenomenon, and all national governments and transnational corporations are aiming at it. The world wants economic prosperity and improved human well-being. Digitalization helps economies expand and develop, which benefits everyone, and therefore it is being done consciously.

D. Bylieva and A. Nordmann (Eds.): PCSF 2022, LNNS 636, pp. 402–412, 2023.
https://doi.org/10.1007/978-3-031-26783-3_33

– Inevitability. Digitalization is an inevitable phenomenon, it will be constantly carried out as one of the components of the overall civilization progress. There can be no end to digitalization, as scientific research is being carried out all the time and new developments are being created. Digitalization is accelerating more and more every year, and the result of all this is obvious. It penetrates in all spheres of people's lives.

The elements of digitalization that can be used in education either are the following:

1. Telecommunication networks (mobile Internet, wireless communication, broadband access).
2. Computer technologies (computers, laptops, tablets, smartphones).
3. Software engineering (operating systems, software).
4. Consequences of using all mentioned above (electronic commerce, media platforms, social networks, educational platforms).

Nowadays, to assess the degree of the countries' population involvement in the digitalization process, indicators are used that indirectly or directly measure it. These include: the Networked Readiness Index (NRI), the Global Innovation Index (GII) and the Digital Economy and Society Index (DESI) [2].

The NRI Networked Readiness Index was developed in 2001. It consists of 53 parameters divided into 3 groups: the conditions' presence of information and communications technologies (ICT) development; the readiness of citizens, business circles and government bodies to use ICT; the level of ICT use in the public, commercial and state sectors [2]. In fact, the indicators reflect the measure effectiveness to improve the life quality in different world's countries: increasing labor productivity, competitiveness, innovations' development and realization, etc. The Network Readiness Index implicitly reflects the digitalization penetration into all spheres of life. The annual results of this index calculation are provided to the World Economic Forum in the report "The Global Information Technology Report" [3].

The second indicator for assessing digitalization is the Global innovation index (GII) [4]. Its aim is forecasting the potential of the country's innovation activity and assessing its result.

Since 2016 the formula for calculating the global innovation index has been calculated as the average of two sub-indices: innovation costs and innovation results.

The innovation expenditure sub-index reflects innovation processes in five main groups:

– an institution;
– a human capital and researches;
– an infrastructure;
– a market development level;
– a business development level.

The level of the country's economy and society inclusion in the digitalization process is intended to be shown by the Digital Economy and Society Index (DESI) [5], calculated according to the European Union methodology based on the 31 parameters' analysis combined into 5 enlarged groups:

1) "Connectivity" (the ability to access communication systems based on digital technologies);
2) Human capital (Human Capital/Digital skills – the level of the population digital skills' possession);
3) Use of Internet by citizens;
4) Integration of Digital Technologies by businesses;
5) Digital Public Services.

The resulting value of the DESI Index, calculated for each EU country, allows us to determine its place in the ranking of the EU countries' digitalization. The overall value of this index for the entire set of EU countries reflects their level of economy and life quality in general.

2   Main Part

Digitalization in education is a process of transition to E-learning system. The main idea of the educational digital transformation is the movement towards the educational process personalization based on the digital technologies' usage. Its important feature is that digital technologies help us in practice to use new pedagogical practices (new models of organizing and conducting educational work), which previously could not take a worthy place in mass education due to the complexity of their realization by means of traditional technologies [6, 7].

The essence of digital transformation is to effectively and flexibly apply the latest technologies to move towards a personalized and results-oriented educational process. It is necessary to solve the following main tasks in digital education:

– The development of material infrastructure. This includes the construction of data centers, the emergence of new communication channels and devices for the use of digital educational and methodological materials.
– The introduction of digital programs, i.e. creation, testing and application of educational materials using machine learning technologies, artificial intelligence, and so on.
– The development of online learning. Gradual phasing out of paper media.
– The development of new learning management systems (LMS). In distance education, LMSs are programs for the administration and training course control. Such applications provide students with equal and free access to knowledge, as well as the flexibility of learning.
– The development of a universal student identification system.
– The creation of educational institution models. To understand where school and university education should go in terms of technologies, we need examples of how it should ideally work: using new LMS, Industry 4.0 tools and devices, and so on.
– Improving the skills of teachers in the field of digital technologies.

Digital technologies are radically changing the content of the taught disciplines and their presentation form [8–10]. Direct connections to electronic databases, news, forums are possible. Publishing houses specializing in educational literature are increasingly switching to electronic versions of their textbooks and teaching aids.

During practical exercises at universities, you may use social networks. Using Skype, Zoom, Webex, messengers, it is possible to participate in a leading specialist expert lesson. The data of using these platforms especially during the pandemic situation are presented in the article (see Figs. 1, 2 and 3).

Fig. 1. The data of using Skype by the users.

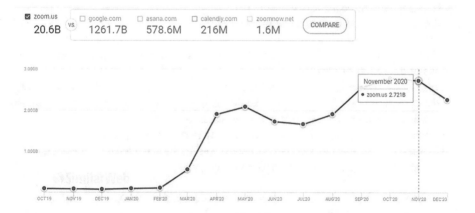

Fig. 2. The data of using Zoom by the users.

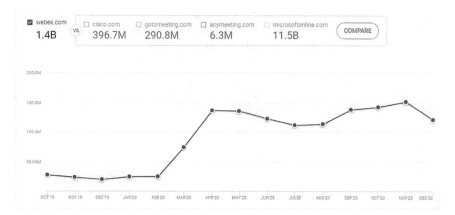

Fig. 3. The data of using Webex by the users.

Massive online education is also becoming very popular nowadays. Massive Open Online Courses (MOOCs) are online courses with free or partially free access that an unlimited number of people can watch. MOOCs were inspired by the Massachusetts Institute of Technology's Open Course Ware service. It was launched in 2001. The materials from various educational programs were made publicly available. The term "massive open online courses" (MOOCs) itself appeared much later. It was first used in 2008. The abbreviation MOOC came into common use only in 2012. It was then that digital platforms appeared that became giants of the modern online education market – Coursera, edX, Udasity. All of them grew out of university online courses, and were led by teachers from these universities [11–13].

Let's take a closer look at the abbreviation:

– Massive – the number of participants in mass courses is not limited and can be in the tens of thousands of people.
– Open – the course has no geographic or physical reference. In addition, it is either free or partially free. For example, viewing is free, but if you want to get a certificate of training, you need to pay (Coursera has such conditions).
– Online – all learning processes take place only in the online space.
– Courses – the course has a certain structure and traditional educational elements, for example, training materials, knowledge testing, and so on.

MOOCs are distance educational and methodological complexes that include video lectures, slide presentations, additional materials for reading or viewing, glossaries, homework in the form of projects, interactive games, simulations, intermediate and final tests, lists of references for the course, useful links, questions for discussion on the forum or in social networks and so on.

There are several advantages of using MOOC, as well as, disadvantages (Table 1).

Table 1. Advantages and disadvantages of MOOC.

Advantages	Disadvantages
Mass and globality	Lack of direct communication
Free access, opportunity of flexible schedules	Technical difficulties
Attracting teachers from the most prestigious universities	The difficulties of building a course
A rich but concise presentation of the theory	Requiring a good level of IT competence for students
Multimedia content of educational materials	Insufficient English proficiency
The latest interactive and information technologies	Problematic enforcement of intellectual property
Mutual verification of tasks	Difficulty in privacy and data protection
Intercultural learning	Course Funding Sources
Gamification (awards, ratings, badges)	The difficulty of student self-organization

It should be noted that the university transformation under the digitalization influence during the knowledge economy formation is a multicomponent process that does not develop equally in its various areas [14–17].

3 Experience

In recent years especially during and after the pandemic situation traditional full-time, online and offline education has been actively discussed in scientific and educational environment. Recently many modern digital E-services have appeared in education.

Online boards are one of the tools through which E-learning is carried out. On the one hand, they keep the online format, on the other hand, they provide the contact between a student and a teacher. As online learning becomes more popular, boards develop and improve. So, the boards are online. Now teachers don't have just boards on which you can write, draw in different colors, choose the thickness of lines, leave notes and comments, but also add any content of your own, some boards at the time of drawing even "recognize" what is drawn and help "finish".

Thanks to the "Screen Sharing" function, with any broadcast from various resources, platforms, you can use your online board to explain the material.

"Mind maps" is a method of organizing ideas, tasks, concepts and any other information. "Mind maps" help visually structure, remember and explain large amounts of information. Psychologist Tony Buzan came up with "Mind maps" in the 1990s, and the first services appeared in the mid-2000s. This is software that helps to create a mental map in a digital form. In most of these services, you can work for free, offline and in a group mode. "Mind maps" is a universal tool that helps to sequentially decompose all tasks, knowledge, questions and plans, and not forget anything.

Padlet is an online whiteboard that you can collaborate on from your computer or smartphone. It is enough to send students a link to the board.

Jamboard is an online whiteboard designed for cross-platform collaboration developed by Google as a part of the G Suite family.

Multifunctional online generator Genial.ly is an intuitive and free online content. Using the service you can create: interactive presentations, video presentations, infographics, interactive games, interactive posters quizzes, web quests, etc. The system offers hundreds of ready-made animated interactive slide templates on any topic. You need to select an appropriate template and fill it with the necessary objects. Among them are text blocks, images, graph and chart templates, infographics, icons, figures, interactive elements (controls for slides and objects on slides), smart objects (objects for demonstrating processes in dynamics). The finished product created in the Genial.ly service can be published on the site, in social networks, or simply share a link to the resource.

You can organize the content of an entire course or a single lecture. You may use "timeline in education", that can be useful in the following cases:

1. Educational material is a history of a phenomenon, fact, event.
2. It is necessary to demonstrate the technology of the process.
3. It is necessary to illustrate the connection between certain elements of the educational material.
4. Visualize the presentation of any educational material.
5. Organize project work for students.

"Sutori" is a service in which informational materials or completed tasks are "strung" on a vertical timeline. It is a service for organizing materials by a teacher or collective student work in an interactive notebook with the ability to insert audio, video, forums and polls. The vertical "timeline" can be displayed as a regular presentation or exported to PDF if necessary.

"Gamification" is a process of introducing game elements, game mechanics, game design methods into originally non-game forms of activity.

"Learningapps.org" is a service for creating interactive teaching aids in various subjects.

"Nearpod" is an innovative interactive online platform that allows you to create educational materials, show them to students and track their progress in real time.

"Prezi" (flying presentation) is a web service for creating presentations, the work of which is based on the method of scaling – zooming in and out of information blocks. Unlike a traditional presentation, where all information is divided into slides, in "Prezi" all its elements are located in one common field.

Studying all these services we can make the conclusion that nowadays teachers can use different services for creating interesting E-lessons at high school [18–21].

During our research work we did a survey about people's attitude to the idea of getting online higher education. The results are shown in Fig. 4.

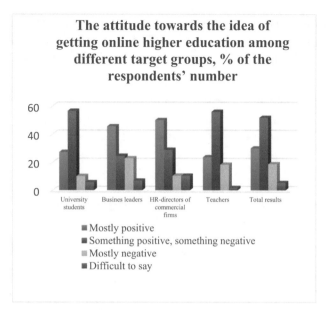

Fig. 4. The attitude towards the idea of getting online higher education among different target groups, % of the respondents' number

During the pandemic situation we also did a survey asking different respondents (especially students (150 people)) who finished MOOCs about their attitude to them.

86% were completely satisfied with the quality of the educational platform. The following characteristics were mentioned in particular: the consideration of the educational student needs (84%); conditions for sharing experience with colleagues in the group (96%); and a partnership style of communication (92%). Some comments were made about the speed of the Internet and some technical difficulties (see Fig. 5).

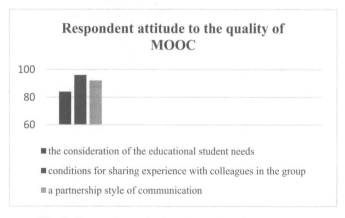

Fig. 5. Respondent attitude to the quality of MOOC in %

When assessing the quality of the educational process, the highest number of fully satisfied respondents was for the following: "Effectiveness of training forms" (92%), "Relevance of training material" (94%), "Practical orientation of training" (88%). (see Fig. 6).

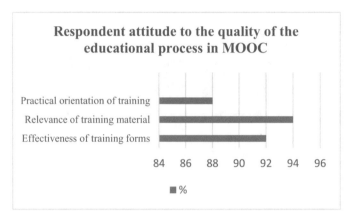

Fig. 6. Respondent attitude to the quality of the educational process in MOOC in %

Among the variety of educational interaction forms, the trainees prefer online discussions (65%), participation in webinars (62%) and personal online consultations (62%). The highest quality score is given to the indicators: "Learning the latest technologies" (92%), "Digital portfolio" (88%) (see Fig. 7).

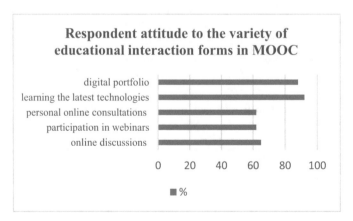

Fig. 7. Respondent attitude to the variety of educational interaction forms in MOOC in %

4   Conclusion

The labor market requires a qualitative specialist. Digitalization requires new, completely different competencies, different from those possessed by graduates of the universities.

The basis of the digital education transformation is the movement towards the educational process personalization through the digital technologies' usage. Digitalization affects the educational content, and its organization. Nowadays a teacher is a navigator who helps to navigate the knowledge bases.

In conclusion, it should be noted that the pedagogical capabilities of digital education, as a means of learning, far exceed the capabilities of traditional means of realizing the educational process, contribute to the educational process improvement, activate and make creative independent and joint work of students and teachers. Thanks to online services, learners learn with great pleasure and their results improve.

No doubt, for a better result we should combine traditional offline education with modern online education.

References

1. Kopylova, N.: The development of an information system for technical university students' teaching and control in English. In: ELEKTRO, pp. 1–4. IEEE, Taormina (2020). https://doi.org/10.1109/ELEKTRO49696.2020.9130317
2. Khalin, V.G., Chernova, G.V.: Digitalization and its impact on the Russian economy and society: advantages, challenges, threats and risks. Manag. Consult. **10**, 46–62 (2018)
3. Index of network readiness information about the research. http://gtmarket.ru/ratings/networked-readiness-index/networkedreadiness-index-info. Accessed 10 June 2022
4. Global Innovation Index. https://vc.ru/flood/44152-global-innovation-indexmesto-rossii-v-mire-innovaciy. Accessed 10 June 2022
5. Ershova, N.Y.: Principles of forming the network learning educational environment. Saratov, University education (2019)
6. Kapranov, V.K., Kapranova, M.N.: Increasing the availability of quality education through the network interaction of school libraries. Open Distan. Educ. **3**(47), 28–32 (2012). (in Rus)
7. Balyshev, P.: The stages of discourse-oriented virtual learning environment modeling. Technol. Lang. **3**(3), 88–105 (2022). https://doi.org/10.48417/technolang.2022.03.07
8. Pozdeeva, E., et al.: Assessment of online environment and digital footprint functions in higher education analytics. Edu. Sci. **11**(6), 256 (2021). https://doi.org/10.3390/educsci11060256
9. Saltanovich, I.: Global-local cultural interactions in a hyperconnected world. Technol. Lang. **3**(2), 162–178 (2022). https://doi.org/10.48417/technolang.2022.02.10
10. Pozdeeva, E., Shipunova, O., Evseeva, L., Kulsariyeva, A.: Systems analysis of the digital agent's role in hybrid social interaction forms. In: Vasiliev, Y.S., Pankratova, N.D., Volkova, V.N., Shipunova, O.D., Lyabakh, N.N. (eds.) System Analysis in Engineering and Control. SAEC 2021. LNNS, vol. 442. Springer, Cham (2022). https://doi.org/10.1007/978-3-030-98832-6_14
11. Kopylova, N.A.: The use of modern e-learning technologies at English lessons in a technical university pedagogical process. In: 9th Mediterranean Conference on Embedded Computing (MECO), pp. 1–6. IEEE, Budva, Montenegro (2020). https://doi.org/10.1109/MECO49872.2020.9134257
12. Bylieva, D., Hong, J.-C., Lobatyuk, V., Nam, T.: Self-regulation in E-learning environment. Educ. Sci. **11**, 785 (2021). https://doi.org/10.3390/educsci11120785
13. Pokrovskaia, N.N., Tyulin, A.T.: Psychological features of the regulative mechanisms emerging in the digital space. Technol. Lang. **2**(2), 106–125 (2021). https://doi.org/10.48417/technolang.2021.02.11
14. Gurakov, A.V.: Technologies of E-learning: tutorial. Electron. text data. Tomsk State University of Control Systems and Radioelectronics, Tomsk (2016)

15. Dyakova, E.A.: Digitalization of education as the basis for training teachers of the XXI century: problems and solutions. Bullet. Armavir Stat. Pedag. Uni. **2**, 24–35 (2019). (in Rus)
16. Hockicko, P.: Nontraditional approach to studying science and technology. Communications **12**(3), 66–71 (2010). https://doi.org/10.26552/com.C.2010.3.66-71
17. Ondrus, J., Hockicko, P.: Braking deceleration measurement using the video analysis of motions by SW tracker. Transp. Telecommun. J. **16**(2), 127–137 (2015). https://doi.org/10.1515/ttj-2015-0012
18. Anas, I., Musdariah, A.: Being an E-Teacher: preparing the ESL teacher to teach English with technology. J. Eng. Lang. Teach. Linguist. **3**(1) (2018). https://doi.org/10.21462/jeltl.v3i1.102
19. Xu, D., Tsai, S.B.: A study on the application of interactive English-teaching mode under complex data analysis. Wirel. Communic. Mob. Comput. **2021**, 2675786 (2021). https://doi.org/10.1155/2021/2675786
20. Bylieva, D., Bekirogullari, Z., Kuznetsov, D., Almazova, N., Lobatyuk, V., Rubtsova, A.: Online group student peer-communication as an element of open education. Futur. Internet **12**, 143 (2020). https://doi.org/10.3390/fi12090143
21. Rusnak, Ì., Nagornij, Â., Vasilik, M.: Peculiarities of use of interactive technologies of teaching English language in training of the future masters-philologists. Studia Gdańskie. Wizje i Rzeczywistość **XV**, 293–305. (2019). https://doi.org/10.5604/01.3001.0014.0482

The Analysis of Successful Teaching University Students in a Multilingual Environment in the Context of Education Digitalization

Evgeniya Khokholeva[1] , Elena Lysenko[1(✉)] , and Svetlana Lipatova[2]

[1] Ural Federal University named after the First President of Russia B. N. Yeltsin, 620002 Yekaterinburg, Russian Federation
e.v.lysenko@urfu.ru

[2] GAOU DPO SO "Institute of Education Development", 620002 Yekaterinburg, Russian Federation

Abstract. In modern historical conditions, there is a unique opportunity for the mass use of remote forms of learning in a virtual environment. Natural, digital languages and languages of signs and technologies simultaneously participate in digital communication. This gives an opportunity to take a fresh look and rethink the potential possibilities of interaction between participants of multilingual communication. The authors, using statistical data and the author's questionnaire, conducted a comparative study of the success of teaching higher school students in a traditional and multilingual environment using remote digital technologies. The results of the study revealed significant multi-level problems: cognitive competence; personal effectiveness; behavioral infantilism. Thus, the multilingual context revealed hidden problems of learning quality. These negative trends as a key task give rise to the need for further research and the formation of multilingual thinking.

Keywords: Multilingualism · Multilingual environment · Digitalization of education · Digital learning · Information educational environment

1 Introduction

Modern communication and interaction of people, groups and communities involves the use of a variety of sign forms of speech, which includes both natural and artificial (symbolic, technical, digital, etc.) languages. This implies adequate perception and decoding of information coming from different sign systems and, in case of inadequate decoding, creates difficulties in decoding and distorting information, which leads to problems in the interaction of subjects of both professional and personal communication. This issue is especially acute in the field of higher education. In the current historical conditions of the pandemic, we experienced a unique opportunity to excessively use distance (digital) forms of education associated with a multilingual context, when natural languages (including foreign languages when teaching foreign students), digital languages (virtual educational environment) are involved in communication at the same time, as well as languages of signs and technologies (pedagogical technologies of distance education). This enables us to take a fresh look and redefine the potential of this form of interaction between participants in multilingual communication. The discussion concerning the implementation of digital technologies in education has been going on since the beginning of the century and is associated with wide use of information technologies in the education. However, in the current context of the pandemic (COVID-19), this problem has become especially relevant. The scale of application of digital technologies has changed: what used to be part of the education system has now become widely used in everyday multilingual practice. All this dictates the necessity to comprehend new learning technologies in a multilingual space.

From the beginning of the 2000s, the possibility of using information technologies at universities has been discussed [1–3].

Studying the problematic aspects of vocational education identified in practice in the developed countries of the West, D. Toropov came to the conclusion that the European educational system is increasingly focused on knowledge acquisition through information processing. Thanks to mobile technologies and the Internet, education is turning from a "lecture" into a "dialogue" between a teacher and a student. Education is transforming from the consumption of knowledge to its production, from an authoritarian educational process to cooperation, from the format of lectures to discussions,, seminars, and strengthening the advisory component in education. Thus, we can state that a transition from at "reproductive" paradigm of education to at "creative" one has occurred: this was possible due to continuous interactions between man, technology, and society [4]. This is how a new multilingual space is formed.

The concept of "developing an information educational environment" is understood as "a holistic pedagogical system integrating basic modalities: 1) modern (innovative) educational technologies aimed at the formation of an intellectually developed, socially significant, creative personality with the necessary level of professional knowledge, skills, and abilities; 2) information educational resources (traditional and electronic media, computer and telecommunication educational and methodological complexes); 3) the means of managing the educational process; 4) psychological and pedagogical conditions that contribute to the creative self-development of students and form their attitudes towards the development of their creative potential" [5]. When speaking of "modern educational technologies", we mainly mean, the use of digital technologies.

Malchikova, Pivovarova, and Kuzmin note that the introduction of teaching methods that implement digital technologies has a positive effect. The authors do not believe that digitalization bears any risks, but also argue for the need to involve the surrounding community in the process of transforming education "Community involvement ensures that the transformation will be based on the real values of this community, provide for the formation of demanded practical skills in students, rely on available resources and become an important component of community life…" [6].

An analytical report by the Ministry of Science and Higher Education of the Russian Federation analyses some aspects that show the real level of readiness of society (in this case, teachers and students) for the transition to digital teaching methods in the multilingual environment. For example, the report provides the following data:

"before the pandemic, 60% of teachers rarely or never conducted lectures and classes in a remote format or in a webinar format" - only "33% of students answered that they like distance learning more than face-to-face" [8]. The report's authors state that the "large-scale retraining of teachers is needed, aimed not only at "passing the PC course", but also at introducing new formats and technologies" [8]. Other Russian scholars have come to the same conclusion. Thus, Perminova believes that "the use of the concept of basic competencies for the digital economy requires not copying them, but a critical attitude and interpretation regarding the content of education at different levels of education and the corresponding training of the teachers—and, above all, the improvement of their methodological, psychological, didactic and technological training" [9].

Semenova and Kazantseva considered the possibilities of mass open online courses in modern education. Such online courses, embodying the features of a new format of education, synthesise education, learning, self-learning and mutual learning. In this case, the emphasis is on the activity and interest of the listener [10]. Wellplanned online learning differs a lot from courses offered online in response to a crisis or critical situation. Colleges and universities striving to provide continuous learning during the COVID-19 pandemic need to understand these differences when evaluating such emergency distance learning [11]. However, the question arises: is there an interest among students for independent academic study? Will students have enough responsibility and discipline to force them to listen to online lectures to the end and complete practical tasks on time without being distracted by extraneous factors?

Whatever answer scholars propose, one thing is clear: the digitalization of education is an inevitable process that has become part of modern reality. The task of the academic community is to identify the problems that society will face when introducing digital teaching methods in a multilingual environment and develop measures to prevent them.

This conclusion is confirmed by P. Lukshin, the author of the project *The Global Future of Education*. He emphasizes that "the modern model of "industrial" education is fundamentally vulnerable: it forms the "skills of the past", not the "skills of the future", and prepares students for a reality that will no longer exist" [12]. According to Lukshin the main trends in the educational sphere over the next twenty years will be the following: all-age education, the spread of network culture values, the pragmatization of education, the automation of routine intellectual operations, and the development of the cognitive ability improvement industry. All this is possible only within the framework of the digitalization of educational processes [12].

In our opinion, all these trends are a vivid demonstration of the formation of a multilingual picture of the world, which requires the formation of a special thinking type—multilingual.

According to Verbitsky, along with the fact that digital learning has huge but little-studied opportunities, there are several limitations associated with their total introduction into the education system [13]. There is an opinion that the use of digital educational technologies threatens the destruction of the higher education system [15]. A detailed analysis of the negative consequences of digitalization "for schoolchildren and students that arise during its implementation, such as the problem of attention, loss of mental abil-ities, decrease in social skills, and the emergence of cancer" is given in modern research (Jarke and Breiter, 2019 [14]; Knox et al., 2020 [15]; Manolev et al., 2019 [16]; Mirrlees and Alvi, 2020 [17]; Suoranta et al., 2022 [18]). Therefore, it is crucially important to find a psychologically, physiologically, pedagogically and methodologically justi-fied balance between the use of computer capabilities and live dialogic communication between the subjects of the educational process—teachers and students. Multilingual thinking creates favorable conditions for this. The digitalization of education will be productive only if it is based on the support of relevant psychological and pedagogical theories [13].

A team of researchers from the Research Centre for Vocational Education and Qual-ification Systems (Blinov, Dulinov, Esenina, Sergeev) note that building a digital edu-cational process is a difficult task that requires scientific justification on the basis of digital didactics, a new branch of pedagogical science [19]. Creating a digital process for vocational education and training based on new didactics will help overcome the prob-lematic situation caused by the digitalization of education, where the dynamic develop-ment of digital technologies and means is combined with the preservation of traditional ("pre-digital") forms of organizing the educational process and learning technologies [19].

2   Methodology

The purpose of the present study is to analyze the impact of the digital format of education on the effectiveness of teaching university students in a multilingual environment during the pandemic.

Objectives:

1) determine the degree of readiness of the student community for the digital format of education as one of the key aspects of multilingualism;
2) conduct a comparative analysis of the effectiveness of teaching university students in online and offline formats;
3) identify the reasons for changing the learning results of university students in the context of online learning as a factor of multilingual educational environment; 4) consider the possibilities for improving the learning results of university students in the context of a forced transition to a digital learning format during the COVID-19 pandemic.

The research program included the following stages: 1) systematization of scien-tific research findings on digitalization in education in a multilingual environment; 2) planning and organizing empirical research and data collection; and 3) processing the obtained results, their discussion and the formulation of conclusions.

The research was conducted using statistical information on final student results for the 2019–2020 academic years (comparative analysis of student performance for the first and second terms—before and during the pandemic, respectively). The author also wrote a questionnaire entitled "Students attitude to the digitalization of education in the conditions of multilingual education", which was structured in four blocks:

1. Attitude towards the digitalization of education;

2. Readiness of students to study digitally;
3. The influence of the online learning process on the effectiveness of student learning;
4. Advantages and disadvantages of the digital form of education.

The study involved 275 respondents.

3   Results

The survey results showed that 46% of students had a neutral attitude towards the digital (online) format, 41% of students had a negative attitude, and only 13% had a positive experience (Fig. 1).

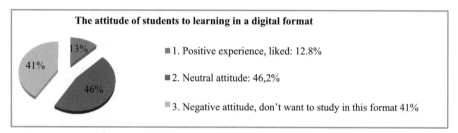

Fig. 1. The attitude of students to learning in a digital format (based on the results of the author's questionnaire "Attitude of students to the digitalization of education in a multilingual environment"

57.5% of respondents have not decided on their readiness to study online in the future, 23.6% are not ready to study in this format in the future, and 18.9% have a positive attitude towards the possibility of studying remotely (Fig. 2).

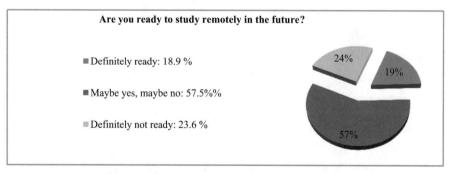

Fig. 2. The readiness of respondents to study in the future in an online format (according to the results of the author's questionnaire "Attitude of students to the digitalization of education").

A comparative analysis of the effectiveness of teaching university students online and offline is presented in Table 1.

Table 1. Evaluation of the effectiveness of the educational process before and during the pandemic on a 5-point scale (where 1 is the minimum result, 5 is the maximum result) (according to the results of the author's questionnaire "Attitude of students to the digitalization of education")

Question	1 point, %		2 points, %		3 points, %		4 points, %		5 points, %	
	Offline	Online	Offline	Online	Offline	Online	Offline	Online	Offline	Online
The quality of professional skills obtained in the study of disciplines	5.1	**17.9**	15.4	**23**	25.6	**38.5**	30.7	**15.4**	23.2	**5.2**
The quality 2.6 of knowledge gained in the study of disciplines		**7.7**	5.1	**15.4**	23.1	**48.7**	43.6	**25.6**	25.6	**2.6**

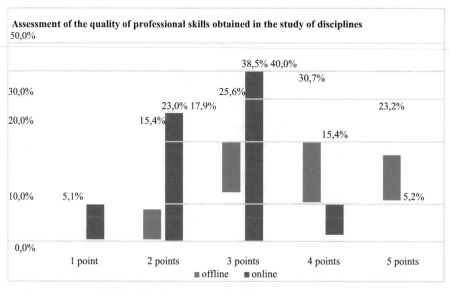

Fig. 3. Assessment of the quality of professional skills obtained in the study of disciplines (based on the results of the author's questionnaire "Attitude of students to the digitalization of education").

Thus, student's subjective assessment of their learning results demonstrates low performance in online learning compared to offline learning (Fig. 3).

54% of respondents positively assessed (awarding "4" and "5" points) the quality of the professional skills acquired in offline mode, while only 20.6% of students give such an assessment to skills acquired in online mode (Fig. 4).

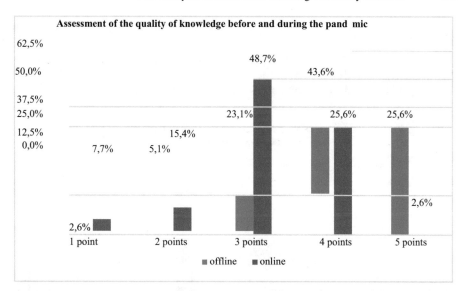

Fig. 4. Assessment of the quality of knowledge before and during the pandemic (according to the results of the author's questionnaire "Attitude of students to the digitalization of education").

69.2% of respondents positively assessed (awarding "4" and "5" points) the quality of knowledge acquired in offline learning mode, while only 28.2% of students give such an assessment to knowledge acquired in online mode.

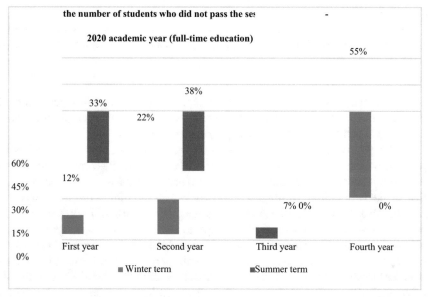

Fig. 5. The ratio of the number of students who did not pass the terms of the 2019–2020 academic years (full-time education).

The obtained results are confirmed by statistical data on student performance for the 2019–2020 academic years (Fig. 5).

Thus, there is an increase underachieving students in years one to three (full-time education) (Fig. 5). There was no increase in failing graduate students (fourth year). A similar trend is also observed among extramural students. (Fig. 6).

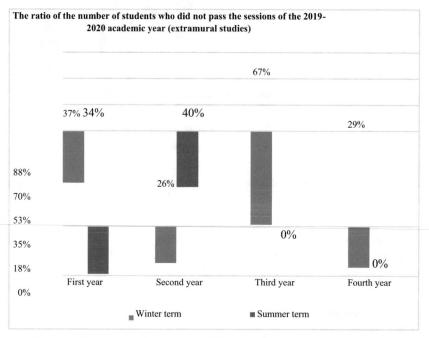

Fig. 6. The ratio of the number of students who did not pass the terms of the 2019–2020 academic years (extramural studies).

For extramural students, graduation takes place in the third and fifth years. It could be assumed that by the end of the course, the level of student responsibility increases and the format (offline, online) does not play such a significant role.

The analysis of academic results based on the results of terms demonstrates the dependence of the quality of learning on the format of learning. It is proven that students in their first years of study (first and second years) need additional control from teachers, which weakens in online learning in comparison with the offline format. It has been established that self-organisation skills in this age group are not sufficiently manifested, so it can be assumed that online learning requires students to be more psychologically (volitional) prepared for learning. These skills were not formed at previous stages of the learning process, and the conditions of the pandemic demonstrate their insufficient level. The online learning format also requires time management skills (the ability to rationally allocate one's time without external control). The questioning of students showed that during the period of distance learning, the number of hours spent on independent work among students increased (Table 2).

Table 2. The ratio of time spent preparing for classes, completing tasks for independent work before the introduction of a distance learning, and during distance learning (according to the results of the author's questionnaire "Attitude of students to the digitalization of education")

Number of hours per week	Before distance learning (%)	During distance learning (%)
Less than 1 h per week	33.1	17.9
1 to 3 h per week	28.4	21.6
4 to 6 h per week	17.2	32.7
More than 7 h per week	21.3	27.8

41% of students noted that when studying online, they significantly increased the preparation time for homework, practical classes, and independent work (Fig. 7).

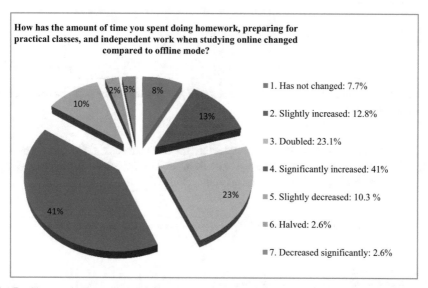

Fig. 7. Changes in the amount of time spent by students on homework, preparation for practical classes and independent work (according to the results of the author's questionnaire "Students' attitude to the digitalization of education").

Interestingly, the respondents experience greater fatigue in the traditional format than in the remote format (Fig. 8), but 44.9% of the respondents did not find a difference in the decrease in concentration due to fatigue (Fig. 9).

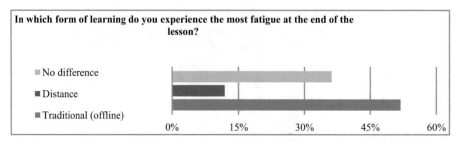

Fig. 8. Evaluation of the degree of fatigue in different forms of learning.

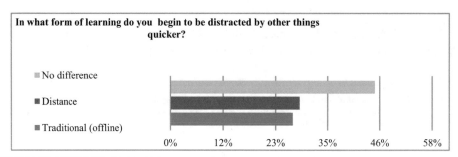

Fig. 9. The answers to the question "In what form of learning do you begin to be distracted by other things quicker?"

71.8% of students consider an online lecture the most successful form of organising online learning, 51.3% prefer online consultations, and 46.2% prefer online knowledge testing (testing and exams) and online seminars (Fig. 10).

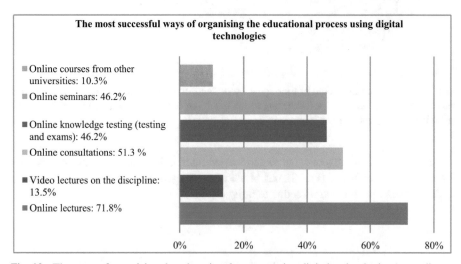

Fig. 10. The ways of organising the educational process using digital technologies (according to the results of the author's questionnaire "Students' attitude to the digitalization of education").

The main attractive factors of online learning (Fig. 11) reduction of travel time (29.7%) and more free time (16.2%); 13.5% of respondents note that online learning is a promising and convenient learning format.

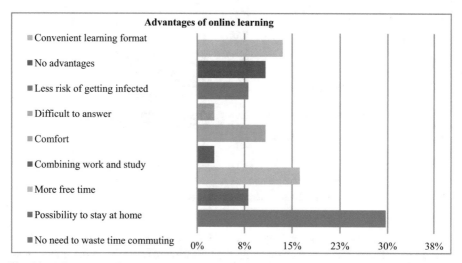

Fig. 11. Advantages of online learning (according to the results of the author"s questionnaire "Attitude of students to the digitalization of education").

The main disadvantage of online learning was the lack of live communication with the teacher (21.1%). The presence of technical problems was indicated by 17.4% of the students surveyed, while organisational problems and a decrease in the level of perception and assimilation of educational material were noted by 16%. 13.2% of students faced a decrease in control, self-control, discipline and self-discipline (Fig. 12). For instance, it was stated that many students connect to an online lecture, but do not listen to the teacher, instead paying attention to their own business.

Thus, this study of the impact of digitalization in education on the academic results of university students showed that a significant number of students are not ready to study online. 41% of students have a negative attitude towards this form of education.

The main reasons are:

1) students lack such qualities like independence, self-control, discipline, and the ability to plan their time. First and second year students are more in need of external control, which weakens in distance learning.
2) some teachers lack online working experience. When giving a traditional lecture, teachers correct and adjust the content of their lecture to the level of student training. In online mode they do not receive enough feedback from students, and, as a result, the online lecture becomes into a monotonous one-and-a-half hour reading session of theoretical material. Because of this, students start to lose attention, which leads to a decline in academic results.

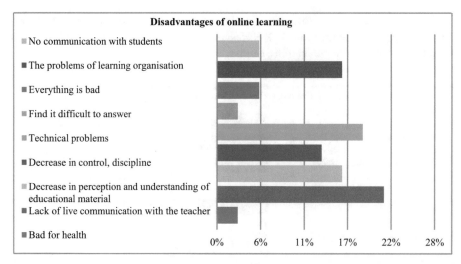

Fig. 12. Disadvantages of online learning.

The comparative analysis of the effectiveness of teaching university students in online and offline formats reveals that with the transition to online learning, the number of students who did not pass tests and exams increased by 21% (first year), 16% (second year), and 7% (third year).

A tendency has emerged: the number of underachieving students declines as the students progress through their courses. Senior students have the most developed sense of responsibility and motivation for learning. Under these conditions, the format of the learning impacts on its effectiveness to a lesser extent.

4   Discussion and Conclusion

Our theoretical analysis of the identified problem allowed us to describe the possibilities for improving the learning results of university students in the context of a forced transition to a distance learning format in a multilingual environment.

The multilingual context revealed the problems associated with the quality of education which did not exist before.

The first of these is the format for structuring and demonstrating lecture materials.

O. Shmurygina suggests dividing online lectures into components of 15–20 min each, followed by checking the learned material. In particular, the author writes: "In a webinar, you can conduct a further survey on some problematic aspects, difficult places in theory. But here a question arises about the technical equipment of each teacher for these purposes [20].

The second problem in online learning is timely feedback from the teacher to the students. In a situation where a teacher checks several dozen students" written works every day, timely feedback is impossible. Auto-testing students" knowledge and the inclusion of mutual testing methods in the educational process, could solve this problem.

The third problem: the use of traditional learning methods reduces the distance learning effectiveness. To work in this format, high-quality teacher retraining is necessary in order to obtain multilingual competencies and be able to work in a multilingual environment. Thus, psychological, technological, methodological support and professional development programs for teachers are vital to minimize the negative impact of rapid changes in the educational process and ensure effective online education [21]. Digital didactics, a new branch of science, needs to be developed. This is not a complete replacement of traditional teaching methods and approaches, but "the introduction of such elements of digital technologies that improve the quality and efficiency of existing formats and facilitate the work of teachers (primarily routine)" [8]. In our opinion, the formation of multilingual pedagogy is the nearest prospect for the development of scientific knowledge about the multilingual world.

The fourth problem is the unpreparedness of students for learning in a multilingual environment. The aforementioned analytical report of the Ministry of Science and Higher Education of the Russian Federation points out the need to teach first-year students "the basics of self-organization in learning, the creating of individual trajectories, and the skill to learn" [8].

Fifth problem: the need for an urgent transition to distance learning due to the COVID-19 pandemic, which escalated the task of forming multilingual thinking as a resource for adapting to changing living conditions and, as a result, to learning conditions.

This study identified the negative impact of digitalization on the learning results of students during a pandemic and the possibility for changing the vector of influence from negative to positive.

The introduction of digitalization into the educational environment requires not only the technological preparedness of teachers and students; the process also affects a large number of personal qualities, and requires a certain level of readiness. The forced digital format of education does not coincide with the student's expectations. All this affects the decrease in the effectiveness of training and requires an integrated approach to solving this problem. The huge potential of digitalization in education is exciting and presents many opportunities and challenges associated with new trends and developments in digital technologies. [22] Changes in the socio-cultural environment towards the formation of a multilingual world, in our opinion, lead to the transformation of three subjective-organizational spatiotemporal layers of human existence: the layer of the subjective (personalized) "image of the world", the layer of communication (group conventionality) and the layer of action. According to Abramova [23, p. 48], "we get grounds for classifying factors that have a broader influence on the formation of the subject's "picture of the world" (not only an individual, but also a social group, ethnic group, nation as a whole), without losing attention to the personality".

Hence, faced with the changed reality, the education system was forced to urgently respond to multilingual challenges. Despite the fact that, in general, the transition was successful, it revealed significant multi-level problems:

1) cognitive-competent success—unpreparedness of teachers and students for fullfledged multilingual communication, including digital (the use of traditional learning models that are not adapted to the changed conditions of the educational process);

2) personal effectiveness—the lack of personal soft-skills necessary for working in a multilingual environment (self-organization, self-discipline, self-motivation, etc.);
3) behavioral infantilism—the inability to transform habitual behavioral patterns in the absence of external control.

These negative trends lead to the need to form multilingual thinking. This key objective should allow adapting to the emerging multilingual environment affecting different levels: socio-cultural, professional-technological, individual-personal.

The immediate goal of scientific research in the field of multilingualism is the search and implementation of innovative technologies for the development of multilingual thinking in the educational environment.

References

1. Bozkurt, A., Sharma, R.C.: Emergency remote teaching in a time of global crisis due to corona virus pandemic. Asian J. Distan. Educ. **15**(1), i–vi (2020). https://doi.org/10.5281/zenodo.3778083
2. Starcic, A.I., Huang, P.-S., Valeeva, R.A., Latypova, L.A., Huang, Y.-M.: Digital storytelling and mobile learning: potentials for internationalization of higher education curriculum. In: Huang, T.-C., Lau, R., Huang, Y.-M., Spaniol, M., Yuen, C.-H. (eds.) SETE 2017. LNCS, vol. 10676, pp. 400–406. Springer, Cham (2017). https://doi.org/10.1007/978-3-319-71084-6_45
3. Daniel, S.J.: Education and the COVID-19 pandemic. Prospects **49**(1–2), 91–96 (2020). https://doi.org/10.1007/s11125-020-09464-3
4. Toropov, D.A.: Digitalization of education: pros and cons, prospects. Pedag. **6**, 109116 (2018)
5. Voevodkin, I.A., Tsaregorodtseva, E.V.: Problems of digitalization of education. Synergy Sci. **33**, 1272–1281 (2019). (in Russian)
6. Malchikova, N.S., Pivovarova, D.M., Kuzmin, E.V.: Research of innovative methods of digital transformation in the field of education. Actual Probl. Human. Nat. Sci. **3**, 131–135 (2017)
7. Teräs, M., Suoranta, J., Teräs, H., Curcher, M.: Post-Covid-19 education and education technology 'solutionism': a seller's market. Postdigit. Sci. Educ. **2**(3), 863–878 (2020). https://doi.org/10.1007/s42438-020-00164-x
8. Stress test lessons. Universities in the conditions of a pandemic and after it. Qual. Educ. **2**, 40–45 (2020)
9. Perminova, L.M.: Digital education in the context of theory and practice. Bull. Vladimir State Univ. Ser. Pedag.Psychol. Sci. **42**(61), 50–65 (2020). (in Russian)
10. Semyonov, V.I., Kazantseva, Y.N.: Mass open online courses as a new format of education. Mod. Probl. Sci. Educ. **6**, 150 (2017). (in Russian)
11. Hodges, C., Moore, S., Lockee, B., Trust, T., Bond, A.: The difference between emergency remote teaching and online learning (2020). https://er.educause.edu/articles/2020/3/thedifference-between-emergency-remote-teaching-and-online-learning
12. Luksha, P.: Educational innovations, or why we need to change education. http://oash.info/download/news/news-4153.pdf. Accessed 6 Jan 2020
13. Verbitsky, A.A.: Cifrovoe obuchenie: problemy, riski i perspektiv [Digital learning: problems, risks and prospects]. Verbitsky, A.A.: Jelektronnyj nauchno-publicisticheskij zhurnal "Homo Cyberus" [Electronic Scientific Journal "Homo Cyberus"] **1**(6) (2019). [Electronic resource]. http://journal.homocyberus.ru/Verbitskiy_AA_1_2019. Accessed 12 Nov 2021. (in Russian)

14. Jarke, J., Breiter, A.: Editorial: the datafication of education. Learn. Media Technol. **44**(1), 1–6 (2019). https://doi.org/10.1080/17439884.2019.1573833
15. Knox, J., Williamson, B., Bayne, S.: Machine behaviourism: future visions of 'learnification' and 'datafication' across humans and digital technologies. Learn. Media Technol. **45**(1), 31–45 (2020). https://doi.org/10.1080/17439884.2019.1623251
16. Manolev, J., Sullivan, A., Slee, R.: The datafication of discipline: ClassDojo, surveillance and a performative classroom culture. Learn. Media Technol. **44**(1), 3651 (2019). https://doi.org/10.1080/17439884.2018.1558237
17. Mirrlees, T., Alvi, S.: EdTech Inc. Selling Automating and Globalizing Higher Education in the Digital Age. Routledge, Abingdon (2020)
18. Suoranta, J., et al.: Speculative social science fiction of digitalization in higher education: from what is to what could be. Postdigit. Sci. Educ. **4**(2), 224–236 (2021). https://doi.org/10.1007/s42438-021-00260-6
19. Blinov, V.I., Esenina, E.Y., Sergeyev, I.S.: Digital didactics of vocational education and training (key theses). Secondary Vocat. Educ. **3**, 3–8 (2019)
20. Shmurygina, O.V.: The educational process in the conditions of a pandemic. Vocat. Educ. Labor Mark. **2**(41), 51–52 (2020). (in Russian)
21. Almazova, N., Krylova, E., Rubtsova, A., Odinokaya, M.: Challenges and opportunities for Russian higher education amid COVID-19: teachers' perspective. Educ. Sci. **10**, 368 (2020). https://doi.org/10.3390/educsci10120368
22. Schmidt, J.T., Tang, M.: Digitalization in education: challenges, trends and transformative potential. In: Harwardt, M., Niermann, P.J., Schmutte, A., Steuernagel, A. (eds.) Führen und Managen in der digitalen Transformation, pp. 287–312. Springer, Wiesbaden (2020). https://doi.org/10.1007/978-3-658-28670-5_16
23. Abramova, M.A.: Multilingualism as a reflection of the implementation of multicultural models of state policy. European multilingual space. In: Collection of materials of the international conference "Eropean Multilingual space: education, Society, law". Arhangelsk (2019). (in Russian)

Techniques and Technologies
for Multilingual Learning

Discursive Approach to Foreign Language Training via Massive Open Online Courses

Artyom Zubkov(✉) 📵

Novosibirsk State University of Economics and Management, 56 Kamenskaya Str,
Novosibirsk 630099, Russian Federation
zubkov_nstu@mail.ru

Abstract. This article considers some aspects of implementation a discursive approach in relation to teaching a foreign language to future professionals. The concepts of discourse and discursive competence are considered. In the course of the study, foreign-language MOOCs of the Coursera platform were selected, the components of discursive competence were identified, a methodological model for the formation of discursive competence of university students was developed, experimental training according to the model developed was carried out, pedagogical conditions for using MOOCs within the framework of a discursive approach for teaching a foreign language in a university were identified, a survey of students and teachers on the effectiveness of using MOOCs in the language classroom of a university was conducted. It is concluded that MOOCs can be an effective tool for the formation of the discursive competence of future professionals.

Keywords: University · Higher education · English · Second language · Students · Discourse · Discursive competence

1 Introduction

The theory and methodology of teaching foreign languages sets itself the main goal of getting to know another culture and the possibility of establishing a dialogue of cultures. In practice, this goal is achieved by the formation of professional foreign language competence of university students. The concept of professional foreign language competence is multicomponent in nature. One of its main components is discursive competence. According to most scientists in the field of linguistics and teaching methods, discursive competence includes knowledge of various types of discourses (e. g. economic, engineering, legal) and patterns of their construction, and in addition, the ability to create and understand them in situations of professional communication. The core of the discursive competence concept is the concept of discourse. Discourse is commonly considered as speech, the process of generating speech, language activity and the way of speaking. The use of discourse concept in the theory and methodology of teaching foreign languages dates back to the 70s of the 20th century. Then the discourse was considered as a functional style with its inherent lexical system and stylistic features. At present, discourse is understood as a complex linguistic phenomenon that includes texts and extralinguistic factors necessary for operating these texts, namely, knowledge about the arrangement of the environment, understanding the goals and attitudes of the addressee.

As for researchers, certain issues concerning the usage of massive open online courses, discourse approach to foreign language training and formation of discursive competence are covered in the works of the following researchers who studied model for the evaluation of oral discursive competence in secondary education [1], using workshop to build postgraduate discursive competences, developing second language oral competence through an integrated discursive approach [2], the impact of role-playing games through second life on the oral practice of linguistic and discursive sub-competences in English [3], the discursive competence at the crossroads of the intralinguistic mediation and the grammatical competence [4], the development of engineering students' foreign language discursive competence [5], ethnolinguodidactic approach to formation of foreign-language discursive competence [6], discursive constructionist approach to narrative in language teaching and learning research [7], discursive practice under conditions of superdiverse linguistic diversity [8], discursive lecturing via student-centered approach [9], teaching of mother tongue reading anchored to discursive approach [10], practical legal cases as an effective method of acquiring the discursive communicative skills of international jurists when learning the professional foreign language [11], formation of the foreign language discursive competence of pedagogical faculties students in the process of intercultural dialogue [12], analysis of written linguistic and discursive competence of first year students in primary education teacher degree [13], teaching foreign language in transport university using massive open online courses [14], increasing effectiveness of foreign language teaching of transport university students in process of online learning [15], implementation of CLIL approach via MOOCs [16], MOOCs in blended English teaching and learning for students of technical curricula [17], foreign language learning environment in transport universities [18], position of foreign language as education for global citizenship [19], development of senior students' writing skills in genres of academic discourse using MOOCs [20], multimedia professional content foreign language competency formation in a digital educational system [21], development of students' polycultural and ethnocultural competences in the system of language education [22], integration of online and offline education [23]. An analysis of the literature demonstrates the majority of existing research works with discursive competence, and much less research is devoted to studying the discursive approach. It should also be noted that the use of digital educational technologies and Internet information resources in a discursive approach is not the focus of researchers' attention. The gap that this study seeks to fill is the answer to the question "What can MOOCs offer for a discursive approach to teaching a foreign language in a university?".

Knowledge of a foreign language by university students is one of the basic general professional qualities. Discursive competence is one of the main components of professional foreign language competence. Therefore, the discursive approach in teaching a foreign language plays an important role in the formation of the discursive competence of students at university. This study proposes to consider the potential of massive open online courses for teaching a foreign language to university students in line with a discursive approach. This goal of the study designated its objectives: 1) to formulate the

pedagogical conditions for the use of MOOCs within the framework of a discursive approach for teaching a foreign language; 2) to build a model for the formation of discursive
competence of university students when teaching a foreign language using MOOCs 3) to
conduct experimental training of university students according to the model developed
and evaluate its results.

2 Materials and Methods

In this study, the following methods were implemented: content analysis of foreign
language massive open online courses of the Coursera platform, analysis of scientific
literature on the research problem, pedagogical modeling, experimental training and
questionnaire.

When selecting educational materials (massive open online courses), we were guided
by the sub-competences of discursive competence—strategic, tactical, genre and textual
[24]. Strategic sub-competence is considered as the ability of an individual to understand his communicative intention and the ability to plan ahead a communicative event
in a communication situation. Tactical sub-competence deals with the analysis of the
communicative situation and the selection of language means for the implementation
of the communicative intention. Genre competence includes the ability to organize discourse according to the characteristics of the operated genre, chosen by the individual
to implement the communicative intention. Possessing the textual sub-competence, the
individual organizes several sentences into a single connected structure—text. The content analysis of the massive open online courses of the Coursera platform was carried
out with the aim of their suitability for the formation of the above sub-competences.
MOOCs from the world's leading universities were selected, the authors of the courses
are native English speakers. The courses selected are demonstrated in Table 1.

Table 1. MOOCs selected for the formation of discursive competence components

Component	MOOC	Developer	Duration
Strategic	Communication Strategies for a Virtual Age	University of Toronto	4 weeks
Tactical	Speaking and Presenting: Conversation Starters	University of Michigan	4 weeks
Genre	High-Impact Business Writing	University of California	4 weeks
Textual	Writing Professional Email and Memos (Project-Centered Course)	University system of Georgia	4 weeks

Pedagogical modeling was aimed at creating a learning environment capable of
effectively using MOOC materials in blended learning within a discursive approach.
The purpose of training is defined by the federal state standard of higher education and
consists in the ability of an individual to use the means of a native and foreign languages
for professional communication, as well as to use the means of a foreign language to

search for new information. In this case, discursive competence is a component of the ability of the individual stated above. Further, according to the stated goal of training, training materials are selected—MOOCs on professional disciplines in English. Further, the foreign language teacher personally completes the MOOCs selected for integration, which subsequently gives him the opportunity to develop supporting materials—language, speech and communication tasks based on the lexical and grammatical minimum, the learning of which will be necessary for university students studying MOOC materials. Next, an introductory seminar is held, at which students are informed about the goals and details of the educational process on the Coursera platform. The educational process takes place in a mixed mode, where the study of supporting materials is integrated into the standard curriculum in the language class, and students take MOOCs independently in their extracurricular time. If necessary, students may need advice from a teacher regarding language or content issues. The result of the work according to the developed methodological model will be the formed discursive competence of students, namely its components. The methodological model for the formation of the discursive competence of university students is demonstrated in Fig. 1.

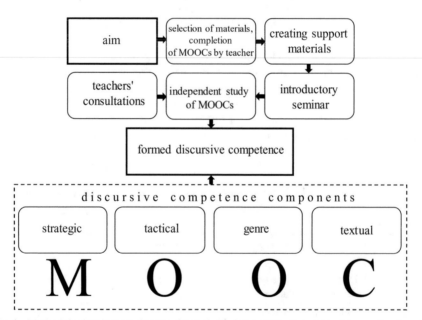

Fig. 1. Methodological model for the formation of the discursive competence of university students

The textbook authored by John Allison, Paul Emmerson "The Business 2.0. Intermediate" (Macmillan, 2013) was used as the main teaching and learning tool in both control and experimental groups of students under the guidance of Russian teachers of English. However, the students of the experimental group additionally used MOOCs as a learning resource for independent learning.

Experimental training was carried out with 2nd and 3rd year students of the Siberian Transport University majoring in International Business for 4 terms, i.e., one MOOC was completed within one term. Placement and final testing to determine the level of formation of the discursive competence of university students in the control and experimental groups were also conducted. The test materials contained questions, by answering which the students had to demonstrate their mastery of communicative business strategies. Also, students of the control and experimental groups were offered a creative task to write a business email and a memorandum. The assessment of the level of formation of discursive competence was carried out jointly and separately for its four components. A survey of students and teachers who participated in experimental training regarding the use of MOOCs to form the discursive competence of university students was implemented.

3   Results and Discussion

The development and implementation of a methodological model for the formation of discursive competence of university students allowed us to formulate the following pedagogical conditions for using MOOCs within the framework of a discursive approach for teaching a foreign language in a university:

1) the concept of discourse occupies a major place in the system of foreign language training of university students;
2) before the implementation of the educational process, it is necessary to select the types of discourse (e. g. technical, economic, managerial) and its genres (e. g. guidance, manual, financial report, business plan), corresponding to the majors implementing in the university;
3) teaching a foreign language should arise with the help of audiovisual presentation (microvideo lectures of massive open online courses).

During the placement and final testing, we determined positive changes in the components of discursive competence of university students, and by calculating the arithmetic mean value, we determined a qualitative change in the general discursive competence (in our assessment, each of the four components of discursive competence makes an equal contribution to the calculation result). Indicators of the formation of discursive competence and its components in the control and experimental groups of university students are demonstrated in Fig. 2 and Fig. 3 respectively.

Before the start of experimental training, after the entrance assessment of the level of discursive competence formation, the control and experimental groups were mixed in composition to achieve the same initial level of formation of discursive competence and amounted to 53% in both groups of students. After experimental training, the level

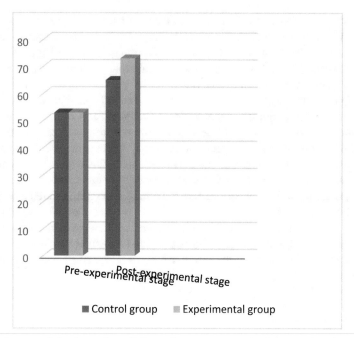

Fig. 2. Indicators of the formation of discursive competence in the control and experimental groups of university students

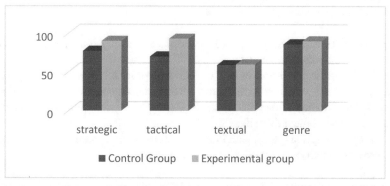

Fig. 3. Indicators of the formation of discursive competence components in the control and experimental groups of university students

of formation of the discursive competence of students in the experimental group was 83%, which is 10% higher than in the experimental group with its 73% indicator. Considering the formation of the components of discursive competence in relation to the control and experimental groups, it should be noted a qualitative increase in all components assessed, while the students of the experimental group always demonstrated higher results (the strategic component is higher by 13%, tactical by 23%, text by 1%, genre by 4%). An insignificant difference of 1% in assessing the text component can be explained by the fact that traditionally foreign language classes at a university (and a transport university is no exception) were aimed at developing linguistic competence, which allowed students in the control and experimental groups to demonstrate relatively the same results. A significant discrepancy of 23% in the tactical component of discursive competence can be explained by the fact that the selected MOOCs, in contrast to department textbooks and teaching aids, are saturated with communicative plots, having studied which, students of a university more skillfully master the ability to select language means for the implementation of a communicative intention. In order to identify the effectiveness of the model developed, the data of the control and experimental groups were processed using the Student's T-test. The data obtained indicate the absence of statistical significance between the results of complex testing between students in the control and experimental groups before the experiment. However, after the experiment, there was a statistically significant difference between the results of the final testing in the control and experimental groups (in the control group $T = 2.61$ at $p = 0.05$; in the experimental group $T = 3.45$ at $p = 0.05$).

Thus, referring to the sources from the literature review section of this study, it should be noted the educational potential of such digital resources as MOOCs for the formation of discursive competence and its components. Answering the question "What can MOOCs offer for a discursive approach to teaching a foreign language in a university?" it is necessary to point out a better formation of discursive competence and all its components among students of a university, provided that the training takes place in accordance with the methodological model developed and the pedagogical conditions for using MOOCs within the framework of the discursive approach in a university are met. The MOOCs selected deal with communication strategies and they are not intended for teaching English. Therefore, in this experiment, these MOOCs are repurposed and become vehicles for learning English in the context of learning about professional economics discourse which, taken together, amounts to a discursive setting for learning English. This dimension of multilingualism and technology might come in through this repurposing.

The results of a survey of students and teachers involved in experiential training state the effectiveness, confidence and interest in MOOCs in the educational process. Students believe that after completing the MOOCs, they can better understand their communicative intentions, plan a communicative event more effectively in a situation of communication in a foreign language. They also feel more confident when choosing language tools and clichés for communication in a foreign language, and are also better versed in the features of the genres of economic discourse—a business email, report, business plan, etc. Teachers also declare positively about the use of MOOCs in teaching a foreign language in the university within the framework of a discursive approach, although they mention the increased labor costs for preparing preliminary teaching materials.

4   Conclusion

The discursive approach is important in the training of future professionals, and, in particular, for the formation of the discursive competence of university students. To use

MOOCs within the framework of a discursive approach, it is necessary to conduct training according to the developed methodological model and subject to the pedagogical conditions for using MOOCs within the framework of a discursive approach in a university. The integration of MOOCs into the traditional curriculum can improve the quality indicators of the discursive competence. However, it is worth mentioning the increased labor costs for the preliminary completing of MOOCs by teachers and the development of tasks for language support. It should also be taken into account that the study was conducted while teaching students of a university of the Russian Federation. This study may be useful for specialists in the field of higher education, foreign language teachers for adults, heads of departments at universities that organize language training for students.

References

1. Núñez-Delgado, P.: Research in language didactics: a model for the evaluation of oral discursive competence in secondary education. Porta Linguarum Rev. Interuniv. Didáctica las Lenguas Extranj (2008). https://doi.org/10.30827/Digibug.24034
2. Bærenholdt, J.O., Gregson, N., Everts, J., Granås, B., Healey, R.L.: Performing academic practice: using the master class to build postgraduate discursive competences. J. Geogr. High. Educ. **34**, 283–298 (2010). https://doi.org/10.1080/03098260903502695
3. Quintín, E., Sanz, C., Zangara, A.: The impact of role-playing games through second life on the oral practice of linguistic and discursive sub-competences in English. In: Proceedings - 2016 International Conference on Collaboration Technologies and Systems, CTS 2016 (2016)
4. Grabowska, M.: La compétence discursive au croisement de la médiation intralinguistique et de la compétence grammaticale: exemple de transposition d'un dialogue en texte narratif. Stud. Rom. Posnaniensia. **44**, 125 (2017). https://doi.org/10.14746/strop.2017.442.008
5. Gritsenko, L.M., Demidova, T.A., Salosina, I.V.: The development of engineering students' foreign language discursive competence: motivational educational factors. Sci. Educ. Today. **9**, 162–181 (2019). https://doi.org/10.15293/2658-6762.1904.10
6. Dzhandar, B.M., Urakova, F.K., Loova, A.D., Tuova, M.R.: Ethnolinguodidactic approach to formation of foreign-language discursive competence of the bilinguals based on the material of adyghe folktales. Mediterr. J. Soc. Sci. **6**, 227 (2015)
7. Prior, M.T., Talmy, S.: A discursive constructionist approach to narrative in language teaching and learning research. System **102**, 102595 (2021). https://doi.org/10.1016/j.system.2021.102595
8. Foldes, C.: Diskursive Praxis unter Bedingungen superdiversiver Sprachenvielfalt: Post-multilingualistische Zugange in der Diskussion, (2020)
9. Einum, E.: Discursive lecturing: an agile and student-centred teaching approach with response technology. J. Educ. Change **20**(2), 249–281 (2019). https://doi.org/10.1007/s10833-019-09341-7
10. Dias Di Raimo, L.C.F., Dearo, T.N.: O ensino da leitura em língua materna ancorado na perspectiva discursiva: uma proposta de leitura de um conto contemporâneo. Calidoscópio. **15**, (2017). https://doi.org/10.4013/cld.2017.151.04
11. Samorodova, E.A., Ogorodov, M.K., Belyaeva, I.G., Savelyeva, E.B.: The study of practical legal cases as an effective method of acquiring the discursive communicative skills of international jurists when learning the professional foreign language (professional French). XLinguae. **13**, 121–138 (2020). https://doi.org/10.18355/XL.2020.13.01.10

12. Ponomarenko, L.N., Zlobina, I.S., Galitskih, E.O., Rublyova, O.S.: Formation of the foreign language discursive competence of pedagogical faculties students in the process of intercultural dialogue. Eur. J. Contemp. Educ. **6**, 89–99 (2017). https://doi.org/10.13187/ejced.2017.1.89

13. Martín, A.M.R., Níkleva, D.: Analysis of written linguistic and discursive competence of first year students in primary education teacher degree. Rev. Signos **49**, 48 (2016)

14. Zubkov, A.: Teaching foreign language in transport university using massive open online courses: pilot study. In: Manakov, A., Edigarian, A. (eds.) TransSiberia 2021. LNNS, vol. 403, pp. 92–100. Springer, Cham (2022). https://doi.org/10.1007/978-3-030-96383-5_11

15. Zubkov, A.: Increasing effectiveness of foreign language teaching of transport university students in process of online learning. In: Manakov, A., Edigarian, A. (eds.) TransSiberia 2021. LNNS, vol. 403, pp. 438–445. Springer, Cham (2022). https://doi.org/10.1007/978-3-030-96383-5_49

16. Zubkov, A.: Implementation of CLIL approach via Moocs: case study of Siberian transport university. In: Manakov, A., Edigarian, A. (eds.) TransSiberia 2021. LNNS, vol. 402, pp. 1002–1010. Springer, Cham (2022). https://doi.org/10.1007/978-3-030-96380-4_109

17. Zubkov, A.D.: MOOCs in blended English teaching and learning for students of technical curricula. In: Anikina, Z. (ed.) IEEHGIP 2022. LNNS, vol. 131, pp. 539–546. Springer, Cham (2020). https://doi.org/10.1007/978-3-030-47415-7_57

18. Komkova, A., Kobeleva, E., Taskaeva, E., Ishchenko, V.: Foreign language learning environment: a case study of STU. In: Manakov, A., Edigarian, A. (eds.) TransSiberia 2021. LNNS, vol. 403, pp. 429–437. Springer, Cham (2022). https://doi.org/10.1007/978-3-030-96383-5_48

19. Almazova, N.I., Kostina, E.A., Khalyapina, L.P.: The new position of foreign language as education for global citizenship. Novosibirsk State Pedagogical Univ. Bull. **6**(4), 7–17 (2016). https://doi.org/10.15293/2226-3365.1604.01

20. Andreeva, S., Khalyapina, L., Almazova, N., Baranova, T.: Development of senior students' writing skills in genres of academic discourse using massive open online courses. In: Anikina, Z. (ed.) IEEHGIP 2022. LNNS, vol. 131, pp. 39–46. Springer, Cham (2020). https://doi.org/10.1007/978-3-030-47415-7_5

21. Khalyapina, L., Kuznetsova, O.: Multimedia professional content foreign language competency formation in a digital educational system exemplified by Stepik framework. In: Anikina, Z. (ed.) IEEHGIP 2022. LNNS, vol. 131, pp. 357–366. Springer, Cham (2020). https://doi.org/10.1007/978-3-030-47415-7_38

22. Almazova, N., Baranova, T., Khalyapina, L.: Development of students' polycultural and ethnocultural competences in the system of language education as a demand of globalizing world. In: Anikina, Z. (ed.) GGSSH 2019. AISC, vol. 907, pp. 145–156. Springer, Cham (2019). https://doi.org/10.1007/978-3-030-11473-2_17

23. Almazova, N., Andreeva, S., Khalyapina, L.: The integration of online and offline education in the system of students' preparation for global academic mobility. In: Alexandrov, D.A., Boukhanovsky, A.V., Chugunov, A.V., Kabanov, Y., Koltsova, O. (eds.) DTGS 2018. CCIS, vol. 859, pp. 162–174. Springer, Cham (2018). https://doi.org/10.1007/978-3-030-02846-6_13

24. Zubkov, A.: Using MOOCs to Teach Foreign Language Writing to University Students. In: Anikina, Z. (ed.) IEEHGIP 2022. LNNS, vol. 499, pp. 23–31. Springer, Cham (2022). https://doi.org/10.1007/978-3-031-11435-9_3

Edubreak with Augmented Reality in a Foreign Language Class at a Non-linguistic University

Elena I. Chirkova⓪, Elena M. Zorina⁽⊠⁾ ⓪, and Elena G. Chernovets⓪

Saint Petersburg State University of Architecture and Civil Engineering, Vtoraya Krasnoarmeiskaya, 4, 190005 Saint Petersburg, Russia
zorinaem@bk.ru

Abstract. The article considers one of the ways of overcoming the psychological characteristics of the generation Z associated with the difficulties of long-term concentration of attention, low motivation for learning and digital addiction to gadgets. The article analyzes the possibility and necessity of holding educational breaks in the process of studying a foreign language at a higher educational institution. It is for students of the digital generation that such pauses take place using augmented reality pedagogical technology. The authors suggested that educational breaks will allow students to maintain their working capacity throughout the lesson, increase motivation to learn a foreign language, improve the level of lexical and grammatical skills of various types of speech activity, and also increase digital literacy. The article provides a comparative analysis of edubreak and spaced learning for similarities and differences. During the experiment, the hypothesis of the researchers was confirmed, and the proposed type of lesson structure (using educational pauses) proved to be effective in teaching students belonging to the digital generation.

Keywords: Foreign language teaching · Augmented reality · Educational pause · Motivation for learning · Digital generation · Concentration of attention · Spaced learning

1 Introduction

All practicing teachers observe situations of weakening the attention of students during a foreign language lesson. These observations are consistent with studies [1] conducted at Johns Hopkins University, USA. Scientists have studied in detail the phenomenon of a break in work and gave some tips for a more productive continuation of activities regarding the need for breaks. Since the studies concerned the activities of office workers, the types of activities during the break were distinguished: general relaxation (for example, resting in a chair), eating, chatting with colleagues, walking in the fresh air, reading the news, checking emails, browsing YouTube, playing on the smartphone). The relevance of such studies is noted not only in biology and psychology, but also in pedagogy, where it is necessary to apply new teaching models.

Despite the fact that the studies of some scientists were carried out on mice [2], and others on humans [3], in both the so-called "spacing effect" of Ebbinghaus [4]

© The Author(s), under exclusive license to Springer Nature Switzerland AG 2023
D. Bylieva and A. Nordmann (Eds.): PCSF 2022, LNNS 636, pp. 440–451, 2023.
https://doi.org/10.1007/978-3-031-26783-3_36

was noted, the essence of which is that the best results are achieved if you practice or revise information with short breaks. Researchers from the National Institutes of Health (NIH), USA [5] experimentally confirmed that during a short pause, the brain can better remember the information about the skill received minutes earlier. According to psychologists, intense mental activity should be alternated with physical activity or a simple short idleness. They propose to establish a rule: every 30–50 min of the lesson, arrange ten-minute breaks.

According to the studies of psychologists and sociologists [6–8], the current generation of university students belongs to generation Z (digital generation). These are people who were born since 2000 and grew up in the era of a technological boom - a digital generation that has a number of psychological characteristics that distinguish them from representatives of other generations. In addition to technological advancement, they are characterized by mosaic thinking, the ability to multitask, restlessness and impatience. We will pay attention to some features of generation Z: difficulties in long-term concentration of attention, reduced learning motivation and the need to introduce digital technologies into the educational process.

Modern educational conditions require the inclusion of new technologies [9], and therefore, to solve the problems of this study, we propose augmented reality (AR) implemented in the form of active links for a mobile device. The term *augmented reality* was coined by Boeing aerospace researcher Tom Caudell in 1990. Virtual reality creates entirely artificial environment, whereas augmented reality uses an already existing environment and imposes new digital information on it [10], sometimes using sensory modalities which usually comprise visual, auditory, tactile, somatosensory, and olfactory ones [11].

Augmented reality (AR), being an innovative technology in the context of the digitalization of society, seeks to become widespread in education [12] which allows it to be called a technology of social significance [13]. It is included in the so-called "passport of the federal project" being the part of the national program "Digital Economy", one of the goals of which is to form a network of educational centres of universities and the development of students along personal trajectories using innovative technologies [14]. Unfortunately, as noted by M. Rumyantsev and I. Rudov [15], nowadays Russian universities utilize AR and VR chiefly when teaching natural sciences.

The discipline "Foreign language" is quite difficult to study, as it is characterized by multilevelness. On the one hand, students need to master various language skills (phonetics, vocabulary, grammar), on the other hand, based on the formed language skills, skills in four types of speech activity should be formed: speaking, reading, writing and listening comprehension. Achieving an adequate level of mastery of language skills and speech skills should lead to the formation of foreign language competence.

As O. Putistina emphasizes [16], it is difficult to qualitatively reconstruct it in an educational institution environment. Moreover, according to the conclusions of V. V. Kotenko [17], the use of augmented reality in teaching a foreign language allows you to simulate communication with native speakers or get to know the culture and history of another country, as well as use excursions, videos, computer educational games and simulators. The formation of the above skills and abilities requires complex brain activity and can be successfully carried out only with the activity of students in the classroom and their high learning motivation. However, the students' oversaturation with digital interactive information can be considered as a negative factor in the use of modern technologies, and traditional teaching aids become boring to them.

Thus, a need for additional study of the didactic potential of augmented reality technology arises, taking into consideration the features of a foreign language teaching, the peculiarities of the educational environment of the university [18]. The aim of the work done is to study the features of applying edubreak with augmented reality in a foreign language class at a non-linguistic university.

Edubreak - (educational break) is a 5–10-min break in a lesson aimed at changing activities, involving active or interactive cooperation of students (possibly gaming) using digital tools (for example, augmented or virtual reality), which increases learning motivation and concentration of attention. Yet, the educational pause involves the emotional students' involvement in the performance of the task, which should have a light form and include the lexical and grammatical material being studied. However, educational breaks should not be just entertainment or relaxation, their methodological goal is training: revision, consolidation of what has been learned.

In addition, during educational breaks, there is a psychological correction of various manifestations of the personality of the digital generation representatives, who prefer an individual set of tasks, do not know how to work in a team, and do not like to lose in front of other students.

Let's return to one of the psychological features of Generation Z students, that is difficulties in long-term concentration of attention. The study of a foreign language, including the language for professional purposes, requires a high concentration of attention from the student. Therefore, one must always remember to create conditions that would avoid overwork and ensure high performance.

The essence of the "spacing effect" of Ebbinghaus (Hermann Ebbinghaus - a German psychologist, is the founder of the psychology of memory) is the alternation of intense and weak learning load, which helps to learn a large amount of information in the shortest possible period. According to the researcher, the optimal load and regularity in learning English is as follows: the most productive approach is when you learn new things during the first 20–30 min from the beginning of the lesson. These first minutes of the lesson are best spent on mastering the most difficult material. This phenomenon is explained by the features of perception, that is, the attention curve, which demonstrates that the concentration of attention drops from 100% to 60% after half an hour from the start of the lesson. Further, it drops to 40% after 45 min, and finally, after an hour from the start of the lesson, the concentration of attention drops to almost 0. It is necessary to return to the information regularly for a certain interval, and between these periods it is important to rest for a while or pause.

The classical system of education at the university does not involve the introduction of this memorization technique. That is why it is advisable to change the type of activity by moving from an academic style of learning to performing emotional tasks that cause a positive attitude [19] or to the application of digital technologies, for example, augmented reality [20]. Such tasks may include: solving humorous crossword puzzles, performing lightweight lexical and grammatical exercises, the work on which would not present real difficulties, watching thematic videos, etc.

Learning using augmented reality during educational pauses is a kind of mobile learning: learning activities using mobile technology serve as a link between the real world of knowledge and the game's visual world [21–25].

2 Methods

To gain the objectives of this study, the methodological ideas expressed in the work of A.V. Grinshkun [26] were realized. We mean that augmented reality helps to maintain the educational motivation and involvement of students at the required level activate the learning process. This technology implies the teacher's utilization of digital tools for presenting material and organizing an educational break, the students' use of the functionality of digital technologies and gadgets in cognitive activity, and the organization of interaction between participants in the process of learning.

In order to prove the effectiveness of the proposed experiment and the type of lesson using educational pauses, a psychological and pedagogical literature analysis and our own experience of pedagogical activity was carried out when obtaining empirical data. The tasks of the study done include the steady attention of students in foreign language classes, to increase learning motivation and the level of learning. The experiment was carried out at the St. Petersburg State University of Architecture and Civil Engineering among 2nd year students of the Faculty of Architecture (3 groups of 15 people).

Its goal is to include educational pauses with augmented reality in the lesson plan, aimed at switching students to another type of activity that provides not only rest (with a decrease in the complexity of the task), but also game and competitive tasks based on the topics studied, how to use the Mobile-Assisted Language Learning (MALL), as well as other features of mobile applications for augmented reality. The students participating in the experiment were distributed between a control group (CG) and two experimental groups (EG1 and EG2) with the same number of students.

For the first and last stages of the experiment, lexical and grammatical exercises of the same type were used, the same for all groups.

For the second (formative) stage of the experiment with augmented reality, several types of tasks were used for the educational break:

- for frontal interrogation:

  - quiz on kahoot.com;
  - survey on mentimeter.com;
  - watching a training video (for example, on YouTube);

- for individual tasks:

  - game exercises on the website learningapps.org;
  - game exercises (plugin "Games") in the training course of the Moodle environment;
  - tasks in augmented reality manuals.

The pedagogical technology of using AR in the study of foreign languages during edubreaks is aimed at switching attention to the next activity. AR improves work efficiency after a pause while continuing to study educational material. AR is a kind of emotional rest, especially for Generation Z, which is comparable to pressing the "RESET" button, that is responsible for rebooting. Sometimes working with the right type of AR tasks allows the student to re-evaluate what he/she did at the beginning of the lesson and why it is important, this is a game-mediated opportunity to find motivation for learning a foreign language, evaluate the positive aspects of the lesson and focus on further work. Edubreaks are also important for reflection and setting goals in future work. They are necessary to relax, since working in AR for the younger generation is a familiar type of activity.

The level of learning motivation and concentration of attention was measured using a questionnaire and Likert scale.

The quality of training was determined by the success of the exercises. The learning effect is considered achieved if its indicator is not less than 70% of the average indicator for the entire group. To calculate it, you need to select the 3 worst results of the group and calculate their average result, and then compare the average of the "worst" with the average score of the entire group.

3   Results

For empirical research, a hypothesis was put forward that the use of educational breaks in foreign language classes will increase the level of educational motivation and learning of students if the proper organization of the educational process using augmented reality is created.

Experiment tasks:

1. To assess the initial level of formation of lexical and grammatical skills among students of three groups of non-linguistic faculty.
2. Develop and test a methodology for using educational pauses with augmented reality to increase the level of motivation for learning and students' training.
3. Evaluate the effectiveness of the implemented methodology for the use of educational pauses with augmented reality to ensure the sustainable attention of students within the whole duration of the lesson.

At the first phase of the experiment, the basic level of formation of lexical and grammatical skills as a type of training was measured. For this, a set of exercises was proposed and offered to the participants of the experiment twice - at the beginning of the lesson and 50 min after it began. All three groups of students showed approximately the same level of initial knowledge.

At the second stage of the experiment in the control group, classes were conducted without pauses, in EG1 the pause was carried out after 40 min from the beginning of the lesson and lasted 10 min. At this time, students could independently relax as they saw fit. It was impossible to make noise and leave the audience. In EG2, the educational break also lasted 10 min and also began at the 40th minute of the lesson. However, it took place in an active change of educational activities and partly in a playful way using students' own gadgets and augmented reality techniques.

During all 5 experimental lessons at the 50th minute of the lesson lexical and harmonic tasks were tested in all groups.

In the 3rd phase of the experiment, the same set of didactic materials was used as in the 1st phase to test the effectiveness of pauses during the lesson. All three groups underwent control testing at the beginning and at the 50th minute of the lesson. The results of testing within the framework of determining the level of training are shown in Table 1.

Table 1. Final cut and level of training

	CG	EG1	EG2
Group mean score	6.75	7.1	9.2
Average score of the 3 worst	4.3	4.8	7.4
Learning effect	63.7%	67.5%	80.6%

During the lesson without the use of a pause (CG), when the concentration of attention was lost, the students became tired, and they completed the learning tasks, spending more time and making more mistakes. During the lesson with the use of uncontrolled rest (EG1), the concentration of educational attention was not restored sufficiently, which led to the fact that the tasks were performed almost the same as in the control group. Students were distracted from learning activities, and it was difficult for them to start actively studying again. During the lesson using the educational pause (EG2), students' attention was restored, the lesson was conducted with the same intensity, and the tasks were completed with fewer errors.

To determine the level of learning motivation among all participants in the experiment, a survey was conducted (Fig. 1).

Judging by the diagram, we can conclude that all the three groups have similar level structure of learning motivation, which allows us to talk about equal initial data. It has a positive effect on the further experiment, allowing us to analyze the data obtained and suggest possible development paths.

As part of determining the concentration of attention, all participants in the experiment were given three statements, to which they showed their attitude using the Likert scale (level of agreement: 0 - completely disagree, 5 - completely agree):

1. During class, I need breaks before starting work that requires concentration of attention (complex exercise);
2. I only use my gadget in class for the learning process.
3. It is difficult for me to concentrate on studying after rest (games, correspondence with friends, listening to music, etc.).

After analyzing the survey, an average level of agreement was revealed separately for groups (Fig. 2).

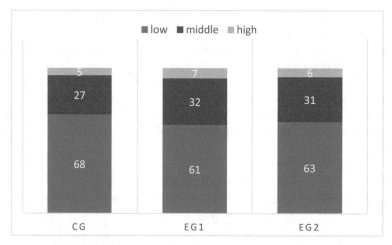

Fig. 1. The level of educational motivation (as a percentage).

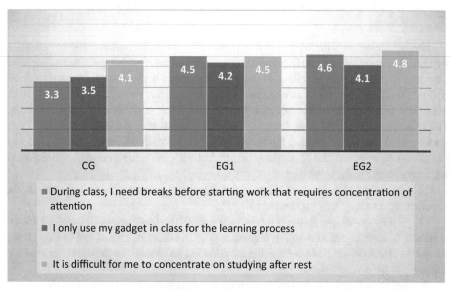

Fig. 2. Average level of agreement to statements about attention span.

In EG2, the following questions were added to the survey:

1. Did you know about the possibilities of augmented reality before the experiment?
2. Did you enjoy the educational break with augmented reality?

The results are presented in two diagrams (Fig. 3 and Fig. 4).

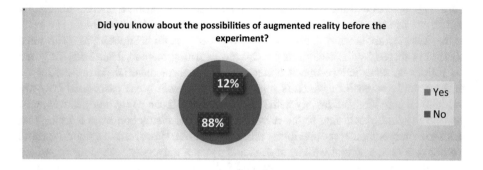

Fig. 3. Answer to the first question.

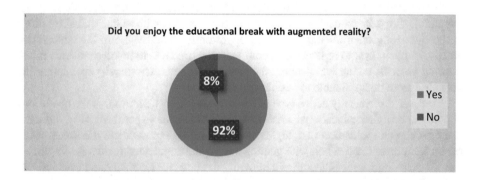

Fig. 4. Answer to the second question.

After analyzing the obtained experimental data, we concluded that the students were satisfied with the educational break with augmented reality, it increased their learning motivation and gave them self-confidence. However, it turned out that most of the students were completely unfamiliar with the possibilities of augmented reality and did not distinguish it from virtual reality. In addition, the survey showed that students' purposeless rest is too relaxing, making it difficult to concentrate on learning. New pedagogical tools related to digital technologies allow to expand and deepen students' knowledge of the subject and improve digital literacy. The experiment showed the possibility of the introduction of educational pauses with augmented reality actively involving students into the learning experiences happening within the classroom.

4 Discussion

Despite the wide range of didactic possibilities of utilizing the augmented reality technology, there are some objective factors that should be taken into consideration while choosing effective methods and tools.

The study of psychological and pedagogical literature shows that researchers do not have a common point of view regarding the need for rest during the lesson. So, David Little [1] believes that the continuation of mental activity during a break is absolutely ineffective as rest. The most effective method of relaxation, in his opinion, is a walk in the fresh air and a short conversation with colleagues on abstract topics. The same opinion is shared by the creator of the Spaced Learning method, Paul Kelly [27] and his followers, who believe that it is necessary to present material during the lesson in three portions, each of which is separated by a 10-min break associated not with educational activities, but mainly with physical activity. It can be rhythmic movements, yoga or origami. According to the researchers, it is the distraction from learning that helps to better remember the previously received material. However, the author's specific method works at a school with teenagers, and not at a university with students, where the academic learning space is not intended for such activity, is not paid attention to. A rather successful implementation of the method was found only for music education, when students study rhythm [28].

We agree with the opinion of researchers from Riga Technical University [29] that integrating the pedagogical benefits of face-to-face learning with technological solutions in the e-learning environment creates an opportunity to add value to the educational process in two ways: to reduce the overall learning time and to involve students in the learning process, taking into account their individual interests, thereby facilitating the transition from unsustainable to sustainable education.

However, a physically active type of activity cannot, by definition, fit the conditions of study at a higher educational establishment. During the experiment, it was proved that educational motivation rose higher precisely in the group where the break in the lesson was educational, albeit with gaming and digital augmented reality. A simple rest was too relaxing for the students, not allowing them to work actively and productively during the rest of the study time. Strong knowledge is directly related to the volitional efforts of students and the application of knowledge in practice [30].

Practice shows that students, representatives of generation Z, quickly decrease their concentration. It is for this reason that a session of 45–60 min does not make much sense, but this time was set long before the advent of the current generation. In universities, the situation is even more difficult due to the fact that "periods" last for 90 min (sometimes with a break of 5 min). Practice shows that few students are able to maintain attention all this time. One of the ways to overcome the students' fatigue in the classroom is to include such types of work that would relieve fatigue in the educational process. One of them may be the inclusion of game breaks in the lesson plan. In our opinion, educational breaks should not exceed 10–15 min. This is also explained by the fact that the academic hours stipulated in the curriculum for the study of the discipline "Foreign Language" by non-linguistic students are not enough. It also requires that edubreak contains not only game moments, but also should be related to the topic of the lesson.

From students' point of view the learning a language can seem just a dull routine, therefore, in order to prepare a specialist of the future, it is necessary to instill an inclination to study a foreign language, a wish to master it at a professional level [16]. In the context of using state-of-the-art technologies (in our study - augmented reality) allows you to diversify a foreign language learning with non-standard methods adding the game elements, which attracts the student, focuses on learning, which is proved by E.A. Plakhova, E.N. Kharapudko, R.R. Nurmieva [31]. Methodologists I.V. Leushina and I.O. Leushin [32] emphasize that the formation of foreign language competence necessarily includes the development of human communicative qualities with state-of-the-art information technologies.

5 Conclusion

The authors came to the conclusion that the experiment conducted proved the hypothesis that edubreaks using augmented reality made some contribution to the qualitative assimilation of language material of all three groups of the students of each group. The offered type of edubreaks helped to remove the psychological problems of students who were unsure of their knowledge and their ability to cope with the challenges of educational problems. Those were the reasons of their low cognitive motivation. The positive educational environment, created due to game pauses with augmented reality contributed to increasing the level of cognitive motivation because during the game pause, the lightweight content of the task was used.

In addition, the use of mobile technologies: reading e-mail, viewing the YouTube service, playing on a smartphone [33] and gaming augmented reality is an important part of the informatization of society associated with the informatization of education. For Generation Z students, smartphones and tablet computers are becoming a major part of their digital lives [34].

Unfortunately, for solving cognitive and educational problems, mobile technologies have not found proper application or are not used enough, but as edubreak they can be fully used in education. Mobile phones are always at hand with the student and are an excellent opportunity to improve the quality and intensity of learning.

The methodically correct holding of game pauses involves the formulation of an activity task that requires interaction from the participants in order to achieve the goal. The game pause should consist of several stages: setting a goal, activities to achieve the goal, analysis of the results. Conducting game breaks involves not only the cooperation of students to complete the task, but also the possibility of moving around the class. Students' motivation for gaming activities is provided by competitiveness, the self-realization and self-affirmation possibilities.

One should stick to the idea that edubreak expands the possibilities and communicative interaction. Mobile devices allow the student to communicate in real time with their classmates and the teacher. This communication neither replaces, nor complements and expands the communicative possibilities of each student. Such educational networking takes place in English, as the English-speaking sector dominates the Internet.

It is the use of an educational pause that opens up new opportunities for interaction in the modern educational space for the teacher and student and makes it possible to think of some other forms of work in a classroom environment. Learning a language is not only memorizing information (for example, vocabulary), but also communication, the ability to respond to uncertain situations correctly. That is why the educational pause can be considered one of the effective pedagogical tools for teaching university students.

References

1. Little, D.F., Zhang, Yu.-X., Wright, B.A.: Disruption of perceptual learning by a brief practice break. Curr. Biol. **27**, 3699–3705 (2017). https://doi.org/10.1016/j.cub.2017.10.032

2. Glas, A., Hubener, M., Bonhoeffer, T., Goltstein, P.M.: Spaced training enhances memory and prefrontal ensemble stability in mice. Curr. Biol. **31**, 4052–4061 (2021). https://doi.org/10.1016/j.cub.2021.06.085

3. Smolen, P., Zhang, Y., Byrne, J.H.: The right time to learn: mechanisms and optimization of spaced learning. Nat. Rev. Neurosci. **17**, 77–88 (2016)

4. Ebbinghaus, H.: Uber das Gedachtnis: Untersuchungen zur Experimentellen Psychologie. Duncker & Humblot, Berlin (1885)

5. Bonstrup, M., Iturrate, I., Thompson, R., Cruciani, G., Censor, N., Cohen, L.G.: A rapid form of offline consolidation in skill learning. Curr. Biol. **29**, 1346–1351 (2019). https://doi.org/10.1016/j.cub.2019.02.049

6. Dutko, Yu.A.: Generation Z: basic concepts, characteristic and current research. Probl. Mod. Educ. **4**, 28–37. (2020). https://doi.org/10.31862/2218-8711-2020-4-28-37. (In Russian)

7. Johnson, S.: Everything Bad is Good for You: How Today's Popular Culture is Actually Making us Smarter. Riverhead, New York (2005)

8. Shamis, E., Nikonov, E.: Generation Theory. Extraordinary X. Synergy University, Moscow (2017). (In Russian)

9. Papagiannis, H.: Augmented Human. How Technology is Shaping the New Reality. O'Reilly Media, Sebastopol (2017)

10. Augmented reality. https://whatis.techtarget.com/definition/augmented-reality-AR. Accessed 2 Oct 2022

11. Hollerer, T.: Augmented Reality: Principles and Practice. Addison-Wesley Professional, Boston (2016)

12. Kameneva, G.A., Bondarenko, T.A.: Pedagogical conditions for enhancing the educational and cognitive activity of students in modern conditions of informatization of education. Sci. Educ. Today **8**(4), 172–186 (2018). https://doi.org/10.15293/2226-3365.1804.11

13. Selwyn, N., Pangrazio, L., Nemorin, S., Perrotta, C.: What might the school of 2030 be like? An exercise in social science fiction. Learn Media Technol. **45**(1), 90–106 (2020). https://doi.org/10.1080/17439884.2020.1694944

14. Vaganova, O.I., Smirnova, Z.V., Kaznacheeva, S.N., Kutepova, L.I., Kutepov, M.M.: Practically-oriented technologies in professional education. In: Popkova, E.G. (ed.) Growth Poles of the Global Economy: Emergence, Changes and Future Perspectives. LNNS, vol. 73, pp. 433–439. Springer, Cham (2020). https://doi.org/10.1007/978-3-030-15160-7_44

15. Rumyantsev, M., Rudov, I.: Project activities of the chair for digital humanities and modern trends in the development of information technology. J. Siber. Fed. Univ. Humanit. Soc. Sci. **9**(7), 1668–1673 (2016). https://doi.org/10.17516/1997-1370-2016-9-7-1668-1673

16. Putistina, O.: Interaction in the concept of autonomous language learning. J. Siber. Fed. Univ. Humanit. Soc. Sci. **8**, 1919–1925 (2015). https://doi.org/10.17516/1997-1370-2015-8-9-1919-1925

17. Kotenko, V.V.: Problems and possibilities of using technologies of augmented and virtual reality in teaching a foreign language. Sci. Notes Univ. Named After P. F. Lesgaft **3**(181), 252–257 (2020). https://doi.org/10.34835/issn.2308-1961.2020.3

18. Sergeeva, N.A., Zakharova, A.N., Tyutyunnik, S.I., Rubleva, O.S.: Features of using methods and means of the augmented reality technology when teaching a foreign language. Perspect. Sci. Educ. **50**(2), 472–486 (2021). https://doi.org/10.32744/pse.2021.2.33

19. Stepichev, P.A.: Pedagogy of surprise as a new means to meet modern challenges in education. In: Bridges, pp. 80–81. Pero, Moscow (2016). (In Russian)

20. Obdalova, O.A., Odegova, O.V.: Intercultural and interlingual communication as a new reality in the context of globalization. Bull. Tomsk State Univ. Philos. Sociol. Polit. Sci. **44**, 70–81 (2018). https://doi.org/10.17223/1998863X/44/7. (In Russia)
21. Kukulska-Hulme, A.: Will mobile learning change language learning? ReCALL **21**(2), 157–165 (2009)
22. Chen, C.M., Hsu, S.-H.: Personalized intelligent mobile learning system for supporting effective english learning. J. Educ. Technol. Soc. **11**(3), 153–180 (2008)
23. Sharples, M.: The design of personal mobile technologies for lifelong learning. Comput. Educ. **34**(3–4), 77–193 (2000)
24. Yamaguchi, T.: Vocabulary learning with a mobile phone. In: Program of the 10th Anniversary Conference of Pan-Pacific Association of Applied Linguistics, Edinburgh, UK. http://www. paaljapan.org/2005Program.pdf. Accessed 2 Oct 2022
25. Yannick, J.: M-Learning: A pedagogical and technological model for language learning on mobile phones. In: Fong, J., Wang, F.L. (eds.) Blended Learning, pp. 327–339. Pearson, New York (2007)
26. Grinshkun, A.V.: Possibilities of using augmented reality technologies in teaching computer science to schoolchildren. Bull. Moscow City Pedag. Univ. Ser. Inform. Inform. Educ. **3**, 87–93 (2014). (In Russian)
27. Kelley, P.: Making Minds: What's Wrong with Education, and What Should We Do About It? Routledge, New York (2008)
28. Nikitchenko, A.E., Minev, E.M.: The use of the practice of rhythmic as part of spaced learning methods in the system of modern jazz education. Sci. Art Cult. **4**(4–1), 162–168 (2014). (In Russian)
29. Kapenieks, J., Kapenieks, J.: Spaced E-learning for sustainable education. J. Teach. Educ. Sustain. **22**(2), 49–65 (2020). https://doi.org/10.2478/jtes-2020-0016
30. Korsun, I.: The formation of learnersí motivation to study physics in terms of sustainable development of education in Ukraine. J. Teach. Educ. Sustain. **19**(1), 117–128 (2017). https:// doi.org/10.1515/jtes-2017-0008
31. Plakhova, E.A., Kharapudko, E.N., Nurmieva, R.R.: Game techniques as a method of the educational process intensification in teaching a foreign language. Human. Soc. Sci. Rev. **7**(6), 38–44 (2019). https://doi.org/10.18510/hssr.2019.769
32. Leushina, I.V., Leushin, I.O.: Foreign language and individualization of student training: realities, trends, options. High. Educ. Russia **28**(3), 147–154 (2019). https://doi.org/10.31992/ 0869-3617-2019-28-3-147-154. (In Russian)
33. Samokhina, N.V.: Using mobile technology in teaching English: the opportunity to develop traditions and searching for the new methodical models. Fundam. Res. Pedagog. Sci. **6**(3), 591–595 (2014). (In Russian)
34. Anikina, Z.: Learner autonomy as an essential component of foreign language teaching in the 21st century. Tomsk State Uni. J. **344**, 149–152 (2011). (In Russian)

Technology and Creativity in Teaching: Audiovisual Translation as a Profession and Method in Multilingual World

Ekaterina A. Samorodova[1](✉) ⓘ, Sofia A. Bakaeva[1] ⓘ, and Elena S. Zakirova[2] ⓘ

[1] MGIMO University, Prospekt Vernadskogo, 76, Moscow 11945, Russia
samorodova.ekaterina.78@mail.ru, s.bakaeva@my.mgimo.ru
[2] Moscow State Linguistic University (MSLU), 38 Ostozhenka St., Moscow, Russian Federation

Abstract. Teaching foreign languages at the university in multilingual world has become a many-sided action. As an experience shows today, this is an interdisciplinary learning process, which integrates different disciplines, educational profiles, methods, and approaches. The teaching a foreign language is considered not as a lexical and grammatical system learning, but the formation of a new professional skills in definite field with the help of new methods and means. The training an audiovisual interpreter is a multi-stage process that requires special means to achieve its goals. The most suitable means for developing the necessary skills and abilities of a future AVT specialist are available electronic learning platforms, playlists of video and audio materials. Within the framework of this study, some methods and techniques for teaching students audiovisual translation will be analyzed and demonstrated from the standpoint of the formation of creative and pragmatic competencies. Moreover, the authors of the article stated that methodic of play list and silent films is efficient to form important professional competences of AVT specialist, speaking different native languages. The authors also stated that the method of audiovisual translation had showed its efficiency in teaching students of humanitarian (non-linguistic) universities a foreign language, in the context of developing creative and professional approaches in multilingual environment.

Keywords: Multilingual world · AVT specialist · Audiovisual translation · Teaching foreign languages · Creative and professional approaches · Playlists · Silent films

1 Introduction

The modern world is filled with information. The growth of technological progress, the development of communication means gives mankind the opportunity to join the culture of another nation without leaving the screen, by watching videos, listening to various information. Can we imagine today's world without cinema, information channels, radio? Can a modern person limit himself or herself to informational, video and audio content only from his or her own country? An information and cultural vacuum could become a reality for any representative of the modern world if humanity had not learned to hear, understand and transmit information from a foreign language to its native language and vice versa. Translation activity, according to the international charter of translators, today is permanent, universal and necessary all over the world; as by making possible intellectual and material exchange between nations, it enriches their life and promotes a better understanding among people. (Charter of Translators) [1].

The translation of foreign language audiovisual content, in other words - films, information channels, commercial media, video games etc. constitutes the main task of the interpreter of the AVT (audiovisual translation). The subject and object of audiovisual translation, in turn, is an audiovisual environment, a discourse that has its own specific features. In addition to the linguistic/verbal component (i.e. text), there is an extralinguistic/non-verbal component in the AVT (i.e. facial expressions, gestures of the speech agent, time frames, emotional coloring) as well as strict compliance with the plot lines of the translated text and the original. Most often, the unit of audiovisual translation is *a film event*, *game event* (*scene*) or *informational event*, which are related to a speech event, but do not fully correspond to it, because within the framework of a film event, for example, quite often verbal speech may be absent. An example would be silent films, individual scenes and frames from films where verbal translation is not required. In addition to the film event, as AVT units can be considered video games, educational platforms, news feeds. Audiovisual translation is a complex cognitive, psychological, empirical and organizational process, the implementation of which requires special professional training of the specialist who performs it.

The audiovisual translator is the most responsible person within the framework of the created mediated communication, where he or she acts not just as a translator, as it happens in the case of simultaneous translation, but as an author for his or her listeners, which in turn have their own cultural, demographic, social and linguistic characteristics. Training in audiovisual translation (AVT) is multi-stage and sophisticated process. It is characterized by the specifics of this professional activity, since AVT combines interpretation and translation, includes fiction, journalistic and special/professional-oriented content.

To prepare a specialist of this level, a special methodological system of AVT teaching for students/undergraduates/postgraduates or graduates of specialized universities is needed, taking into account all the features of this profession/specialization. Audiovisual translation is considered as a conglomeration of several disciplines and its teaching takes place based on the general and special didactic principles, from different sciences points of view. *Apparently, the developed system of education should be based on an interdisciplinary or integrative approach* [2] within which there is an interconnection of different approaches. Such an interconnection of various systems and approaches within the framework of teaching audiovisual translation may allow us to consider AVT as a special method of a foreign language teaching.

Within the framework of this study, some methods and techniques for teaching students audiovisual translation will be analyzed and demonstrated from the standpoint of the formation of creative and pragmatic competencies. The authors also consider teaching students of humanitarian (non-linguistic) universities a foreign language by the method of audiovisual translation, in the context of creative and professional approaches.

2   Materials and Methods

The theoretical basis of this article includes the work of Russian translators and method-ologists who considered audiovisual translation and teaching it from the standpoint of activity and functional approaches: Krupnov [3], Minyar-Beloruchev [4], Khaleeva, [5]. The scientific researches show that AVT is a labor-intensive and multifaceted process. Possession of fundamental knowledge of a foreign and native languages remains the main requirement and necessity for the implementation of high-quality, correct transla-tion. It should be noted that the works of Tareva, Schlabach, Hufeisen, Bylieva, Hong, Lobatyuk, Nam, Krasnoshchekov, Rubtsova, Wang, Aronin, Krylov, Vasileva et al. [7–11] are devoted to the problems of linguodidactics, features of the language competencies formation in a multilingual world. The teaching foreign languages through technological tools is highlighted in the works by Balyshev, Bylieva, Moccozet, Chirkova, Chernovets, Zorina, Ramming, Aurora, Krylov, Vasileva, Pozdeeva Shipunova et al. [12–17]. The works of researchers of the Russian school of translation such as Vygotsky, Winter, Leontiev, Dashinimaeva [18–21] develop the concept of active and psycholinguistic approaches in AVT teaching. It should be noted the work of Talyzina, Latyshev, Retsker, Hell, Schweitzer, Chaume, Delisle [22–27] which are very important for this article due to their studying of audiovisual translator professional competencies. It is known that initially only two types (oral and written) of translation have been identified: (Komis-sarov [28] Minyar-Beloruchev [4]). One of the types of oral translation is translation by ear. Performing such a translation, the translator perceives information in the language of the original text orally and transmits it orally as well as to the recipient. The works of Minyar-Beloruchev, in which the author emphasizes the subcategories of oral (syn-chronous, sequential) translation and oral-visual translation (sight translation), while the author notes that "oral-visual translation is performed without prior reading" [4]. As for written translation, we usually mean translation in written form, i.e. in the form of a linked text of the original work intended for reading. In written translation, the translator has the opportunity to repeatedly return to the original text, while in oral translation this cannot be done. Returning to the above mentioned, we emphasize once again that all types of speech activity are involved in the AVT process; it requires the specialist to have special translation and professional competencies. According to translation researchers (Schweitzer, [29], one of the most important and relevant competences of a translator is intercultural communicative competence. This is due to the fact that creating a translated text in a foreign language or in a native language based on the understanding its mean-ing in oral or written form the translator must be guided by the principle of respect for his or her target audience taking into account its cultural and social norms. According to the international Charter of Translators, we confirm the abovementioned statement: *the correct translated content contributes to the rapprochement and understanding of nations and peoples.* The intercultural communicative competence of the audiovisual translator/interpreter, in turn, includes the following components:

– linguistic component (knowledge of the original language and the target language, including both general language knowledge, lexical and grammatical features, style, and genre features of the speech of various target audiences, the ability to spontaneously select suitable language constructions, etc.);

- pragmatic component (the ability to convey all the necessary information to the target audience, while calculating the time, while determining the desired degree of transcreation, selecting all the necessary language tools);
- discursive component (knowledge of the discourse of translation).
- It is important to note that we mean a specialized, professionally/oriented language, i.e. the translation of a film on medical topics or detective story, where professional vocabulary is widely used. In this case, we mean a film event only, but not an international professional conference, in which simultaneous or consecutive translation is done.

Undoubtedly, all the components of the intercultural communicative competence of the audiovisual translator/interpreter are closely interconnected and interact; these are implemented in AVT process, which takes place in several stages.

For the purposes of this study, we will consider the activities of an audiovisual translator on the example of preparing the translation of a feature film for dubbing and voice-over.

The first stage of the AVT is based on the study and careful work on the original text, which can be presented in the form of cut sheets (when it is a film/movie event or subtitles for soundtracks in news events). It is not always possible for an audiovisual translator to access both the written text and the audio at the same time. There are a number of legal reasons related to property rights, video rental of the film, etc. In some cases, the translator has only a video sequence, and then the specialist is faced with the task of compiling a film script. This is one of the most difficult tasks in this activity.

The second stage of the AVT is voice-over/dubbing. This stage is characterized as pragmatic and organizational one. In addition to correcting the phrases of dialogues and monologues from the standpoint of linguistic content, the length of the phrase, styling are verified. This is a very crucial moment, since the length of the phrases of the original language and the translation in many cases (not always) are very different. This is due to the linguistic, stylistic features of the language. For comparison:

- Я не могу тебе этого обещать (ru) – 12 words.

– I can't promise you that (en) – 5 words.

The length of the film in the translation should not exceed the original version, while the acceleration of the tempo of voice-over is not allowed.

The dubbing text of the film also requires very thorough preparation in terms of compliance with the idea of the film, style, genre, features of its characters and emotional coloring.

The characters' replicas, facial expressions in the translated video sequence must fully correspond to the original. It is considered that dubbed translation is a special, separate art [4]. When compiling the dubbing text, the translator is guided, along to other things, by the rules of phonetics, more precisely, prosody. The emotional perception of the film depends on the peculiarities of phonetic modeling. It would be rather strange to see a situation where the film actor articulates a closed syllable, while the dubbing actor articulates an open one. For example:

- Ура! Ура! / здорово!(ru.)

– Cool! (en.)

Such inaccuracy is common and require careful training of specialists.

The third stage of film preparation deals with the work with dubbing actors. The directors carefully select candidates for voicing roles. Thus, audiovisual translation of film events is a complex, multi-stage process in which labor groups and teams can be involved. The researchers in AVT, depending on the tasks performed, distinguish the following subcategories of audiovisual interpreter professions:

A. **The audiovisual interpreter into native language and into foreign language**
This specialist is required to have fundamental knowledge both of native and foreign languages, knowledge of the culture and socio-cultural characteristics of the target language, including the skills and abilities of audiovisual translation of any content and the performance of any translation task: from voice-over to subtitling. A translator with such training is responsible for the final version of the translated text, for transcreation as well as for transculturation (Kozulyaev [2]).

   Leonardo Bruni *(1370–1444) wrote his famous Treatise on Correct Translation in 1426. His main quote is entirely consistent with the idea of audiovisual translation. We agree with the assessment that "similar to someone after the model of a picture, paint another picture, borrow from the model both the figure, and the pose, and the forms of the whole body, intending not to do something himself or herself, but to reproduce what another has done, - so in translations, the best translator conveys all the content, and the spirit, and intention of the author of the original, reincarnates to the maximum extent possible the figures of his or her speech, posture, style and outlines of the original, trying to recreate them all"* [21].

B. **The translator of dubbing, composing the text for dubbing/dubbing director**
This profession of a translator is really considered as the most creative one. The dubbing translator creates not just a text but a picture of the film, taking into account the transcreation of the picture, the prosodic features of the dialogues, the emotional idea of the film and its stylistic coloring. In fact, the dubbing translator is also the director of the film.

C. **Subtitle translator**
This specialist has in addition to linguistic knowledge the definite special skills in working with professional programs. The subtitle translator ensures the creation of subtitle tracks in accordance with the necessary legal requirements of national and international legislation. There are rules which establish the length of a track, its duration and so on.

D. Audiovisual translator for the target audience

– Tefflo- translator/Tefflo - commentator

Most often, a specialist of this training level is engaged in compiling translation texts for people with developmental disabilities, cognitive disabilities and vision problems. It should be noted that psycholinguistic approach is used to implement such translation, since the text of the translation should be focused not only on its linguistic components, but on the characteristics of the psycho-emotional structures of the audience. The word turns into an image and creates a picture in the mind of the listener; the speech from a system of signs and sounds is transformed into a concept. One of the types of intersemiotic translation is audio description; the term "audio commentary" is used in Russia. In practice, this is the creation of additional subtitling tracks for hearing-impaired people and it is the creation of additional audio tracks with comments for visually impaired, blind people with special cognitive development. Currently, audio commentary is used in the translation of foreign performances/films/information channels into the language of the target audience due to which the listeners' perception becomes more complete.

– Audiovisual translator for children's content

A specialist of this profile, in addition to his or her basic foreign language competencies both of the native language and the target language must have deep knowledge of children's content: cartoons, video games, educational and gaming platforms. The children's animation and film industry is a special language environment that requires close attention and detailed study.

The foresaid thoughts allow us to conclude that for the training of specialists with the appropriate qualifications, certain methodological techniques and complexes are required for the formation of professional competencies at the training stage.

Some methodological techniques aimed at implementing the audiovisual translator's professional skills and competencies, particularly intercultural and special (information-organizational, strategic) competencies within the framework of interdisciplinary and creative approaches will be analyzed in this article.

3   Results and Discussion

We consider the translation text within a film event is the very important task for the audiovisual translator. When compiling the final version of the translation, the translator must follow certain rules. Moreover, the text must comply with certain requirements: sociolectism, pseudo-orality, conciseness, genre [2]. In order to teach students the pragmatic component of audiovisual translation, the authors of the article have developed and offered a set of exercises aimed at the developing of the necessary skills and using ICT. The widespread use of such methods in the teaching of foreign languages has largely justified them as very effective ones. According to the average indicator, more than 40 percent of students consider learning foreign languages using computer technology to be very productive. The modern world has received a special impetus to the development of educational platforms and e-learning tools based on new ICT methods during the pandemic. Certainly, digital education cannot and will never be able to replace the human factor and the human-to-human dialogue. Nevertheless, a skillful methodological combination of such means to achieve the set goals can become effective. In addition, the use of computer technology in the study of foreign languages is a key principle of audiovisual interpreter/translator training: this is the principle of informatization and technical support (Kozulyaev). Moreover, in didactics the following principles of AVT teaching are distinguished: *the principle of a personality-oriented nature of learning* [2], *the principle of intercultural communication* [6], *the principle aimed at the formation of the creative process, the principle of independence in learning.*

The profession of audiovisual interpreter/translator is multi-tasking, as mentioned above. Accordingly, a variety of means aimed at solving such problems should be involved in the training of such a specialist. It should be noted that overloaded academic programs in an educational institution do not always allow practicing the necessary translation skills in the classroom. A lot of time is spent for studying the theoretical disciplines as well as for professional practice, group work, which are also necessary and useful for a future specialist. In most cases, the student independently works on improving the translation skills and competencies necessary for his or her development as a specialist. The purpose of any technique is the formation and improvement of professional knowledge and skills of a specialist. We consider that the process of teaching audiovisual translation being characterized as widely interdisciplinary one with a large set of different components require the technologies aimed at the development of the students' creative abilities. It should be noted that methodical educational technologies using the Playlist have been practiced on the educational market for a long time. There are many video courses offering various programs for teaching general language competencies. As our experience shows a AVT student is a student with completely different educational intention. On the one hand, exercises and techniques intended for such students should be clearly aimed at achieving a specific task bringing certain skills to automatism, and the interconnection of these skills on the other hand.

We offer to consider the practical examples of such skills formation (Table 1).

Lesson 1

Table 1. Main goals and objectives of the lesson

The purpose of the lesson	Knowledge	Ability	Skills
The development of pragmatic/creative competence in the context of translating a passage/phrase	The lexical and grammatical components of the translated text, the meaning of words, terms (if necessary), phraseological expressions, etc	To apply the necessary constructions for translation, determine the level of transcreation	To create the translation text taking into account certain requirements: duration the text/its volume/speed of speech

The exercise is offered in the Play list on the computer screen. This can be done in the form of an audio track and text, or in the form of an image and an audio track (Table 2).

Table 2. Exercise

Original text (French) Playing time: 00.00.03	- Ah! T'es voila! Dis, donc, où tu étais ?/(Here you are ! Where have you been?)

Tasks

1 Listen carefully to the text. Determine the genre of speech.
   -The task is focused on observing the properties of audiovisual content: genre/plot.
2 Translate the text as close as possible to the meaning of the original.
3 Translate the text, expressing the emotional coloring and idea of the text. Determine the level of transcreation and prosodic.
   The exercise is aimed at the formation of a creative approach to translation, at the transmitting the property of pseudo-orality.
4 Translate the phrase keeping/not exceeding the time duration of the original text.
   The exercise is aimed at observing conciseness as the properties of audiovisual content. The text should be logical and short. Do not exceed the original text.
5 Speak your translation into the microphone (Table 3).

Table 3. Results

Original text (French) Playing time: 00.00.03	- Ah! T'es voila! Dis, donc, ou tu étais ? ?/(Here you are ! Where have you been?)
Translation text (Russian) Playing time: **00.00.03**	(students answers) A. А! Вот и ты! Где была, скажи-ка? B. Пришла! Где бродила? C. А! Вот и ты! И где же ты был?

We consider that the working with the Play list is very convenient and efficient for many reasons. First of all, it is effective training due to accessibility to the material: the student can stop and continue his or her translation at any time. Really, it is possible to work in pairs or in group with partners: in this case the text can be read by one of the participants. This type of work is quite often used in the classroom to practice consecutive or simultaneous translation. It should be noted that the text on electronic media sounds with the same intonation and speed, while a native speaker can speed up the rate of speech and change intonation, so it is best to use audiovisual means for learning purposes. Moreover, in AVT teaching it is necessary to see and hear the text at the same time. It is unacceptable to reduce intentionally the duration of a phrase by means of physically its shortening and thus distorting the meaning. The text must be read very carefully. Accordingly, electronic programs in this case are the best option for independent work. The above exercise is aimed at developing practically all the most basic skills of a AVT interpreter.

The Lesson 2 exercise will demonstrate the formation of a AVT interpreter's professional skills to compose the translation text in time (Table 4).

Lesson 2

1. Listen to the text on the screen. Mark the length of the sound/number of pronounced syllables.
2. Translate the text with a rate exactly matching with speech longitude.
3. Speak your translation into the microphone
4. Record your result in a table on the screen

Table 4. Exercise

Original text	Playing time / number of syllables	Translation text	Playing time / number of syllables
- Vas -y, ne sois pas paresseuse !	00.00.02/ 8 words.	- Давай! Не лентяйничай!/Go on! Don't be lazy!	00.00.02 / 7 сл.
-C'est à moi que tu parles comme ça? - Comment?	00.00.04/ 10 words.	- Со мной так разговариваешь?- Как ?/ Are you talking to me so?	00.00.04 / 10 сл.
-Je le jure devant Dieu, si tu fais encore un pas, entre nous, c'est fini.	00.00.08/16 words.	- Клянусь Богом! Еще один шаг и между нами все кончено! /I swear to God! One word,more and nothing between us!	00.00.04 /18 сл.

When performing this exercise, it is undesirable to speed up the rate of speech during translation. The rate of speech when reading the original text and the translated text must match but small differences are acceptable during training. It is proposed to record the results in a table for self-control at the end of the exercise.

Lesson 3

This lesson suggests the use of silent films or films without sound to train technical, pragmatic competence, taking into account the priority of the visual series over the verbal one. The student is asked to watch a certain short excerpt from the video, focusing on the facial expressions of the character, and then try to find the most suitable cue, synchronized with the movement of the lips and gestures of the hero. Thus, the student trains not only such aspects of AVT as lip-sync, prosodic, phrase length, dynamic equivalence, but also situational attention, social perception, instant character and scene mapping, creative approach to translation.

Exercise 1: Pragmatic Competence Training

1. Look carefully at the silent scene on the screen. Do you understand the emotion expressed by the character? Describe the emotional map of the scene in 1–2 words.
2. Carefully watch the silent scene on the screen again, focus on the movement of the lips, facial expressions, and gestures of the character. Find the cue that best fits this articulation. Write it down.
3. Speak out your chosen cue in parallel with the video clip (Table 5).

For this exercise, it is important to choose an episode that represents the most under-standable short situation, an emotionally colored scene, where the viewer unambiguously reads the feelings of the characters (love, anger, fear, repentance). By this way we "canal-ize" the student and constrict the student's search space for a suitable cue, focusing his attention not so much on the content part, but on the technical part.

In such cases as 1) the emotional concept of the episode is calculated (for example, the hostess's sincere gratitude to her maid); 2) the "skeleton" of the scene is deter-mined (the heroine thanks the maid and asks her to leave) then the student can focus on the practical, technical part of the task i.e. to choose a cue corresponding to facial expressions/lip movements/gestures, analysis of openness/closeness of syllables. Thus, the student trains the technique of combining text and video series, monitors the length of the phrase and pauses, works on the synchronization of speech movements in the intended original text and the translated text. We consider the guessing facilitates the development of a student's translation sensitivity, sharpening simultaneously several channels of information perception while relying only on a visual source. Since working with a silent film does not involve self-examination the voiced version this exercise gives space for creativity, search, more variations.

Exercise 2. Training of Pragmatic Competence with Self-examination

The student views the selected movie clip without sound, maps the scene and the speak-ing character, determines (as far as it is real) the emotional background of the episode, analyzes the articulation, makes an intuitive guess about the possible replica that the character says. This exercise differs from the previous one in that at the last stage it

Table 5. Examples of exercises for guessing and putting short lines on the lips

Source 1:	
The Kid (1921), w. Charlie Chaplin	
https://youtu.be/LQE0c1Zugx8 (accessed 18.08.2022)	
Studied scene time	**Possible replica**
11:35–11:36	Oh my God !
14:31–14:33	Right, papa ?
21:33–21:35	Where to put flowers ?
Source 2:	
The Haunted House (1921), w. Buster Keaton	
https://www.youtube.com/watch?v=cKOoOK09HvQ (accessed 18.08.2022)	
Studied scene duration	**Possible replica**
2:08–2:16	(female character)
	Oh, my God, I have to go!… No, please.. Please, please !
Source 3:	
Camille (La dame aux camélias) (1921), w. Rudolph Valentino	
https://www.youtube.com/watch?v=QcZTUnFOD-E (accessed 19.08.2022)	
Studied scene duration	**Possible replica:**
7:56–8:02	Thanks… You can go. Go!

becomes possible to listen to the voiced text in the original language and check your guess. Then the second part of the exercise involves searching for a suitable cue in the target language, which directly trains translation competence. When practicing this exercise, the teacher and students jointly discuss the options for replicas, the lip technique, speech toning, prosodic, pseudo-orality, localization, transcreation, analyze errors and more successful sentences (Table 6).

Exercise 3. Description of Actions: Training Audio Description

The use of selected excerpts from silent films is aimed at training the most accurate and understandable, but concise description of what is happening on the screen in a foreign language. Such an exercise develops a number of competencies required for AVT for the target audience (in particular, audio commentary) (Table 7).

In addition, silent films are recommended to be used to compose possible dialogues in a foreign language, as a training for spontaneity in plot and situationality, as well as for the development of linguistic and extralinguistic skills. Another way to develop extra-linguistic and AVT skills, as well as transcreation and transculturation training, can be, for example, the analysis of poems translation that are included in the film text. This exercise requires special preparation of the teacher and students, a thorough search for practical material, as well as additional theoretical sources, since it involves not only working out the technical details of the AVT, but also studying the cultural (in this case, literary) context. The exercise is mainly an analysis of existing translation examples.

Table 6. Exercise

Source:	The studied scene duration:	Brief description of the scene / emotional card:	Articulation analysis (number of open and closed syllables)	Possible replica of the character (*student options*):	Original character replica:	Translation of the character's remark (student options):
Le Dîner de cons (1988), Francis Veber. VFF. https://youtu.be/K-tZhMH3IiE (accessed 19.08.2022)	4:13 -4:17 00:02.90)	Le désarroi, la confusion, l'inquiétude		1) - Eh, vous savez, je ne comprends pas tout, ça va, mais.. 2) - …c'est un peu partout… 3) Excusez-moi, c'est vous qui êtes un peu partout, j'aimerais comprendre, mais…	- Excusez moi, j'avoue que je suis un peu perdu, j'essai de comprendre, mais…	- Простите, признатьс я, я немного растерян, я пытаюсь понять, но…/ I'm sorry to admit, I'm a little confused, I'm trying to understand, but … /

Table 7. Example exercise

Source	The studied scene duration
A Night in the Show (1915), w. Charlie Chaplin https://www.youtube.com/watch?v=EIOVo6DrBMQ	8:27–8:45

Description (student options)

The audience is in the hall. The couple is sitting in the first row. The man is very nervous. He can't sit on his seat. Finally, he is changing his place. The woman is angry with him

Source	The studied scene duration
The Kid (1921), w. Charlie Chaplin https://www.youtube.com/watch?v=V_M3Jw3zoPY	0:03–0:35

Description (student options)

The child is making pancakes. He is trying a little bit and he is very pleased with it. His father is lying on his bed reading a newspaper and smoking a cigarette. The boy is taking a plate with pancakes and putting it on the table. He is asking for his father to eat. But his father is going on reading. Then the boy is coming to him and taking his newspaper away

An Example of an Exercise. For analysis, let's take an episode from W. Wyler's film "Roman Holiday" (1953), where the heroine O. Hepburn quotes the beginning of the poem "Arethusa" by the English poet Shelley (Table 8).

Table 8. Exercise

Source:	The studied scene duration:	Original text (English)
Roman Holiday (1953), William Wyler	0:26:38-0:26:48	Arethusa arose From her couch of snows In the Acroceraunian mountains
Official film translation into Russian (dubbing):	**Official literary translation into Russian by Balmont:**	
Ариадна поднялась в снегу. В Акроцианиранских горах/ *Ariadne rose in the snow in the Acrocyaniran mountains*	Аретуза проснулась, На снегах улыбнулась, Между Акрокераунских гор/ *Arethusa woke up smiled in the snow, between the Akrokeraunian mountains*	

The system of the above exercises was proposed by the authors of the article to students of philological faculties with translation as the professional direction in their education.

For a certain time, 150 students of 2nd-3rd year studying in linguistic universities participated in a pedagogical experiment, the purpose of which was to form the professional competencies of an audiovisual translator through the use of technology [30].

Under the creative competence (translator) in the framework of this study is understood as a set of special, personal, professional, creative skills and abilities of future specialists in creating a new translation text using linguistic and extralinguistic means of communication in multilingual world. Translation the texts differ from each other in their content, character and coloring. The creative competence of a translator is one of the fundamental professional advantages in the modern world, as it implies a flexible approach to work, going beyond the boundaries of stereotypes, the ability to apply an innovative approach within the framework of translation, aimed at achieving the best result when working with a foreign language text.

Under pragmatic competence in the context of this article, the authors meant the totality of the necessary skills and abilities to build a translation text in accordance with certain conditions and compliance with the rules required by a given situation: discourse, target audience, time frame, and more.

For a certain time (6 months), students were asked to work independently with the system of exercises proposed by the authors of the article, with the implementation of organized control by the teacher during classroom sessions, in which the problematic aspects of the work were considered in detail. Evaluation of the results and progress was carried out on the basis of a comparative-analytical method.

The following table presents professional competencies (pragmatic competency/creative competency). It should be noted that the author's methodology and the results of the research work have been aimed at the formation of these competencies. In percentage terms, progress is indicated in the course of the research work.

Table 9. Results of experiment

Methodology	Pragmatic competency	Creative competency	Work duration
Working with playlists	23%	34%	3 months
Working with silent films	25%	56%	3 months

During the experiment, an overt progress was noted among students who regularly perform exercises according to this method. As the Table 9 shows, the average level of vocational training has increased by 23%. A positive result was achieved mainly due to the high motivation of the students and the desire to master professional competencies. It should be emphasized that AVT students noted the exercises with silent films to a greater extent. Particular attention of the students was paid to the tasks aimed at developing audiovisual commentary skills. Often the video series is not accompanied by any sound, except for the music. The viewer of the film sees everything on the screen. As mentioned above, for a special audience, which is represented by people with a complete lack of vision, it is absolutely impossible to understand the picture without sound accompaniment. It is evident that such a viewer sees the film through the audio translator's eyes. It should be noted that the number of such an audience is growing every year and such specialists are needed more. We should emphasize that audio translation or audio commentary is a special creative process that requires special training. As it has been mentioned above silent films are considered to be an interesting and effective didactic tool aimed at developing not only professional translation skills, but also creative competencies that are very necessary for specialists in this profession.

The main goal of the experiment held by the authors was to test the AVT methods aimed at forming students' professional and creative competencies. It should be noted that not only AVT students but the students of pedagogical and philological universities studying linguodidactics and linguistics took part in the experiment. Moreover, the students of legal faculties studying the international law and often acting as interpreters at international legal conferences took part in the experiment too. The students of international journalism faculties the future activities of which are often closely related to the translation of media and video materials took part in the experiment too. It should be emphasized that in the training of specialists in these areas and professions, audiovisual translation acts as a special method of teaching foreign languages aimed at the formation of professional and creative competencies.

During the experiment the students were offered the exercises with playlists and silent films. After the experiment, the students had to fill in the Table 10 and reflect in percent which method seemed to them the most effective and interesting one.

The experiment showed that students in the areas of humanitarian profile, such as methods of teaching a foreign language, journalism, philology, gave their preference to silent films. This is due to the fact that the creative process is the basis of these professions. It is important for a future teacher to master the skills of presenting a text in a foreign language and retelling it, which is very useful in this profession, since the teacher must be able to clearly and exactly explain new material to the students. A consistent description of movie heroes' actions in silent films i.e. auditory commentary helps to logically

Table 10. Professional competencies (PC), Creative competencies (CC).

Students	Playlists		Silent films	
	Professional competencies (PC),	Creative competencies (CC)	Professional competencies (PC),	Creative competencies (CC)
Students of pedagogical and philological universities	25%	15%	23%	37%
Students of international journalism faculties	25%	20%	25%	30%
Students of philological universities	25%	16%	22%	37%
Students of International law faculties	47%	23%	15%	15%

frame the narrative and describe the actions in detail. In addition, audiovisual translation of cartoons and films is also an interesting exercise for students, and a future teacher can create his or her translation since being a student of the university. The work with playlists proposed above will help the future teacher to master pragmatic competence, particularly in lesson planning, in timing, and in proper distribution of language material.

As mentioned above, the methodology is aimed at the formation of professional competencies. The profession of international lawyers is dominated by the need to work with legal documentation, negotiating in a foreign language, and interpreting conferences. An international lawyer rarely encounters creativity in his profession; he operates with facts. The methodology proposed by the authors will help the future specialist to cope successfully with the translation of audiovisual legal content while working at international conferences. Students of this profile are often offered educational films and playlists as educational material based on video documents from meetings of the UN Court, international organizations, etc. As for the students whose future profession connected with journalism deals with AVT on a daily basis, especially future international journalists, many of whom later become audiovisual translators. Accordingly, the exercises proposed by the authors of the article are interesting and effective for the students.

4 Conclusion

Teaching foreign languages at the university has become a many-sided action. As practice shows today, this is an interdisciplinary learning process, which integrates different disciplines, educational profiles, methods and approaches. The teaching a foreign language is considered not as a lexical and grammatical system learning, but the formation of a new professional in definite field with the help of new methods and means. The training an audiovisual interpreter is a multi-stage process that requires special means to achieve its goals. The most suitable means for developing the necessary skills and abilities of a future AVT specialist are available electronic learning platforms, playlists of video and audio materials. Taking into account the realities of the modern world, independent educational work often becomes a priority. Therefore, the creation of effective methodology that support and ensure such a format is the main task for foreign language teaching at the universities. The exercises and methodology developed by the authors and presented in the article will help future audiovisual interpreters/translators improve some professional skills and abilities. Teaching a foreign language to humanitarian students (except future interpreters/translators) through the techniques of audiovisual translation will contribute the formation of some professional competencies, among which creative competence is one of the very important.

References

1. Charter of interpreters. https://studfile.net/preview/12505691. Accessed 28 July 2022
2. Kozulyaev, A.V.: Teaching dynamically equivalent translation of audiovisual works: experience in developing and establishing innovative methods within the framework of the School of Audiovisual Translation. In: Problems of linguistics and pedagogy, vol. 3 (13), pp. 3–24. Bulletin of PNRPU, Moscow (2015)
3. Krupnov, V.N.: Leveled development of translator's actions in the educational process as a principle of teaching translation. In: Problems of teaching translation of the English language, vol. 203, pp. 183–198. Publishing House of the Moscow State Pedagogical Institute of Foreign Languages named after M. Torez (1982)
4. Minyar-Beloruchev, R.K.: General theory of translation and interpretation. Military Publishing Hous, Moscow (1980)
5. Khaleeva, I.I.: Fundamentals of the theory of teaching the understanding of foreign speech. Higher school, Moscow (1989)
6. Tareva, E.G.: A system of culturally appropriate approaches to teaching a foreign language. Lang. Cult. **40**, 318–336 (2017). (in Rus)
7. Schlabach, J., Hufeisen, B.: Plurilingual school and university curricula. Technol. Lang. **2**(2), 126–141 (2021). https://doi.org/10.48417/technolang.2021.02.12
8. Bylieva, D., Hong, J.-C., Lobatyuk, V., Nam, T.: Self-regulation in E-learning environment. Educ. Sci. **11**, 785 (2021). https://doi.org/10.3390/educsci11112078
9. Aronin, L.: Multilingualism in the age of technology. Technol. Lang. **1**, 6–11 (2020). https://doi.org/10.48417/technolang.2020.01.02
10. Bylieva, D., Krasnoshchekov, V., Lobatyuk, V., Rubtsova, A., Wang, L.: Digital solutions to the problems of Chinese students in St. Petersburg multilingual space. Int. J. Emerg. Technol. Learn. **16**, 143–166 (2021). https://doi.org/10.3991/ijet.v16i22.25233
11. Krylov, E., Vasileva, P.: Convergence of foreign language and engineering education: opportunities for development. Technol. Lang. **3**, 106–117 (2022). https://doi.org/10.48417/technolang.2022.03.08
12. Balyshev, P.: The stages of discourse-oriented virtual learning environment modeling. Technol. Lang. **3**, 88–105 (2022). https://doi.org/10.48417/technolang.2022.03.07

13. Bylieva, D., Moccozet, L.: Technology-mediated communication for educational purposes (in Russia and Switzerland). Technol. Lang. **2**, 75–88 (2021). https://doi.org/10.48417/techno lang.2021.03.06

14. Chirkova, E.I., Chernovets, E.G., Zorina, E.V.: Enhancing the assimilation of foreign language vocabulary when working with students of the digital generation. Technol. Lang. **2**, 89–97 (2021). https://doi.org/10.48417/technolang.2021.03.07

15. Ramming, U.: Calculating with words: perspectives from philosophy of media, philosophy of science, linguistics and cultural history. Technol. Lang. **2**, 12–25 (2021). https://doi.org/10.48417/technolang.2021.01.02

16. Aurora, S.: Natural language as a technological tool. Technol. Lang. **2**, 86–95 (2021). https://doi.org/10.48417/technolang.2021.02.08

17. Pozdeeva, E., et al.: Assessment of online environment and digital footprint functions in higher education analytics. Edu. Sci. **11**(6), 256 (2021). https://doi.org/10.3390/educsci11060256

18. Vygotsky, L.S.: Psychology of human development. Eksmo, Moscow (2005). (in Rus)

19. Zimnyaya, I.A.: Psychological analysis of translation as a type of speech activity. In: Questions of the theory of translation, vol. 127, pp.37–49. Publishing House of the Moscow State Pedagogical Institute of Foreign Languages named after M. Torez, Moscow (1978). (in Rus)

20. Leontiev, A.A.: Fundamentals of psycholinguistics, 3rd edn. Smysl, Moscow (2003). (in Rus)

21. Dashinimaeva, P.P.: Translation theory. Psycholinguistic approach: textbook. Publishing House of Buryat State University, Ulan-Ude (2017). (in Rus)

22. Talyzina, N.F.: Management of the process of knowledge acquisition. Military Publishing House, Moscow (1979). (in Rus)

23. Latyshev, L.K.: Technology of translation: textbook. Manual for the preparation of interpretors (from German). Academy, Moscow (2007). (in Rus)

24. Retsker, Y.I.: Translation theory and translation practice: essays on the linguistic theory of translation, 3rd ed. Valent, Moscow (2007). (in Rus)

25. Schweitzer, A.D.: Interdisciplinary Status of Translation Theory. In: Translator's Notebooks, pp. 20–31. MGLU Publishing House, Moscow (1999)

26. Chaume, F.: Audiovisual Translation: Dubbing. Routledge, London (2012)

27. Delisle, J.: Translation: An Interpretative Approach = L'analyse du discours comme méthode de traduction: part I. University of Ottawa Press, Ottawa, Canada, London, England (1988)

28. Komissarov, V.N.: Theoretical foundations of the methodology of teaching translation. REMA, Moscow (1997). (in Rus)

29. Schweitzer, A.D.: Translation theory: status, problems, aspects. Nauka, Moscow (1988). (in Rus)

30. Pidbutska, N., Knysh, A., Chebakova, Y., Shtuchenko, I., Babchuk, O.: Motivation behind the preference of distance education of higher education students. J. Educ. Cult. Soc. **13**(1), 201–210 (2022). 201–210. https://doi.org/10.15503/jecs2022.1.201.210

International Students' Creative Thinking Development via the TRIZ Method in a Multilingual Audience

Svyatoslava Bozhik[1]([⊠]) [iD], Ekaterina Bagrova[1] [iD], and Ekaterina Osipova[1,2] [iD]

[1] Peter the Great St. Petersburg Polytechnic University, Saint Petersburg 195251, Russian Federation
bozhik_sl@spbstu.ru
[2] St. Petersburg State University, Saint Petersburg 199034, Russian Federation

Abstract. The article validates the possibility and efficiency of improving creative thinking ability while teaching English via the TRIZ (theory of inventive problem solving) method to international students. The objectives of the current research are (a) to investigate the lingoudidactic potential of TRIZ problem-solving method in the process of teaching English to international students, (b) to describe the algorithm of international students' creative thinking development via the TRIZ method, (c) to check the effectiveness of the devised algorithm. In our research TRIZ is considered as a set of methods for solving logical tasks, consisting of the following steps: general preparation, situation analysis, questions preparation, hypotheses generation, hypotheses selection, hypotheses testing. The ambitions behind the devised algorithm are to teach in-depth analysis of the situation, to intensify critical and logical thinking in a non-standard situation, to overcome stereotypical thinking, to stimulate teamwork and communication abilities. The methodology has been tested on the second-year international students from Peter the Great St. Petersburg Polytechnic University majoring in Public Relations in the year 2021. The methods of this research include both theoretical (assessment of the key pedagogical principles of the TRIZ Method integration and generalization of facts and concepts, data analysis) and empirical ones (methodological algorithm development, a set of experiments: educational and control, evaluation of educational outcomes). We have obtained satisfactory results demonstrating that the problem-based learning algorithm devised leads to an improvement in students' creative thinking ability.

Keywords: Creative thinking · TRIZ method · Intercultural communication · International students · Foreign language learning · Multilingual audience

1 Introduction

The scale of change is increasing, and new lines of development are creating new requirements for education, characterized by high creativity. In order to fulfill the above-mentioned demand, the Russian educational system is actively renovating. The educational environment is fundamentally transforming into a creative, open, accepting, and stimulating environment [1].

© The Author(s), under exclusive license to Springer Nature Switzerland AG 2023
D. Bylieva and A. Nordmann (Eds.): PCSF 2022, LNNS 636, pp. 469–479, 2023.
https://doi.org/10.1007/978-3-031-26783-3_38

The urgency of the problem of developing students' creative thinking stems from the social functions of education, that is, "reliance on talent, creativity and initiative of a person as the most important resource of economic and social development" [1, p. 892]. The most significant requirement for educational results is to demand the number of creative competences [2].

The current educational objective is to help students find a new way of thinking to develop both the critical and creative potential required for the development of a personality [3]. These conditions are particularly remarkable in the higher education system., since "self-reliance, creativity, mobility, responsibility, ability for both personal and professional creative self-fulfillment become fundamental for any professional and indicate one's competitiveness in the job market" [4, p. 737].

A serious problem for English language teachers is the search for educational technologies that aim not only to assimilate students to some knowledge but also to develop their creative thinking abilities. One of the concepts that allows this balance to be maintained is the teaching system of the theory of inventive problem solving (TRIZ), which is based on the problem search method and independent creative activity of students. This issue is particularly relevant for teaching English to international students in a multilingual audience, who tend to suffer from the great complexity of the English syllabus and the lack of ways of its adaptation to the needs of students [5].

The purpose of the research is to design the algorithm to teach international students English based on the TRIZ method and check its effectiveness in international students' creative thinking development.

2 Literature Review

TRIZ is a systematic approach for understanding and solving any problem and a method for innovation and invention. With the help of increasing pressure for innovation in a globally expanding and competitive world, many have been driven to apply TRIZ as a method for in absolutely different spheres of science and work from engineering and design [6–8], robotics [9], biomedicine [10, 11] to problem solving and product innovation [12]. Nowadays TRIZ and its principles have also become the source of new ideas and solutions in and post-COVID-19 time [13–15]. The basis for this development and application originates from the TRIZ history. The beginning of TRIZ dates back to 1946, but it was not supported by the government. Later the experience was learnt and used in the USA. One of the most successful approaches to teaching TRIZ has been the utilization of 40 principles by Altshuller [6].

The principles of TRIZ "offer a systematic approach for problem solving, still it requires creativity abilities" [16, p. 191] Due to its potential to generate new ideas and look at the situations from an unknown angle, TRIZ has begun to be used as a tool for the development of creativity in education [17, 18]. Despite its ability to stimulate students in search of innovative solutions, TRIZ methods have the following teaching problems: (a) the advanced and complex structure is challenging for new students, (b) confusion due to multiple versions due to advances being made, (c) long time is needed for students in industry and universities to become capable, (d) disparate collection of tools which further complicates teaching further, (e) difficult to apply manually, (f) no simple process to solve problems of intermediate complexity [19].

In the field of education TRIZ is connected with problem-based learning (PBL), which has been used successfully in various situations [18, 20]. A novice starts with peculiar cases and later, when some progress has been made, will bring out by abstraction, and adopt intensively advanced principles. Contrariwise, the know-how mastering way would be paved through abstract rules and ideas to certain cases [21]. In this principle, it is essential to create a trusting communication between the teachers and the students. The mixture of modeling and problem-based learning gives students an opportunity to work with different options and knowledge in secure and reliable atmosphere. "This merger cultivates an awareness of creative thinking, critical analysis, and decision-making abilities from extrapolating and relating the theoretical and practical knowledge..." [20, p.142]. The main goal of PBL is to provide opportunities to students to use the gained knowledge.

One more well-known approach included into the TRIZ method is project-based learning. Project-based learning makes learners to study "important and meaningful questions through a process of investigation and collaboration" [22, p.274]. The main pluses of project-based learning are that students learn the main concepts via projects, students build their own approach to the curriculum, and work on their own pace.

Project-based learning and problem-based learning are quite similar in their results and in their multidisciplinary orientation. However, some differences could be noted: (a) project tasks are closer to professional reality and consequently need more time for execution than problem-based tasks, (b) the purpose of project work is more to apply knowledge, whereas problem-based learning is more aimed at acquiring knowledge, (c) likewise, project work is more self-directed than problem-based learning, since the problem does not guide the learning process to the extent that the project can [23].

To solve the tasks of the current research, we apply problem-based learning only, although project-based approach might become a part of our further investigation of the TRIZ method.

3   Methodology

Participants. Totally 22 s-year bachelor students from Peter the Great St. Petersburg Polytechnic University majoring in Public Relations participated in the study in the year 2021. The participants are representatives of different countries: China (2 students), Vietnam (4 students), Peru (1 student), Columbia (1 student), Kazakhstan (10 students), Turkmenistan (4 students). According to the curriculum international students choose either English or Russian as a foreign language. The participants studied English at home countries and came with their own proficiency level. The average level of English of the participants was from Pre-Intermediate (B1) to Intermediate (B2).

Procedure. The experimental teaching lasted for 7 weeks, four hours per week (totally 28 academic hours) and consisted of two steps: 1) the educational experiment, 2) the control experiment.

The educational experiment aims at developing creative thinking potential via the TRIZ Method in EFL classroom. The TRIZ-based training was used as the final block on the topic under study and was carried out in the following stages: 1) preparatory work – students were asked to prepare for the lesson and find information on a given question, topic (the areas of the questions were engineering, art, culture, history, law, climate, etc.); 2) work in class – students were offered a logical task to solve in a team, for this purpose students were divided into groups, received the task and discussed possible solutions, to generate as many hypotheses as possible students used a devised pattern with the algorithm steps; 3) the final stage – all groups offered their hypotheses, discussed, voted for the best one and the teacher represented the correct answer. The example of the logical task and the devised pattern are suggested below.

The control experiment is needed to check the effectiveness of the suggested algorithm and compare the level of creative thinking and interactive abilities in pre-training and post-training stages.

At the control experiment international students were asked to solve four logical tasks. 20 min per task were given to analyze the situation, generate possible hypotheses, discuss with the group and come up with the most plausible solution. The set of tasks was suggested to students twice before the TRIZ-based training (pre-training stage) and after (post-training stage). The only difference is that students at the post-training stage should use the devised pattern with the algorithm steps. At the stage of analyzing and generating hypotheses, students were writing notes in their patterns. At the stage of discussing, they worked with the hypotheses from the patterns.

To assess the changes in students' creative thinking abilities, we devised the following criteria:

(a) productivity – the number of hypotheses generated per task,
(b) originality – the number of non-standard, non-stereotype ideas and associations suggested per task,
(c) curiosity – the number of prepared and asked questions per task,
(d) accuracy – the presence of a correct hypothesis or a hypothesis logically close to the correct answer.

The average rate of productivity, originality, curiosity, and accuracy was calculated by formula:

$$K = \frac{N1}{N2} \tag{1}$$

$$X = \frac{(K1 + K2...K22)}{n} \tag{2}$$

K – the average rate of productivity/originality/curiosity/accuracy of one student,
N1 – the number of ideas,
N2 – the total number of logical tasks, offered to students.
X – the average rate of the productivity/originality/curiosity/accuracy,
n – the number of students in a group.

To assess the changes in international students' interactive abilities we devised the following criterion: interactivity – the ability to teamwork, to listen to interlocutors, to express ideas, to develop other team members hypotheses, to argue.

To measure the average rate of interactivity we distinguish 4 levels: (a) silence (non-interaction), (b) low level (partial interaction), (c) medium level (average interaction), (d) high level (intensive interaction). Students get 0, 1, 2, 3 points for each of the levels accordingly. The number of students with abovementioned points gives the percentage ratio of students in a group owning one of the interactivity level.

Methods. The following methods have been used in our research: a) methods of a theoretical analysis (assessment of the key pedagogical principles of the TRIZ Method integration and generalization of facts and concepts, data analysis); b) empirical methods (methodological algorithm development, a set of experiments: educational and control, evaluation of educational outcomes); c) a tabular and graphical presentation of information.

Materials. The theory of inventive problem solving or TRIZ is a set of methods for solving technical problems. Nowadays it includes the solving of logical tasks and is widely used for educational purposes. The classical TRIZ algorithm is a complex tool that requires special skills. A simplified version adopted for discussions in EFL classes is offered to international students. It consists of the following 6 steps: (1) general preparation, (2) situation analysis, (3) questions preparation, (4) hypotheses generation, (5) hypotheses selection, (6) hypotheses testing. The logical tasks are chosen and designed with the following purposes: to teach in-depth analysis of the situation, to intensify critical and logical thinking in a non-standard situation, to overcome stereotypical thinking, to stimulate teamwork and communication abilities. Here is an example of a logical task.

Read and analyze the situation. Follow the steps in the pattern and be ready to present your hypotheses: On Manfred Mann's Earth Band album envelope «Watch» there is a man running down a runway with his arms outstretched. Despite his pose, good perspective photos of the airfield, the taking off effect is not so evident.
Question: How to enhance the feeling of taking off?
Key: The shadow of a running man has the shape of an airplane.

4   Results

4.1   The Results of the Educational Experiment

At the educational experiment stage, international students were trained to solve logical tasks via the suggested algorithm with the help of a devised pattern. We would like to mention that to solve a logical problem students had to turn to knowledge from other scientific fields, conduct independent work on finding the necessary information, which allowed them to consolidate the material on the topic, expand their understanding of the topic under study, diversify lexical units, it allowed to have a conversation on the topic when discussing answer options and generating ideas. This method contributes not only to the development of creative thinking but intercultural communication, the ability to speak a foreign language, to increase erudition, in-depth knowledge of the discipline through searching and working with information, the development of logical thinking and teamwork. On the other hand, we understand that the number of international students taking part in the educational experiment was not so large and we could not extend the measured criteria. It seems to us our further investigation on the topic.

Here is an example of the pattern and step-by-step analysis of the abovementioned task conducted together with the international students. It includes description of steps and verbatim students' answers (Table 1).

Table 1. The algorithm of generating hypotheses via TRIZ method

Steps	Algorithm	Answers
Step 1	**General Preparation** Read the terms of the task. Rephrase the problem conditions in your own words and write them down.	**You are Given:** Manfred Mann's Earth Band album envelope, on the cover a man runs down the runway with his arms outstretched **You should Find (Explain):** how to create a feeling of take-off, separation from the ground, weightlessness, lightness
Step 2	**Situation Analysis** *Analyze the conditions of the task:* 1. What is the main object in this task? What parts or elements does it consist of? 2. What objects are located around the main object? What objects and how does it interact with? 3. What processes are taking place in the object itself, with its participation and around it?	**Situation Analysis** *Analyze the conditions of the task:* 1. Album cover. Airfield. The runway. A man on the runway. 2. Air. Space. Sky. Horizon. Planes. The sun. 3. A man is running along the runway; his arms are spread out like an airplane. There is a perspective.
Step 3	**Questions Preparation** Questions I can ask myself and my interlocuters concerning the task, its conditions, etc.	What's wrong with the pose? Why is there no feeling of a flight? What prevents you from creating this feeling? How do photographers usually create a feeling of a flight, air in a photo or picture?
Step 4	**Hypotheses Generation** Think about how the phenomena listed below could contribute to obtaining the desired result. Formulate hypotheses. *List of phenomena:* mechanical, acoustic, thermal, chemical, electrical, magnetic, optical, nuclear, biological, social.	**Hypotheses** 1. Put wings on your hands 2. Change into a bird 3. Change into a plane 4. A man jumps up and is photographed in the air, in a jump 5. A man jumps with a parachute and is photographed landing on the strip 6. A flock of birds is flying in the sky with wings spread like a man's 7. A plane takes off from the runway, a man runs after him with outstretched arms, repeating the movement of the plane

(continued)

Table 1. (*continued*)

Step 5	**Hypotheses Selection** Select the most plausible hypotheses and arrange them in descending order of plausibility	**More plausible hypotheses** ✓ hypotheses 6 and 7 are more plausible, synchronicity of movement can create a feeling of a flight ✓ hypothesis 4 can be plausible because a person will not touch the ground, there is a feeling of a flight, movement in the air ✓ hypothesis 5 is less plausible because the parachute will be an extra element, in this case the person lands, not takes off ✓ hypotheses 1-3 are implausible, will create frivolity, makes the photo heavier **Less plausible hypotheses**
Step 6	**Hypotheses Testing** Suggest experiments, including mental ones, to test each plausible hypothesis or perform appropriate calculations. (If possible) You can surf the Internet and find the right answer.	Calculations in these conditions are unclear how to perform, but a search on the Internet gave the result, the students found the album cover and the author's idea – the shadow of a man was made in the form of an airplane taking off.  Hypothesis 7 was close to being solved.

4.2 The Results of the Control Experiment

The results of the experimental learning via the TRIZ Method and control experiment are presented in the graphs below. The results of changes in international students' creative thinking abilities (Fig. 1) illustrate the growth of all criteria in post- and pre-training stages that let us talk about the positive change in creating thinking development. The results of the interactivity level change (Fig. 2) also illustrate the positive dynamics and improvement.

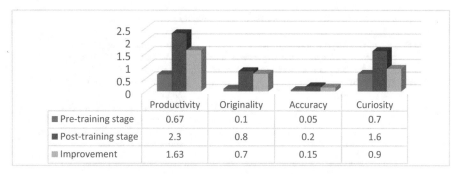

Fig. 1. The creative thinking abilities changes

Fig. 2. The interactivity level changes

5 Discussion and Conclusion

Training international students using the TRIZ method helps increase student creativity. TRIZ is a method which can be applied to the theory of creativity for the following conditions: (a) being systematic, (b) taking the lead for the placing of the ideal solution in very wide solution space, (c) being repeatable and reliable, (d) reaching creative information, (e) adding creative acknowledge [24].

Creativity associated with the TRIZ method was measured in other research studies. For the assessment of the use of TRIZ, Livotov [25] developed an approach to assess TRIZ's efficiency in process intensification and by analyzing compliance with objectives, the impact of the secondary problems, and the diversity and efficiency of ideas generated as a solution. Chang [26] showed the advantages of using TRIZ in a study of 121 university students, improving their creativity and developing and implementing better new ideas. Belski [27] found that self-efficacy in solving problems between the units used by TRIZ and a conventional engineering unit improved. The documented advantages of using TRIZ justify its use for education to engineering students.

However, our research has proved that the TRIZ method, originally created for the training of technical specialists, can be successfully applied in the humanitarian specialties in language training of international students in a multilingual audience and is considered as a universal unique tool for creative thinking development. Moreover, simultaneously with the language development, this method helps remove the cultural border between different countries representatives while working in a team at one logical task suggesting different ideas, knowing each other better. In addition, we should mention that a task content allows giving information on various topics including historical and cultural facts that also gives a chance to a multilingual audience to learn more about each other's cultural background [28].

The priority goal of modern education is not only the transfer of knowledge, skills and abilities from a lecturer to a student, but the full-fledged formation and development of the student's abilities to independently outline the educational problem, formulate an algorithm for solving it, control the process and evaluate the result [29]. This research focuses on the implementation of the TRIZ method in foreign language training and proves that the problem-based learning algorithm devised leads to an improvement in students' creative thinking ability. Furthermore, it suggests the idea of using the algorithm with the help of digital technologies [30, 31]. The present findings might help motivate both teachers [32, 33] and the students, let them look for the knowledge they required, and let them be proactive in the education process. Future research should consider the potential effects of the project-based approach in the TRIZ method on the development of students' creative thinking.

References

1. Barysheva, T.A., Gogoleva, V.V., Zyabkina, T.F., Maksimova, E.V.: Development of student's creativity by means of reflective technologies in educational information environment. In: Anikina, Z. (ed.) IEEHGIP 2022. LNNS, vol. 131, pp. 891–903. Springer, Cham (2020). https://doi.org/10.1007/978-3-030-47415-7_96
2. Kuzminov, Y., Frumin, I.: Russian education – 2020: model of education for economics based on knowledge. In: Proceedings of the IXth International Scientific Conference "Modernization of Economics and Globalization", pp. 32–64. SU HSE, Moscow (2008)
3. Lubart, T., Zenasni, F.: A new look at creative giftedness. Gifted Talented Int. **25**, 53–57 (2010). https://doi.org/10.1080/15332276.2010.11673549
4. Sigal, N.G., Linyuchkina, E.G., Plotnikova, N.F., Zabolotskaya, A.R., Bagmanova, N.I.: Academic environment for the development of creative fulfillment of innovative students. In: Anikina, Z. (ed.) IEEHGIP 2022. LNNS, vol. 131, pp. 737–744. Springer, Cham (2020). https://doi.org/10.1007/978-3-030-47415-7_78
5. Stroganova, O., Bozhik, S., Voronova, L., Antoshkova, N.: Investigation into the professional culture of a foreign language teacher in a multicultural classroom from faculty and international students' perspectives. Educ. Sci. **9**(2), 137 (2019). https://doi.org/10.3390/educsci90 20137
6. Altshuller, G.: 40 Principles: TRIZ keys to innovation. Technical innovation center, Inc. (2002)
7. Sharaf, H.K., Ishak, M.R., Sapuan, S.M., Yidris, N.: Conceptual design of the cross-arm for the application in the transmission towers by using TRIZ–morphological chart–ANP methods. J. Mater. Res. Technol. **9**(4), 9182–9188 (2020)
8. Canbulut, F., Demitras, B.: Using TRIZ and brainstorming in design: with the case study of a bed base scissors lift system. GU J. Sci. Part C **7**(3), 614–626 (2019)

9. Wan, H., Weng, S., Du, H., Dong, D., Wang, B., Yu, T.: Innovative design of compact heavy-load independent transfer device for nuclear engineering system. Hindawi Sci. Technol. Nucl. Installations **2022**, 5256808 (2022). https://doi.org/10.1155/2022/5256808

10. Chen, S., Kamarudin, K.M., Yan, S.: Analyzing the synergy between HCI and TRIZ in product innovation through a systematic review of the literature. Hindawi Adv. Hum. Comput. Inter. **2021**, 1–19 (2021). https://doi.org/10.1155/2021/6616962

11. Tan, D.W.H., Ng, P.K., Noor, E.E.M.: An assimilation of TRIZ in dissecting the statistical outcomes of tactile sensitivity, pinch force and endurance among elderly people. Cogent Eng. **8**(1), 1891710 (2021). https://doi.org/10.1080/23311916.2021.1891710

12. Dwh, T., Pk, N., Eem, N.: A TRIZ-driven conceptualisation of finger grip enhancer designs for the elderly (2021). https://doi.org/10.12688/f1000research.51705.1

13. Shao, P., Tan, R., Peng, Q., Zhang, L., Wang, K., Dong, Y.: Problem-solving in product innovation based on the cynefin framework-aided TRIZ. Appl. Sci. **12**, 4157 (2022). https://doi.org/10.3390/app12094157

14. Chang, D.-S., Wu, W.-D.: Impact of the COVID-19 pandemic on the tourism industry: applying TRIZ and DEMATEL to construct a decision-making model. Sustainability **13**, 7610 (2021). https://doi.org/10.3390/su13147610

15. Wang, C.-N., Tran, K.-M., Huang, C.-C., Wang, Y.-H., Dang, T.-T.: Supporting luxury hotel recovered in times of COVID-19 by applying TRIZ method: a case study in Taiwan. Systems **10**, 33 (2022). https://doi.org/10.3390/systems10020033

16. Bertoncelli, T., Mayer, O., Lynass, M.: Creativity Learning Techniques and TRIZ. Procedia CIRP **39**, 191–196 (2016). https://doi.org/10.1016/j.procir.2016.01.187

17. Wessel, W., Wits Tom, H.J., Vaneker Valeri Souchkov.: Full immersion TRIZ in education. In: Rizzi, C. (ed.) TRIZ Future Conference 2010, Proceedings of the TRIZ Conference, pp 269–276. Bergamo University Press, Bergamo (2010)

18. Nakagawa, T.: Education and training of creative problem solving thinking with TRIZ/USIT. Procedia Eng. **9**, 582–595 (2011). https://doi.org/10.1016/j.proeng.2011.03.144

19. Coates, D.: Tech 61095/33095, TRIZ: Theory of Inventive Problem Solving, Kent State University (2007)

20. Murphy, S., Hartigan, I., Walshe, N., Flynn, A., O'Brien, S.: Merging problem-based learning and simulation as an innovative pedagogy in nurse education. Clin. Simulat. Nurs. **7**(4), e141–e148 (2011)

21. Sire, P., Haeffelé, G., Dubois, S.: TRIZ as a tool to develop a TRIZ educational method by learning it. Procedia Eng. **131**, 551–560 (2015)

22. Frank, M., Lavy, I., Elata, D.: Implementing the project-based learning approach in an academic engineering course. Intern. J. Tech. Design Educ. **13**(3), 273–288 (2003)

23. Mills, J., Treagust, D.: Engineering education - is problem-based or project-based learning the answer? Australas. J. Engin. Educ. **3**(2), 2–16 (2003)

24. Ekmekci, I., Koksal, M.: Triz Methodology and an application example for product development. Procedia Soc. Behav. Sci. **195**, 2689–2698 (2015)

25. Livotov, P., Mas'udah, Chandra Sekaran, A.P.: On the Efficiency of TRIZ Application for Process Intensification in Process Engineering. In: Cavallucci, D., Guio, R., Koziołek, S. (eds.) Automated Invention for Smart Industries. TFC 2018. IFIP Advances in Information and Communication Technology, vol. 541, pp. 126–140. Springer, Cham (2018). https://doi.org/10.1007/978-3-030-02456-7_11

26. Chang, Y., Chien, Y., Yu, K., Chu, Y., Chen, M.: Effect of TRIZ on the creativity of engineering students. Thinking Skills Creativity **19**, 112–122 (2016). https://doi.org/10.1016/j.tsc.2015.10.003

27. Belski, I., Baglin, J., Harlim, J.: Teaching TRIZ at university: a longitudinal study. Intern. J. Eng. Educ. **29**, 346–354 (2013)

28. Almazova, N., Rubtsova, A., Eremin, Y., Kats, N., Baeva, I.: Tandem language learning as a tool for international students sociocultural adaptation. In: Anikina, Z. (ed.) IEEHGIP 2022. LNNS, vol. 131, pp. 174–187. Springer, Cham (2020). https://doi.org/10.1007/978-3-030-47415-7_19

29. Osipova, E., Bagrova, E.: Corpus linguistic technology as a tool to improve creative thinking in the interpretation of English language idioms. In: Bylieva, D., Nordmann, A. (eds.) PCSF 2021. LNNS, vol. 345, pp. 948–962. Springer, Cham (2022). https://doi.org/10.1007/978-3-030-89708-6_76

30. Bylieva, D., Krasnoshchekov, V., Lobatyuk, V., Rubtsova, A., Wang, L.: Digital solutions to the problems of Chinese students in St. Petersburg multilingual space. Int. J. Emerg. Technol. Learn. **16**, 143–166 (2021). https://doi.org/10.3991/ijet.v16i22.25233

31. Bylieva, D., Nordmann, A.: Technologies in a Multilingual World. Technol. Lang. **3**(3), 1–10 (2022). https://doi.org/10.48417/technolang.2022.03.01

32. Almazova, N., Eremin, Y., Kats, N., Rubtsova, A.: Integrative multifunctional model of bilingual teacher education. In: IOP Conference Series Materials Science and Engineering, vol. **940**, pp. 012134 (2020). https://doi.org/10.1088/1757-899X/940/1/012134

33. Rubtsova, A.V, Almazova, N.I., Bylieva, D.S., Krylova, E.A.: Constructive model of multilingual education management in higher school. In: IOP Conference Series Materials Science and Engineering, vol. **940**, pp. 012132 (2020). https://doi.org/10.1088/1757-899X/940/1/012132

A Communicative Approach for Foreign Language Learning via Social Media

Yulia Petrova[✉] [iD]

Rostov State University of Economics, Bolshaya Sadovaya, 69, Rostov-on-Don 344002, Russia
julia-pp@yandex.ru

Abstract. Our research focuses on the meaning of social media platforms and social messengers for foreign language learning, on the example of English. The urgency of the issues of education on the whole and the theory and practice of language learning, including related to the current trends of global changes, increasing the role of the educational sphere, applying new methods of teaching and modern technologies. This is a decisive international trends in its development, which comes down to the processes of transfer of knowledge, on which the educational system relies, accordance with the realities of our days. This kind of reflection allowed us to make a hypothesis that social media platforms are a powerful resource for language learning for both personal and professional purposes. Which gives a strong motivation, active students participation and engagement in educational process, involving both teachers and students into multilingual technology environment. The study consists of two parts. The first part is a systematic review of the scientific literature focused on social media platforms as an integral part in English learning among students. The second part of the study contains a conclusion drawn based on an analysis of social media platforms users' preferences for foreign language learning. Social media platforms are presented in the article as an additional resource for foreign language learning for general and professional purposes, motivating teachers and students take actively participate and engage in a multilingual technology environment.

Keywords: Social media platforms · Multilingual technology environment · English language · Second learning language · Language barrier

1 Introduction

The language barrier is a problem for people all over the world. But even if languages are learned from early childhood, language barriers with native speakers may still remain. Immersion in the language environment (study, work, courses) promotes the acquisition of conversational skills for several months. Why not link learning via social media platforms for breaking the barrier from the beginning of the study without immersing yourself in the language environment of another country.

Information and communication technologies have eliminated communication barriers, transformed forms of communication, created a global virtual community. The list of Internet phenomena, which includes elements of online learning and teaching modified by technology, is long indeed. It is apparent that social media platforms, being part of the innovation process in modern society, have had and continue to have an impact on people's foreign language learning, especially among young people.

The widespread use of smartphones, iPhones, laptops, tablets, and PCs in the learning environment in a multilingual communicative space to integrate linguistic diversity for students in both linguistic and non-linguistic specialties has been suggested as a useful tool in the student learning process. Students participate in publication activities in international scientific journals, have classes with native language teachers, using Skype, Zoom – video conferencing programs etc. As education and technology interact, subject matter experts share their knowledge and experience. Their blogs and social media channels are numerous and increasing. Students successfully use Multilingual Technology Environment (MTE), they create, attend course in social media platforms, make video presentations in YouTube. Also, many technologies have been developed to provide free translation software programs as: Google Translate, Yandex Translate, Promt, Reverso Context etc., which give an extra opportunity for students to plunge into virtual reality multilingual environment, for improving different forms of written and oral communication. Some of these programs use electronic systems based on text pattern recognition algorithms, others can translate the text using speech recognition technology [1].

In recent years, after numerous Covid-19 lockdowns, the transition to a distance learning format, with the use of the latest modern technologies have transformed education from full-time to distance learning [2]. This type of transformation allowed students to access and send information regardless of location and time through various virtual learning platforms such as Moodle, iSpring Learn, Teachbase, etc. using modern information and communication technologies [3] within the modern communication paradigm of multilingualism [4].

In the research, we conducted a social survey to determine the role of ICT in multilingual learning via social media platforms on the example of English as a foreign language (EFL). This allowed us to build a framework model using modern social messengers as: WhatsApp, Telegram, and VK Messenger to receive and share information and instructional materials for teachers and students. This study confirms the possibility of social media platforms to integrate the linguistic diversity as a useful tool in the learning process.

2 Research Methodology

For our study, a systematic review of research on technology in teaching EFL was conducted. We believe the choice of a systematic review of studies is appropriate because it aims to present a comprehensive synthesis of the scientific literature and to demonstrate the relevance of the topic as modern innovations deliver the necessary high-quality learning outcomes [5].

The use of technology in curriculum development creates a personalized experience in teacher-student interaction [6] to increase students' real use of technology in language learning [7], digital media combining interactive language functions and archive writing functions [8], and spoken forms, because language is more than just grammatical patterns [9]. Regardless of the mode of instruction, students themselves value the usefulness and ease of use of technology in English language learning [10].

Social media is free and easy to use, giving teachers and students the ability to communicate, create, share information and learn. As a result, social media platforms in day-to-day learning become an integral part of the methods used in the learning process [11]. Researchers consider changes in digital integration in using of the Internet to access authentic materials, communicate with other students or native speakers of English [12]; in using of open courses on online platforms: Coursera, Udemy, EdX, MasterClass, etc., with the goal of continuing to study the material in video format [13, 14]; in using of video games as supplemental texts in EFL classes to support reading learning [15]; in using of mobile learning applications where the motivation to EFL learning is much higher than in traditional learning [16, 17]; in using of modern ICT to create video content for students in a specific thematic environment in English to complete the tasks set by the teacher [18]; in using smart learning environments with ICT tools for online English teaching, such as: YouTube [19], where students can watch different learning videos; in using social media platforms in EFL learning by creating Facebook groups, which are even more preferred by students than face-to-face groups, and show a higher level of student involvement and motivation [20]. The visualization provided by social media to users enhances group collaboration among students and teachers, which ultimately leads to better academic performance [21].

However, it should be noted, that several years ago, not all teachers had the desire to use social media platforms in their work with students [22]. The reason for this could be that teachers lack the competence to use new technologies for learning and teaching [23]. Today, after removing forced restrictions related to COVID-19, social media platforms remain a tool to improve learning. The social nets can be used to complement and improve learning process, becoming a powerful resource that promotes education, especially when students are engaged in higher order thinking in EFL learning. Such a mixed form of learning is now becoming an integral part of higher professional education and is a priority for universities around the world. The main advantage of this approach is the possibility to save time and achieve better results than expensive face-to-face methods, which are tied to the place and time of practical classes and lectures. Blended learning offers a wide range of flexibility and allows students to access materials from any location and at any time, becoming a powerful resource for advancing education, especially when students are thinking along with higher-level learning EFL [24].

Modern education is both effective and productive when using technological applications, MTE in education by modern social messengers is considered by us in the educational system in the context of a social super-system, where new arguments about the effectiveness of technology in the learning process and the effectiveness of interaction as teachers with students foreign language learning and between the students themselves, leading not only to a productive exchange of information, but also to a rapid transfer and processing of the studied material. As education and technology are interconnected, subject matter researchers have gained unlimited opportunities to share their knowledge and experience by creating blogs and social media channels. Today it is very numerous and diverse, which can only cause interest in EFL learning and thus stimulate academic achievement.

Rethinking the concept and role of language as the collective macro-bases of understanding between individuals [25], get people interested in foreign language learning. The collective representation proposed by E. Durkheim [26] is considered by us as the communication between members of certain groups and communities, for example, students (foreign language learning) and university teachers using modern social messengers such as WhatsApp, Telegram and VK Messenger. In this environment, verbal and visual organization is done via texts, images, audio and video files, which creates a MTE. All these factors have allowed us to create a framework model (see Fig. 1).

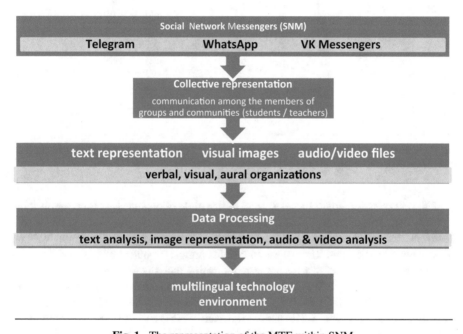

Fig. 1. The representation of the MTE within SNM.

Our study also used a data collection method that was conducted using a link to an electronic social survey and sent to bachelor students via messengers during September 2022. Students who agreed to take part in the survey filled out online questionnaires. Students were informed that their participation was voluntary and confidential and that they had the right not to participate in the survey. Out of 168 participants, 149 completed the questionnaire in full by selecting the appropriate answers from the three questions provided. There was no time limit for completing the questionnaire. Storing filled-in questionnaires was no longer than five days.

3 Survey Results

(See Fig. 2).

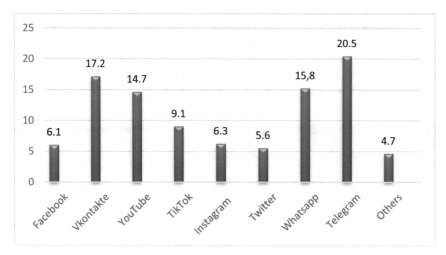

Fig. 2. The answers of the respondents on the question: What kind of social media platforms do you use most of all for improving EFL knowledge?

The first question could be responded by multiple choice answer (maximum three from the list). And it showed that more than 50% of students prefer such social media platforms and social chat groups for improving EFL knowledge, as Telegram, VK and WhatsApp. YouTube and TikTok accounted for almost a quarter of the selection, the remaining social media platforms are less interesting than the top five.

Originally launched in Russia in 2006, the VK social media platforms service was an analogue to Facebook [27]. In 2022, the monthly Russian audience size reached about 100 million people, where 21,8% of users are from 18–24 years old [28]. The preference on online social media for improving EFL learning in VK was chosen by 17,2%, which is on the second place after Telegram – 20,5%.

The ninth point in the survey marked as – "others" variant, gave an opportunity to type personal choice of the student. Among 4,7% of the answers, the following social media platforms for improving EFL were marked: Babbel – the application for Android and iOS devices, provides EFL learning and 13 languages more, from beginner to advanced level. This visual dictionary allows a student to repeat a word behind a voice recorder, record it and play it in the right context [29]; Italki service that offers already developed crowdsourcing live chat rooms and text checking [30]; Lang-8 for work on written speech [31] (Fig. 3).

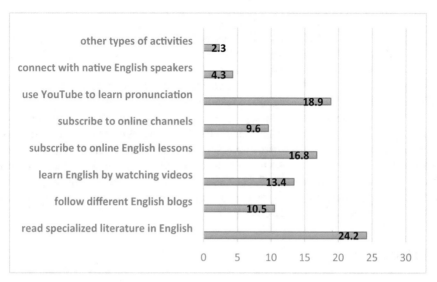

Fig. 3. The answers of the respondents on the question: What kind of social media activities help students to improve EFL?

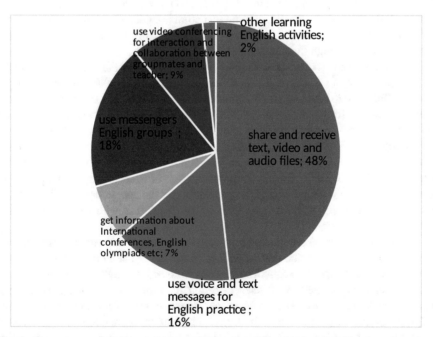

Fig. 4. The answers of the respondents on the question: What kind of activities for using group chats on messengers do find most useful for learning English in educational process?

When working independently using social media platforms, students place more emphasis on reading. To develop their writing skills, expand their vocabulary, and improve their pronunciation, are preferably selected by more students than those who prefer live conversation with native speakers.

The responses to this question show diversity in the choice of additional models for EFL learning, as there is no single zero response (Fig. 4).

Half of the surveyed students use group chats on messengers (WhatsApp, VK, Telegram), based on the answers which were on the second question and prefer sharing and receiving files, which provides quick access to the information received at any time and the ability to send a file to a classmate or teacher and vice versa. More than 40% use group chats on messengers with the purpose of language practice both in group and individually, and only 7% of respondents receive information content about upcoming events (conferences, Olympiads, competitions).

4 Discussion

Specifically, we examined how the experience of EFL online with learning chats on messengers is related to various online activities, digital skills, and the perceived usefulness of social media for language learning among students.

The results show that perceived usefulness of the Internet strongly predicts the usefulness of learning in terms of both interest and satisfaction. Interest in learning describes how students respond to a particular topic that attracts students' attention [32], and students who are more interested in learning are more likely to rate training outcomes positively and show high satisfaction.

As our research was conducted on a small scale, it therefore did not allow us to make strong generalizations based on the use of online social media platforms for language learning. Taking into account the observations made in this article, our current research opens the way for further research, especially regarding the optimal and appropriate use of social media platforms in teaching and foreign language learning. Further research with a large number of students is needed to provide more in-depth knowledge in this area. The interest in researching the use of social media for studding or improving EFL level is explained by such statistics as the percentage of users among young people in Russia, which accounts for 6.5% of the population for the age group 18 to 24 years, in total with 129,8 million active Internet users in Russia in 2022, with 106.0 million users of social media platforms [33].

We rely on statistical data to offer a microscopic model that explains the growth, dynamics of social media and our experience in the MTE. On an individual level, our research reflects the interest of modern students, who acquire information and communication technologies, the way the users distribute their interactions among groupmates in social media platforms, chat groups in messengers during classes. From the macroscopic side, the study reproduces the key topological and dynamic characteristics of social media platforms: a wide distribution of levels and types of activities for EFL learning.

The emergence and development of the Internet has influenced the intersection of social and spatial relations in the modern world, led to the birth and flourishing of social media platforms, which not only unite like-minded friends in online communities, but unite all students around the world with the same interest, wishing to know EFL. So students can watch videos, read articles, communicate with native speakers, which provide the opportunity to immerse into language environment.

Every year, the number of people who subscribe to and use social media not only for language learning, but for language teaching, – is growing. This emergence of a large number of Internet applications is the basis for a new research in this area, which does not lose its relevance.

5   Conclusion

A foreign language learning via social media promotes cohesion between individuals in education via modern social messengers and social media platforms where students can determine their own learning pace and get access to educational material from any place where there is an Internet. It is equally true to say that for many students of today's generation, the boundaries between online and offline learning are blurred. Despite the end of strict pandemic barriers, many students continue to use different social media platforms to improve and develop their language skills. Both for self-development and career advancement, but also as a resource for professional training at colleges and universities. Networks provide an online space that is certainly very convenient and understandable in using. In a communicative approach for foreign language learning via social media, there are no time restrictions for attending lectures and practical classes, no geographical limitations, no transportation costs, no rent for premises, etc.

Sharing and receiving learning materials via social media platforms and using various digital tools have shown how students can improve word comprehension in context, increase language levels using text, audio, and video files, and develop skills not only verbally but also in writing via group chats on messengers. This constant language practice certainly improves the quality of learning in general.

Learning from communication with native speakers makes it possible to hear and understand pronunciation, learn informal and everyday expressions and start speaking English without fear of making mistakes. Thus, a communicative approach seems to be the most effective way for learning languages. Social media platforms provide an opportunity to immerse students in the language environment without overcoming geographical barriers.

Learning English via social media has been shown as a method that requires efforts to organize the learning process, but it is the effective and productive cooperation between teachers and students, both for self-development and future career advancement, and also as a resource for professional training at colleges and universities. This study may be of interest to foreign language teachers, students, people who want to learn or improve their foreign language, and also have a continuation of the study in further papers devoted to the role and importance of social media platforms in EFL teaching and learning.

References

1. Gauvain, J.L., Lamel, L.: Structuring broadcast audio for information access. Eurasip. J. Adv. Signal Process. **2003**, 642019 (2003). https://doi.org/10.1155/S1110865703211033

2. Muftah, M.: Impact of social media on learning English language during the COVID-19 pandemic. PSU Rev. Res. (2022). https://doi.org/10.1108/PRR-10-2021-0060

3. Yu, Z., Yu, L., Xu, Q., Xu, W., Wu, P.: Effects of mobile learning technologies and social media tools on student engagement and learning outcomes of English learning. Technol. Pedag. Educ. **31**(3), 381–398 (2022). https://doi.org/10.1080/1475939X.2022.2045215

4. Petrova, Yu.: Meme language, its impact on digital culture and collective thinking. In: E3S Web Conference, vol. 273, p. 11026 (2021). https://doi.org/10.1051/e3sconf/202127311026

5. Serdyukov, P.: Innovation in education: what works, what doesn't, and what to do about it? J. Res. Innov. Teach. Learn. **10**(1), 4–33 (2017). https://doi.org/10.1108/JRIT-10-2016-0007

6. Costley, K.C.: The positive effects of technology on teaching and student learning. Arkansas Tech University (2014). https://files.eric.ed.gov/fulltext/ED554557.pdf. Accessed 10 Oct 2022

7. Murphy, K., DePasquale, R., McNamara, E.: Meaningful connections: using technology in primary classrooms. Young Child. **58**(6), 12–18 (2003). https://www.learntechlib.org/p/101494/. Accessed 4 Oct 2022

8. Warschauer, M., Matuchniak, T.: New technology and digital worlds: analyzing evidence of equity in access, use, and outcomes. Rev. Educ. Res. **34**(1), 179–225 (2010). https://doi.org/10.3102/0091732X09349791

9. Nesmeyanov, E., Petrova, Yu.: Cross functional effect between language and culture in the understanding of the context in the educational process. In: SHS Web Conference, vol. 70, p. 05006 (2019). https://doi.org/10.1051/shsconf/20197005006

10. Thang, S.M., Nambiar, R.M.K., Wong, F.F., Mohd Jaafar, N., Amir, Z.: A clamour for more technology in universities: what does an investigation into the ICT use and learning styles of Malaysian 'digital natives' tell us? Asia Pac. Educ. Res. **24**(2), 353–361 (2014). https://doi.org/10.1007/s40299-014-0185-2

11. Halim, A.: Importance of digital technology in English language learning. J. Linguist. Eng. Lang. Teach. **1**(1) (2021). https://jlelt.org/index.php/jlelt. Accessed 12 Oct 2022

12. Marull, C., Kumar, S.: Authentic language learning through telecollaboration in online courses. TechTrends **64**(4), 628–635 (2020). https://doi.org/10.1007/s11528-020-00488-2

13. Kleftodimos, A., Evangelidis, G.: Using open source technologies and open internet resources for building an interactive video based learning environment that supports learning analytics. Smart Learn. Environ. **3**(1), 1–23 (2016). https://doi.org/10.1186/s40561-016-0032-4

14. Lima, C., Bastos, R.C., Varvakis, G.: Digital learning platforms: an integrative review to support internationalization of higher education. Educ. Rev. **36** (2020). https://doi.org/10.1590/0102-4698232826

15. Lawrence, A.M., Sherry, M.B.: How feedback from an online video game teaches argument writing for environmental action. J. Lit. Res. **53**(1), 29–52 (2021). https://doi.org/10.1177/1086296X20986598

16. Sun, J.-Y., Chang, K.-Y., Chen, Y.-H.: GPS sensor-based mobile learning for English: an exploratory study on self-efficacy, self-regulation and student achievement. Res. Pract. Technol. Enhanc. Learn. **10**(1), 1–18 (2015). https://doi.org/10.1186/s41039-015-0024-y

17. Yu, Z., Xu, W., Sukjairungwattana, P.: Motivation, learning strategies, and outcomes in mobile English language learning. Asia-Pac. Educ. Res. (2022). https://doi.org/10.1007/s40299-022-00675-0

18. Tai, Y., Ting, Y.-L.: New aspect of technology adoption: a case study of students' self-made English-learning video. Asia Pac. Educ. Rev. **17**(4), 663–675 (2016). https://doi.org/10.1007/s12564-016-9464-3

19. Alobaid, A.: Smart multimedia learning of ICT: role and impact on language learners' writing fluency—YouTube online English learning resources as an example. Smart Learn. Environ. **7**(1), 1–30 (2020). https://doi.org/10.1186/s40561-020-00134-7

20. Akbari, E., Naderi, A., Simons, R.-J., Pilot, A.: Student engagement and foreign language learning through online social networks. Asian-Pac. J. Second Foreign Lang. Educ. 1(1), 1–22 (2016). https://doi.org/10.1186/s40862-016-0006-7

21. Saqr, M., Viberg, O., Vartiainen, H.: Capturing the participation and social dimensions of computer-supported collaborative learning through social network analysis: which method and measures matter? Int. J. Comput.-Support. Collab. Learn. 15(2), 227–248 (2020). https://doi.org/10.1007/s11412-020-09322-6

22. Conole, G., Alevizou, A.A.: Literature review of the use of Web 2.0 tools in Higher Education. Milton Keynes: The Open University (2010). http://oro.open.ac.uk/23154/. Accessed 20 Sept 2022

23. Ifenthaler, D., Spector, J.M., Sampson, D.G., Isaias, P.: Advances in cognitive psychology, educational technology and computing: an introduction to the special issue. Comput. Hum. Behav. 32, 290–291 (2014). https://doi.org/10.1016/j.chb.2013.08.018

24. El Boghdady, M., Ewalds-Kvist, B.M., Alijani, A.: A review of online platforms in training and surgical education. Eur. Surg. 51(2), 41–48 (2019). https://doi.org/10.1007/s10353-019-0569-x

25. Tada, M.: Language and imagined Gesellschaft: Émile Durkheim's civil-linguistic nationalism and the consequences of universal human ideals. Theory Soc. 49, 597–630 (2020). https://doi.org/10.1007/s11186-020-09394-1

26. Crittenden, B.S.: Durkheim: sociology of knowledge and educational theory. Stud. Philos. Educ. 4, 207–253 (1965). https://doi.org/10.1007/BF00395481

27. Kaustubha, K., Sant, A., Sant, M.A.: Understanding World Media. N.p., K.K. Publicat. 230–231 (2021)

28. Similarweb. vk.com (2022). https://www.similarweb.com/ru/website/vk.com/#overview. Accessed 12 Oct 2022

29. Nushi, M., Eqbali, M.H.: Babbel: a mobile language learning App. TESL Reporter 51(1), 110–111 (2018)

30. Saadatara, A.: Italki: www.italki.com Review. TESL Reporter 52(2), 81–84 (2021)

31. Buendgens-Kosten, J.: Blogging in the target language: review of the "Lang-8" online community. SiSAL J. 2, 92–99 (2011). https://doi.org/10.37237/020207

32. Hong, S., et al.: Complement and microglia mediate early synapse loss in Alzheimer mouse models. Scien. 6(352), 712–716 (2016). https://doi.org/10.1126/science.aad8373

33. Digital 2022 the Russian Federation. Datareportal (2022). https://datareportal.com/reports/digital-2022-russian-federation

Students' Making Test-Format Questions in English for Specific Purposes Course as Means of Improving Reading Skills

Marina S. Kogan[✉] [iD], Ekaterina V. Kulikova[iD], Elena K. Vdovina[iD], and Nina Popova[iD]

Peter the Great Saint Petersburg Polytechnic University, Polytechnicheskaya 29, 195251 Saint Petersburg, Russia

kogan_ms@spbstu.ru

Abstract. Development of reading skills in the English for specific purposes course is an important problem, which is often underestimated. To comprehend specialist texts in a foreign language students have to use different cognitive practices and have solid background and linguistic and vocabulary knowledge. The paper describes the results of the 10-week experiment on involvement of second year IT students in making test-format questions as a part of their home assignment. At the beginning of the term, the academic group with a lower average placement test score (though statistically insignificant) was selected as the experimental one for the pedagogical experiment, with the second group becoming the control one. Both groups received instruction based on the same teaching materials for the following 10 weeks. The only difference was that students in the experimental group prepared test-format questions as a part of their home assignment and worked in small teams testing each other at the end of each class. These activities resulted in improving students' reading skills and better memorization of target vocabulary items during the experiment. The final test revealed statistically significant difference in the knowledge of vocabulary and grammar structures between the experimental and control groups. The students' reaction to making test-format questions was overwhelmingly positive. The paper discusses the findings and outlines further research. The proposed technology of improving reading comprehension skills in English may be extrapolated on teaching other foreign languages in our multilingual world.

Keywords: English for specific purposes · ESP · Teaching foreign language reading · Target vocabulary items · Reading comprehension assessment · Students' making test-format questions

1 Introduction

In today's information society, reading plays a key role in the acquisition of information and knowledge. Despite the fact that a whole set of different literacies has been suggested in research literature including digital, technological, academic, and functional literacies

D. Bylieva and A. Nordmann (Eds.): PCSF 2022, LNNS 636, pp. 490–505, 2023.
https://doi.org/10.1007/978-3-031-26783-3_40

in addition to emotional, visual and communicative ones, the development of the reading literacy both in the native and in foreign languages remains a priority of educational policy of most countries [1–4]. Target language reading comprehension is a foreign language skill of particular importance, mastering professionally-oriented reading skills is considered a priority for learners at the level of undergraduate and graduate programs [1]. Intending to teach professionally-oriented reading comprehension, teachers realize the fact, though, that learners do not always possess sufficient knowledge and skills to fulfill specially developed reading tasks. Problems arising in the ESP course are often low level of language proficiency among the students of non-linguistic majors and mixed level learner groups. Also, among the factors are insufficient academic and research competence of language teachers, as well as educational standards limiting the number of hours in Master's and Postgraduate programs where ESP courses are especially in demand. Therefore, the aim of this paper is to present both a theoretical and a practical basis for the preparatory learner-centered ESP technology with the focus on reading comprehension development.

The term "technology" is currently used not only to describe digital technologies, but is also widely used in the field of education [5]. The concept of "teaching technology" in pedagogical terms is a related term with respect to the concept of "teaching method-ology", which is a set of teaching methods. The teaching technology proposed by us to improve learner's reading skills represent several definite stages of the learning pro-cess, guaranteeing the achievement of a certain standard of learning and reproducibility by other teachers if needed. The possibility of repeating our way of enhancing student reading competence may be considered to be one of the examples of the "teaching tech-nology" implementation for the sake of achieving effectiveness in terms of learning results.

To some degree the topicality of the research is supported by the importance of reading skills when a new foreign language is learnt, especially in the sphere of profes-sional language acquisition. Many researchers involved in studying transfer of reading skills in the native language and first foreign language on a third language conclude that the better reading skill is in the first foreign language the easier it is transferred on a new one especially for genetically related languages [6, p. 42]. In the modern world, mastering a few foreign languages in the lifelong learning format contributes to one's professional growth and participation in international projects [7, 8]. Experts in mul-tilingual education call for unconventional approaches to multilingual teaching future engineers.

2 Literature Review

2.1 The Role of Reading in ESP Course

According to Chmelíková, when studying a foreign language at university, the impor-tance of developing skills in reading special texts is massively underestimated [2]. Her study conducted in a number of universities in Slovakia shows that students of technical specialties read little, often superficially, and have too low level of linguistic knowledge to be able to understand domain-specific foreign language texts. In addition to background knowledge of the topic and linguistic knowledge of predicting, making connections, visualizing, inferring, questioning, and summarizing, the author suggested a number of prerequisites of more successful reading, such as knowledge of different reading models and strategies, concluding that the efficiency of reading domain-specific texts can be improved in a relatively short time – two months in her study [2, pp. 65–66].

According to Korotkina, any English for Special Academic Purposes (ESAP) course should be focused on academic skills, which are based on academic reading and writing competencies [1, p. 129]. Justifying her point of view, the author argues that for the students who study in their native language environment, the speaking and listening skills, though important, are secondary in comparison with the importance of reading skills. The author argues that education at a modern university is impossible without processing a large amount of specialist literature in a foreign language [1, p. 132]. Dealing with authentic texts related to their major allows students to solve many problems associated with writing, since it facilitates understanding of academic writing discourse, written text structures, correct use of references, etc., thus familiarizing with the main international academic writing conventions. The researcher favors students working in groups, firstly, because students demonstrate a wide range of different cognitive practices and background knowledge which they can share in their collaboration and, secondly, their cognitive capacities may not correlate with their level of language proficiency benefiting group work [1, p. 133].

One more important component of reading is emphasized in 2018 year research in which the authors point out that reading, although classified as a receptive skill, initiates active, productive, and constructive mental processes [5, p. 246]. They refer to the ESP university course for graduate students in Arts with a low level of German language fluency, students work with authentic texts from the very beginning. Students are given tasks to recognize and explain vocabulary and grammatical constructions such as word order, functions of relative clauses, etc. and later work on their own consolidating their knowledge of these grammatical items. The researchers maintain that the lack of attention to the grammatical structures and lexical details can prevent students from understanding domain-specific texts when reading them on their own. Therefore, the teacher in the classroom should facilitate the process of constructing students' own knowledge using a deductive method alongside with the use of active forms of pair and small group work [9, p. 247]. Moreover, Burdumy and Bohlander emphasize that students should reach a threshold of understanding grammar structures corresponding with the high-level cognitive skills such as "Analysis" and "Synthesis" in Bloom's taxonomy [10]. Otherwise they will be unable to read special texts effectively on their own in the future. Researchers also find it important to address pronunciation in courses aimed at developing reading skills, because when reading in a foreign language, students pronounce the words. Due to the discrepancy between the morphological and phonological systems of the English language, which has deep or opaque orthography [11], the pronunciation of unfamiliar words must be given attention in the classroom, including reading aloud activity [9, p. 247].

Most researchers believe that knowledge of vocabulary and syntax plays a key role in teaching foreign language reading [4, 12, 13]. Recently, attempts have been made to assess the vocabulary size for effective independent reading of the texts in different genres. For English, the following estimates are given: an optimal threshold, which is the knowledge of 8,000 word families yields the coverage of 98% and a minimal threshold of 4,000–5,000 word families results in the coverage of 95% [14].

2.2   Reading Assessment Tests

Standardized tests such as the Stanford Diagnostic Reading Test [15] are used to assess reading skills. The Stanford Diagnostic Reading Test, which has sections on comprehension, vocabulary and scanning, is widely used in US secondary schools to measure the reading skills of low achievers. Another example of a standardized test used to assess reading skills is the English part of the Psychometric University Entrance Test used in Israel [14]. Its goal is to predict the academic success of university applicants, and English is the language of academic texts. The English part of the test includes about 60 questions which tap the learners' comprehension of academic English. Some questions focus on understanding words or sentence structure, others on understanding global textual information, both explicit and implicit.

In European universities, preference is given to the IELTS test, which simultaneously measures the development of language and meta-language/linguistic skills [16, p. 108]. At the moment, the IELTS English language assessment system is considered to be one of the most objective diagnostic methods and is used as a means of monitoring the formation of language and meta-language skills within the discipline "English language" and "English for Academic Purposes" in a number of Russian universities. Murtazina and Petkova conducted a study of the section which tests reading skills in the IELTS test consisting of four groups of tasks which correlate with various levels of Bloom's taxonomy from lower- to higher-order thinking skills [16, p. 113]. The first group includes the tasks aimed at finding particular information in the text, which correspond to the lower levels of knowledge, understanding and recognition. The second group includes true/false tasks which involve the analysis of the data. The third group includes tasks aimed at generalizing information, based on the analysis and synthesis. The fourth group of tasks is aimed at data visualization, and the implementation of these tasks involves the mechanisms of visual thinking.

Having analyzed the results of the test done by 204 participants of the study, the authors concluded that the most difficult tasks are those which they ascribed to synthesis, the sixth and the highest level in Bloom's taxonomy, for example, tasks to match headings and paragraphs or tasks to fill academic abstracts with the words from the list. In the latter type of tasks participants made 556 mistakes [16, p. 119].

In the 2016's review of studies on the effect of self-questioning in developing reading comprehension in secondary schools, the authors indicate that, in addition to standardized tests, various tasks are used to assess reading comprehension, most typical ones being multiple choice, true/false, short answer, fill in the blank, and maze assessments in which every seventh word in a reading passage is deleted and replaced with three word or word collocations choices. In addition, a commonly used method of checking text comprehension is the formulation of the central idea of the text and some supporting details [15, pp. 158–159].

2.3 Students as Test-Makers

The idea of involving students in developing test items is not new. The first publications appeared in the 1980s and they described positive results of this activity in enhancing student motivation and performance across disciplines. At the same time, other studies, for example, published by Merrill Swain demonstrated that traditional pre-/post-testing was insufficient for determining what students have really learned due to the teacher's intervention (see in [17, pp. 8–9].

F.-Y. Yu, Professor of the National Cheng-Kung University and one of the leading experts in this field[1], published fifteen papers dedicated to students test writing and analyzed thirteen papers with the same focus written between 1985 and 2013 by authors from different fields of knowledge [18, p. 90, 101]. Most of her own research has been done at the elementary and middle level school, including the study of the impact of the test item development on the progress of Taiwanese elementary school learners of the English language [19]. According to Yu's research, a majority of related studies deal with mathematics and natural sciences while papers exploring the effects of test writing in language instruction are much more rare.

However, the bibliography of research articles in the field can be extended by including papers written in the languages other than English, for example, in the Russian language [20, 21] and papers on the impact of students' test development in the course of English as a foreign language [22–25].

A relative novelty of the discussed type of the learning activity might explain the fact that researchers use different terms for the methodological concept such as *student-generated questions; problem posing; student item construction; student-developed assessment items*, etc. Yu and the co-authors prefer the term *student-question generation* (SQG), defining it as "a learning activity during which students generate a set of questions corresponding to specific previous instruction or experiences they deem educationally important and relevant for self- and peer-assessment purposes" [18, p. 89]. In their study of 2015, Yu and the co-authors reported about 54 Taiwanese sixth graders who practiced writing test items including multiple choice, matching, and yes/no questions based on the material covered in the online learning system called QuARKS (Question-Authoring and Reasoning Knowledge System) developed at the National Cheng-Kung University. For eight weeks, students worked in small groups for 20 min at the end of each English language lesson. The students were streamed according to the entrance test results: mini-groups with the English proficiency above average, below average and average. The results of the final test showed a statistically significant higher achievement among the students from the SQG groups in comparison with their peers from the control group who studied English using the same textbook and a traditional methodology, mainly aimed at memorizing and repeating (Drill and Practice, using the authors' terminology). The results of the questionnaire also showed that the SQG group was more motivated to learn English than the students from the control group [19, p. 13].

---

[1] Distinguished Professor Fu-Yun Yu's page on the National Cheng-Kung University's website https://www.ed.ncku.edu.tw/en/team_detail.asp?nid=4.

One of the pioneers of this approach in teaching English, L.M. Kaufman, described the experiments carried out in the 1990s at the University of Puerto Rico in his publications [22, 23]. He maintained that turning the test-taker into a test-maker can transform the process into an imaginative, exciting game involving both students and the teacher in which the learning and assessment go in hand. Nowadays this idea gains popularity as approaches to assessing student achievement are changing, and less formal formative assessment is gaining more and more importance in the educational process [26, 27].

In the experiment described by Kaufman, two groups had different levels of foreign language proficiency: intermediate-level ESL students, and advanced-level ESL teacher trainees. The procedure included work in small groups of 3–4 students throughout the term making up lesson plans, teaching their classmates, developing questions, exercises, and tests, and taking the tests themselves [23, p. 80]. The teacher also participated in the preparation of test/quiz items. In general, the ratio of teacher-made to student-made quiz or exam items was 50:50. Thus, all students were aware that the test would include both types of items: teacher-developed and student-written items. A survey conducted at the end of the term indicated that motivation increased in both groups; both groups found the experience of writing their own tests beneficial as the activity increased the level of understanding of the language input and the level of confidence before the test and relieved some pre-test anxiety.

Research in cognitive sciences, conducted in the 21st century, suggests that these implications occur due to the mobilization and actualization of various cognitive processes, for example, a reflection on student learning experience, making sense of the learning materials, searching one's existing knowledge reservoir for possible connections and appropriate use of the language, seeking additional information [19, p. 13]. Nevertheless, despite the positive results of the experiment, Kaufman points to some limitations of the method which arise due to a number of factors including the characteristics of the learners, their proficiency level, and specific tasks which the students undertake [23, p. 83]. Yu et al. also recommend instructors to examine the influence of the SQG strategy in different second/foreign language learning contexts, learners' language learning experience, different educational levels, and cultural differences [19, p. 14].

In SPbPU, the instructors occasionally involved students in making test-format questions before. According to a preliminary observation, Master's students majoring in language teaching can effectively contribute to the development of different test-format tasks such as multiple-choice questions, matching exercises, true/false questions, etc. for ESP courses working in pairs or small groups, under the supervision of their university instructor [28, 29].

3   The Experiment and Its Goals

In response to Yu et al. advise we decided to apply the described learner-centered approach to increase the efficiency of the educational process and motivation in the ESP course for 2nd year students of the St. Petersburg State University of Aerospace Instrumentation (SUAI).

3.1   Research Questions

We were specifically interested in the following research questions:

1. Does the Students' Making Test-format Questions (SMTFQ) technology contribute to better memorization of new vocabulary and grammatical structures?
2. Does the SMTFQ technology contribute to the improvement of reading comprehension?
3. Does the SMTFQ technology improve language skills among students with lower level language competence?
4. Does the SMTFQ technology increase students' motivation to learn English?

3.2   Participants and Procedure

Participants. The experiment was carried out in the fall semester of the 2020–2021 academic year. The experiment involved students of two academic groups of the second year. A total of 44 students aged 20–23: 38 male students and 6 female students, who learned English at secondary school, having studied English for more than 6–8 years. According to the placement test conducted at the beginning of the experiment, students in both groups have the same English language proficiency mainly level A2 – B1 (according to CEFR), with some students having B2 and some A1 level. It was decided to conduct the experiment in the group with slightly lower average score of entrance test (although statistically insignificant). The experimental training lasted 10 weeks, culminating in a final test. Mini-groups were formed taking into account the results of entrance test so that in each mini-group of 3–4 students there were students whose level of English is above average, below average, and the average level of English proficiency in this academic group.

Materials. The training in both groups was carried out according to the course manual compiled by the teachers of the English language department A Book of Science and Computers [30]. The complexity of the texts, as measured by the Versatext Profiler[2], corresponds to the B2 level. The manual contains a glossary of terms consisting of 15–20 words and phrases at the beginning of each lesson. These lists were required to be memorized in every lesson.

Procedure. The students of the control group were instructed in a traditional way: warm-up activities at the beginning of the lesson, pre-reading exercises, familiarization with key vocabulary, reading and analyzing the text, answering questions from the textbook, doing lexical and grammatical exercises working in pairs and small groups, etc. At the end of each lesson, students of both groups wrote a short summary of the text from the previous lesson using ten or more key words from pre-reading vocabulary lists as an additional incentive to memorize the topic-focused vocabulary. The summaries were used for measuring their weekly progress.

---

[2] The URL of the Versatext Profiler developed by J. Thomas https://versatext.versatile.pub/.

In addition to what the control group did for each lesson within the course study, the students of the experimental group got an additional home assignment to make up some test-format questions on grammar issues and key vocabulary considered in the course unit. The experimental group used their test-format questions in a 15-min pair work at the end of the lesson before writing a summary of the text.

As the experimental group students had had no experience in compiling test-format questions before, they were instructed how to make them and practiced four specific types of test-format question, one type per a lesson: multiple choice, matching, gap filling, and true/false test format-question. That is why short training sessions were allocated at the end of the first four lessons in the experimental group. Table 1 summarizes the content and the output of these training sessions.

Table 1. The types and the sequence of different test format questions during the training sessions

Lesson number	Type of a test-format question targeted at the lesson	Type of test-format questions made up by the students
1	Matching	–
2	Multiple choice	Matching
3	Gap filling	Multiple choice, Matching
4	True/False statements	Gap filling, Multiple choice
5	–	True/False statements
6–10	–	Students generate test-format questions of their choice (any of the four types)

Students submitted their test-format questions to the teacher electronically for evaluation and possible corrections where necessary.

Instruments to Measure Students' Achievements. The following instruments were designed to measure changes in the English language performance: a placement test (a standardized PET test) and an academic achievement test conducted in both experimental and control groups at the beginning and end of the course. Students in both groups wrote texts summaries at the end of each class, which were served as the instrument to test students' text comprehension, grammar revision, and target vocabulary acquisition. The post-experiment questionnaire was offered to the experimental group.

The instructor-developed academic achievement test was composed of 30 questions based on the textbook and included 16 multiple-choice questions, 8 true/false questions, and 6 matching questions.

Data Collection. The following four types of data were collected during the experiment:

- a collection of test-format questions composed by the students of the experimental group;

- summaries of 10 texts written at the end of each class by the students in both groups;
- the placement and final academic achievement tests;
- questionnaires completed by students of the experimental group.

4 Results

4.1 Test-Format Questions Made by Students

In total, 1,680 test items were received from the experimental group. The number of test-format questions of each type is shown in Table 2.

Table 2. The number of test-format questions of each type

Type	Number of test-format items	Percentage
Multiple choice	960	57.1
Matching	370	22.0
Gap filling	240	14.3
True/False statements	110	6.6

Table 2 shows that students gave preference to multiple choice questions for testing grammar structures (57.1% of all the test items). Matching test-format questions (22.0%) included the following types: matching words and their Russian language equivalents, topical words and their definitions which students looked up in dictionaries. Sentence gap-filling with target vocabulary items was used in 14.3% of the test items. Finally, true/false statements as test-format questions were definitely the least popular type (6.6%). The latter might have been found by the students as a time consuming activity.

4.2 Text Summaries

The number of the target vocabulary items (TVIs) in each summary was counted. The results are presented in Fig. 1. The results were processed using T-test for independent samples with the help of the statistical software available at the Compleat LexicalTutor website[3].

The analysis reveals that, despite having clear instructions to use at least 10 words/collocations from the target vocabulary list in the summary, students in both groups equally failed to meet the requirement writing a summary of Unit 1 text. The difference between the control and experimental groups appeared statistically insignificant in summaries of Unit 1 text, with Texp being less than Ttable (p = 0,05, *Ttable = 2,018 for df = 42. Texp = 0,17).

---

[3] https://www.lextutor.ca/stats/t-test/.

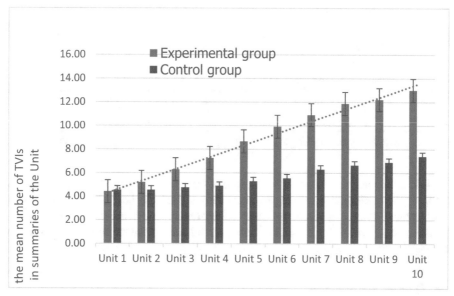

Fig. 1. Changes in the quantity of the target vocabulary items in students' summaries during the ten-week experiment

However, Fig. 1 shows that students in the experimental group started including more TVIs in their text summaries in the following Units than the students in the control group, with the gap between the two groups increasing after Unit 3 so that the difference changed into statistically significant. As we can see, the experimental group closely approached the required number of 10 TVIs in their summaries in Unit 6 and in later summaries (Units 7 to 10) exceeded this quantity in their summaries.

Though we can observe a tendency to increase TVIs use in the control group's summaries, the control group students, on average, never achieved the level of required vocabulary use during the experimental time, with the mean number of TVIs in summaries based on Unit 10 text rising to 7.4. Eventually, none of the control group students exceeded the required quantity of TVIs.

The Cohen's d effect size measure shows that for summaries for Unit 5 and 10 the difference between the groups is of particular importance (medium effect size). It is worth mentioning that there were no spelling errors in TVIs, which can be explained by the IT specialists' professional attention to spelling as even tiny sign error in a program code may result in long hours spent on debugging the program later.

4.3 ESP Course Academic Achievements

Table 3 demonstrates the data concerning the number of correct answers in pre- and post-tests. The experimental group showed better results than the control group in the post test (final term test), and this difference is statistically significant with small effect size.

Table 3. The number of correct answers in pre- and post-tests

	Experimental group (n = 24)		Control group (n = 20)		T exp*	Cohen's d effect size
	Mean	SD	Mean	SD		
Pre-test	12.3	4.45	13.4	5.26	0.76	1.29
Post-test	25.08	3.34	20.65	3.54	4.27	

Significance level p = 0.05 (5%), *T table = 2,018 for df = 42.

Taking into account that the pre-test results were slightly worse in the experimental group at the beginning of the experiment, we assumed the possibility of their lower level language competence, and we can conclude that the proposed technology may also be effective in achieving the goal of levelling out the difference in students' English language proficiency level.

4.4 Questionnaire Results

The questionnaire was designed to find out students' opinion on test item writing, language learning motivation, and students' attitude to making test-format questions technique. It was only administered in the experimental group which was involved in this learning activity.

The questionnaire included eight statements, and the respondents were asked to respond by agreeing or disagreeing. The answers were ranked using the simplified Likert-type scale of three levels of agreement/disagreement from "disagree" through "neither agree nor disagree" to "agree". In the final item of the questionnaire, the students of the experimental group were asked to write names of academic disciplines in which they believe the SMTFQ technology might be useful.

In the questionnaire, we were interested in the respondents' opinion on the SMTFQ technology to facilitate language learning and the motivation to learn English. The list of statements follows.

1. This technology is effective for understanding grammar rules.
2. This technology is effective for applying grammar rules in practice (in oral and written communication).
3. This technology is effective for learning vocabulary.
4. This technology helps to succeed in the final test at the end of the course.
5. This technology helps to better understand written texts.
6. I became more confident in my knowledge of the English language.
7. This technology increases motivation to learn a foreign language.
8. This technology can be used in the study of other academic disciplines. If you chose "Agree", give some examples of these disciplines.

According to the results, students expressed a rare unanimity. More than 95% chose "Agree" for all eight questions, and, at the end of the questionnaire, instead of writing the names of disciplines, the respondents generalized by writing *"Any discipline"* or *"all disciplines"*.

5 Discussion and Further Research

The experiment demonstrated that the students of the experimental group more actively use the target vocabulary and grammar structures in their summaries. When compiling test-format questions targeting lexical and grammar issues from the text the student analyzes the text grouping words within sentences and using the resulting groups as semantic milestones (context cues), which helps him to determine the sentences necessary for further use in test format. He compares the sentences in order to make the right choice of grammar patterns and simplifies them by segmenting in order to underline a certain grammar rule. During this kind of intellectually intensive work students' receptive lexical and grammatical reading skills are trained. Students' involvement in the textual content becomes more profound than in case of skimmed reading typical for preparing regular homework. When students read the text analytically, searching for the grammar and lexical patterns, their way of reading becomes closer to scanning, that is reading with the purpose of finding the required information. So scanning the text for choosing the targeted examples of grammar and lexical items' use is hypothetically more instrumental for mastering guided reading skills than skimming.

A side effect of close reading is that students from the experimental group regularly used the exact word collocations and grammar structures from the texts they read in different units of the course book, as the analysis of the summaries showed. This can be explained as follows:

Firstly, they followed the instruction to use a given number of TVUs more often than students in the control group due to the attention paid to the experimental group by the teachers. Secondly, this group repeatedly dealt with the texts when making test-format items and discussing them with their peers and team members, which, as a result, lead to the involuntary memorization of parts of the texts they used in their summaries. And finally, students are likely to underestimate paraphrasing and generalization in their summary writing, the techniques requiring special training and more attention to. One of the suggestions could be to formulate the main idea in the Russian language, and then to translate this secondary text into the English language, thus avoiding text copying. This technique is practiced by some researchers who believe that two-way translation leads to better comprehension of the main idea of the text and to the development of written communication skills, which is important for the ESP course [31, 32]. Obviously, these aspects of writing a summary should be a focus of future research.

The questionnaire showed that the students really liked the role of test-makers. A positive attitude of the students to the SMTFQ technology can be explained by students' interest in what test-format questions other students may bring for them to do in their team work. Also, in the second half of the experiment, we noticed a growing level of competitiveness since students tried to compose more interesting and challenging test-format questions than their peers. In the course of the experiment, we observed that the students' interaction in the classroom and in preparation for it was growing and becoming more effective, which might be considered as indicators of students' growing motivation to participation in this activity, in particular, and language learning in general.

The experiment was conducted in a classroom-format, but we believe that this technology can be successfully used not only in classroom training, but also in a blended or distant learning format since modern webinar platforms such as Microsoft Teams and Zoom provide facilities to organize students' cooperation online in small groups in synchronous communication, and LMS support for asynchronous communication modes. It is a well-known fact that meaningful communication in online courses is an indispensable prerequisite for successful learning in any discipline [33].

The unanimous approval by the participants of the experiment of the proposed technique points to its educational potential which can be used when reading and processing professionally-oriented authentic texts in foreign language studies, and when studying other disciplines.

Within the framework of the experiment, its effectiveness was proved by better results of the final test shown by the students of the experimental group. This is especially important given that the students of this group had shown slightly worse score (though statistically insignificant) at the entrance test than the students of the control group. We have good grounds to believe that this active, learner-centered approach can be used to level out, to some extent, the foreign language proficiency of students with lower level language competence in multi-level groups. Even though the teacher's time spent on preparing for classes increased by about 20% due to the necessity of regular checking test items compiled by the students, we consider it worthwhile for enhancing students' foreign language skills.

Thus, the experiment has allowed us to give affirmative answers to the research questions. In particular, the SMTFQ technology contributes to better memorization of new vocabulary and grammatical structures and facilitates reading comprehension. Test making activities enhance students' motivation and improve language skills of the poorer students and increase their chances to catch up with students having a better command of English.

We also think that the SMTFQ technology can be used to develop other language skills within the ESP course, in particular, listening comprehension, but this requires experimental verification. We are also sure that SMTFQ technology can be successfully projected on other foreign languages and teaching Russian as a foreign language. Foreign students will hopefully, given further research is conducted, appreciate the way of mastering their reading skills through making test-format questions for grammar and vocabulary comprehension, as it appears to be didactically more motivating. Also, as their answers to the final survey show, students can use this testing technique on their own for better comprehension of other disciplines they study in Russian, thus transferring skills developed at ESP classes to other disciplines. We did not ask students whether or not they were planning to learn new foreign languages in the future. Taking into account a general trend towards multilingualism [34] we are risking assuming that some of them will do it in the future. We have good grounds to think that the proposed technology of improving text comprehension, vocabulary and grammar structures acquisition can help students to learn new languages. However, this assumption as well as other statements formulated here need experimental verification, thus forming the outline of future research.

The SMTFQ technology used in the classes can provide one more benefit. Test items written by the students can be accumulated in an electronic databank, which can be used by teachers in further test preparation. The tests can also be included into the LMS Moodle system for use when necessary. If the educational institution does not have an LMS, tests items can be turned into an interactive electronic format using popular resources such as LearningApps[4].

Learning to read in an ESP course is a very important aspect. It should prepare future specialists for reading authentic specialist scientific literature and technical documentation in a foreign language. Practical teaching shows that performing pre-text and post-text tasks from ESP textbooks is not enough to form a thoughtful reader of specialist texts who can understand all important information. Besides the revision of the target vocabulary and grammar structures, it is important to help students become more active readers of domain-specific literature. For a high level of reading comprehension, students need various cognitive skills, which can be ensured by involving them in the activities which develop these skills. Students' making test-format questions can be exploited for all these purposes.

Acknowledgements. The research is partially funded by the Ministry of Science and Higher Education of the Russian Federation under the strategic academic leadership program 'Priority 2030' (Agreement 075-15-2021-1333 dated 30.09.2021).

References

1. Korotkina, I.B.: Transdisciplinary approach to the development of a course on academic reading of specialized texts on the example of public policy. In: Bagramova, N.V., Smirnova, N.V., Schemeleva, I.Yu. (eds.) Teaching Reading in a Foreign Language at a Modern University: Theory and Practice, pp. 123–142. Zlatoust, St Petersburg (2016). (in Russian)
2. Chmelíková, G.: Possibilities of improving reading of subject-specific texts. J. Teach. Eng. Spec. Academ. Purp. **9**(1), 61–70 (2021). https://doi.org/10.22190/JTESAP2101061C
3. Watkins, P.: Teaching and Developing Reading Skills. Cambridge University Press, Cambridge (2017)
4. Burt, M., Peyton, J.K., Adams, R.: Reading and adult English language learners: a review of the research. Center for Applied Linguistics, Washington, DC (2003). https://eric.ed.gov/?id=ED505537. Accessed 27 Aug 2022
5. Galskova, N.D.: New education technologies in the context of a modern education theory in the field of foreign languages. Иностранные языки в школе. **7**, 9–15 (2009)
6. Puig-Mayenco, E., González, A.J., Rothman, J.: A systematic review of transfer studies in third language acquisition. Second Lang. Res. **36**(1), 31–64 (2020). https://doi.org/10.1177/0267658318809147
7. Prokhorova, A.A., Bezukladnikov, K.E.: Multilingual training of technical university students: sound arguments. Lang. Cult. **52**, 215–231 (2020). https://doi.org/10.17223/19996195/52/14. (in Russian)
8. Prokhorova, A.A.: Multilingual approach to power engineering students language teaching. Eur. J. Nat. Hist. **4**, 41–45 (2019)

---

[4] LearningApps: https://learningapps.org/.

9. Burdumy, A., Bohlander, C.: Designing and delivering a program of reading skills classes to postgraduate students. J. Teach. Eng. Spec. Academ. Purp. **6**(2), 245–252 (2018). https://doi.org/10.22190/JTESAP1802245B

10. Anderson, W., Krathwohl, D.R.: A Taxonomy for Learning, Teaching, and Assessment. A Revision of Bloom's Taxonomy of Educational Objectives. Addison Wesley Longman, Inc., Boston (2001)

11. Seymour, P.H.K.: Early reading development in European orthographies. In: Snowling, M.J., Hulme, C. (eds.) The Science of Reading: A Handbook, pp. 296–315. Blackwell Publishing, New Jersey (2005). https://doi.org/10.1002/9780470757642.ch16

12. Grabe, W., Stroller, F.L.: Teaching and Researching Reading. Pearson Education, London (2011)

13. Rahimi, M.A., Rezaei, A.: Use of syntactic elaboration techniques to enhance comprehensibility of EST texts. Engl. Lang. Teach. **4**(1), 11–17 (2011). https://doi.org/10.5539/elt.v4n1p11

14. Laufer, B., Ravenhorst-Kalovski, G.C.: Lexical threshold revisited: lexical text coverage, learners' vocabulary size and reading comprehension. Read. Foreig. Lang. **22**(1), 15–30 (2010). http://www2.hawaii.edu/~readfl/rfl/April2010/articles/laufer.pdf

15. Joseph, L.M., Alber-Morgan, S., Cullen, J., Rouse, C.: The effects of self-questioning on reading comprehension: a literature review. Read. Writ. Quarterly **32**(2), 152–173 (2016). https://doi.org/10.1080/10573569.2014.891449

16. Murtazina, P.A., Petkova, E.N.: Research of the formation of intellectual skills of Russian-speaking students in academic reading in English (based on the analysis of the performance of tasks in academic reading in the format of the IELTS exam). In: Bagramova, N.V., Smirnova, N.V., Schemeleva, I.Yu. (eds.) Teaching Reading in a Foreign Language at a Modern University: Theory and Practice, pp. 101–122. Zlatoust, St Petersburg (2016). (in Russian)

17. De Jesus Sales, A.: The output hypothesis and its influence in the second language learning/teaching: an interview with Merrill Swain. Interfaces Brasil/Canadá. Florianópolis/Pelotas/São Paulo **20**, 1–12 (2020). https://doi.org/10.15210/INTERFACES.V20I0.18775

18. Yu, F.-Y., Wu, C.-P.: The effects of an online student-constructed test strategy on knowledge construction. Comput. Educ. **94**, 89–101 (2016). https://doi.org/10.1016/j.compedu.2015.11.005

19. Yu, F.-Y., Chang, Y.-L., Wu, H.-L.: The effects of an online student question-generation strategy on elementary school student English learning. Res. Pract. Technol. Enhanc. Learn. **10**(1), 1–16 (2015). https://doi.org/10.1186/s41039-015-0023-z

20. Halimova, N.M.: Pedagogical testing as a factor in the success of students' learning: on the example of technical disciplines. Doctoral dissertation. Krasnoyarsk State Pedagogical University, Krasnoyarsk (1999). (in Russian)

21. Evdokimova, L.A.: Test making as a means of consolidating educational material. Quest. Meth. Teach. Univer. **3**(17), 444–446 (2014). (in Russian)

22. Kaufman, L.M.: Students writing their own tests: an experiment in student-centered assessment in two cultures. In: The 27th Annual Convention of TESOL, Atlanta, GA (1993). https://files.eric.ed.gov/fulltext/ED362060.pdf. Accessed 27 Aug 2022

23. Kaufman, L.M.: Student-written tests: an effective twist in teaching language. J. Imagination Lang. Learn. 80–85 (2000). https://files.eric.ed.gov/fulltext/ED476595.pdf. Accessed 27 Aug 2022

24. Murphy, T.: Tests: learning through negotiated interaction. TESOL J. **4**(2), 2–16 (1994)

25. Ashtiani, N.S., Babaii, E.: Cooperative test construction: the last temptation of educational reform? Stud. Educ. Eval. **33**, 213–228 (2007). https://doi.org/10.1016/j.stueduc.2007.07.002

26. Bullock, D.: Assessment for Learning. British Council. https://www.teachingenglish.org.uk/article/assessment-learning. Accessed 27 Aug 2022

27. Antonova, K., Tyrkheeva, N.: Formative assessment of critical reading skills in higher education in Russia in the context of emergency remote teaching. J. Teach. Eng. Spec. Academ. Purp **9**(2), 137–148 (2021). https://doi.org/10.22190/JTESAP2102137A

28. Almazova, N.I., Popova, N.V., Evtushenko, T.G.: Organizational and methodological aspects of creating professionally oriented didactic resources in a foreign language at a technical university. Bullet. Kemerovo State Univ. Series: Humanit. Soc. Scien. **4**(1), 1–11 (2020). https://doi.org/10.21603/2542-1840-2020-4-1-1-11. (in Russian)

29. Popova, N.V., Almazova, N.I., Evtushenko, T.G., Zinovieva, O.V.: Experience of intra-university cooperation in the process of creating professionally-oriented foreign language textbooks. High. Educ. Rus. **29**(7), 32–42 (2020). https://doi.org/10.31992/0869-3617-2020-29-7-32-42. (in Russian)

30. Gromova, I.I. (ed.): A Book of Science and Computers. SPbGUAP, St Petersburg (2010)

31. Leonardi, V.: Teaching business english through translation. J. Lang. Transl. **10–1**(March), 139–153 (2009)

32. Mažeikienė, V.: Translation as a method in teaching ESP: an inductive thematic analysis of literature. J. Teach. Eng. Spec. Academ. Purp. **6**(3), 513–523 (2018). https://doi.org/10.22190/JTESAP1803513M

33. Vajndorf-Sysoeva, M.E., Grjaznova, T.S., Shitova, V.A.: Methodology of Distant Teaching: Textbook. Jurajt, Moscow (2018). (in Russian)

34. Council Resolution of 21 November 2008 on a European strategy for multilingualism. 2008. The Council of the European Union. November 21. https://eur-lex.europa.eu/legalcontent/EN/ALL/?uri=CELEX:32008G1216(01). Accessed 27 Aug 2022

Project-Based Learning in Higher Education as a Tool for Integrated Hard, Soft Skills and Engineering Language Development

Liudmila V. Daineko$^{(\boxtimes)}$, Inna I. Yurasova , Natalia M. Karavaeva ,
and Tatiana E. Pechenkina

Ural Federal University named after the first President of Russia B.N. Yeltsin, 19 Mira St.,
620002 Yekaterinburg, Russia
`l.v.daineko@urfu.ru`

Abstract. Modern multilingual world provokes university graduates to master not only natural languages, but also the languages of signs and technologies. Multilingualism, as a new form of literacy, implies mastering other ways of communication, not only verbal-linguistic ones. The paper considers the impact of project-based learning on the formation of traditional hard skills, necessary soft skills, and mastering of visual-image engineering language, which implies publicly available presentation of the results of project activities. The experience of Ural Federal University in organizing project competitions on behalf of employers' partners aimed at solving urgent problems of the society is considered. Close interaction between employers and the university allows employers to participate in training future employees for their company, solving their production problems, and universities to improve the quality of the educational process and the demand for graduates in the labor market. The use of the project-based learning method allows students on the one hand to consistently explore and delve into the details of the problem under study, using the hard skills available. Group work on projects develops soft skills that involve communication and collaborative project work in a project team. The need to present the results of one's work enhances engineering language proficiency.

The results of the survey of the participants of the project activities are presented. It is concluded that the educational tool used is successful. The obtained research results have practical significance and can be used in the organization of the educational process in higher educational institutions.

Keywords: Engineering language · Project-based learning · Hard skills · Soft skills · Educational organization · Higher education

1 Introduction

Modern reality requires every person to adapt quickly to a multilingual environment [1], and multilingualism in this case is not considered exclusively in the format of communication with native speakers of different languages. The perception of the surrounding multilingualism as a new form of literacy, involving interaction with the outside world not only through language, but also through the use of various applications [2], computer programs, without which it is impossible to imagine modern digital world [3], encourages universities to provide their graduates with new competencies. Traditionally, universities prepared graduates with solid professional skills. Assuming mastery of known technologies, standard ways of solving problems, ability to use equipment and technical means. Then a trend appeared in education aimed at the necessity of additional formation of soft skills allowing students to adapt to the requirements of their environment by solving non-standard tasks and communicating with other people. The rapid integrated development of hard and soft skills can be realised through the use of project-based learning methodology. The need to present the results of project work in an accessible way enables students to learn engineering language in a practical way. In the future, graduates' mastery of visual-image engineering language, which is an international language of professional communication, will enable them to communicate with foreign partners in the best and most intuitive way and present the results of their professional activity without verbal support. High level of hard, soft skills and engineering language proficiency will ensure high competitiveness of specialists in the international labour market.

2 Problem Statement

Understanding that a language is a set of symbols used for communication [4] and facilitating it, allows for using project-based training as a means to increase the level of proficiency in all languages of the multilingual world for greater expressiveness in communication. The importance of timely exchange of professional information for an unambiguous understanding of the situation, including for partners from different countries, has led to the need for university graduates to possess linguoprofessional competence [5]. The engineering language, or the language of technology, is equally understood by all. That is why it is so important to "translate" the results of one's work into engineering language [6], including to assess the feasibility and cost of the project. Of course, with the rapid development of information technology, this process has also accelerated. Thus, the traditional geometric and graphic training of university students is being replaced by 3D modelling, which increases the productivity of the work, the level of visualisation quality, its variability and visibility [7]. Another important aspect of good engineering proficiency is the ability to visualise according to the dimensionality of the object being modelled, qualitatively changing the technology and ideology of geometric modelling [8]. Thus, the wide range of tasks set by society [9] leads to the need to transform the role of engineering language in communication. The ease of use of engineering language allows the development of the aesthetic level of the professional [10], his creative and creative skills, the ease of communicating with team members, that is, what is called soft skills. Soft skills are personal qualities, goals, motives and preferences that are valued in the labor market. Researchers believe that soft skills are the main factor in a successful person's life, and mastering them by students should become an important task supported by the state [11]. However, there are a number of contradictions between the state regulation of education and the ability to meet these requirements [12].

Schulz, in his study on education beyond knowledge, notes the importance of soft skills, which play an important role in the formation of a person's personality, effectively complementing hard academic and technical knowledge [13]. Wats & Wats note that hard skills contribute to success by only 15%, while soft skills (interactive, communicative and human) provide 85% of success [14]. Moreover, a direct link between soft skills and the performance of work functions by graduates has been proven [15].

The experience of working with students in the form of project-based learning at the university is quite large: teachers work with both bachelors [16] and masters [17, 18], actively mastering new competencies, including digital ones [19]. The main reasons for the introduction of project-based learning in university programs is to increase the activity of working with partner employers [20] to increase the compliance of the competencies obtained by graduates with the modern labor market [21]. According to many authors, the technology of project-based learning contributes well to the development of soft skills, encouraging the acquisition of engineering language. Thus, Kechagias, in his study on the measurement and evaluation of soft skills, notes that the acquisition and development of "soft" skills is important for advancement in life and work, and very often, these skills are not sufficiently developed among young people [22]. Musa et al., considering instilling soft skills in students, believe that employers highly value many skills acquired during project-based training, including the ability to interact with team members, the ability to cope with interpersonal conflicts, make informed decisions, and solve complex problems [23], including ones in the field of sustainable development [24], ability to improve business processes based on data analysis [25], to increase the efficiency of various production activities [26]. Vogler et al. investigating the hard work on soft skills in interdisciplinary project-based training, found that the interdisciplinarity of a project develops many aspects of learning that cannot be achieved within the framework of traditional training. Researchers note the importance of interdisciplinary collaboration, which pushes students to go beyond the disciplines studied [27]. In a study of the conceptual foundations of project-based learning for the integration of soft skills to students, Dogara et al. noted that project-based learning by its nature uses the potential of students in the field of interpersonal communication skills formation [28]. According to Chassidim, Almog & Mark, a project-oriented educational environment is a necessary factor for students to achieve soft skills, combining academic topics, self-study and the implementation of social skills, the development of creativity and other necessary competencies [29]. According to Tadjer et al., most universities face the fact that it is necessary to improve simultaneously the cognitive skills and the soft skills of students [30]. The importance of soft skills in professional activity is noted by Ravindranath who emphasizes the need to master them in curricula at all levels of study [31]. Nealy believes that active forms of learning developed in response to business needs can be a good tool for developing soft skills [32]. Sudana et al., investigating the possibilities of managing the assessment of soft skills in the process of studying at a vocational school, noted that about 10% of graduates cannot find a job due to the fact that the results of their education are poorly evaluated. This is mainly due to the fact that the educational process adheres to the assessment of hard skills and ignores the assessment of 4C (creativity, critical thinking, communication and cooperation) [33]. Shakir, in his study of soft skills in higher education, believes that the development of human capital through social skills should be included in bachelor's degree programs [34]. And according to Claxton and colleagues, an important aspect of human capital development is not only a person's abilities, but also his perception and inclination to use these abilities correctly in appropriate situations [35], as well as in the formation of corporate culture [36].

However, researchers have not sufficiently addressed the issue of assessing students' integrated mastery of hard and soft skills during project-based learning and the use of engineering language to present the results of group project work.

3   Materials and Methods

In order to increase the efficiency of the educational process, develop hard and soft skills in students, including mastering engineering language, it was decided to organize group work on projects, according to the applications of partner employers. Group work on projects allows students to gain new practical competencies, to establish communication within a previously unfamiliar team, as well as to master the engineering language that is the language of visualization of the work. The opportunity offered to students to present the results of their projects by any means serves the task of mastering the engineering language at a high level, including through the development of creative abilities and the development of soft skills.

The project-based method of teaching is not new. However, this does not reduce its relevance in the learning process. The project-based learning process allows students to experience the practical relevance of theoretical knowledge acquired during traditional lectures. The competences acquired during project work will allow future graduates to feel more confident in the production process. The production process often involves communication between a large number of unfamiliar people, not always speaking the same language, but united by the fulfilment of a common task.

While working on a project, students get to know the characteristics of other team members, build communication with each other, solve arising problems, learn to agree with each other, divide responsibilities, reflect on the solution of the task, gather necessary information, learn to plan their own and others' time, choose ways to visualize their project solution. It should be noted that nowadays there are international students in almost every academic group at the university, for whom it can be difficult to communicate with their classmates. Therefore, university projects can be seen as a rehearsal for post-graduated life, involving hard and soft skills, as well as the use of engineering language tools. It is of particular interest to work on real projects together with employer partners. The involvement of potential employers in the project increases the students' interest in the project results and their involvement in the work. Public defense of projects allows, on the one hand, to show the results of the work done and, on the other hand, to receive feedback on its results. An important aspect of the presentation of the results is the visualization of the developed project, which allows demonstrating the skills of engineering language-visualization. Competition in the project competition is an incentive for students to improve their work. Moreover, looking at the same object of research from a different perspective enables new ideas, stimulates communication between different teams, discussion of proposed solutions, and possibly the emergence of new project team members. The comparison of one's vision of the project described in engineering language with the projects of other teams allows to develop the content of one's language, to expand the "lexicon" of engineering language.

269 undergraduate students studying construction (Institute of Construction and Architecture) and construction management (Institute of Economics and Management) in the Ural Federal University took part in the projects, being divided into teams of 3 to 6 people. As a result, students presented 55 projects. The projects were defended in two stages. At the first stage, all projects were presented in academic groups with the participation of invited experts, from which 22 projects considered to be the best were selected by a closed vote of the group's students and experts, then presented to experts and partner employers in the framework of the competition "Integrated Development of territories and urban projects in the context of sustainable development". Integrated development of territories is a trend of urban development in Russia and includes strategic urban planning and implementation of large-scale residential development. Often, such projects involve redevelopment of ex-industrial or abandoned areas.

The Administration of the Berezovsky City District and the Administration of the Sysertsky City District of the Sverdlovsk region offered students land plots for the creating of territorial development projects. The value of such requests is to provide students with the opportunity to gain practical experience in implementing their theoretical knowledge in a real socially significant project. In turn, the opportunity to get the opinion of young people about the direction in which the territory of real estate should be developed is valuable for customers.

On the territory of Berezovsky, students designed:

- Development of the territory of the "Historical Square" in order to build a historical complex with elements of reconstruction;
- Renovation of the territory of the "Park of Entrepreneurs" in order to build a new amusement park "Berezovsky Park";
- Development of the "Green Valley" territory in order to form a new affordable housing market and provide comfortable living conditions, with highly developed infrastructure, social and household facilities;
- Development of the territory of the "Sports Complex" in order to build a sports complex, including a swimming pool and an ice arena.

On the territory of the Sysert City District:

- Construction of a new zoo near the village of Patrushi;
- Renovation of the territory of the Uralgidromash plant for the purpose of integrated development of the territory and housing construction;
- Construction of the residential district "Solnechny" with an area of 71.9 hectares near the village of Patrushi on former agricultural lands;
- Development of the territory of three land plots in Sysert city for the purpose of housing construction and integrated development of the territory;
- Construction of the cottage settlement "Bazhov's Tales" on the territory of the Sysert City District.

According to the project assignment, students had to design a project for the development of the territory of the municipality based on a comprehensive socio-economic analysis of the urban district, taking into account the appearance of potential growth points in the territory under consideration for a period of 5 to 20 years. The results were to be presented in the form of an explanatory note and visualization of the proposed project in any form, including three sections:

1. Analytical, consisting of an analysis of the socio-economic development of the city, including a SWOT-analysis to identify the reasons affecting the attraction of new residents and job creation, an assessment of the budget structure and the main sources of replenishment and expenditure of budget funds.
2. Socio-demographic, containing an analysis of the demographic indicators of the city (population change, gender, age and employment structure classifications), the results of a sociological survey of the city's population for satisfaction with the quality of life and activities of the territory, the state of the urban environment, the results of an assessment of the entrepreneurial activity of the city, etc.
3. The project plan, which includes the main directions of spatial, social and economic development of the projected territory developed on the basis of the results of the analysis, including infrastructure analysis, the main directions of rational nature management and ensuring environmental safety of the projected territory, the developed concept of territory development from the point of view of a comfortable living environment.

The customers highly appreciated the students' projects, the best of which were awarded prizes in the competition.

4   Results and Discussion

In order to improve the efficiency of work on real projects, the authors conducted a survey of students who participated in the project work in May 2022 after summarising the results of the competition. 113 people out of 269 project participants (42%) responded to the survey. The survey was conducted using the Yandex Forms service, the survey results were processed using the Vortex program[1]. 63 boys (55.8% of respondents) and 50 girls (44.2%) answered the questionnaire, of which 91 students of the Institute of Construction and Architecture (80.5% of respondents) and 22 students of the Institute of Economics and Management (19.5%), most of whom are 19 (42.5% of respondents) and 20 years old (41.6%), the rest of the respondents are 21 years old and older (15.9%).

When asked about new discoveries during project work, the majority of respondents (62.8%) replied that they learned a lot of information about the small town of the Sverdlovsk region. It should be noted that most of the students had never been to the territory of the cities of Berezovsky and Sysert before the project. 60.2% of respondents said that they got acquainted with the methodology of assessing the socio-economic condition of the territory in practice. Most of the participants in the project activity are

---

[1] Program for processing and analysis of sociological and marketing information Vortex10. – URL: https://www.vortex10.ru/main.

students of the Institute of Construction and Architecture, and practical work in the field of assessing the socio-economic condition of the territory is not the purpose of their training. However, the understanding that in the future they will be engaged in the investment and construction sector makes this practical experience extremely important, with 43.4% of respondents noting that they have an understanding that they can really influence the development of a particular territory, and 35.4% - that they have learned how to compile and process public opinion polls. From the point of view of the development of soft skills, it is significant that 57.5% of respondents said that they learned how to work in a team. Only 5 respondents (1.7%) found it difficult to answer this question and 4 people (1.3%) replied that they did not learn anything new.

When asked what digital tools were used when working on the project, the majority of students (93.8%) noted standard office programs, which is expected, since an explanatory note was required. 34.5% performed drawings in special programs, 30.1% used manual drawings, 23% used computer games to visualize the project (The Sims, Minecraft), and 11.5% used video. It should be noted that the experts and other participants of project activities greeted the results of the projects presented not just by means of the familiar PowerPoint presentations, but also in the form of manual drawings or in special programs, or through video visualization created in gaming apps, positively, showing the effectiveness of using engineering language in presenting the results of the project work.. Moreover, when communicating with students after the competition, many of them mentioned that they took note of the idea of presenting the project results at their discretion, which indicates the success of the chosen tool for the development of students'

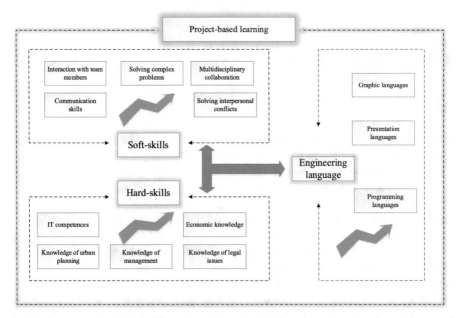

Fig. 1. Process of hard, soft skills and engineering language development in project-based learning

multilingual competencies and increasing students' interest in mastering engineering language. The process of hard, soft skills and engineering language development in project-based learning is presented in Fig. 1.

Answering the question about evaluating their work and the work of their team, the opinions were distributed, as shown in Table 1.

Table 1. Distribution of respondents' answers (%) according to their evaluation of the project

Scores/Performance evaluation	Evaluating level of own professional work (hard skills)	Team performance evaluation (soft skills)	Evaluation of team project visualization (mastery of engineering language)	Team performance evaluation (soft skills)
5	51,3	64,6	63,7	54,9
4	33,6	22,1	25,7	34,5
3	13,3	11,5	8,8	6,2
2	0,9	1,8	1,8	4,4
1	0,9	–	–	–
Total	100	100	100	100

Thus, there is a critical attitude of students to the results of their work and recognition of the role of the team in achieving the goal set for the team. Low marks for their work, visualisation and team performance performances will allow students to avoid similar project defense results in the future by focusing on the need to improve their hard, soft competences and engineering language skills.

When asked if it were interesting to take part in the work on territorial development projects, 64.6% of students answered positively, 15.9% said that they were forced to take part in joint work in order not to let their team down, 14.2% found it difficult to answer and only 5.3% answered that they were not interested in project work.

When asked how many times before the spring of 2022 they had taken part in scientific competitions and conferences, the majority of students (41.6%) replied that they had never done it, and the majority of those who answered (40 people) were students of the Institute of Construction and Architecture. 19.5% of respondents took part once, 24.8% two or three times, 13.3% more than five times, 0.9% more than 10 times (one respondent).

Answering the question of whether they plan to take part in scientific competitions and conferences in the future, respondents chose "yes, of course", "yes, if it is the teacher's requirement" and "I find it difficult to answer" were distributed almost equally - 30.1%, 28.3%, 26.5%, respectively. 5.3% of respondents (6 people) answered negatively and 9.7% (11 people) replied that they did not know where to find information about such events.

Interesting results were given by a question with a request to share the emotions and feelings from project work and presentation. Only 13 respondents (11.5%) did not answer the question, and five of them inserted positive "smiles" in the answer field. Only two respondents wrote that they did not understand why this project work was carried out. The rest of the answers contain a positive assessment of the project activity and the project competition. Students noted that project activity is an interesting and unusual experience for them. Moreover, if initially the attitude towards the task was negative, in the process of working on the project, the attitude towards it has changed significantly. The students noted their responsibility to the citizens, for whom territorial development projects will be implemented in the future, their excitement during the performance, the fighting spirit from teamwork. Another important point for students was the opportunity to compare different development projects of the same territory, which allowed them to take a different look at the features of these projects. The students noted the value of teamwork, appreciated the logic of step-by-step work on the implementation of the projects and the reality of the project assignment related to their future professional activities, although they noted that the development of the territory of Yekaterinburg would be more interesting for them. Some of the students took information visualization ideas from other teams, confirming their interest in improving their personal engineering language skills. However, some students did not have detailed enough (in points) evaluation results of their project to improve further projects, and some decided to approach the issue of project team composition more carefully in the future. Several students expressed their desire to participate in such competitions in the future and noted that they had been afraid of such events for no reason.

Summing up the results of the project competition, the curators of the project teams agreed with the positive assessment of the students, and developed directions for improving the organization of students' project activities in the form of a project competition to learn hard, soft skills and engineering language.

The project customers, who acted as experts, requested all student projects for further study of the possibilities of implementing ideas for the development of territories.

5 Conclusion

A multilingual world implies proficiency not only in foreign languages and a set of hard skills, but also in soft competences, engineering language, and the language of technology. One of the priority components of a graduate's professional competence is a high level of proficiency in visual engineering, an international language of professional communication that can be understood without verbal support.

Teaching students on real projects commissioned by partner employers allows them to assess the level of professional training in solving practical problems in the field of urban planning, economic justification of design solutions, construction organization and project management, i.e. mastery of hard skills. While working together on projects, students gain soft skills by developing creative potential, building teamwork, and solving issues that arise. A public project defence involving the visualisation of project results enhances engineering language skills. The expansion of the boundaries of the personal world through the expansion of language boundaries [37] for a more complete expression of one's thoughts, the results of one's work and the results of one's creativity can be forced by joint teamwork on a project.

In working on a real-world project, students learn hard and soft skills in an integrated way, as well as the engineering language for presenting the results of group project work, thus achieving a result that is feasible to apply in practice.

Work on real projects for the development of territories stimulates the intensification of scientific research on topical issues of the development of the regional economy, including the field of investment and construction activities.

The success of using project-based learning is due to the practical orientation of the training, implemented on real examples and evaluated by practitioners as experts. Involving students in solving the pressing problems of the city makes it possible to increase the social responsibility of young people and the competitiveness of graduates in the labour market.

References

1. Bylieva, D.S., Lobatyuk, V.V., Rubtsova, A.V.: Information and communication technologies as an active principle of social change. In: IOP Conference Series: Earth and Environmental Science, vol. 337, no. 1, p. 012054 (2019). https://doi.org/10.1088/1755-1315/337/1/012054
2. Bylieva, D., Lobatyuk, V., Ershova, N.: Computer technology in art (Venice Biennale 2019). In: Proceedings of Communicative Strategies of the Information Society (CSIS 2019), Article no. 18, pp. 1–6. ACM, New York (2019). https://doi.org/10.1145/3373722.3373785
3. Reshetnikova, O.: Effective learning tools in e-learning. In: Proceedings of the 20th European Conference on e-Learning 2021 (ECEL), pp. 387–393. Academic Conferences and Publishing International Limited, Berlin (2021). https://doi.org/10.34190/EEL.21.091
4. Pelz, P.: Good engineering design – design evolution by languages. Technol. Lang. 1(1), 97–102 (2020). https://doi.org/10.48417/technolang.2020.01.20
5. Pershin, V., Makeeva, M., Tcilenko, L.: Lingvoprofessiogramma inzhenera. High. Educ. Rus. 5, 162–163 (2004). (in Russian)
6. Plotnikov, M.A.: Kak perevesti trebovaniya potrebitelya na inzhenernyj yazyk i ocenit' zatraty. Qual. Manag. Meth. 11, 36–43 (2007). (in Russian)
7. Amirdzhanova, I., Vitkalov, V.G.: The current state of the development of geometric culture and the competence of future specialists. Sci. Vector Togliatti State Univ. 2(2), 26–31 (2015). (in Russian)
8. Rukavishnikov, V.: Geometric-graphic training of an engineer: the time of reforms. High. Educ. Rus. 5, 132–136 (2008). (in Russian)
9. Lojko, A.I.: Discourse analysis of the institutional language of modern engineering. In: Professional Communicative Personality in Institutional Discourses, pp. 58–61. Belarusian State University, Minsk (2018). (in Russian)
10. Matvejkina, Zh.V., Kalinin, A.A., Semencov, M.N.: Descriptive geometry and engineering graphics are the main language of the engineer. In: Modern Scientific Research: Problems and Prospects: Proceedings of the IV International Scientific and Practical Conference, Pero, Zernograd, pp. 268–272 (2019). (in Russian)
11. Heckman, J.J., Kautz, T.: Hard evidence on soft skills. Labour Econ. 19(4), 451–464 (2012). https://doi.org/10.1016/j.labeco.2012.05.014
12. Zaitseva, E.V., Zapariy, V.V., Asryan, G.G.: The results of reforms in the field of training highly qualified personnel in the late XX – early XXI century: scientific personnel. Hist. Mod. Perspect. 4(2), 65–75 (2022). https://doi.org/10.33693/2658-4654-2022-4-2-65-75

13. Schulz, B.: The importance of soft skills: education beyond academic knowledge. J. Lang. Commun. **2**, 146–154 (2008). https://www.bcsgea.org.bd/wp-content/uploads/2019/11/The-Importance-of-Soft-Skills-Education-beyond-academic-knowledge.pdf

14. Wats, M., Wats, R.K.: Developing soft skills in students. Int. J. Learn. **15**(12), 1–10 (2009). https://doi.org/10.18848/1447-9494/CGP/v15i12/46032

15. Voronina, L., Zaitseva, E., Gaisina, E., Izmozherova, N.: Informational-communicational and digital competence of teachers of Russian universities. In: 13th International Conference on Education and New Learning Technologies, pp. 7336–7340. EDULEARN (2021). https://doi.org/10.21125/edulearn.2021.1483

16. Daineko, L.V., Yurasova, I.I., Karavaeva, N.M.: Creative projects as a link between theory and practice. In: Bylieva, D., Nordmann, A. (eds.) PCSF 2021. LNNS, vol. 345, pp. 558–572. Springer, Cham (2022). https://doi.org/10.1007/978-3-030-89708-6_47

17. Daineko, L., Davy, Y., Larionova, V., Yurasova, I.: Experience of using project-based learning in the URFU hypermethod e-learning system. In: 18th European Conference on e-Learning, pp. 145–150. Academic Conferences and Publishing Limited, Valencia (2019). https://doi.org/10.34190/EEL.19.066

18. Daineko, L.V., Reshetnikova, O.E.: Project method – an effective instrument for developing competencies of future professionals. Eur. Proc. Soc. Behav. Sci. **98**, 221–230 (2020). https://doi.org/10.15405/epsbs.2020.12.03.23

19. Voronina, L., Zaitseva, E., Gaisina, E., Izmozherova, N.: Personnel potential of teachers of Russian universities in the conditions of digitalization. In: 13th International Conference on Education and New Learning Technologies, pp. 7470–7476. EDULEARN (2021). https://doi.org/10.21125/edulearn.2021.1517

20. Daineko, L., Goncharova, N., Larionova, V., Ovchinnikova, V.: Experience of introducing project-based learning into university programmes. In: 13th International Conference on Education and New Learning Technologies, pp. 8038–8043 (2021). https://doi.org/10.21125/edulearn.2021.1632

21. Daineko, L.V., Goncharova, N., Larionova, V.A., Ovchinnikova, V.A.: Fostering professional competencies of students with the new approaches in higher education. Eur. Proc. Soc. Behav. Sci. **98**, 231–239 (2020). https://doi.org/10.15405/epsbs.2020.12.03.24

22. Kechagias, K.: Teaching and Assessing Soft Skills. 1st Second Chance School of Thessaloniki, Neapolis (2011)

23. Musa, F., Mufti, N., Latiff, R.A., Amin, M.M.: Project-based learning (PjBL): inculcating soft skills in 21st century workplace. Proc.-Soc. Behav. Sci. **59**, 565–573 (2012). https://doi.org/10.1016/j.sbspro.2012.09.315

24. Goncharova, N.V., Pelymskaya, I.S., Zaitseva, E.V., Mezentsev, P.V.: Green universities in an orange economy: new campus policy. In: Bylieva, D., Nordmann, A. (eds.) PCSF 2021. LNNS, vol. 345, pp. 270–284. Springer, Cham (2022). https://doi.org/10.1007/978-3-030-89708-6_23

25. Antonova, A., Aksyonov, K., Ziomkovskaia, P.: Development of method and information technology for decision-making, modeling, and processes scheduling. AIP Conf. Proc. **2425**(1), 130003 (2020). https://doi.org/10.1063/5.0082100

26. Khalyasmaa, A., Zinovieva, E., Eroshenko, S.: Performance analysis of scientific and technical solutions evaluation based on machine learning. In: Proceedings – 2020 Ural Symposium on Biomedical Engineering, Radioelectronics and Information Technology, USBEREIT 2020, pp. 475–478. IEEE, Yekaterinburg (2020). https://doi.org/10.1109/USBEREIT48449.2020.9117774

27. Vogler, J.S., Thompson, P., Davis, D.W., Mayfield, B.E., Finley, P.M., Yasseri, D.: The hard work of soft skills: augmenting the project-based learning experience with interdisciplinary teamwork. Instr. Sci. **46**(3), 457–488 (2017). https://doi.org/10.1007/s11251-017-9438-9

28. Dogara, G., Saud, M.S.B., Kamin, Y.B., Nordin, M.S.B.: Project-based learning conceptual framework for integrating soft skills among students of technical colleges. IEEE Access. **8**, 83718–83727 (2020). https://doi.org/10.1109/ACCESS.2020.2992092

29. Chassidim, H., Almog, D., Mark, S.: Fostering soft skills in project-oriented learning within an agile atmosphere. Eur. J. Eng. Educ. **43**(4), 638–650 (2018). https://doi.org/10.1080/030 43797.2017.1401595

30. Tadjer, H., Lafifi, Y., Seridi-Bouchelaghem, H., Gülseçen, S.: Improving soft skills based on students' traces in problem-based learning environments. Interact. Learn. Environ. (2020). https://doi.org/10.1080/10494820.2020.1753215

31. Ravindranath, S.: Soft skills in project management: a review. IUP J. Soft Skil. **10**(4), 16 (2016)

32. Nealy, C.: Integrating soft skills through active learning in the management classroom. J. College Teach. Learn. (TLC) **2**(4) (2005). https://doi.org/10.19030/tlc.v2i4.1805

33. Sudana, I.M., Apriyani, D., Suryanto, A.: Soft skills evaluation management in learning processes at vocational school. In: Journal of Physics: Conference Series, vol. 1387, no. 1, p. 012075. IOP Publishing (2019). https://doi.org/10.1088/1742-6596/1387/1/012075

34. Shakir, R.: Soft skills at the Malaysian institutes of higher learning. Asia Pac. Educ. Rev. **10**(3), 309–315 (2009). https://doi.org/10.1007/s12564-009-9038-8

35. Claxton, G., Costa, A., Kallick, B.: Hard thinking about soft skills. Educ. Leadersh. **73**(6), 60–64 (2016)

36. Zaitseva, E., Zapariy, V.: Role of the university corporate culture in university management. In: 10th International Days of Statistics and Economics, pp. 2089–2095. Melandrium (2016)

37. Wittgenstein, L.: Tractatus Logico-Philosophicus. Routledge & Kegan Paul (1922)

An Interdisciplinary Project as a Means of Developing Digital Skills

Yevgenia Vorontsova(ID), Anna Grishina(ID), Alexandra Dashkina(ID),
and Nina Popova(✉)(ID)

Peter the Great Saint Petersburg Polytechnic University, Polytechnicheskaya 29,
195251 Saint Petersburg, Russia
ninavaspo@mail.ru

Abstract. The article gives insights into developing students' individual abilities to use information and communication technologies in certain contexts using project technology. Its application is not only conducive to developing students' critical thinking, but also teaches them to work in a team and to find creative approaches to solving problems, inter alia by using digital tools. The article illustrates how economic majors improved their digital skills by participating in a project referred to as "Breaking News". The students participating in the project used *Google Sites* to create a web-site on business and economic news coverages. The project was interdisciplinary since it integrated three components: economics, the English language and digital literacy. At the initial stage of the project the participants formed groups and familiarized themselves with the digital tools. Then each student had to prepare a presentation in English on a recent event, make a glossary of the key terms used in the upcoming presentation and then give a talk in class. At the final stage, the site was discussed in class. A final poll was aimed at assessing the extent to which the students' digital skills had improved by the end of the project. The poll illustrated that 100% of the participants improved their digital skills; 96% of the students felt that working on the project motivated them to study English, and 76% of the respondents claimed to have improved their knowledge of economics. Thus, similar interdisciplinary projects can be included in the course of teaching English for Special Purposes.

Keywords: Digital skills · Project · Interdisciplinary · Website · Poll · Presentation

1 Introduction

At present digitization has brought about drastic changes in social, political and economic spheres; art, culture and education. Tertiary education is in a state of flux since the conventional educational system as well as teaching methods is now being replaced with a new innovative model. The idea of digital university implies combining all educational services in a single digital space, which will make it possible to develop the synergy capable of bringing the teaching and learning process to a new level [1, p. 58]. Achieving this synergy is only possible if all the stakeholders of the educational process are willing to adapt to these changes. Teaching techniques can go digital in a variety of ways depending on the university, but it is the digitization of higher education that can really put universities on the map.

Digitization can turn the future professional into a valuable asset since digital solutions integrated into the educational process are based primarily on particular problems that students have to address in the course of studying their professionally-oriented subjects. On the other hand, digitization creates the demand for university graduates capable of solving practical problems in accordance with a particular situation, thinking critically, working in teams and working independently when it comes to doing assignments that are essential to individual and professional development.

It is vital that modern specialists should be multilingual since in their professional communication they will find themselves in the situations in which foreign language is a tool of intercultural interaction. To this end, educational technologies should be aimed at training professionals capable of working in the multilingual digital environment. Consideration of educational and digital technologies leads us to an understanding of their gradual convergence, a qualitative graph of which is presented in Fig. 1.

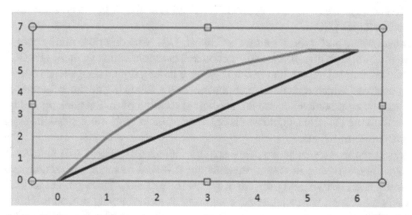

Fig. 1. Qualitative graph of the convergence of educational and digital technologies in modern higher education

This graph shows that over time (vertical axis), there is a gradual improvement in digital technology (red line above the conditional horizontal axis of achievements), which is accompanied by a simultaneous increase in the number of educational technologies (green line). Due to the fact that at present practically no educational technology can do without the use of digital resources, we can conclude that these two trends in higher education are convergent.

The use of problem-based learning technologies, in particular project ones, involved in vocational education is observed on the world scale. The use of digital project technologies in a modern university is especially useful for improving the efficiency of training students majoring, inter alia, in management and economics.

Noting the convergence of digital and educational technologies in the modern university educational process, we consider it important to note their positive impact on the educational process and continue further research into them. The modern system of tertiary education is aimed at creating the conditions that are conducive to the integral personal development of graduates, who will be knowledgeable not only in their employment field, but also related subjects [2]. Today, researchers stress that transferrable and soft skills in higher education are as important as studying vocational subjects. Thus, university education should be aimed at building such skills [3].

According to the results of research in the sphere of language education, a foreign language is a subject that makes it possible to build communication skills, transferrable skills and personality traits that are necessary for engineering students, such as the pursuit of self-improvement, flexibility in the course of solving professional problems, self-discipline and the ability to make the most of their potential [4].

When it comes to developing transferrable skills, it is worth mentioning that if a foreign language is taught on the basis of innovative techniques, technologies and digital tools, the students acquire digital skills. Future professionals need to acquire digital skills since they enhance their employability in the changing market conditions. In order to build a successful career, they need to be capable of continuous lifelong learning, and using the emerging digital tools inter alia. Different scientists have defined digital skills in different ways. Petrova, V. S. defined digital skills as the ability to take advantage of new technologies [5].

However, we prefer the definition by Gileva, T.A., who describes digital competencies as individual abilities to use information and communication technologies (ICT) in a certain context [6]. According to the DigComp model, digital competence comprises five main areas, which include 21 components in the following fields: information and data literacy; communication and collaboration; creation of digital content; safety and problem solving [7]. Thus, digital skills combine a number of various subskills essential for a modern professional.

Project technology helps students acquire these skills, which will enhance their employability. Working in teams, students are involved in social interaction through discussion and clarifying any uncertain issues [8; p. 1249]. Since project technology is closely associated with teamwork, it contributes to developing critical thinking and teaches students to adopt creative and unconventional approaches to solving problems. To this end, digital tools can be applied.

Researchers and educationalists recognize the fundamental role of interdisciplinary interaction in the practical implementation of the competency-based approach. In addition, in the university education, interdisciplinary links ensure continuity and integrity of learning, the interaction of the subject matter and procedural components of the disciplines involved being considered [9].

Interdisciplinary projects in education are relevant since they ensure the integrity of the learning process by incorporating the subject matter and procedural aspects of related disciplines when it comes to solving interdisciplinary problems.

2 Literature Review

Scholars are still doing extensive research into application of the project technology in teaching foreign languages. By participating in a project, students do not only take part in a discussion, but develop a final product, for instance, a real project that could be applied in education [10; p. 15]. This method can be successfully applied in teaching a foreign language for special purposes in universities, as it is aimed at obtaining practical results, thus increasing students' motivation to improve their knowledge of the subject in question and in related fields.

In modern pedagogical science, project technology is defined as a pedagogical technology that teaches students to put their existing knowledge into practice and acquire new knowledge. In this paper, we adhere to the definition given by Mezentseva and Popova, according to which the project method is students' independent work facilitated by a teacher. It is aimed at achieving practical results and developing students' critical thinking and creative potential [11; p. 54].

At the same time, interdisciplinary projects should be integrated into foreign language teaching since they develop foreign-language communicative and cognitive skills. The authors define these skills as an integrative characteristic of a person, which indicates whether a would-be specialist is able to make informed decisions in various situations and to participate in professional communication in a foreign language using digital technologies in the course of communication. In our opinion, it is advisable to talk about the integration of interdisciplinary projects into the process of teaching a foreign language. The research suggests that such educational projects contribute to the development of cognitive skills, integrative knowledge and allow learners to gain practical experience in their professional field.

In the scholarly papers dedicated to the integration of interdisciplinary projects into the process of teaching a foreign language, foreign researchers define an interdisciplinary project as a method of collaborative learning. They highlight the following characteristics of this educational technology:

1. Students' active involvement in the educational process;
2. Independent learning and teamwork;
3. Integration of knowledge and skills into professional areas;
4. Students' independence and awareness;
5. Activation of critical thinking;
6. Positive attitude to the subject of research;
7. Students' satisfaction with the result of the project work [12; p. 329].

In this definition, an interdisciplinary project is characterized by the independence of the learning process. However, in our opinion, the teacher should coordinate and facilitate the students' work and check their performance at each stage of the project.

In the course of working on an interdisciplinary project, students integrate information, data, methods, tools, concepts and/or theories from two or more subjects to launch intellectual products, explain phenomena or solve problems in the ways that would be unattainable on the basis of a "single-subject approach" [13; p. 54].

The authors developed an interdisciplinary educational project and ran it in the classroom and in the course of students' independent work as part of the discipline "Foreign language for special purposes". Since the project includes the content components of such subjects as foreign language, economics, and digital literacy, it represents a three-component interaction or an interdisciplinary triad. Being a conditional universal component of interdisciplinary synthesis [9], computer science is actually identical to the concept of digital literacy. Therefore, we will use this term to further describe the triadic interaction of the disciplines in our project, especially since the students in the surveyed sample have digital literacy in their curriculum.

According to the theory of interdisciplinary links, in the interdisciplinary triad two substantive components of the interacting disciplines vary, whereas the third, digital component is preserved as a universal procedural or activity component which provides for interdisciplinary connections [14–16].

According to the classification proposed by Polat, the interdisciplinary project described in this paper combines the informational and research types of educational projects. It enables students to develop their analytical skills and is conducive to individualization of educational and scientific-cognitive activities [17].

The method of teaching a foreign language through an interdisciplinary project has a number of characteristics: 1. The student becomes the subject of the educational process, who independently sets tactical goals and determines what information is needed in accordance with the concept of the project; 2. The individual pace of students' work on the educational project ensures its smooth running; 3. Since students acquire profound basic knowledge of several subjects, they can apply it in various social and professional situations [18].

Thus, in the interdisciplinary project described in this paper, the teachers of a foreign language and economics act as mentors/facilitators and consultants, respectively, whereas the development and implementation of the educational project are geared towards the students' independent extracurricular work. Being able to work independently is a key transferrable skill that needs developing in the context of digitization of the educational environment [19–21]. The efficiency of independent work is closely linked with digital skills, which were earlier defined as the students' individual abilities to use information and communication technologies to their advantage [22].

3　Results and Discussion

In a number of studies, the use of interdisciplinary projects in ESL classes is also considered as a way to encourage students who do not specialize in linguistics to study a foreign language [23]. It was substantiated by the results of a survey which was conducted in Peter the Great Polytechnic University in a group of third-year students majoring in economics.

The authors of this paper launched the "Breaking News" project aimed at developing digital skills in sophomores majoring in economics at the Institute of Industrial Management, Economics and Trade of St. Petersburg Peter the Great Polytechnic University. The project involves the use of advanced digital technologies, and by the end of the semester its participants are expected to create a website using the Google Sites service. The site is supposed to contain brief news updates prepared by the participants of the project and presented by them in foreign language classes.

This project meets the requirements for interdisciplinary educational projects and enables the students to develop foreign-language communicative, cognitive and digital skills, as well as improve their knowledge in the field of economics. The final output of the project is the development of an informational website containing English-language news updates prepared by students and posted on the Google Sites platform. This assignment is to be part of students' independent homework.

The Breaking News can be referred to as an interdisciplinary educational project since it involves studying economics as the core discipline. Thus, economics is a disciplinary component that forms vertical retrospective links with the discipline of foreign language. Therefore, since Economics is studied earlier as part of a professionally-oriented foreign language course, the interdisciplinary connections implemented in this project help the students retrieve the background knowledge from their memory and develop the skills that will allow them to put their theoretical knowledge into practice. Figure 1 shows the disciplines that are conducive to interdisciplinary interaction in the Breaking News project, where computer science is a conditional disciplinary component aimed at developing the students' digital literacy, which will enable them to create a website on the Google Sites platform.

Interdisciplinary links in project "Breaking News" are shown in Fig. 2.

Fig. 2. "Interdisciplinary links in the Breaking News educational project"

The essential aspects of interdisciplinary synthesis in the course of implementing the project are illustrated in Table 1:

Table 1. Implementation of the interdisciplinary links in the Breaking News project

Disciplines	Interdisciplinary interaction
Foreign Language + Economics	Retrieving from memory background knowledge in the field of economics to determine the topic and objectives of the project in the Foreign Language

(continued)

Table 1. (*continued*)

Disciplines	Interdisciplinary interaction
Foreign Language + Economics + Digital Literacy	The search and analysis of professionally-oriented information on English-language sites teaches the students to use various Internet resources, analyze and systematize the professionally-oriented material and highlight the key facts
Foreign Language + Economics	Summarizing the information given in a foreign-language professionally-oriented text boils down to reducing the volume of the article while significantly preserving its main content. This activity develops cognitive skills
Foreign Language + Economics + Digital Literacy	Creating a website containing classified sections with foreign-language news updates develops the students' digital literacy and improves their cognitive skills

This table illustrates the vertical interdisciplinary connections between the three disciplines: Foreign Language, Economics and Digital literacy. Their didactic potential has been successfully fulfilled in the interdisciplinary educational project. The combination of Foreign Language and Economics provided the subject matter in our project, since all English-language news updates covered economic topics. By studying the discipline Digital literacy, the students improved their procedural skills, which helped them implement the project task.

The interdisciplinary project Breaking News has a number of general characteristics. Its purpose is to create a group news site by mainstreaming the interdisciplinary links between the Foreign Language, vocational subjects and the disciplines of the basic cycle. The following tasks are to be accomplished in the course of the interdisciplinary project:

1. The students' knowledge in the field of economics is consolidated and put into practice;
2. The students make progress in studying the foreign language and extrapolate these results to other disciplines;
3. The students understand the meaningful links between individual disciplines in the course of working on the interdisciplinary project.

The project consists of the following stages:

1. The research topic as part of the discipline "Foreign language. Basic course" is defined and the project teams are formed in the ESL class. The topic and the team membership are approved by the teacher;
2. The information on the chosen topic is selected (the students with a poor command of the foreign language should use the sources selected by the teacher);

3. The students retell (render) the articles, compile a glossary of professional terms and ask questions on the content of the article. These assignments are to be completed in the course of the students' extracurricular independent work;
4. The students create a website containing classified sections by using the service *Google Sites*. They analyze, sort out and categorize the news articles on the site. At this stage the students carry out the assignments independently as part of their homework after having the progress tutorial with the teacher.
5. The students present the project, discuss its results, determine the team accomplishments and individual achievements of each team member in an ESL class.

The project comprises organizational, individual, team and final stages of work:

1. At the organizational stage the students are fully briefed on what they are supposed to achieve. The research area is determined (the subject matter of the news updates should correspond to the area the students specialize in), teams are formed, the participants' roles are defined, and each of them is given an individual assignment. The students familiarize themselves with the digital tools selected by the teacher that will help them complete the project;
2. Individual stage: at this stage, each student prepares a presentation in English about one of the latest world events, compiles a glossary of the terms used in the presentation, gives a presentation of the selected news in class, expresses their opinion on the problem and answers questions from the audience;
3. At the team stage the participants of a small subgroup use the Google Sites service to create a news site on the basis of the materials prepared at the stage of individual work. At the team stage the students are also given counselling by the teachers of vocational subjects;
4. At the final stage, the students report on the results of the project. The created news site is presented, discussed and evaluated; the performance is assessed; the achievements of the team as well as the individual accomplishments are considered.

Since the project is interdisciplinary, it is carried out in close collaboration with the teachers of vocational subjects. The semester project involves weekly interim reporting. It enables the teacher to evaluate each student's individual performance at different stages of the project and, if necessary, to adjust the work of the project team. The study shows that the vast majority of the students pointed out that they were more willing to study the foreign language while they were taking part in the interdisciplinary project. They also felt that they had become more interested in applying digital technologies.

In our opinion, digital literacy (a student's ability to work efficiently with a variety of digital tools, programs and applications) can be regarded as an essential component of digital competence.

Before the students started working on the project, they were asked to take part in a survey and assess their own digital literacy. The survey comprised 7 sections, which, in turn, consisted of 10 questions, except for the first section, which included 5 questions. The survey was aimed at assessing the ability to work:

i.   with the Windows operating system;

ii.   in the Word text editor;
iii.   with browsers;
iv.   with e-mail programs;
v.   with the Power Point presentation program;
vi.   with software for processing and recording information in audio and video formats;
vii.   in virtual learning environments, inter alia, in Moodle;
viii. in blogs and social networks.

Table 2 illustrates the scale for evaluating the survey results.

Table 2. Digital literacy assessment scale

Level	Evaluation score
Basic level	0–45 points
Intermediate level	46–70 points
Advanced level	71–75 points

Initial levels of digital literacy (before the project) are illustrated in Fig. 3.

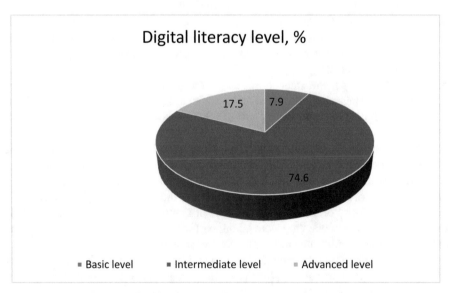

Fig. 3. Initial levels of digital literacy (before the project)

The bar chart illustrates that the vast majority of the respondents highly appreciated their own level of digital literacy. It means that the students are able to work with the software described above, and they are also capable of performing different kinds of specific tasks within each block.

However, the detailed analysis of the survey results allowed us to ascertain that despite such high results of self-assessment, the students pointed out some skills that needed improving in each block. The assignments aimed at their development were included in the list of the project tasks (Table 3).

Table 3. Characteristics of the main project stages and the tasks aimed at the development of digital literacy

#	The name of the stage	The tasks aimed at the development of digital literacy
1	Solo	• Search for specific information on the internet in accordance with a certain topic • Search for illustrations (using free-of-charge stock graphic resources); creating a website template in a Word document, working with headers and footers, text formatting, designing the layout of relevant images, compiling tables • Communication via the team email; setting up a group mailing system
2	Team	• Discussing ideas in the Moodle-based forum • Creating a preliminary Power Point presentation of the project • Adding animation to the presentation • Adding an audio recording (made with the Audacity program) to the presentation • Adding a video (made with the Movie Maker program) to the presentation • Creating a website using Google services • Adding or deleting pages • Sorting out information • Adding images, a banner, a logo

The solo stage of the project develops the learners' cognitive skills and self-discipline, since the student has to use English-language Internet portals to find, select and analyze the information relevant to the topic of the curriculum, as well as to express his/her own opinion about the event. In addition, at the solo stage of the project, the students learn to use various internet resources, analyze and sort out the material, highlight the key facts. Thus, this stage is conducive to the development of the students' digital literacy. Moreover, at the solo stage, students learn to plan their own activities and their cognitive interest grows, since they have to select the information and resources relevant to their assignment without being given any instructions.

The students also develop their receptive and productive language skills by making a presentation in the classroom and expressing their own opinion about the event. They improve their pronunciation and grammar, as well as replenish their vocabulary by compiling glossaries with word definitions. They also learn to choose and use adequate language forms and means of communication, and practice using verbal and non-verbal means of communication. This classroom activity contributes to the development of communication skills.

At the team stage the students create a news site by working together in the digital environment and using the Google Sites service. The Google Sites service has a number of features that enable students to work on the site together remotely in real time, maintaining communication in the digital environment. At the same time, the teacher can monitor the progress and evaluate each student's individual performance.

Apart from text files, the following elements of a website can be created by using the service: layouts in the form of ready-made sections with several elements inside (usually a combination of text and photos); a table of contents; a carousel of images; YouTube videos; marks on the map; documents, presentations, tables, forms and diagrams made by using other Google services. At this stage, students learn to work in a team and also develop their digital skills, which can be defined as the willingness and ability to apply information and communication technologies confidently, efficiently and safely in various spheres of daily life. An example of the website created by the students in the project is shown below (Fig. 4).

Fig. 4. Interface of the *Breaking News* website created by the students

In order to get feedback on teaching a foreign language for special purposes by applying the project technology, we surveyed 63 third-year students at the Institute of Industrial Management, Economics and Trade of St. Petersburg Peter the Great Polytechnic University. At the end of the semester, the students who participated in the project were asked 12 questions to confirm our hypothesis about the feasibility of using interdisciplinary educational projects in teaching a foreign language. Table 4 presents the results of the survey at the end of the Breaking News project.

The students who participated in the survey also mentioned the problems they faced while they were working on the project. A high percentage of the respondents (42%) claimed that they had difficulty selecting relevant news articles in English-language internet sources. This problem was mainly mentioned by the students with a low level

Table 4. The results of the student survey at the end of the Breaking News project

Question	Yes	No
Have you improved your knowledge of the foreign language by taking part in the project?	96%	4%
Have you improved your knowledge of economics/business by taking part in the project?	76%	24%
Have you improved your digital literacy by taking part in the project?	100%	–
Has the Breaking News project helped you improve your self-discipline and time management?	96%	4%
Have you personally succeeded in achieving the project's objectives (creating a news website and a glossary of special terms)?	100%	–
Does your personal assessment of the work under the Breaking News project agree with the teacher's assessment?	100%	–
Do you think that this format of independent work is conducive to finding new solutions and tools for its implementation?	92%	8%
Do you think that this teaching technology increases the students' motivation to learn a foreign language?	96%	4%
Do you think that this teaching technology stimulates the students' cognitive activity?	100%	–
Do you think that incorporating projects into foreign language classroom activities can improve the students' background knowledge of economics (their core discipline)?	76%	24%

of foreign language proficiency. Thus, it is essential to offer students a selection of news websites with current news updates in a foreign language.

According to the results of the survey, all the respondents were interested in working on the project since digital technologies were integrated into the learning process. They commented that the presentation of news reports in online classes and developing a website in teams were the most interesting project assignments. All the participants pointed out that they had more incentive to learn the foreign language while working on the project, and also described it as an effective means of activating cognitive activity. Most of the respondents (76%) also regarded the Breaking News project as an effective means of improving their cognition [19]. They expressed a desire to continue learning a foreign language by carrying out interdisciplinary educational projects.

After the students designed the website, we conducted a survey to gauge the extent to which their digital literacy had improved. They were asked the same questions they had answered before they started working on the project (Fig. 5).

After completing the project, the students with low levels of digital literacy made remarkable progress, while the students with adequate levels of digital literacy (according to their own estimates) showed just a modest improvement.

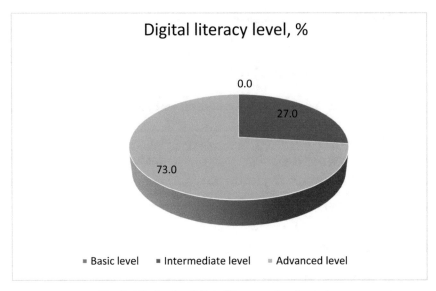

Fig. 5. The levels of digital literacy after the project

4 Conclusion

The results of this study showed that students of a technical university should acquire digital skills. We defined digital skills and considered their crucial role in enhancing a modern graduate's employability. This paper describes the educational project aimed at improving these skills. Project-based learning can be integrated into the course of the discipline "Foreign language for special purposes", which is on the curriculum at the Institute of Industrial Management, Economics and Trade of St. Petersburg Peter the Great Polytechnic University. The project can be implemented by applying distance learning educational technologies.

Through the involvement in the interdisciplinary educational project, students acquired analytical skills by studying and interpreting the information available on the internet and deciding on the best way of doing an assignment. The students' digital literacy also improved on the solo stage of the project in the course of searching, selecting and processing information in English.

The use of problem-based learning technologies, in particular, interdisciplinary project ones, can be regarded as an effective means of developing learners' digital skills. The educational project technology used by us for teaching students majoring in economics can be projected to learners of any other profile. Thus, the convergence of educational and digital technologies is confirmed.

References

1. Gromova, N.: Digitalization of education in Russia: from modern technologies to an innovative model of the educational environment. In: Nazarov, A. (ed.) 2nd International Scientific

and Practical Conference on Digital Economy (ISCDE 2020), vol. 156, pp. 54–58. Atlantis Press, Dordrecht (2020). https://doi.org/10.2991/aebmr.k.201205.009

2. Popova, N.: Interdisciplinary paradigm as the basis for the formation of integrative skills of students at a multidisciplinary university (illustrated by foreign language teaching practices): abstract of doctoral dissertation. Saint Petersburg (2011). (in Russian)

3. Dashkina, A., Nam, T.: Prospects and experience of integrating basic and non-formal humanitarian education in a technical university. World Acad.: Cult. Educ. **5**, 34–50 (2020). (in Russian)

4. Popova, N., Pyatnitsky, A.: Forming the universal skills of processing foreign-language scientific information by teaching students abstract translation and commenting. Foreig. Lang. Sch. **4**, 2–12 (2016)

5. Petrova, V., Shcherbik, E.: Measuring the level of formation of digital competencies. Moscow Econ. J. **5**(3), 237–244 (2018). https://qje.su/otraslevaya-i-regionalnaya-ekonomika/moskov skij-ekonomicheskij-zhurnal-5-2018-114/. (in Russian). Accessed 10 June 2022

6. Gileva, A.: Competencies and skills of the digital economy: development of a personnel development program. Herald of Ufa State Oil Tech. Univ. Sci. Educ. Econ. Series: Econ. **2**(28), 22–35 (2019). https://cyberleninka.ru/article/n/kompetentsii-i-navyki-tsifrovoy-ekonomiki-razrabotka-programmy-razvitiya-personala. (in Russian). Accessed 10 June 2022

7. Carretero, S., Vuorikari, R., Punie, Y.: DigComp 2.1. The Digital Competence Framework for Citizens. With Eight Proficiency Levels and Examples of Use. Publications Office of the European Union, Luxembourg (2017). https://doi.org/10.2760/38842

8. Guilland, A., Liliya, L., Nieminen, S.: Teaching and learning transversal competences in the higher education. Learnings from Erasmus + Socces-project. In: Proceedings of INTED2017 Conference, Valencia, Spain, pp. 1248–1254 (2017). https://doi.org/10.21125/inted.2017. 0044

9. Popova, N.: Interdisciplinary integration as a basis for designing the educational process in higher school. Univ. Sci. J. Publishing house: St. Petersburg Univ. Sci. Consort. **3**, 98–109 (2012)

10. Biasutti, M., EL-Deghaidyp, H.: Interdisciplinary project-based learning: an online wiki experience in teacher education. Technol. Pedag. Educ. **24** (3), 1–17 (2014). https://doi.org/10. 1080/1475939X.2014.899510

11. Mezentseva, M., Popova, N.: Modern trends in developing the method of project technologies at school and university. Sci. Tech. Bull. St. Petersburg State Polytechnic Univ. Soc. Commun. Educ. **4**, 56–68 (2019). https://doi.org/10.18721/JHSS.10406

12. Thuan, P.: Project-based learning: from theory to EFL classroom practice. In: Proceedings of the 6th International Open TESOL Conference, pp. 327–339. OpenTESOL, Ho Chi Minh City (2018). https://www.researchgate.net/publication/331071691_project-based_learning_ from_theory_to_efl_classroom_practice. Accessed 25 June 2022

13. Boix Mansilla, V., Dillon, D., Middlebrooks, K.: Building Bridges Across Disciplines: Organizational and Individual Qualities of Exemplary Interdisciplinary Work (2002). https://static1.squarespace.com/static/5c5b569c01232cccdc227b9c/t/5ee78feba7b1 db5317ee793c/1592233963722/BUilding+Bridges.pdf. Accessed 25 June 2022

14. Akopova, M., Popova, N.: Interdisciplinary connections in higher education: textbook. Peter the Great SPbPU, St. Petersburg (2014)

15. Krylov, E., Vasileva, P.: Convergence of foreign language and engineering education: opportunities for development. Technol. Lang. **3**, 106–117 (2022). https://doi.org/10.48417/techno lang.2022.03.08

16. Balyshev, P.: The stages of discourse-oriented virtual learning environment modeling. Technol. Lang. **3**, 88–105 (2022). https://doi.org/10.48417/technolang.2022.03.07

17. Polat, E.: Project method: history and theory of the issue. Sch. Technol. **6**, 43–47 (2006)

18. Ivanova, M., Kuznetsova, E.: Technology of teaching foreign languages on the basis of an interdisciplinary approach. Societ. Social. Psychol. Pedag. **2**, 160–162 (2016). (in Russian)
19. Volodarskaya, E., Pechinskaya, L.: Improving the efficiency of independent work in the study of a second foreign language by undergraduates. In: Anikina, Z. (ed.) IEEHGIP 2022. LNNS, vol. 131, pp. 530–538. Springer, Cham (2020). https://doi.org/10.1007/978-3-030-47415-7_56
20. Shipunova, O., Pozdeeva, E., Evseev, V., Romanenko, I., Gashkova, E.: University educational environment in the information exchange agents evaluations. In: Rocha, Á., Fajardo-Toro, C.H., Rodríguez, J.M.R. (eds.) Developments and Advances in Defense and Security. SIST, vol. 255, pp. 501–511. Springer, Singapore (2022). https://doi.org/10.1007/978-981-16-4884-7_42
21. Bylieva, D., Hong, J.-C., Lobatyuk, V., Nam, T.: Self-regulation in e-learning environment. Educ. Sci. **11**, 785 (2021). https://doi.org/10.3390/educsci11120785
22. Grishina, A.: Interdisciplinary project on the basis of the discipline foreign language as a form of development of students' educational and cognitive activity. Mod. Probl. Sci. Educ. **3** (2021). https://doi.org/10.17513/spno.30920. (in Russian)
23. Berezovskaya, I., Shipunova, O., Kedich, S., Popova, N.: Affective and cognitive factors of internet user behaviour. In: Bylieva, D., Nordmann, A., Shipunova, O., Volkova, V. (eds.) PCSF/CSIS -2020. LNNS, vol. 184, pp. 38–49. Springer, Cham (2021). https://doi.org/10.1007/978-3-030-65857-1_5

Checklist as a Working Tool for the Formation of Digital Literacy of BIM-Specialists in the Multilingual World

Daria Shalina(✉) [iD], Vladislav Tikhonov [iD], Natalia Stepanova [iD],
and Viola Larionova [iD]

Ural Federal University (UrFU), 19 Mira St., Ekaterinburg 620002, Russia
`d.shalina2011@yandex.ru`, {`n.r.stepanova,v.a.larionov`}`@urfu.ru`

Abstract. The article is devoted to the analysis of the problem of personnel short-age of BIM (Building Information Modeling) specialists and the search for ways to solve it. Today, the training of BIM-specialists is considered as a constant improvement of their digital literacy in order to adapt to constantly changing conditions to the needs of the enterprise. However, current educational requirements make it difficult to attract successful students to work in international companies. BIM-specialists should develop the ability to solve professional tasks in a multilingual world. The reasons for the personnel shortage are considered, such as a lack of understanding of the need for BIM-specialists, lack of places for BIM training and difficulty in training. The relevance of BIM technologies is proved by statistics and correlation analysis. The directions of training BIM-specialists at universities are given. It is proposed to use transforming checklists in training for the requests of the enterprise with instant translation into the language chosen by the student to familiarize himself with vacancies and master the program. This is a modern working tool of an effective multilingual world, which allows to increase digital literacy in professional training and training of BIM-specialists, depending on the actual production problems at different stages of planning the work of an enterprise in the construction industry and the real estate market.

Keywords: Digital literacy · BIM-specialist · Selection · Training · Digitalization · Transformation

1 Introduction

1.1 Justification of the Problem

Today, one of the main trends in the development of Russian economic sectors is digitalization. Currently, it is planned to modernize almost all types of production through the switching to digital technologies and the introduction of scientific and technical developments [1–3].

Increasing the level of modernization and transformation of production directly entails increasing the level of knowledge and competencies of employees through adaptation and development programs and, of course, the selection and training of students to work at enterprises. It follows that together with the concept of "digitalization", such a concept as "digital literacy" appeared, i.e. the skills necessary for living and working in the digital world [4]. The need for qualified specialists who possess this digital literacy, own and use digital technologies in their professional activities is constantly increasing [1].

At the same time, the changing living conditions of people affect the requirements for working in international teams on projects. Globalization and digitalization have contributed to the emergence of the phenomenon of a multilingual world, where people speak several languages [5–7]. However, if there is no advanced knowledge in the specifics of owning special documentation on project activities, the problem of linguistic inequality in understanding issues arises. Therefore, we need a universal algorithm of actions for all members of the working team, which we must create.

As part of the digitalization of the construction industry, it is necessary to achieve the use of modern information technologies, such as Building Information Modeling (BIM), at all stages of the life cycle of construction objects [2]. BIM is the main digital tool in the field of construction production [8]. For the implementation and application of BIM, full resource security is necessary – these are the technologies and specialists themselves. Both cause difficulties.

Information modeling technologies involve the purchase of software and equipment for work, which requires significant costs [9]. However, even if the necessary technological resources are acquired, another question arises: will there be specialists capable of implementing these technologies and work in a multilingual world?

The objective problem of digitalization of construction is the actual problem of personnel shortage of BIM-specialists [2, 10, 11], i.e. their training and retraining in a multilingual world.

1.2 Relevance and Practical Significance

The relevance of the research is due to the modernization of the system of professional training and development of employees to meet the key trends of the construction industry [11–13]:

– The combination of engineering and linguistic activities as a challenge to the multilingual world [14–16];
– Complication of the design and construction of newly erected and reconstructed buildings and structures;
– Introduction of innovative technologies, including BIM;
– High competition in the real estate market;
– Reduction of the cost of construction due to high technological efficiency.

All of the above forces specialists working in the construction industry to constantly study, know technical and linguistic languages and have a minimum of up-to-date knowledge on the basics of digital literacy, constantly updating them in accordance with the trends of the time.

1.3 Purpose and Objectives of the Study

The purpose of this study is to form ways to solve the problem of personnel shortage of BIM-specialists in the field of construction based on the development of algorithms for actions in the selection process and preparation for the requests of enterprises. We suggest using checklists as a working tool. Such checklists for the adaptation or professional development of employees and students will help students to meet the constant changes and successfully transform to meet the new challenges of the enterprise.

To achieve this goal, the following tasks were identified:

– Study the theoretical foundations of BIM and classify BIM-specialists;
– Analyze the causes of the personnel shortage of BIM-specialists;
– To study the best practices of adaptation to digital reality, including the use of checklists;
– To form ways of solving the identified problem through the training practice of training and retraining of personnel for the needs of the enterprise;
– To propose using a system of transforming checklists for the needs of the enterprise as algorithms of actions with specified boundary conditions as a training tool;
– Provide for three planning horizons – short-term, medium-term and strategic.

2 Research Methodology

2.1 Theoretical Foundations

Digital literacy is the possession of a set of functional knowledge and skills for the competent and effective use of digital technologies [1, 4, 17, 18]. Digital competence is the next level after digital literacy, when a specialist has a professional set of knowledge and skills in a certain area and is able to apply them reasonably and carefully to optimize their activities and to increase its efficiency [17]. A digital citizen is a subject of the digital economy that has a certain set of developed competencies for productive work with digital technologies, development and realization of its potential in a digital society [1].

Projecting the term "digital citizen" into the construction sphere, we get the characteristic of a specialist with digital literacy in construction, i.e. the ability to work effectively with digital tools in the implementation of construction projects.

Now the main digital tool of the construction industry is BIM or information modeling. This tool allows you to create a three-dimensional information model of the object and fill it throughout the entire life cycle of the project, starting with the justification of investments and ending with operation.

The main idea of BIM is to improve quality. Thanks to the accuracy of design in BIM, it is possible to identify errors and make an operational decision to eliminate them. Such predictive analytics saves time and costs [12].

The use of BIM is declining from the initiative and design stage to the construction and operation stage. This phenomenon is explained by the difficulty of transferring data to the next stages of the life cycle and depends on the level of development of BIM in the company. The degree of adaptation of the organization to the implementation of BIM is shown by 4 levels of maturity of the model [2, 19, 20]:

- The first level (3D). 3D model appears that contains design documentation, engineering analysis, 3D visualization, content inspection, collision detection, spatial coordination, and on-site logistics planning.
- The second level (3D+4D). 4D involves linking the model to time and visualizing the graph. With this change, you can view the construction sequence and track production. Phased coordination and planning of logistics is achieved.
- The third level (3D+4D+5D). 5D represents the cost-linked model. By drilling down the information model, you can extract the exact quantity of required materials, predict costs, and manage orders. The presence of 5D promotes cost management through cash flow analysis and payroll analysis.
- The fourth level (3D+4D+5D+6D). The next dimension is 6D (building management). In the 6D information model, it becomes possible to manage assets, space and objects. The information model is integrated with the building management system.

To get the most out of the application of BIM, appropriate specialists are needed. In Table 1, we present a classification of BIM-specialists with explanations of their main tasks.

Table 1. Classification of BIM-specialists [21, 22].

Profession	Goal	Work tasks
BIM-manager	Ensure the use of BIM throughout the life cycle of the construction object	Formation of BIM training methods Development of internal regulations for creating BIM and working with it Formation of the BIM terms of reference Conducting an audit of the information model
BIM-coordinator	Coordinate the work of the project team with BIM	Development of the BIM terms of reference Regular audit of the information model Checking for collisions
BIM-modeller	Designing the BIM model section	Creating component libraries Reproducing data from 2D drawings into a 3D model

BIM-manager, BIM-coordinator and BIM-modeller are specialists who must have a sufficient level of digital literacy in the field of construction. Also, BIM-specialists should develop the ability to solve professional problems in a multilingual international environment [14].

2.2   The Reasons for the Personnel Shortage of BIM-Specialists

Analyzing the problems of personnel shortage, the question arises: why is there a lack of a sufficient number of BIM-specialists? As an answer to the question, we have identified three possible reasons for the personnel shortage at the training level:

– Lack of vision of the real need for BIM-specialists;
– Lack of places for training in BIM-specialties;
– The complexity of selecting candidates and their BIM training.

In the course of the study, we presented these reasons in the form of questions and answered them.

3   Research Results and Discussion

3.1   Question No. 1. What is the Reason for the Need to Study BIM?

According to Rosstat statistics [25], the volume and pace of construction are increasing (see Fig. 1). The larger the volume of construction, the more resources are required for the implementation of projects.

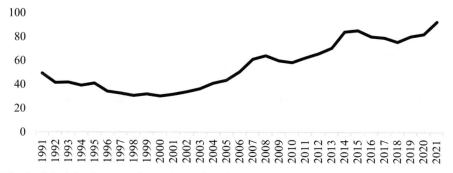

Fig. 1. Schedule of construction volumes from 1991 to 2021 in the Russian Federation, million sq.m [25]

With significant construction rates, technologies such as BIM are required, as they allow [8, 9, 24, 26]:

– To reduce the risks of increasing the cost of projects;
– To generate a more accurate forecast of estimated costs;
– To reduce the number of design errors and the likelihood of their occurrence;
– To effectively plan and make an accurate construction schedule;
– To develop a detailed model;
– To improve construction process management systems;
– To work together with different specialists in a single model.

These advantages are internal motivations for the implementation of BIM. In theory, developers are using BIM more with the growth of construction volumes.

To confirm this judgment, a correlation analysis was carried out, where the relationship between the volume of construction and the percentage of use of BIM, as well as between the volume of construction and the level of maturity of BIM in the company was analyzed.

In total, there are 10 areas of BIM application [27]:

1. Justification of investments;
2. Architectural solutions;
3. Constructive solutions;
4. Engineering, engineering solutions;
5. Budget of the project;
6. Working documentation;
7. Construction schedule;
8. Control of the construction progress;
9. Executive documentation;
10. Operation.

For correlation analysis, the weight of each zone in the amount of 10% was taken. The level of development of the BIM company was indicated as follows:

Level 1 – 3D;
Level 2 – 3D and 4D;
Level 3 – 3D, 4D and 5D;
Level 4 – 3D, 4D, 5D and 6D.

The initial data for correlation analysis are presented in Table 2. The total number of observations is 20.

Table 2. Initial data of correlation analysis [27, 28].

Company name	Under construction, sq.m	Share of the life cycle of the project where BIM is applied (from investment justification to the operation stage), %	The level of development of BIM
PIK Group of Companies PJSC	5 999 312	90%	3
GC LSR	2 631 085	50%	2
GK Airplane	2 968 307	60%	1
Setl Group Holding	1 847 180	70%	3

(*continued*)

Table 2. (*continued*)

Company name	Under construction, sq.m	Share of the life cycle of the project where BIM is applied (from investment justification to the operation stage), %	The level of development of BIM
GC FGC	1 371 525	60%	3
GC Don stroy	1 181 961	50%	2
GC Etalon	1 127 416	70%	3
GC Granel	957 724	10%	1
GC MR Group	933 313	70%	1
INGRAD Group	893 111	60%	1
CCG Group of Companies	872 309	50%	2
Glavstroy Group of Companies	845 827	30%	1
Lingonberry Group	757 621	80%	1
GC A101 DEVELOPMENT	690 297	40%	1
Aquilon Group	683 299	10%	1
GC KORTROS	653 828	30%	3
GC MIC	554 657	30%	1
GC Leader of Groups	468 585	30%	1
GC Development	358 222	10%	1
GC Geometry	348 638	10%	1

The results of the correlation analysis are presented in Table 3.

The results of the correlation analysis show that there is an average relationship between the proportion of the life cycle where BIM is applied and construction volumes. At the same time, the relationship is high and significant, since the p is 0.007, which is less than the significance level of 0.01. With the growth in construction volumes, developers are applying BIM to more stages of project implementation.

Table 3. Results of correlation analysis.

Indicators	Under construction, sq.m	Share of the life cycle of the project where BIM is applied (from investment justification to the operation stage), %	The level of development of BIM
Under construction, sq.m	1		
Share of the life cycle of the project where BIM is applied (from investment justification to the operation stage), %	0,59	1	
The level of development of BIM	0,45	0,48	1

The relationship between the level of BIM development and construction volumes is medium and significant ($p < 0.05$). The level of development of BIM in the company increases with the volume of construction. Consequently, knowledge and skills in BIM are becoming more relevant and necessary for modern construction.

3.2 Question No. 2. Where Can You Learn BIM Technologies?

According to the VUZOPEDIA website, 113 universities in Russia have training areas in the field of BIM. In most cases, the programs do not use the word "BIM", but are called as "Construction of buildings, structures and development of territories" (UrFU, Yekaterinburg) or "Industrial and civil construction of unique buildings and structures" (SPbPU, St. Petersburg). It is assumed that these areas of training teach design in BIM [23].

Thus, students have the opportunity to study BIM-competencies at universities, which can increase the flow of BIM-specialists in the labor market.

3.3 Question No. 3. In What Ways Can the Process of Training BIM-Specialists be Simplified?

The process of teaching digital literacy is complex and complex and requires the use of new approaches, where the desire is caused by an internal urge.

Training in such digital professions as BIM-specialists should take place in a comprehensive and systematic manner. At the same time, it is important to take into account for which generation the training will be. In this case, potential BIM learners are representatives of generation Z, who have clip thinking and strive not to miss any benefit.

Concretization and accuracy of actions are important to them. In this regard, the study suggests using a modern learning tool – checklists.

One of the popular practices of effective learning is gamification [29]. Digital technologies are used in a playful way to make strategic decisions. Another practice that is gaining popularity is the development of transforming checklists. Here, digital technologies are taught using a pre-prepared algorithm of actions.

The checklist is a list of sequential actions with a note of completion. After completing all the actions, you move to the next level, where a different algorithm of actions has been developed [30]. Completing checklists is like going through the stages of a game with different levels.

Potential BIM-specialists are young people of generation Z who have their own psychological and behavioral characteristics [30–32]:

– Generation Z has clip thinking and perceives information visually with frequent (clips);
– Representatives of generation Z switch between tasks quickly and without problems. At the same time, it is important for them to break down large tasks into small stages;
– Generation Z finds it difficult to implement long-term projects. It is necessary that the work be interesting and not boring;
– It is important to save time when completing tasks, so representatives of generation Z are motivated to use virtual tools;
– Generation Z controls a large amount of information, so it studies it superficially;
– Generation Z adjusts its questions so as to receive specific information for the task.

Taking into account the psychology of generation Z, it is important to present information briefly and clearly. The checklist is a suitable working tool for improving digital literacy for the multilingual world today.

4   Applied Aspects of Research

In the classification of BIM-specialists, there are three specialists: BIM-manager, BIM-coordinator and BIM-modeller.

In the case of a BIM-specialist, the workflow is accompanied by human-machine communication (technical language describing engineering processes) and human-human communication (linguistic language). Technical languages include programming language, technical drawing and engineering elements (crossbar, floor slab, grid of columns, etc.) [33]. Knowledge of several languages contributes to the assimilation of information beyond the language representations [14].

A BIM-specialist must be multilingual, i.e. understand the context of communication and be able to decipher information in various communication situations. It is important to use the methods of multilingualism (co-learning of different languages) in the training exercises and the learning process [35]. This is effectively implemented using a set of checklists that are accessible and understandable to everyone.

Currently, three training formats are actively practiced: full-time, online and hybrid (full-time + online).

We will also divide the levels of training into three parts: the first level is minimal knowledge, understands what it is and how it works; the second level is the formation of digital competence, direct work with technology; the third level is a digital expert who himself becomes a mentor.

Based on the above, a matrix of checklists for improving digital literacy was formed (see Table 4).

Table 4. Matrix of checklists for improving digital literacy.

Type of mastering digital literacy	Face-to-face format	Online format	Hybrid format (face-to-face + online)
First level (short-term)	Min knowledge + live communication	Min knowledge + flexible schedule	Min knowledge + live communication + flexible schedule
Second level (medium-term)	Digital competence + live communication and application in practice	Digital competence + flexible schedule and application of knowledge remotely	Digital competence + flexible schedule with live communication + application in practice and remotely
Third level (strategic)	Digital Expert + Mentoring	Digital Expert + Online mentoring	Digital Expert + Hybrid Mentoring

Checklists for BIM specialists will differ in professional knowledge and skills and be modified to meet the needs of the enterprise.

As an example, we will present a developed checklist for a BIM manager (see Fig. 2) at the first level, where training is carried out in a hybrid format using the 365done service [36]. The purpose of such a program for students is to get a basic set of theoretical knowledge in the field of BIM and master the algorithms for their adequate use.

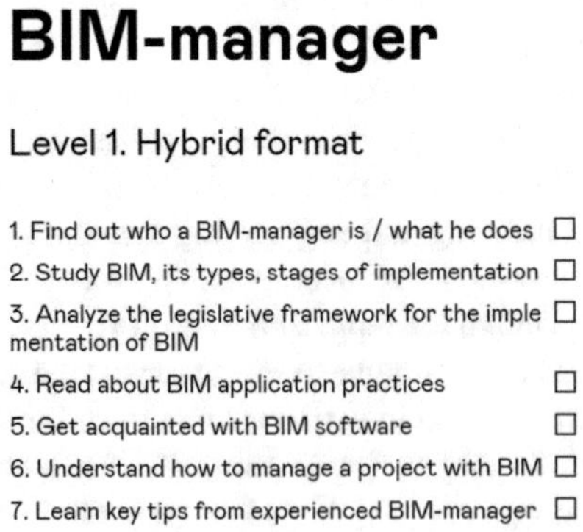

BIM-manager

Level 1. Hybrid format

1. Find out who a BIM-manager is / what he does ☐
2. Study BIM, its types, stages of implementation ☐
3. Analyze the legislative framework for the imple ☐ mentation of BIM
4. Read about BIM application practices ☐
5. Get acquainted with BIM software ☐
6. Understand how to manage a project with BIM ☐
7. Learn key tips from experienced BIM-manager ☐

Fig. 2. Checklist for a BIM manager. Level 1. Hybrid format.

5   Conclusions

BIM-manager, BIM-coordinator and BIM-modeller are relevant professions in modern construction. Their value and relevance is due to the growth of construction volumes. Also, with the increase in construction production, developers are implementing BIM at more stages of project implementation. This is explained by the growing demand for BIM specialists and, hence, the growth of applicants wishing to improve their qualifications.

Current educational requirements make it difficult to attract successful students. Training for such digital professions as BIM-specialists should be comprehensive and systematic. At the same time, it is important to consider for which generation the training will be and what languages are needed for work. In this case, potential BIM learners are representatives of generation Z, who have clip thinking and strive not to miss any benefit. They need specificity and accuracy of actions. Also, a BIM-specialist must speak at least English (to understand the program interface) and a special technical language (to work in the program). With import substitution, there is a need for knowledge of other foreign languages.

In this regard, the study suggests using a modern learning tool – modified checklists as an algorithm of actions for the multilingual world when working in groups on international projects. The corresponding checklist in training and work will help modern students to assimilate the material, gain knowledge and skills through an exact sequence of actions that is simply translated into different languages.

Thus, the study suggests using checklists as a training tool as algorithms of actions with specified boundary conditions for training and development of a specialist while improving digital literacy in multilingual learning. This approach is distinguished by taking into account the characteristics of generation Z and the combination of technical and linguistic languages, which makes the learning process effective and efficient. Moreover, according to the scale of solving urgent problems at the request of the enterprise, all key components and parameters of digital literacy training are determined by calculation formulas, have a target value and can be compared with the current ones. The principle of decomposition in training is also observed (a lower level allows you to reach the upper level). And segmentation in learning is possible, i.e. setting your key indicators for your specific segment according to the requests of the enterprise for a specific task based on the modification of checklists.

The next study on this topic will be of an applied nature and contain ready-made checklists for BIM-specialists for multifunctional objects in the construction industry (for example, combining sports, culture and tourism).

References

1. Toktarova, V., Rebko, O.: Digital literacy: concept, components and evaluation. Bull. Mari State Univ. **2**(42), 165–177 (2021)
2. How is the digitalization of the construction industry. https://ria.ru/20220812/minstroy-180 7198851.html. Accessed 20 Sept 2022
3. Aladyshkin, I.V., Kulik, S.V., Odinokaya, M.A., Safonova, A.S., Kalmykova, S.V.: Development of electronic information and educational environment of the university 4.0 and prospects of integration of engineering education and humanities. In: Anikina, Z. (ed.) IEEHGIP 2022.

LNNS, vol. 131, pp. 659–671. Springer, Cham (2020). https://doi.org/10.1007/978-3-030-47415-7_70

4. Berman, N.: On the issue of digital literacy. Rus. J. Educ. Psych. **6**(2), 35–38 (2017)

5. Malykh, L.: On the content of the concept of "multilingual education" in the Russian education system. Bull. Udmurt Univ. Philos. Psych. Pedag. **31**(1), 108–119 (2021). https://doi.org/10.35634/2412-9550-2021-31-1-108-119

6. Hufeisen, B., Nordmann, A., Liu, A.W.: Two perspectives on the multilingual condition - linguistics meets philosophy of technology. Technol. Lang. **3**(3), 11–21 (2022). https://doi.org/10.48417/technolang.2022.03.02

7. Bylieva, D., Nordmann, A.: Technologies in a multilingual world. Technol. Lang. **3**(3), 1–10 (2022). https://doi.org/10.48417/technolang.2022.03.01

8. Vozgment, N., Astafyeva, O.: Advantages of BIM modeling in the investment and construction sector in the context of digital transformations of the industry. Bull. Univ. **7**, 58–66 (2021). https://doi.org/10.26425/1816-4277-2021-7-58-66. (in Russian)

9. Yushkin, I., Alamidi, S., Stashevskaya, N.: Problems and advantages of BIM implementation at construction industry enterprises. Constr. Mech. Eng. Struct. Struct. **18**(2), 172–181 (2022)

10. Sanzhina, O., Zharkaya, G.: Key trends of digitalization of the construction industry. In: VIII International Conference Problems of Mechanics of Modern Machines, pp. 627–632. East Siberian State University of Technology and Management, Ulan-Ude (2022)

11. BIM technologies (Russian market). https://www.tadviser.ru/index.php. Accessed 20 Sept 2022

12. Fontokina, V., Savenko, A., Samarsky, E.: The role of BIM technologies in the organization and technology of construction. Bull. Eurasian Sci. **14**(1), 1–6 (2022)

13. Leśniak, A., Górka, M., Skrzypczak, I.: Barriers to BIM implementation in architecture, construction, and engineering projects—the Polish study. Energies **14**, 2090 (2021). https://doi.org/10.3390/en14082090

14. Bezukladnikov, K., Prokhorova, A.: Methodological system of multilingual education of future engineers: an empirical study. Bull. Tomsk Stat. Univ. **466**, 158–164 (2021)

15. Krylov, E., Vasileva, P.: Convergence of foreign language and engineering education: opportunities for development. Technol. Lang. **3**, 106–117 (2022). https://doi.org/10.48417/technolang.2022.03.08

16. Krylov, E., Khalyapina, L., Nordmann, A.: Teaching English as a language for mechanical engineering. Technol. Lang. **2**(4), 126–143 (2021). https://doi.org/10.48417/technolang.2021.04.08

17. Korshunov, G., Kroitor, S.: Digital literacy as a key factor of successful adaptation of a person and society to digital realities. Soc. Econ. **1**, 38–58 (2020). https://doi.org/10.31857/S020736760008037-9

18. Bylieva, D., Hong, J.-C., Lobatyuk, V., Nam, T.: Self-regulation in e-learning environment. Educ. Sci. **11**, 785 (2021). https://doi.org/10.3390/educsci11120785

19. Marco, L., Manuele, C., Davide, T., Benedetta, B.: BIM level of detail for construction site design. Proc. Eng. **123**, 581–589 (2015). https://doi.org/10.1016/j.proeng.2015.10.111

20. Moses, T., Heesom, D., Oloke, D.: Implementing 5D BIM on construction projects: contractor perspectives from the UK construction sector. J. Eng. Des. Technol. **18**(6), 1–28 (2020). https://doi.org/10.1108/JEDT-01-2020-0007

21. Nabiev, R., Iglina, N., Luneva, T.: Management of training of specialists in the investment and construction sector in the conditions of digitalization of the economy. Bull. Astrakhan State Tech. Univ. Series: Econ. **2**, 50–59 (2020)

22. BIM professions: how to become a BIM manager. https://digital-build.ru/bim-professii-i-ih-otlichiya. Accessed 20 Sept 2022

23. Nicał, A., Wodynski, W.: Enhancing facility management through BIM 6D. Procedia Eng. **164**, 299–306 (2016). https://doi.org/10.1016/j.proeng.2016.11.623

24. Golovin, K., Kopylov, A., Tomilova, B.: Analysis of the current state of BIM technologies in the construction market. Proc. Tula State Univ. Tech. Sci. **12**, 278–282 (2020)
25. Construction. Federal State Statistics Service. https://rosstat.gov.ru/folder/14458. Accessed 20 Sept 2022
26. Shalina, D., Larionova, V.: Building Information Modeling (BIM) as a way to reduce the risks of increasing the cost of the project. Fundam. Res. **12**, 215–222 (2021). https://doi.org/10.17513/fr.43179
27. The level of application of TIM by developers of the Russian Federation in the construction of residential facilities. Competence Center for TIM. Digital Academy. HOUSE.RF. https://наш.дом.RF/technologies-information-modeling. Accessed 20 Sept 2022
28. Rating of the TOP developers of the Russian Federation. A single resource of developers. https://erzrf.ru/top-zastroyshchikov/rf?topType=0&date=220901. Accessed 20 Sept 2022
29. Larionova, V., Stepanova, N., Shalina, D.: Management games: organizational behavior and emotional intelligence. In: 20th PCSF and 12th CSIS-2020. European Proceedings of Social and Behavioural Sciences, vol. 98, pp. 1–12 (2020). https://doi.org/10.15405/epsbs.2020.12.03.27
30. Badikova, I.: The use of checklist technology for the organization of research activities of students in the field of pedagogy and psychology. Bull. Voronezh Stat. Univ. Series: Probl. High. Educ. **3**, 168–173 (2018)
31. Stillman, D., Stillman, I.: Generation Z at work. How to understand him and find a common language with him. Mann, Ivanov and Ferber, Moscow (2018)
32. Razinkina, E., Zima, E., Pozdeeva, E., Evseeva, L., Tanova, A.: Convergence of employers' and students' expectations in the educational environment of the agricultural university. In: E3S Web Conference, vol. 258, p. 10019 (2021). https://doi.org/10.1051/e3sconf/202125810019
33. Pelz, P.: Good engineering design. Design evolution by languages. Technol. Lang. **1**(1), 97–102 (2020). https://doi.org/10.48417/technolang.2020.01.20
34. BIM manager: which universities teach where to enroll in Russia. https://vuzopedia.ru/professii/432/vuzy?page=2. Accessed 20 Sept 2022
35. Baranova, T., Kobicheva, A., Tokareva, E.: Application of the multilingual approach in training students of a technical university. Mod. High Technol. **6**, 129–135 (2021). https://doi.org/10.17513/snt.38710
36. Checklist and list constructor 365done. https://my.365done.ru. Accessed 20 Sept 2022

Author Index

Printed in the United States
by Baker & Taylor Publisher Services